Governing Innovation: A Global History of Patent-Driven Technological Progress

Qiang Yue • Fei Yang

Governing Innovation: A Global History of Patent-Driven Technological Progress

Springer

Qiang Yue
Beijing, China

Fei Yang
Beijing, China

ISBN 978-981-95-5569-7 ISBN 978-981-95-5570-3 (eBook)
https://doi.org/10.1007/978-981-95-5570-3

Jointly published with Publishing House of Electronics Industry

The print edition is not for sale in China (Mainland). Customers from China (Mainland) please order the print book from: Publishing House of Electronics Industry.

Publishing House of Electronics Industry

Translation from the Chinese Simplified language edition: "《全球专利科技发展史》" by Qiang Yue and Fei Yang, © Publishing House of Electronics Industry Co., Ltd 2025. Published by Publishing House of Electronics Industry Co., Ltd. All Rights Reserved.

This Springer imprint is published by the registered company Springer Nature Singapore Pte Ltd.
The registered company address is: 152 Beach Road, #21-01/04 Gateway East, Singapore 189721, Singapore

Preface

Patents record, in exhaustive detail, the modernization of applied technology. Yet, they have rarely been examined systematically as historical sources. This absence is striking. As practitioners in the field of patents, we undertook this study in our spare time, with the aim of reading patents as historical texts—documents that illuminate not only technological change but also the deeper logics of innovation, strategy, and governance.

In this book, patents are treated as a lens for exploring the rise and fall of companies and industries, tracing the patterns and transformations that shape their trajectories. By following the histories revealed in patent filings and litigation, we hope to offer new perspectives for intellectual property scholars, corporate strategists, and policymakers: to reflect on how protection and innovation interact, and to consider how promises of innovation are pursued, implemented, or deferred.

The book is organized into four parts—Rise and Fall, Disruption and Revolution, Core and Frontiers, and Governance and Administration. Each section explores how leading firms such as IBM, Kodak, Sony, Apple, Matsushita, Siemens, Disney, and Bosch deployed patents as part of their strategies. Drawing on data from patent filings, litigation, and corporate R&D histories, we analyze how companies responded to waves of technological change, how patents reflected processes of market expansion, replacement, and supersession, and how innovation outcomes were shaped by both successes and failures in strategic decision-making.

Rise and Fall examines the two-hundred-year histories of IBM and Kodak, highlighting critical patent disputes and the role of patents in shaping corporate destinies. Disruption and Revolution turns to the consumer electronics and IT revolutions, analyzing cases such as Kodak, and Apple to show how shifting technologies and consumer preferences—often through the supersession of the old by the new—became inscribed in patent records. Core and Frontiers considers companies such as Matsushita, Siemens, Disney, and Bosch, focusing on how established industrial leaders used patents in their attempts, sometimes successful and sometimes not, to expand into new technologies and markets. Finally, Governance and Administration reflects on the broader relationship between innovation, patents, and institutional design. For knowledge-intensive enterprises, our recommendation is clear: patents should not be seen only as shields protecting existing advantages, but as platforms for renewal—springboards for new business models and continuous innovation.

True innovation requires both a stable foundation of patent-supported advantage and a capacity to continually adapt and expand patent portfolios. Companies that fail to do so risk being overtaken by competitors. Intellectual property professionals, likewise, must recognize that their responsibility is not to encourage hoarding or opportunistic litigation, but to evaluate how patents can genuinely foster innovation. In this way, the principles of governance, the protection of intellectual property, and the pursuit of innovation can be aligned.

We believe that technology must ultimately be people-centered, and that patents, innovation, and corporate development should serve human welfare. The cases examined in this book demonstrate that only innovations responsive to social needs produce patents of lasting value. The true significance of patents lies in safeguarding exceptional products and services, protecting firms that excel in serving the public, and affirming that the right to serve humanity must be preserved. This, we argue, is the rightful path for corporate governance, innovation strategy, and patent strategy.

Beijing, China

Yue Qiang
Yang Fei

Competing Interests The authors have no competing interests to declare that are relevant to the content of this manuscript.

Contents

1 **Introduction** 1

Part I Rise and Fall

2 **From Looms to Computers** 25

3 **From Patent Examiner to CTO** 35

4 **From Unicorn to Monopoly** 49

5 **Thrill of Achieving Monopoly via Patents** 63

6 **The Fall of a Legendary Patent** 73

7 **From Cards to Disks** 83

8 **From Opponent to Champion of Software Patents** 103

9 **From Paper Mill to Patent Factory** 111

10 **From Mainframes to Microcomputers** 121

11 **Holding Patents for Quick Profits** 131

12 **A Supplementary Note: The Yellow Giant** 155

Part II Disruption and Revolution

13 **Introduction** 185

14 **Tracing the Origins of Magnetic Tape Technology** 199

15 **The Birth of the Cassette Tape** 205

16 **The Sudden Rise of the Walkman** 215

17 **Products as Art** 221

18 **Sony for the Young** 241

19 Philips in the Ivory Tower ... 245

20 Sony at the Crossroads of the Digital Music Era ... 251

21 The Dream of Total Domination ... 255

22 The Petty King of Patents ... 263

23 A New Generation Surpasses the Old ... 267

24 The Failure of Betamax ... 271

25 The Birth of the CD ... 287

26 A Hesitant Revolution ... 291

27 Unhelpful Allies ... 299

28 Time-Shifting for Others' Gain ... 307

29 Philips and the Battle of Formats ... 317

30 The Endgame of the Format War ... 321

31 Beginning with *Star Trek* ... 331

32 Newton and General Magic ... 335

33 From iPod to iPhone ... 349

34 A Brief History of Multi-Touch Patents ... 353

35 Twilight at the Moment of Transition: Sony and Apple from a Patent Perspective ... 359

36 Summary ... 369

Part III Core and Frontiers

37 Introduction ... 373

38 The Long Road from Scientific Innovation to Productization (Part I) ... 383

39 The Long Road from Scientific Innovation to Product (Part II) ... 403

40 The Formation of the Bosch Core ... 415

41 Bosch's Expansion Beyond Its Core ... 421

42 New Frontiers: From Hair Clippers to Power Drills ... 433

43 The Formation of Xerox's Core Zone: From Copying Techniques to Xerography ... 439

44 Xerox's Expansion at the Frontier: Tried, Struggled, and Failed ... 459

45 Hewlett-Packard's Boundary Expansion: Instruments, Computers, and Printers 469

46 Building a Core of Light and Shadow 485

47 Blurring the Lines Between Illusion and Reality 495

48 Legacy and Innovation in the Digital Age 505

Part IV Governance and Administration

49 Introduction 521

50 Innovation and Patents: A Fundamental Connection 531

51 Innovation Protection: Evolution of Rules 537

52 Intellectual Property Rules: Tao of Application 573

Chapter 1
Introduction

The people are most precious, the state is secondary, and the ruler is least important. Mencius

Samuel P. Huntington once made a famous assertion: "*The West won the world not by the superiority of its ideas or values or religion but rather by its superiority in applying organized violence. Westerners often forget this fact; non-Westerners never do.*"

The Age of Discovery in the fifteenth century evolved into a history of European colonization across Asia, Africa, and Latin America. During this period, European adventurers led the charge with muskets, followed closely by missionaries with Bibles. First came military suppression, then ideological conversion—a dual assault on the bodies and minds of indigenous peoples in the Asia, Americas and Africa. For a long time, colonized people believed that white men were inherently superior in intelligence and morality. Even today, this mindset lingers subconsciously in some minds like an ideological branding iron.

In a similar way, in the realm of intellectual property, many Chinese tend to believe that Westerners have a stronger awareness of IP and a more mature patent system, hence their greater capacity for innovation and superior products. Therefore, to enhance Chinese companies' innovation capabilities, it's often suggested that the Chinese must strengthen their IP awareness and align institutional frameworks with those in Europe and the United States. Such thinking was not surprising when China had just joined the World Trade Organization and opened its doors. At the time, many Western companies entered the domestic market with undeniably excellent products and innovative IP strategies. National brands like Robust (乐百氏), Maxam (美加净), and Zhonghua Toothpaste were acquired and then shelved by foreign firms. Patent pool owners of technologies like DVD and MPEG collected enormous licensing fees without offering concrete, manufactured products. The shock felt by the Chinese may have been akin to what Native Americans experienced when facing volleys of musket fire.

Today, China has become the world's factory. More and more long-established Western companies have abandoned manufacturing and even R&D, outsourcing production to Chinese firms and relying on brand licensing to capitalize on their

Q. Yue, F. Yang, *Governing Innovation: A Global History of Patent-Driven Technological Progress*, https://doi.org/10.1007/978-981-95-5570-3_1

century-old reputation as "tech companies" to make quick profits. When these tactics no longer work, they turn to their patent portfolios for one last harvest. Some companies, already defeated in the market, have transferred decades of accumulated patents to NPEs (Non-Practicing Entities). As Chinese tech companies expand into international markets, they often encounter patent ambushes from these two types of companies.

We believe that the further a company is from its production line and end users, the weaker its product capabilities, and its patents inevitably become weak. When legacy firms attempt to enforce patents that have no supporting products, their behavior borders on patent trolling. Such patents are not inherently threatening, but because of the high litigation costs, unfamiliar languages, and jurisdictions, Chinese companies do face a natural disadvantage in overseas patent disputes—an undeniable reality.

In this regard, we agree with Huntington's view: this advantage does not arise from the West's superior "*ideas or values or religion*," but from its greater ability to extract tangible wealth from patents—an ability cultivated over more than two centuries.

To many, concepts like "patent monetization" and "patent trolls" appear to be modern phenomena. Historically speaking, the term "patent troll" only began to appear in the 1990s. From a business practitioner's viewpoint, patent trolling gained traction in the U.S. after the burst of the dot-com bubble in 2001, when a wave of bankruptcies led to patent fire sales, and NPEs seized the opportunity to hoard patents. However, this impression is misleading. In the United States, the patent system has a history spanning more than 200 years. Large-scale patent litigation, strategic patent operations, and sophisticated portfolio planning emerged more than a century ago—not merely in the past few decades.

The first wave of large-scale patent litigation in American history began in the late eighteenth century. Eli Whitney, an American, invented the cotton gin, enabling large-scale cotton processing. He applied for a patent in 1794 and received authorization (see Fig. 1.1), granted by none other than one of America's founding fathers—Thomas Jefferson. Many patent professionals may know that Jefferson was the first U.S. patent examiner in history. However, in truth, Jefferson reviewed only a few applications, serving in a substantive examination capacity for merely 3 years (1790–1793). In 1793, the U.S. abandoned substantive examination, requiring only a sworn declaration of originality from the applicant. If formalities were met, patents were granted.

Whitney found a partner and opened a factory to manufacture cotton gins, targeting plantation owners in the American South. However, his business model was immature: production couldn't keep up with demand, and his pricing was exorbitant—he asked customers to pay him one-third of the cotton profits generated by his machine. Moreover, the gin's design was simple and its mechanism obvious to the trained eye. In Whitney's time, it was unrealistic to expect people to respect "intellectual property rights." Very soon, Southern plantation owners began to copy his machine. Whitney then launched a series of patent infringement lawsuits to defend his rights. But being unfamiliar with the South, and facing delays or outright

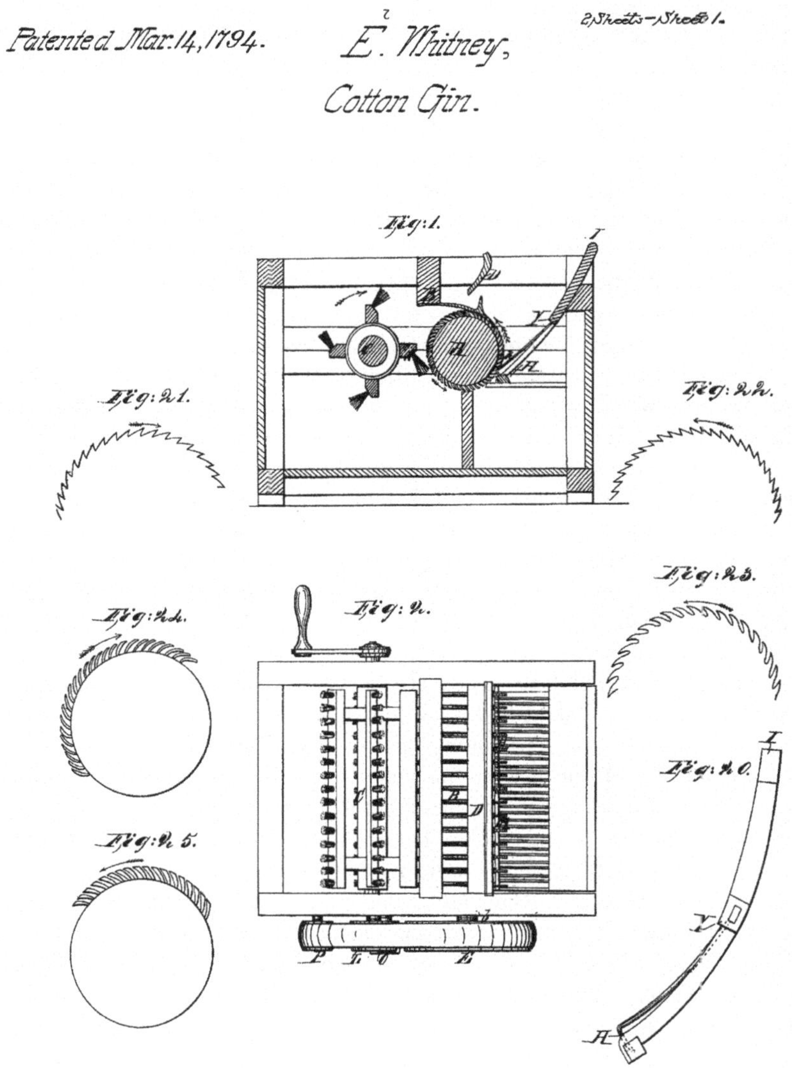

Fig. 1.1 Illustration of Whitney's 1794 patent

rejections in court, he found little success. Some states didn't even recognize patent rights. Though Whitney won some legal battles, the resulting damages barely covered his legal expenses. He famously lamented, "*An invention can be so valuable as to be worthless to the inventor.*"

His statement was painfully accurate—Whitney's cotton gin transformed cotton into an ideal crop for the South's slave-based economy, while the textile boom fueled industrial growth in the North. The resulting economic divergence helped shape the national structure and ultimately contributed to the outbreak of the Civil War. Yet the man whose invention reshaped a nation's economy earned almost nothing from it. Whitney continued inventing, but due to the bitter experience with the cotton gin, he never filed another patent application.[1]

However, the failure of the cotton gin patent earned Whitney broad recognition. He later secured a major federal contract to manufacture 10,000 muskets for the U.S. Army.[2] He proposed the use of interchangeable parts, completed the order efficiently, and made a substantial profit.[3] The interchangeable parts method took root in Northern industry but failed to take hold in the slave-driven Southern economy—and eventually became one of the decisive factors in the North's victory during the Civil War. In this sense, Whitney got his revenge.

After Whitney's failed attempt to enforce his rights, it took several more decades before successful cases of patent monetization began to appear. In 1818, American inventor Thomas Blanchard designed an automatic lathe capable of processing irregular objects like gunstocks, shoe lasts, and axe handles. He submitted a patent application, which was immediately challenged by a competitor. At the time, under America's "first to invent" system, such challenges (known as interference proceedings) required both parties to prove who had first completed the invention. After trial, Blanchard was granted the patent. However, customers still questioned its stability. So Blanchard withdrew his application, rewrote the specification, and was granted a reissued patent on January 20, 1820 (see Figs. 1.2 and 1.3).

By the 1830s, Blanchard had developed several lathes capable of producing interchangeable parts, which were covered by patents US3 through US9. Anyone familiar with American patent history would recognize the significance of these single-digit numbers: they were among the first patents granted under the revised 1836 U.S. Patent Act. Prior to 1836, patents were identified by date, not number.[4] Blanchard's most important patent was in fact this reissued one, which had no patent publication number. In the 1820s, lathes produced using this patent sold for $400, and users paid $100 annually in licensing fees.

From the 1830s to 1850s, Blanchard launched dozens of infringement lawsuits and won most of them. His success owed not just to patent quality or legal counsel, but to a major reform in U.S. patent law: if a patentee could prove they hadn't received "reasonable remuneration" during the patent term, and the cause wasn't due to their own negligence or fault, they could apply for an extension. Blanchard

---

[1] Resis, R. (2006). History of the patent troll and lessons learned. Intellectual Property Litigation, 17(2), 1–6. Retrieved January 22, 2024 from https://bannerwitcoff.com/_docs/library/articles/HistoryOfPatentTroll.pdf

[2] Robert O. Woods. (2010). How the Cotton Gin Started the Civil War. Retrieved January 22, 2024 from https://www.asme.org/topics-resources/content/how-the-cotton-gin-started-the-civil-war

[3] Randy Alfred. (2009). Oct. 28, 1793: Whitney's Cotton Gin Patent Not Worth Much. Retrieved January 22, 2024 from https://www.wired.com/2009/10/1028whitney-cotton-gin/

[4] Dobyns, K. W. (1994). A History of the United States Patent Office: The Patent Office Pony. Retrieved January 22, 2024 from http://www.myoutbox.net/popch16.htm

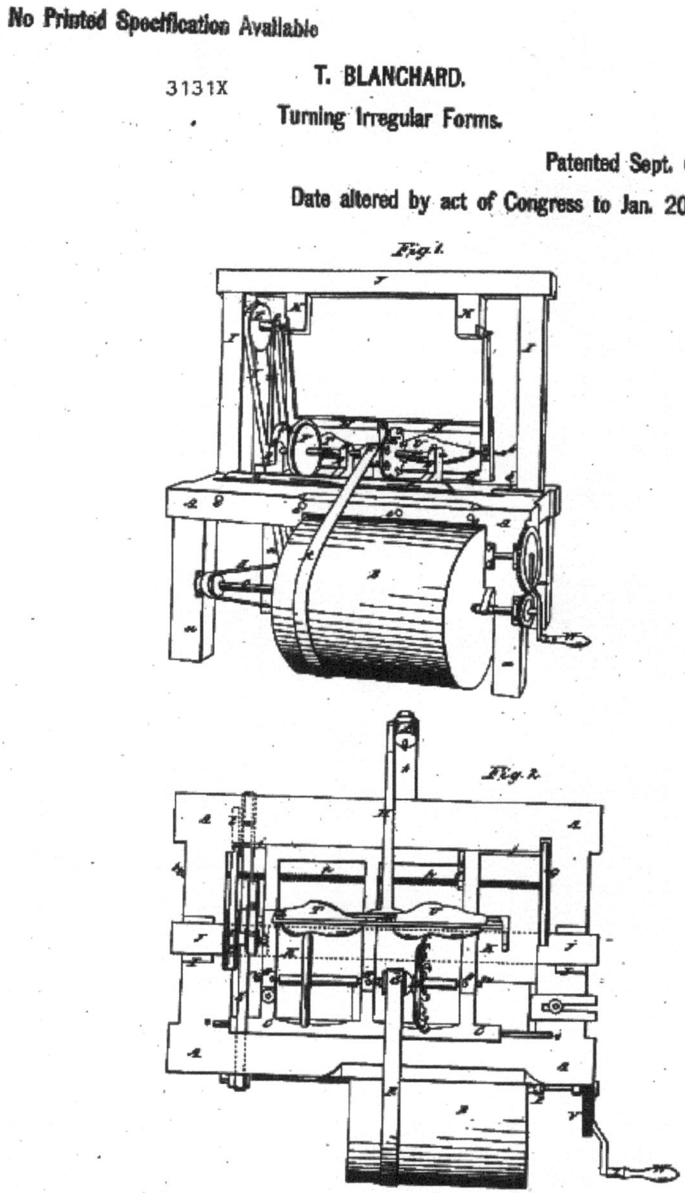

Fig. 1.2 Patent drawing of Blanchard's 1820 patent

Fig. 1.3 Original Blanchard Lathe, 1822

became one of the biggest beneficiaries of this reform. At the time, most defendants wouldn't even consider designing around decades-old patents.

Around 1834, Blanchard's 1820 patent for the lathe was approaching expiration. He began lobbying the United States Congress, hoping it would again grant him a special extension. He argued that the patent, had it been properly commercialized, could have earned him $215,440 in revenue. However, he claimed that during the patent's effective term, he had been diligently working at the Springfield Armory in service to the country, which prevented him from practicing the patent effectively. As a result, over the 14 years of protection, he earned only a modest sum—a total of $3665.84, down to the last cent.

Faced with such precise and compelling evidence, and after persistent lobbying, Congress passed a bill in June 1834 granting a 14-year extension for Blanchard's 1820 patent.

This episode illustrates just how rudimentary the U.S. patent system was at the time—far less sophisticated than it is today. Because Blanchard's patent had officially expired in January 1834, and Congress didn't approve the extension until June, there was nearly a 6-month window in which the patent had no legal effect. During this gap, when Blanchard sued competitors for copying his lathe, they countered by claiming they had made the copies during the window of patent invalidity, shifting the burden of proof onto Blanchard to demonstrate otherwise.

Adding to the complications, a clerical error by Congress mistakenly recorded the patent's grant date as January 12 instead of January 20. As mentioned earlier, before 1836, American patents had no numerical identifiers and were typically referenced by their issue dates. Therefore, getting the date wrong was not a trivial matter. One defendant even argued that the congressional extension applied to a patent that technically did not exist, and should be considered null and void. Blanchard had to return to Congress and request a correction through another act. Fortunately, these minor bureaucratic hurdles didn't stop him from collecting substantial licensing fees and compensation.

Another 14 years later, as the extended patent again neared expiration, Blanchard came prepared. This time, he offered Congress a lavish "gift." At the time, photographic technology was still in its infancy, and high-ranking officials still followed medieval customs of commissioning portraits or busts. Blanchard acquired a number of plaster busts of members of Congress and used his lathe to carve marble replicas. He then shipped them to Washington, D.C., for exhibition. The gesture flattered many legislators, and in 1848, Congress passed yet another bill extending Blanchard's patent protection for another 14 years. This time, the paperwork was error-free, and the extension seamlessly bridged the existing term with no lapse in protection.

As a result, the United States witnessed a truly unprecedented event in its patent history: a single patent that enjoyed a protection term of 42 years.[5]

Blanchard became one of the earliest inventors in the nineteenth century to actively enforce his rights through patent litigation and reap significant financial rewards. However, he was neither the first nor the only one to do so.

Beginning with Blanchard, many inventors began filing lawsuits after securing patents and amassed considerable wealth through licensing. According to research by legal historian Christopher Beauchamp at Brooklyn Law School, the U.S. District Court for the Southern District of New York handled 381 patent infringement cases in 1880.[6] That number exceeded the annual caseload of any U.S. federal district

---

[5] Cooper, C. C. (1991). Social Construction of invention through patent management: Thomas Blanchard's Woodworking Machinery. Technology and Culture, 32(4), 960–998.

[6] Beauchamp, Christopher, The First Patent Litigation Explosion (December 6, 2015). *125 Yale Law Journal 848, 2016*, Brooklyn Law School, Legal Studies Paper No. 428. Retrieved January 22, 2024, from https://ssrn.com/abstract=2699964

court for patent disputes in 2010, and even in 2014, it would have ranked among the top three.

Further research shows that more than 650 patent infringement lawsuits were filed in federal courts across New York State in 1880 alone—more than in any U.S. state in 2011.

As the table shows, an inventor named Oliver H.P. Parker initiated as many as 150 patent lawsuits in the 1850s—a number that few enterprises in modern-day China could rival. The second- and third-most litigious parties were both connected to the prolific inventor Charles Goodyear. Goodyear invented vulcanized rubber, a milestone in industrial history, and the famous Goodyear Tire & Rubber Company was named in his honor. Over his lifetime, Goodyear was granted 28 U.S. patents.[7] His business partners relied heavily on these patents, launching hundreds of lawsuits to control the early American rubber industry. The cases listed in Table 1.1 only scratch the surface.[8]

The Goodyear Dental Vulcanite Company, ranked second in Table 1.1, was one of Goodyear's partners. Starting in 1866, this company launched over 2000

Table 1.1 presents the top entities involved in patent litigation from 1840 to 1910 in the Southern District of New York and the Eastern District of Pennsylvania

Name/entity	Number of suits filed	Location	Field	Peak decade(s)
Oliver H. P. Parker	150	Philadelphia, PA	Water Power	1850
Goodyear Dental Vulcanite Co.	65	New York, NY	Dental Rubber	1870, 1880
Charles Goodyear, Joined by Various Licensees	49	Connecticut	Rubber Goods	1850, 1860
American Bell Telephone Co. (with Local Licensees)	36	Boston, MA	Telephones	1880, 1890
George H. Wooster	33	New York, NY	Garments	1880
Farbenfabriken of Elberfeld Co.	26	New York, NY	Pharmaceuticals	1900, 1910
George C. Roberts	21	New York, NY	Refrigeration	1870
Columbia Motor Car Co. & George B. Selden	20	Hartford, CT & Rochester, NY	Automobiles	1910
George Gregerson, Assignee of John Lightner	20	Roxbury, MA	Railroad Equipment	1860
Richard Imlay	20	New York, NY	Railroad Equipment	1860

[7] Jo Ann Calzonetti. Christopher J. Laursen. (2010). Patents of charles goodyear: His international contributions to the rubber industry September 2010 Rubber Chemistry and Technology 83(3):303–321

[8] Cai Guise-Richardson, (2010). Redefining Vulcanization: Charles Goodyear, Patents, and Industrial Control, 1834–1865. Technology and Culture, 51(2), 357–387. https://doi.org/10.1353/tech.0.0437

enforcement actions against dentists across the United States, making a profound societal impact.

In the nineteenth century, plantation economies in Central and South America and the southern United States significantly boosted sugarcane production. As a result, sugar became cheap and widely accessible, turning into a daily staple for ordinary people. This rapid development meant that few realized the need for sugar consumption moderation. Dental disease skyrocketed, making tooth decay one of the most prevalent ailments in North America. Simultaneously, advances in anesthesia made tooth extraction and denture fitting more routine, and dentists became highly sought-after professionals. This was a key reason why the Goodyear Dental Vulcanite Company aggressively targeted the dental profession.

Readers familiar with American Westerns or gangster films might know that mid-nineteenth century America was lawless and unruly—a time when private detectives and bounty hunters roamed free. Even the U.S. Patent Office was burglarized twice, making headlines at the time. Aside from storing patent specifications, the office also kept and exhibited models of patented inventions. The model exhibition hall even displayed important national artifacts, such as the original Declaration of Independence, the Constitution, and diplomatic gifts from foreign governments—including valuable gold and jewels.

In December 1841, a thief broke into the Patent Office's model exhibition hall and stole a diamond-encrusted gold snuffbox gifted by the Russian Tsar, a pearl necklace from the Imam of Muscat, and a golden sword sheath from the Viceroy of Peru. The stolen items were eventually recovered after considerable effort. Then in November 1848, the Patent Office was burglarized again—largely the same set of treasures was stolen a second time.[9]

Given this backdrop, it is unsurprising that professional lawyers of the time employed all manner of tactics to gather evidence. The Goodyear Dental Vulcanite Company used a wide range of underhanded methods, including bribing or coercing the servants and neighbors of their target dentists. According to secondary sources citing *The New York Times,*[10] their most frequently used tactic was allegedly a honey-trap. Once a dentist was identified as a target, the company would hire a beautiful woman to visit the clinic under the pretense of ordering rubber dentures. Despite being highly educated professionals, many dentists were caught off guard in the presence of an attractive client. After receiving the dentures, the woman would take them directly to the company's legal team, who would then file a lawsuit in court.

In 1872, dentists finally saw the expiration of Goodyear's original patent. But the Goodyear Dental Vulcanite Company quickly brought out another patent—U.S. Patent No. 43009A, granted in 1864 to a dentist named John Cummings—and resumed their nationwide campaign to collect licensing fees.

---

[9] Dobyns, K. W. (1994). A History of the United States Patent Office: The Patent Office Pony. Retrieved January 22, 2024 from http://www.myoutbox.net/popch21.htm

[10] Beauchamp, C. (2016). The first patent litigation explosion. *Yale Law Journal, 125*(4), 848–933. https://www.yalelawjournal.org/article/the-first-patent-litigation-explosion

Little is known about John Cummings beyond the fact that he was a dentist. He held no other patents, and his name appears in no other patent records. Cummings had filed his application in 1855, but it was initially rejected. At the time, rudimentary forms of rubber dentures were already in use, and similar patents existed in Britain. For unknown reasons, the U.S. Patent Office eventually approved his application in 1864, and the patent was immediately transferred to the Goodyear Dental Vulcanite Company.[11]

The patent specification consisted of only two pages—one containing drawings (Fig. 1.4), and the other a brief specification and claims (Fig. 1.5). The invention was exceedingly simple: it replaced metal with vulcanized rubber as the material for dental plates. Despite its simplicity and prior rejections, the patent survived at least two invalidation challenges in court.

Later, dental associations discovered that in one of those challenges, the Goodyear Dental Vulcanite Company had secretly funded the defendant—the party arguing the patent was invalid—and even encouraged them to appeal. The suspiciously rushed and sloppy nature of the appeal proceedings raised serious doubts.[12] Whether the company bribed the defendant or the court remains unknown today, but the fact remains: Cummings's patent survived litigation and was upheld.

The Goodyear Dental Vulcanite Company typically offered licensing terms of $45 per year plus $1 to $2.50 per denture.[13] However, when signing a license, dentists were often required to pay a lump sum covering ten or more years of fees, amounting to hundreds of dollars—a full year's income for many practitioners. Dentists, harassed and frustrated, banded together to establish foundations to fight back and lobbied local legislatures to oppose patent extensions, all to no avail.

This historical episode unfolded shortly after the American Civil War, just a few decades before the era depicted in the film *Django Unchained*. In that movie, one of the main characters, Dr. Schultz, is a dentist-turned-bounty-hunter who poses as a slave owner to help Django rescue his enslaved wife. In the film's climactic scene, Schultz negotiates the purchase of Django's wife with Calvin Candie, a sadistic plantation owner played by Leonardo DiCaprio. Although Schultz remains composed throughout Candie's insults and provocations, he ultimately snaps when Candie demands a handshake to seal the deal. Unable to stomach the gesture, Schultz shoots Candie point-blank, triggering a bloody gunfight and sacrificing his own life.

The Goodyear Dental Vulcanite Company eventually encountered a real-life dentist who, like Schultz, could no longer tolerate their aggressive tactics—and that confrontation ultimately brought the company's legal onslaught to an end. A dentist

[11] Goodyear Dental Vulcanite Company v. Davis, 102 U.S. 222 (1880)

[12] John Marr. (2015). History's Forgotten "Denture Patent War" Spawned A Truly Bizarre Murder. Retrieved February 1, 2024, from GIZMODO: https://gizmodo.com/historys-forgotten-denture-patent-war-spawned-a-truly-1716425468

[13] John Marr. (2015). History's Forgotten "Denture Patent War" Spawned A Truly Bizarre Murder. Retrieved February 1, 2024, from GIZMODO: https://gizmodo.com/historys-forgotten-denture-patent-war-spawned-a-truly-1716425468

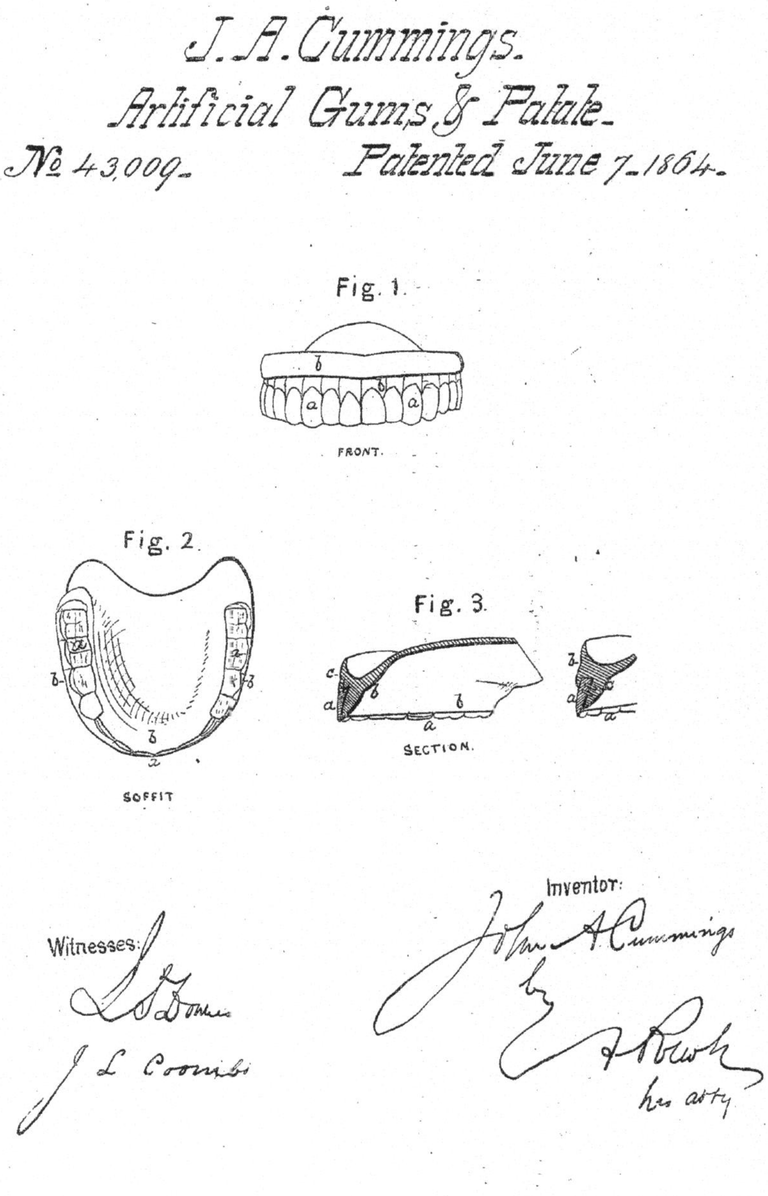

Fig. 1.4 Patent drawing of Cummings' 1864 patent

UNITED STATES PATENT OFFICE.

JOHN A. CUMMINGS, OF BOSTON, MASSACHUSETTS.

IMPROVEMENT IN ARTIFICIAL GUMS AND PALATES.

Specification forming part of Letters Patent No. **43,009**, dated June 7, 1864.

To all whom it may concern:

Be it known that I, JOHN A. CUMMINGS, of Boston, in the county of Suffolk and State of Massachusetts, have invented certain new and useful Improvements in the Manner of Forming Artificial Palates and Gums Used for Inserting Artificial Teeth; and I do hereby declare that the following description, taken in connection with the accompanying drawings, hereinafter referred to, forms a full and exact specification of the same, wherein I have set forth the nature and principles of my said invention, by which it may be distinguished from others of a similar class, together with such parts as I claim and desire to have secured to me by Letters Patent.

The figures of the accompanying plates of drawings represent my improvements.

Figure 1 is a front elevation of a set of artificial teeth and plate formed by my improved method. Fig. 2 is a view of the lower side of the plate and teeth. Fig. 3 is a central vertical section of the same.

The method hitherto practiced of attaching artificial teeth to a metallic plate fitting to the roof of the mouth is attended with many objections and inconveniences, as the plate, being hard and unyielding, naturally hurts the mouth and in some degree impedes mastication and perfect articulation, while the expense of the metal employed prevents many from availing themselves of the advantages of artificial teeth.

The present invention consists in forming the plate and gums to which the teeth are attached of rubber or some other elastic material, so indurated as to be rigid enough for the purposes of mastication and pliable enough to yield a little to the motions of the mouth and in one piece, the teeth being embedded in the elastic material while the said material is in a soft condition and then baked with the gums and plate, so that the teeth, gums, and plate will all be connected, forming, as it were, one piece.

a a a and *c* in the drawings represent the artificial teeth prepared in the usual manner.

The plate and gums *b b* are formed in one piece, and of rubber or of rubber and the compounds commonly employed therewith, or of gutta-percha, or, in fact, of any elastic substance which can be reduced to a soft condition and then vulcanized or hardened sufficiently to answer the purpose. The rubber or other material used is first molded to fit the shape of the mouth and the gums formed, and while soft and pliable the teeth *a a* are embedded in the gums and as an additional security fastened to the same by small pins *c c*, Fig. 3. The teeth, gums, and plate, being thus connected, are then baked until the elastic material becomes sufficiently vulcanized, when the process is completed, the plate being hard and rigid enough for the purposes of mastication, and yet by its elastic nature yielding to the motions of the mouth, while the teeth are firmly embedded and secured in the gums.

The plate and gums can be colored in any proper manner to imitate nature.

Having thus described my improvements, I shall state my claim as follows.

What I claim as my invention, and desire to have secured to me by Letters Patent, is—

Forming the plate and gums in which the teeth are inserted in one piece of hard rubber or vulcanite—*i. e.*, an elastic material which can be hardened sufficiently for the purpose of mastication, and retain a portion of its elasticity so as to yield a little to the motion of the mouth—as herein set forth, and for the purposes specified.

JOHN A. CUMMINGS.

Witnesses:
A. POLLOK,
J. L. COOMBS.

Fig. 1.5 Specification and claims of Cummings' patent

named Samuel P. Chalfant refused to pay licensing fees to the company. To avoid litigation, he closed his practice in Delaware and relocated to St. Louis, only to flee again to San Francisco when the company's lawyers followed him. No matter where he moved, they would inevitably track him down.

In 1879, during a negotiation with a high-ranking company executive, Chalfant snapped and pulled a gun, fatally shooting the man on the spot.[14] The executive, Josiah Bacon, was a key decision-maker behind the company's aggressive litigation strategy. After his death, the company's extensive campaign of lawsuits gradually died down.

Remarkably, the court showed leniency toward Chalfant, sentencing him to just 10 years in prison. While incarcerated, he was treated relatively well—he worked as the prison dentist, avoided manual labor, and didn't have to wear a prison uniform. Occasionally, he was even allowed to treat patients from outside the prison for extra income. The dental community viewed him as a hero, often visiting him and campaigning for his pardon.

One female admirer even disguised him for an escape attempt, helping him board a train to flee. Unfortunately, while eating on the train, his fake beard fell off, arousing suspicion and leading to his recapture. Despite this bold disregard for the law, Chalfant was eventually pardoned and released after serving just 6 years.[15]

This entire episode illustrates how controversial patent enforcement was in nineteenth-century America—and how deeply unpopular "patent trolls" were, even at the dawn of the patent system.

Among the entities listed in Table 1.1, another key player was the American Bell Telephone Company, the predecessor of AT&T. As a pioneer in high technology, Bell played a central role in U.S. patent litigation history, with many landmark cases centered on its innovations. In the twentieth century, other household names such as Thomas Edison and the Wright brothers would become formidable patent holders, intimidating competitors through aggressive enforcement. For much of the twentieth century, AT&T and IBM consistently topped the U.S. Patent and Trademark Office's rankings for most patents granted.

Of course, we shouldn't dismiss these inventors and companies as mere "trolls." One of the core principles of patent law is to enable inventors to profit from their creations. As mentioned, the sudden rise in patent litigation in the latter half of the nineteenth century largely stemmed from the 1836 Patent Act, which allowed for term extensions. These extensions breathed new life into aging patents, giving them decades of renewed enforcement power—leaving most businesses with no time or ability to redesign around them. The lesson is clear: American patent lawyers and technology firms have been honing their litigation strategies for more than two

[14] David J. Owsiany. (2016). 150 years ago the dental profession fought against denture patents. Retrieved February 1, 2024, from Ohio Dental Association Website: https://www.oda.org/news/150-years-ago-the-dental-profession-fought-against-denture-patents/

[15] John Marr. (2015). History's Forgotten "Denture Patent War" Spawned A Truly Bizarre Murder. Retrieved February 1, 2024, from GIZMODO: https://gizmodo.com/historys-forgotten-denture-patent-war-spawned-a-truly-1716425468

centuries. Whether in courtroom argument, legal writing, procedural lobbying, or underhanded tactics like seduction and espionage, they combine historical depth with modern sophistication. When China joined the World Trade Organization, its first generation of patent professionals found themselves up against this formidable legacy. And clearly, this power has little to do with "ideas, values, or religion."

As mentioned earlier, starting in 1793 when Thomas Jefferson ceased his role in substantive patent examination, the U.S. Patent Office handled only formal reviews. To obtain a patent, applicants merely needed to submit the necessary documents, drawings, and models, along with a $30 registration fee—granting was almost guaranteed.[16] By 1836, the total number of U.S. patents issued had reached 10,000.[17] Since the Patent Office lacked the capacity for substantive review, patent validity could only be challenged in court. The ever-growing burden of patent litigation eventually overwhelmed the judiciary, prompting Congress to reform the system in 1836. The revised law reinstated substantive examination, requiring patents to demonstrate originality, utility, and importance.

That same year, the Patent Office appointed its first full-time examiner, Charles Keller. In 1837, the number of examiners increased to two; by 1861, to 12; and by 1870, to 22 full-time examiners supported by 44 assistant examiners. The creation of this modern examination system directly gave rise to the patent agency profession.

In fact, less than 4 months after the 1836 reform took effect, a man named Thomas Jones saw a business opportunity and opened the first patent agency in Washington, D.C. Jones had previously taught school, worked briefly at the pre-reform Patent Office, and served as an editor for the *Journal of the Franklin Institute*, which held considerable influence among engineers at the time. His agency's very first advertisement appeared in that journal, earning him early clients.

In 1837, the Patent Office hired him back as an examiner. However, perhaps dissatisfied with the meager government salary, Jones resigned again in 1838 and returned to his agency.

Jones was not unique—many early examiners left the Patent Office in the 1840s to start their own agencies.[18] These included Charles Keller (the first full-time examiner) and Charles Page (the fourth), among others, many of whom went on to enjoy considerable success. This revolving door between government and private practice created a perfect public-private pipeline.

After leaving the Patent Office, Charles Keller, for instance, partnered with former colleagues to help a client obtain a reissued patent worth over a million dollars—a major scandal at the time. These accounts underscore that the patent industry was already entering a golden age, with rapid growth in patent prosecution, litigation, and monetization.

---

[16] Kara W. Swanson, (2009). The Emergence of the Professional Patent Practitioner. *Technology and Culture*, 50(3), 519–548. https://doi.org/10.1353/tech.0.0298

[17] Cooper, C. C. (1991). Social Construction of invention through patent management: Thomas Blanchard's Woodworking Machinery. *Technology and Culture*, 32(4), 960–998. P6

[18] Kara W. Swanson, (2009). The Emergence of the Professional Patent Practitioner. *Technology and Culture*, 50(3), 519–548. https://doi.org/10.1353/tech.0.0298

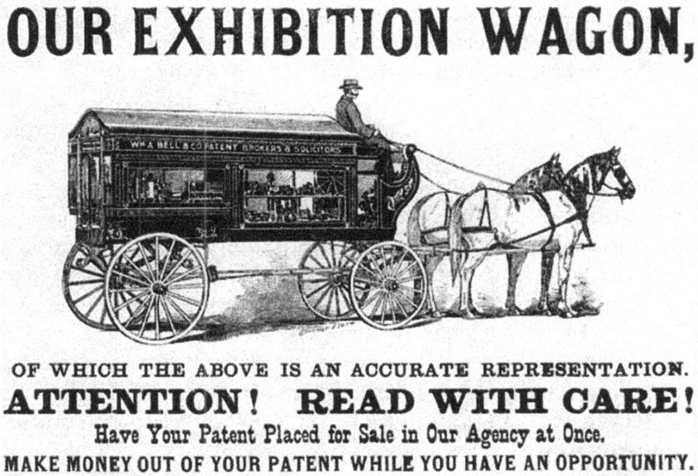

Fig. 1.6 Advertisement of patent agency William A. Bell & Co., 1891

Figure 1.6 shows an 1891 advertisement from the William A. Bell & Co. patent agency, featuring a horse-drawn carriage they used to promote patents in rural counties. The wagon's exterior was painted with diagrams of patented inventions, and inside it carried documents and models for nearly 150 patents. The agency promised to display its clients' patents at trade shows and expositions, and to connect them with manufacturers and investors. The bold headline read:

> ATTENTION! READ WITH CARE! Have Your Patent Placed for Sale in Our Agency at Once. MAKE MONEY OUT OF YOUR PATENT WHILE YOU HAVE AN OPPORTUNITY.

This advertisement is now part of the Smithsonian's National Museum of American History. Records show that the Bell Company's wagon toured the Midwestern states in the summer and fall of 1891 and traveled through the southern states in the winter of 1892.[19]

This may well be the earliest recorded instance of "bringing intellectual property to the countryside"—a practice more than 130 years old, and a phrase frequently used by the Chinese Communist Party and familiar to many Chinese IP professionals..

By the early twentieth century, patent monetization had become a mature business model, and some cases from that era continue to exert influence today. One such example is Thomas Edison's 1893 invention of the *Kinetoscope*—a coined term, typically translated as "motion picture viewer." The *Kinetoscope* is widely recognized as a precursor to modern motion picture technology.

---

[19] Cooper, C. C. (1991). Social Construction of invention through patent management: Thomas Blanchard's Woodworking Machinery. *Technology and Culture*, 32(4), 960–998. P6

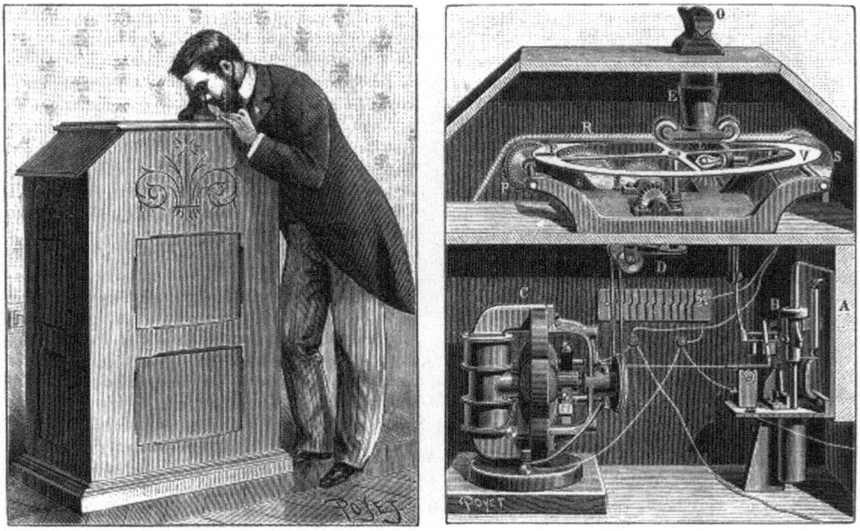

Fig. 1.7 The *Kinetoscope*

From a product perspective, the *Kinetoscope* allowed a single viewer to watch low-resolution moving images in a rather awkward manner (see Fig. 1.7, with the left image showing how it was viewed and the right showing its internal mechanism). It bore little resemblance to today's cinematic experience. However, the scope of Edison's patent was broad enough to encompass the very essence of projected motion pictures.

U.S. Patent No. 589168A, issued for the *Kinetoscope* (or *Kinetographic camera)*, included the following in its Claim 1:

> *An apparatus for effecting by photography a representation, suitable for reproduction, of a scene including a moving object or 0bjects,comprisingameansforintermittently projecting at such rapid rate as to result in persistence of vision images of successive positions of the object or objects in motion, as observed from a fixed and single point of View, a sensitized tape-like film, and a means for so moving the film as to cause the successive images to be received thereon separately and in a single-line sequence.*

Claim 5 read:

> *An unbroken transparent or translucent tape-like photographic film having thereon equidistant photographs of successive positions of an object in motion, all taken from the same point of View, such photographs being arranged in a continuous straight-line sequence, unlimited in number save by the length of the film, substantially as described.*

It is clear that Edison's patent claims covered the basic principle behind film projection—that is, the phenomenon of persistence of vision created when a sequence of images on film moves past the viewer's eyes at constant speed. The medium and venue of projection were irrelevant to the scope of the patent (Fig. 1.8).

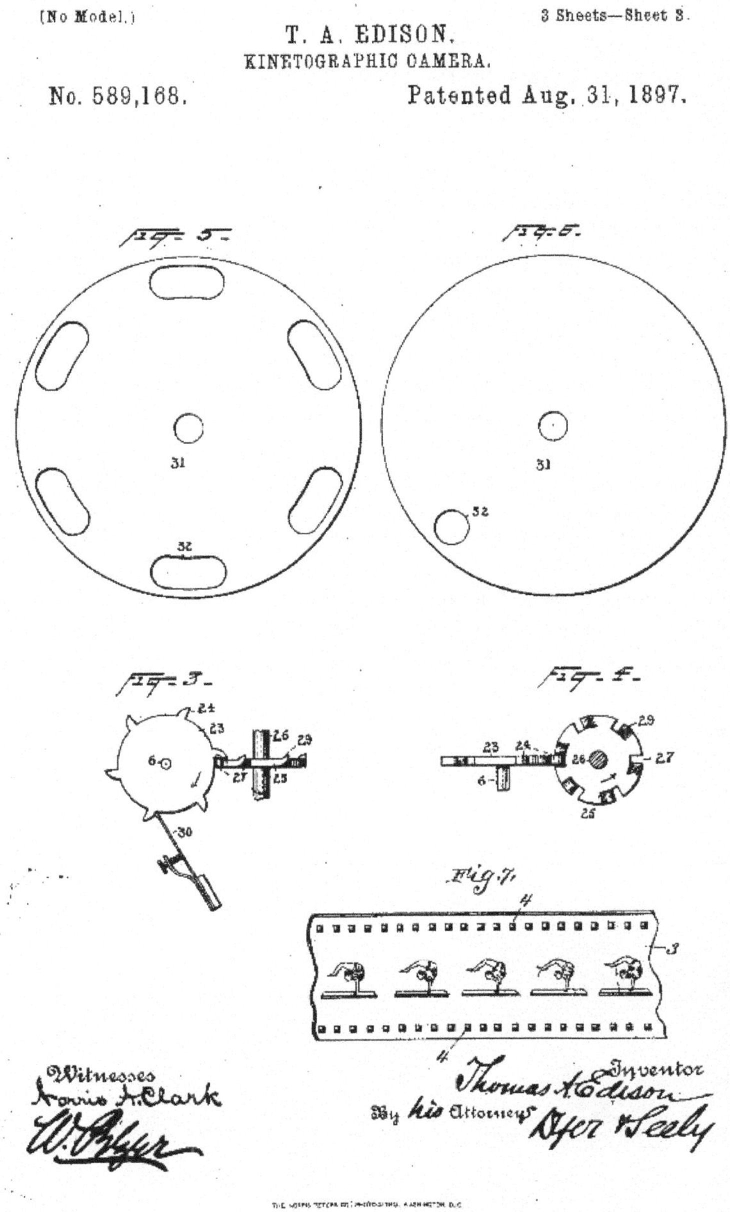

Fig. 1.8 Illustration of Edison's Kinetoscope patent

A few years later, the Lumière brothers in France developed film projection technology based on Edison's groundwork. On December 28, 1895, they held a public screening of projected films in Paris, which caused a sensation. Although Edison had filed his patent earlier, the world generally recognizes this date as the birth of cinema.

Thus, while motion pictures stepped out of Edison's viewing box, they never escaped the legal shadow of his patents.

As the film industry rapidly evolved after the Lumière brothers' exhibition, Edison took notice. Beginning in 1902, he began threatening film distributors and theaters across the U.S. with patent litigation unless they adopted his equipment. In 1907, he launched a series of lawsuits that drove several competitors out of the business.

At that time, the only viable alternative to Edison's cameras was equipment made by the Biograph Company, which was based on a slightly different mechanical concept. Edison attempted to invalidate Biograph's patents through interference proceedings but failed. He then invited Biograph to join forces, and together they established the Motion Picture Patents Company (MPPC), a cartel that monopolized all key patents in film production, exhibition, and film stock.

Kodak became MPPC's exclusive supplier of raw film stock (see Fig. 1.9, featuring Kodak founder George Eastman with Edison).

MPPC aggressively targeted independent filmmakers—through litigation, industrial espionage, and even hired thugs to sabotage production sites. Although Kodak sometimes sold film stock to independents under the table, technical constraints of

Fig. 1.9 George Eastman and Thomas Edison

the time meant that no camera design could evade the combined force of Edison's and Biograph's patents.[20]

At the time, Edison's headquarters were in New Jersey. Initially, the nascent American film industry had been concentrated in the New York area. But due to intense pressure from the MPPC, many independent producers fled across the country to the West Coast, eventually settling in Los Angeles, California.

Los Angeles had mountains, oceans, and even deserts—ideal for diverse film locations. More importantly, in an era of slow transportation and limited communication, it was simply too far from New Jersey for Edison's reach. And so, Hollywood was born.[21] A 2021 article in the venerable American magazine *The Saturday Evening Post* even referred to Edison as an "unintentional founder of Hollywood".[22]

Among Edison's most important patent attorneys were Lemuel Serrell, George Dyers, and Richard Dyers. In Western countries, law schools have long histories, but back then, there were no specialized programs in intellectual property or patent law. These attorneys gained their expertise primarily through family apprenticeship: Lemuel Serrell's father was among the first generation of patent lawyers after the 1836 U.S. patent law reform. George Dyers and Richard Dyers were father and son; after George grew old, Richard inherited the role and continued working for Edison.

Early patent practitioners in the U.S. often passed down their skills through families and navigated a "revolving door" between government and private practice, accumulating practical experience along the way. Within this historical context, patent departments began to emerge in American companies—some scholars even consider the patent department to be one of the earliest internal corporate departments in U.S. business history.[23]

In short, by the time China joined the Berne Convention and the World Trade Organization to officially become a member of the global intellectual property community, American patent professionals had already been battling on the patent frontlines for over two centuries. Whether in drafting, mining, litigating, or monetizing patents, their experience was deep and extensive—far beyond the reach of China's first generation of IP practitioners.

To be fair, throughout the 200-year history of U.S. patent law, countless dedicated legal professionals have worked hard to secure rights for inventors and the public alike, bridging legal gaps and shaping doctrines that still benefit society. But alongside them were numerous individuals and entities that operated on the fringes of the law—leveraging public power for private gain, suppressing rivals, or

[20] Bowers, Q. D. (1997). *Thanhouser Films: An Encyclopedia and History*. Retrieved February 1, 2024, from https://www.thanhouser.org/tcocd/Narrative_files/c2s5.htm

[21] Bach, S. (1999). Final cut: Art, money, and ego in the making of heaven's gate, the film that sank united artists. Newmarket Press. ISBN: 9781557043740

[22] Garrett O'Brien. (2021). Thomas Edison: The Unintentional Founder of Hollywood. Retrieved February 1, 2024 from Saturday Evening Post Website: https://www.saturdayeveningpost.com/2021/03/thomas-edison-the-unintentional-founder-of-hollywood/

[23] Drahos, P., & Braithwaite, J. (2017). *Information feudalism: Who owns the knowledge economy*. Routledge. pp45

undermining innovation, all in direct contradiction to the original intent of the patent system.

At the same time, the American academic community has engaged in deep reflection on the legitimacy, values, and ethics of patents. These discussions have fostered a rich body of thought, making it easier for people to believe that the American IP system is superior in terms of ideals and principles.

In our view, the true value of patent professionals lies in their ability to safeguard innovation. Their principles should be grounded in upholding the law and serving the public. For companies, the value of patents is realized through the implementation of technical solutions—by turning them into real products or services. Patents must not be elevated above the products they protect. Broadly speaking, an entrepreneur's mission is to improve lives and change the world, not to terrify competitors through legal intimidation. More specifically, customer satisfaction is the bedrock of a company's survival and growth.

Yes, patents are both swords and shields, and their importance cannot be dismissed. But as ancient Chinese wisdom says, "Weapons are instruments of ill omen; only when absolutely necessary should they be used." While empires may be won from horseback, they cannot be governed from the saddle.

Over the years, dazzled by the sophisticated patent maneuvers of Western tech companies, many in China have come to view patent monetization as a lofty and elite business activity—often at the expense of focusing on products themselves. A common saying goes: "Top-tier companies sell standards, mid-tier companies sell technology, and bottom-tier companies sell products." But since Venice first implemented a patent system in 1474, is there a single company in history that has achieved centuries of success purely through patent monetization?

When it comes to both patent strength and longevity as a technology enterprise, no company embodies both qualities better than IBM. Thus, our story begins with IBM.

In the entire history of human technological progress, IBM stands out as a giant deserving detailed study. Between 1970 and 1990, six IBM employees won Nobel Prizes for their work—contributions that left an indelible mark on humanity.

In 2021, IBM was granted 8682 patents by the United States Patent and Trademark Office (USPTO), marking its 28th consecutive year at the top of the agency's annual patent rankings.[24] It wasn't until 2022 that IBM relinquished its crown to Samsung.

This enormous patent portfolio gave IBM an unbeatable position in litigation. Though occasionally targeted by NPEs or so-called patent trolls, few companies with real operations dared to bring a patent infringement suit against IBM. As prominent patent attorney Clegg Ivey once said, "*IBM is perhaps the most formidable IP litigation opponent imaginable, not least because of its thermonuclear*

[24] IFI Claims. (2022). 2021 Top 50 US Patent Assignees. Retrieved March 30, 2024, from https://www.ificlaims.com/rankings-top-50-2021.htm

patent portfolio."[25] In a 2013 research report,[26] Villanova law professor Michael Risch noted, "*(IBM) is rarely, if ever, sued for patent infringement by its competitors. The reason is simple: any company that might sue IBM for infringing a patent would face counterclaims for infringing several IBM patents. The result is either no action or a cross-license agreement between IBM and the other party.*" Jerry Rosenthal, then a senior executive in IBM's IP department, proudly stated, "*IBM engineers simply design the best product possible without worrying about patent concerns.*"[27]

Patents, of course, are not just shields—they are also swords. IBM maintains the world's largest patent arsenal not just to defend itself, but to assert its power when necessary. As early as 2000, IBM's annual income from patent licensing had already exceeded $1 billion, setting it apart from the rest of the industry. By 2017, IBM earned $1.466 billion from patent licensing alone, accounting for more than a quarter of its total annual profit that year (approximately $5.753 billion, reduced that year due to U.S. tax reform).

These earnings are undeniably impressive. But even for IBM, a company defined by its vast IP holdings, it was products—not patents—that built its empire and determined its fate.

[25] Parchomovsky, G., & Wagner, R. P. (2005). Patent portfolios. *U. Pa. L. Rev.*, *154*, 1. Retrieved March 30, 2024 from https://scholarship.law.upenn.edu/cgi/viewcontent.cgi?article=1294&context=penn_law_review

[26] Risch, M. (2013). Patent Portfolios as Securities. *Duke LJ*, *63*, 89.

[27] Parchomovsky, G., & Wagner, R. P. (2005). Patent portfolios. *U. Pa. L. Rev.*, *154*, 1. Retrieved March 30, 2024 from https://scholarship.law.upenn.edu/cgi/viewcontent.cgi?article=1294&context=penn_law_review

Part I
Rise and Fall

Chapter 2
From Looms to Computers

The first two inventions that achieved commercial success and could be considered "computers" were both born in France. Their inventors were Charles-Xavier Thomas de Colmar and Joseph Marie Jacquard, respectively.

De Colmar once served under Napoleon, where he was in charge of auditing the army's logistics system. The French army at the time was teeming with mathematical talent; Napoleon himself was a well-known enthusiast of mathematics. He often hosted academic salons, inviting top mathematicians of the day to solve math problems for intellectual entertainment. Laplace was his mentor, while Lagrange and Monge were regulars in his salons. Fourier even held important posts in both his government and military. Thanks to these world-class mathematicians, the French army achieved exceptional precision in everything from ballistics and troop deployment to logistics management. This was one key factor in Napoleon's military success. As a senior logistics officer, De Colmar frequently dealt with complex calculations, which inspired him to design mechanical tools to aid computation.

After Napoleon's fall, De Colmar retired from the military and entered the business world, focusing on the insurance industry, where he did quite well. He founded two insurance companies: one called "Le Soleil" ("The Sun") and the other "L'Aigle" ("The Eagle"). The former was a symbol of the Sun King, Louis XIV, while the latter represented Napoleon. Following the restoration of the Bourbon monarchy, public sentiment remained mixed. Those nostalgic for Napoleon would choose to insure with L'Aigle, while royalist aristocrats and loyal citizens opted for Le Soleil. This strategy of hedging bets allowed De Colmar to profit from both sides. Before his death, the Soleil-Aigle Group became the largest insurance company in France. The choice of company names reveals that, had he lived today, De Colmar would surely have been a branding-savvy entrepreneur.

In his spare time, De Colmar refined the design of his mechanical calculator and filed two patent applications in 1820 and 1822. Figure 2.1 shows the diagram from his 1820 patent application.

De Colmar amassed significant wealth in his lifetime, mostly from his insurance companies. His mechanical calculator served his professional needs and acted as a

Q. Yue, F. Yang, *Governing Innovation: A Global History of Patent-Driven Technological Progress*, https://doi.org/10.1007/978-981-95-5570-3_2

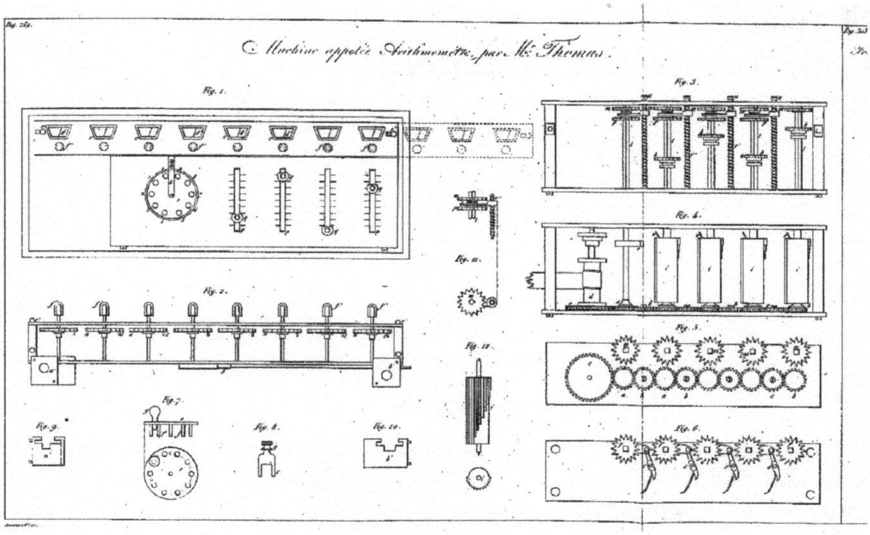

Fig. 2.1 Patent drawing of De Colmar's 1820 patent

personal calling card. But since he was financially secure, he didn't feel compelled to develop it into a serious commercial product—until 1842, when an unexpected humiliation changed his mind. That year, a clockmaker created a similar mechanical calculator that won a gold medal at a Paris exhibition. De Colmar's own calculator, showcased at the same event, only received a silver medal. This loss wounded his pride and spurred him to market his invention more actively.[1] He lavishly adorned the exterior of his calculators to make them as exquisite as works of art and gifted them to royalty and dignitaries. He also made further improvements, filing new patents in 1849, 1850, and 1865. By embracing aesthetics and top-down marketing, De Colmar established his calculator's reputation in the business world. When he died in 1870, his factory had sold about 1000 units[2]—half of them after 1865[3] (Fig. 2.2 shows a surviving example). This modest commercial success signaled that, on the eve of the electrical revolution, the task of "calculation" could already be handed over to machines. Humanity was getting closer to computers.

[1] Alain Guyot. (2021). Research, IT, industry: In Voiron a revolutionary calculator was invented in 1842 (1/10—year 2021) | ECHOSCIENCES. Retrieved March 30, 2024, from https://www.world-today-news.com/research-it-industry-in-voiron-a-revolutionary-calculator-was-invented-in-1842-1-10-year-2021-echosciences/

[2] Jeremy Norman. (2004). The Thomas Arithmometer, the First Commercially Produced Mechanical Calculator. Retrieved March 30, 2024 from Jeremy Norman's HistoryofInformation.com: https://www.historyofinformation.com/detail.php?id=448

[3] Alain Guyot. (2021). Research, IT, industry: In Voiron a revolutionary calculator was invented in 1842 (1/10—year 2021) | ECHOSCIENCES. Retrieved March 30, 2024, from https://www.world-today-news.com/research-it-industry-in-voiron-a-revolutionary-calculator-was-invented-in-1842-1-10-year-2021-echosciences/

Fig. 2.2 De Colmar's mechanical calculator. Photograph by the author at the Computer History Museum (USA)

Slightly earlier than de Colmar's time, there was another Frenchman, Joseph Marie Jacquard, who holds special significance in the history of computing. His invention had little to do with computation itself, but it was closely related to the core logic of today's computers: he invented a loom that used holes in punched cards to control the weaving of various patterns with needle and thread. In a sense, it was a "programmable" loom. As shown in Fig. 2.4, the punched cards (1) were strung together into a chain and wrapped around a wooden box (2) that had perforations on all four sides. The side of the box covered by the punched cards would move toward several control rods (3). If a particular position on the card had no hole, the rod would be pushed toward the side with a spring (9); if there was a hole, the rod would enter it and remain unmoved. This setup enabled the mechanical recognition of 1 s and 0 s purely through physical means. The movement of the rods determined whether the metal hook (5) would lift the thread (4), thus allowing the loom to weave complex patterns.

Jacquard applied for his first loom patent in 1800 and was later granted authorization (Figs. 2.5 and 2.6 show his patent drawings and the physical machine, respectively). During his lifetime, the United States remained relatively stable domestically, while France was mired in turmoil, experiencing seven coalitions against France, the fall of the First Empire, the Hundred Days rule, and two Bourbon restorations. Jacquard could not, like the Americans, retreat behind closed doors and run his own business in peace. Fortunately, successive governments placed high value on his technology, and Napoleon even personally intervened in the matter of his patent royalties. Thanks to this, Jacquard—despite his humble origins—was able to enjoy a comfortable life.

On April 13, 1805, as Napoleon passed through Lyon, he made time during his busy itinerary to summon Jacquard and requested to see his loom. Two days later, Napoleon issued an imperial decree stating that "from the date of this order and for

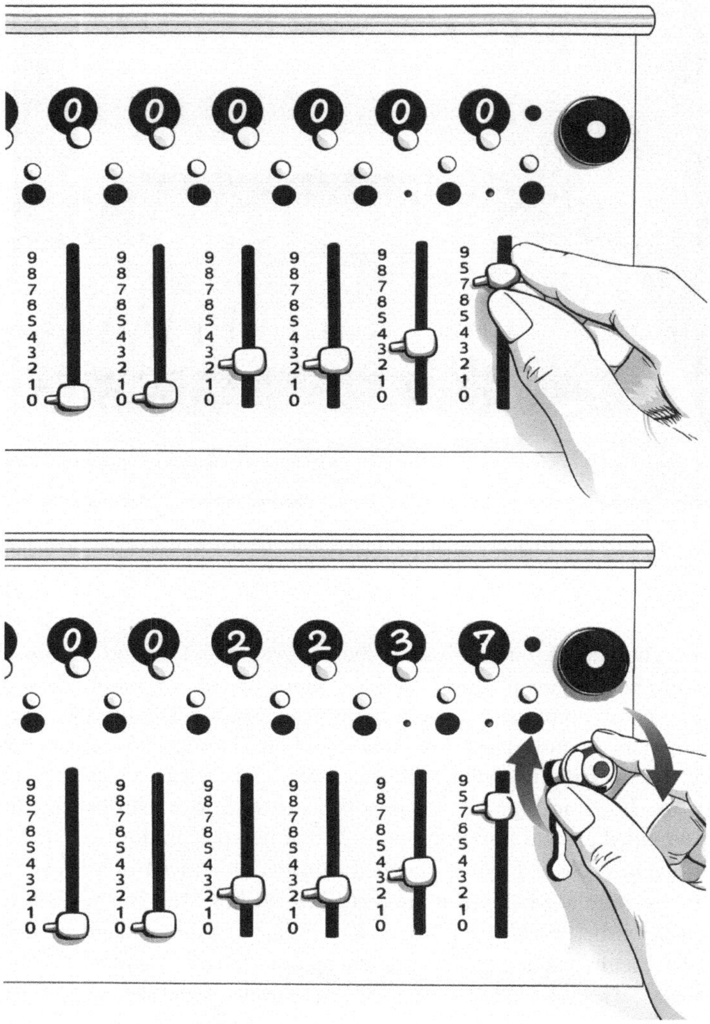

Fig. 2.3 Schematic illustration of the operation of De Colmar's mechanical calculator. As an example of addition, the user slides the levers to select the digits in each position and then turns the crank at the lower right corner of the panel. The entered number is displayed in the black output windows at the top of the panel. The user then slides the levers again to input another number and turns the crank once more. The black output windows will now show the sum of the two numbers. The operation procedure is illustrated in Fig. 2.3.

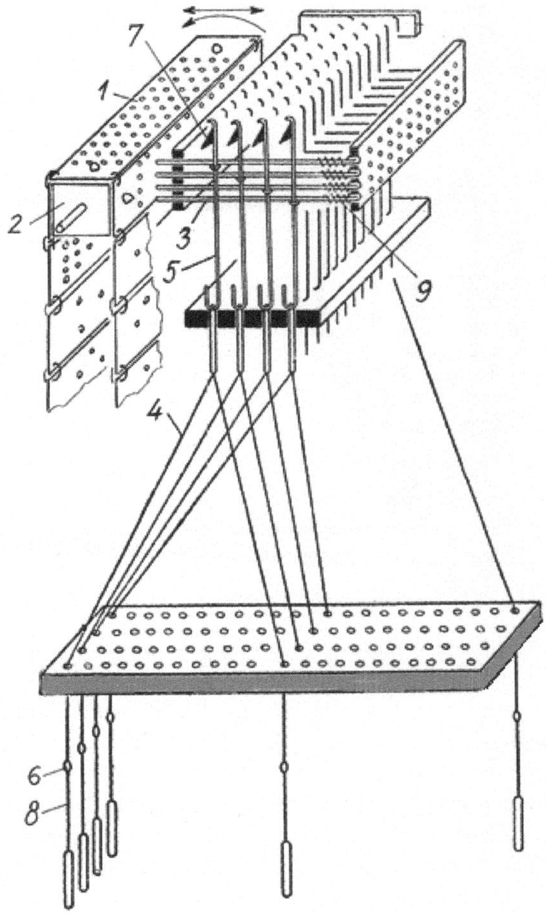

Fig. 2.4 Schematic diagram of the principle of the Jacquard loom. (*Source: Ryj. (n.d.). Jacquard loom punch card sample* [Image]. Wikimedia Commons. https://commons.wikimedia.org/wiki/File:Jacquard_12.png. This image is shared under the Creative Commons CC0 1.0 Universal license (public domain dedication))

the next six years, for each loom delivered for use, a payment of 50 francs shall be made (to Jacquard)." In 1806, Napoleon issued another decree instructing the Lyon municipal government to pay Jacquard an annual pension of 3000 francs and directed him to dedicate all his time and work to the City of Lyon.[4] According to historical records, Jacquard delivered at least 41 looms to the city in 1806 and 57 more in 1811. Figure 2.7 shows an intricate fabric sample produced by the Jacquard loom.

After Napoleon's fall, the Bourbon monarchy did not treat Jacquard as a remnant of Napoleon's regime. On the contrary, they continued to honor him with generous

[4] Marc Allégret. (2001). Jacquard, Joseph Marie, (1752–1834), Mécanicien-Tisserand. Retrieved March 30, 2024 from Napoleon.org: https://www.napoleon.org/histoire-des-2-empires/biographies/jacquard-joseph-marie-1752-1834-mecanicien-tisserand/

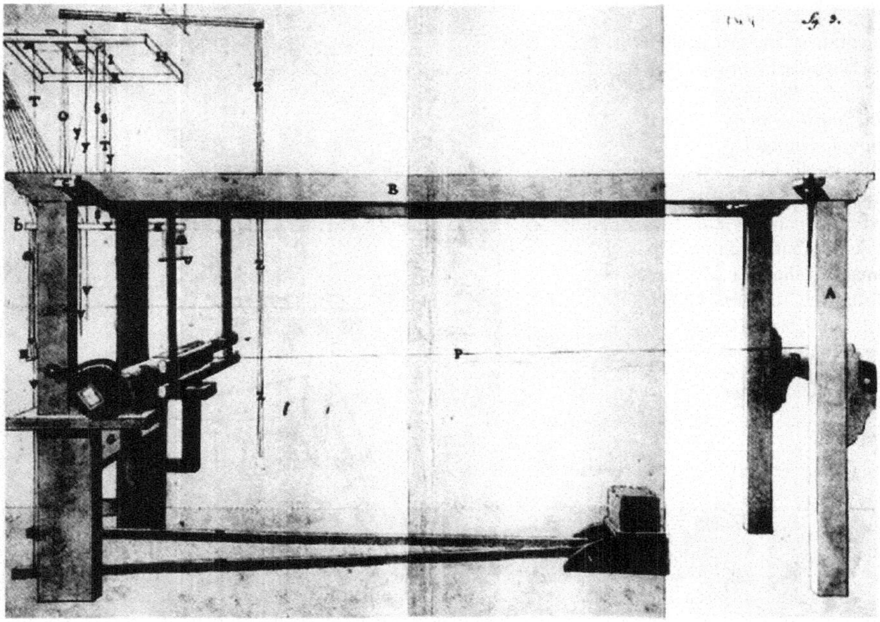

Fig. 2.5 Patent drawing of the loom patented by Jacquard in 1800

treatment. In 1819, King Louis XVIII awarded him the Legion of Honor. The total number of Jacquard looms installed in France exceeded 10,000, significantly reducing the need for manual labor and triggering serious unemployment. Outraged unemployed workers targeted Jacquard personally. During one public uprising, Jacquard, as the inventor of the automatic loom, was beaten by a mob. Had the police not arrived in time, he might have been thrown into the Rhône River.[5]

As mentioned earlier, the Jacquard loom had no direct relation to "computation," but it inspired a true pioneer of computing: British mathematician Charles Babbage. Unlike the insurance magnate de Colmar or the industrialist Jacquard, Babbage was a purer kind of scientist. Inspired by the Jacquard loom, he proposed the concept of the Difference Engine—a device designed to compute polynomial functions. However, Babbage's demands for mechanical precision far exceeded the technological capabilities of his time. After 20 years of work, he had not only exhausted the £17,000 provided by the British government, but also spent most of his own fortune, ultimately completing only a small portion of the prototype (Fig. 2.8 shows the unfinished Difference Engine).

Jules Verne's famous novel *Around the World in Eighty Days* was published in 1872, about a year after Babbage's death. In the novel, the protagonist Phileas Fogg

[5] Marc Allégret. (2001). Jacquard, Joseph Marie, (1752–1834), Mécanicien-Tisserand. Retrieved March 30, 2024 from Napoleon.org: https://www.napoleon.org/histoire-des-2-empires/biographies/jacquard-joseph-marie-1752-1834-mecanicien-tisserand/

Fig. 2.6 Actual Jacquard loom controlled by punched cards, with a far greater number of holes than in Fig. 2.4 (Source: Ulf Malmros. (n.d.). *Jacquard loom sample display* [Image]. Wikimedia Commons. https://commons.wikimedia.org/wiki/File:Jacquardv%C3%A4vstol_Norrk%C3%B6pings_stads-museum.jpg. This image is shared under the Creative Commons CC0 1.0 Universal license (public domain dedication))

possesses a total fortune of approximately £40,000. He stakes £20,000 on a wager that he can travel around the world in 80 days—a move regarded by many wealthy gentlemen as a reckless gamble. Fogg's entire expense for the journey—including the high cost of purchasing a ship to cross the Atlantic—also happens to total £20,000. In other words, the amount of money Babbage spent would have been enough for him to live a refined, leisurely, and aristocratic life like Fogg's, or even to throw away fortunes traveling around the world. But Babbage was not interested in any of that. He insisted on pouring everything into scientific research that offered no foreseeable return. Such unwavering pursuit of dreams and truth by scientists deserves the admiration of every one of us.

In the process of developing the Difference Engine, Babbage went even further and proposed the concept of the Analytical Engine. This time, Babbage's vision was even more ambitious. In his conception, the Analytical Engine would be a large machine powered by a steam engine, equipped with purely mechanical arithmetic

Fig. 2.7 Complex fabric woven by the Jacquard loom (photograph by the author at the Computer History Museum, USA)

logic units and memory, and using punched cards for both input and output. This grand blueprint was far ahead of its time. Like the Difference Engine, the Analytical Engine was never completed during Babbage's lifetime (Fig. 2.9 shows the unfinished Analytical Engine).

Babbage's purity of purpose won him many followers. Ada Lovelace, the only daughter of the great poet Lord Byron, was drawn to this purity and became Babbage's close confidante. She proposed a method for using the Analytical Engine to compute Bernoulli numbers. This visionary idea earned Lovelace the title of "the world's first programmer." Unfortunately, this passionate group of idealistic math enthusiasts, driven by pure love for the subject, ultimately failed to realize Babbage's dream.

Fig. 2.8 Babbage's Difference Engine No. 1, 1824–1832. (Source: Mrjohncummings. (n.d.). *Babbage's Difference Engine No. 1, 1824–1832* [Image]. Wikimedia Commons. https://commons.wikimedia.org/wiki/File:Babbages_Difference_Engine_No_1,_1824-1832._(9660573845).jpg. This image is shared under the Creative Commons CC BY-SA 2.0 license)

Fig. 2.9 Babbage's Analytical Engine, 1834–1871. (Source: Mrjohncummings. (n.d.). *Babbage's Analytical Engine, 1834–1871* [Image]. Wikimedia Commons. https://commons.wikimedia.org/wiki/File:Babbages_Analytical_Engine,_1834-1871._(9660574685).jpg. This image is shared under the Creative Commons CC BY-SA 2.0 license)

The entrepreneur Thomas de Colmar made a fortune through his invention and mathematical skills, while the scientist Charles Babbage exhausted most of his life and wealth in pursuit of a dream. De Colmar's calculator could only perform basic arithmetic operations. Though expensive, it was still affordable for most businesses. In contrast, Babbage's envisioned machine was conceptually much closer to a modern computer, but its construction complexity far exceeded the technological capabilities of the time—even the combined efforts of Britain's top talent couldn't bring it to life. As patent practitioners, they met with very different fates—something well worth our reflection.

Chapter 3
From Patent Examiner to CTO

As previously mentioned, by the mid-nineteenth century, patent litigation and licensing had already become important tools for technology enterprises. In the rise of IBM, patents likewise played a crucial role.

When people talk about IBM, what they often recall is the rapid rise under Thomas J. Watson Sr., the dominance under Thomas J. Watson Jr., and the turnaround led by Lou Gerstner. Many also know that before Watson Sr. took over IBM, the company's core technological foundation was the tabulating machine invented by Herman Hollerith, also known as the punched card tabulator. However, few have closely studied the invention process and patent history of this tabulating machine. As a result, even those highly familiar with IBM's history may not realize: in matters of patents, Hollerith's level of professionalism far surpassed that of most inventors of his era.

Hollerith had worked as an assistant examiner at the United States Patent Office and later started his own firm as a patent agent. Even when he was fully immersed in inventing and focused on developing the tabulating machine, patent work remained a steady source of income and support. Before his invention achieved commercial success, he had also gone to great lengths to court his future wife, Lucia Talcott. Reportedly, her mother once gave the following assessment of her prospective son-in-law:

> *He has staked everything—all his money, his time and his thoughts—upon this invention, and it would be a terrible disappointment to him if it failed. But should it fail, he can easily make a living with his patent practice. He can have as much or as little of that as he cares to undertake.*[1]

Today, many founders of tech companies, like Hollerith, possess both unique technical insight and strong business acumen and leadership. However, during the difficult early stages of a startup, these companies often lack the conditions to consult

[1] Austrian, G. D. (1982). *Herman Hollerith: Forgotten giant of information processing*. Columbia University Press.

Q. Yue, F. Yang, *Governing Innovation: A Global History of Patent-Driven Technological Progress*, https://doi.org/10.1007/978-981-95-5570-3_3

with top-tier patent professionals, and the quality of early-stage patents is typically not high. Only after reaching a certain scale do companies establish dedicated patent team or begin working with high-quality patent firms. But Hollerith was different from the average tech entrepreneur: he was not only a technical expert but also capable of drafting his own patents. As a result, even before IBM was officially founded, it already had high-quality patents and a mature patent strategy.

Coincidentally, another company that, like IBM, long dominated the top of the U.S. patent grant rankings—AT&T—also had a strong patent expert at its inception. The predecessor of AT&T was the Bell Telephone Company, founded by Alexander Graham Bell, the inventor of telephone technology, who contributed his inventions as equity but had no formal background in patent law. However, another co-founder, Gardiner Greene Hubbard, was different. He was a seasoned attorney and a well-known public figure in both political and commercial circles, with substantial experience in the field of patents. He married his daughter to Bell, became one of Bell's major investors, and personally served as the first president of the Bell Telephone Company. Readers familiar with American patent history will know that Bell rushed to the patent office immediately after inventing the telephone, filing his application just 4 hours ahead of his competitor, securing a patent of epoch-making importance. This was, in large part, the result of his lawyer father-in-law Hubbard's repeated advice and guidance.[2] Hubbard was a formidable figure—between 1876, when Bell's telephone patent was granted, and its expiration in 1894, the Bell Telephone Company was involved in over 600 patent lawsuits and did not lose a single case.[3]

AT&T's history includes many stories related to patents, which we'll discuss in later chapters. For now, let's return to Hollerith. In 1879, he graduated from Columbia School of Mines and later followed one of his professors to work at the U.S. Census Bureau, where he gained practical experience in statistics and computation. At the time, census statistics in the United States were performed almost entirely manually. The only mechanical aid available was a tabulating device invented by Charles W. Seaton, the bureau's chief clerk. We know that the machine which made Hollerith famous—commonly known as the "tabulator"—offered basic computational and statistical functions and is widely regarded as a precursor to the modern computer. In contrast, Seaton's tabulating device was little more than a tool for making tables. Its main function was to help clerks draw grids on continuous sheets of paper and then cut them into individual forms (Fig. 3.1 shows the patent diagram for Seaton's tabulating device). This crude wooden box was granted a patent in 1872 (U.S. Patent No. 127435A), and the Census Bureau purchased the patent from Seaton for $15,000, using the device in the 1880 census. As mentioned earlier, 1872 was also the year *Around the World in Eighty Days* was published. At the time, one British pound was roughly equivalent to five U.S. dollars. So, $15,000 was

[2] Villanueva Garcia, B. (2011). Alexander Graham Bell and the Telephone: The Invention That Changed Communication. p42

[3] Chandler Jr, A. D., & Cortada, J. W. (Eds.). (2000). A nation transformed by information: How information has shaped the United States from colonial times to the present. Oxford University Press. p93

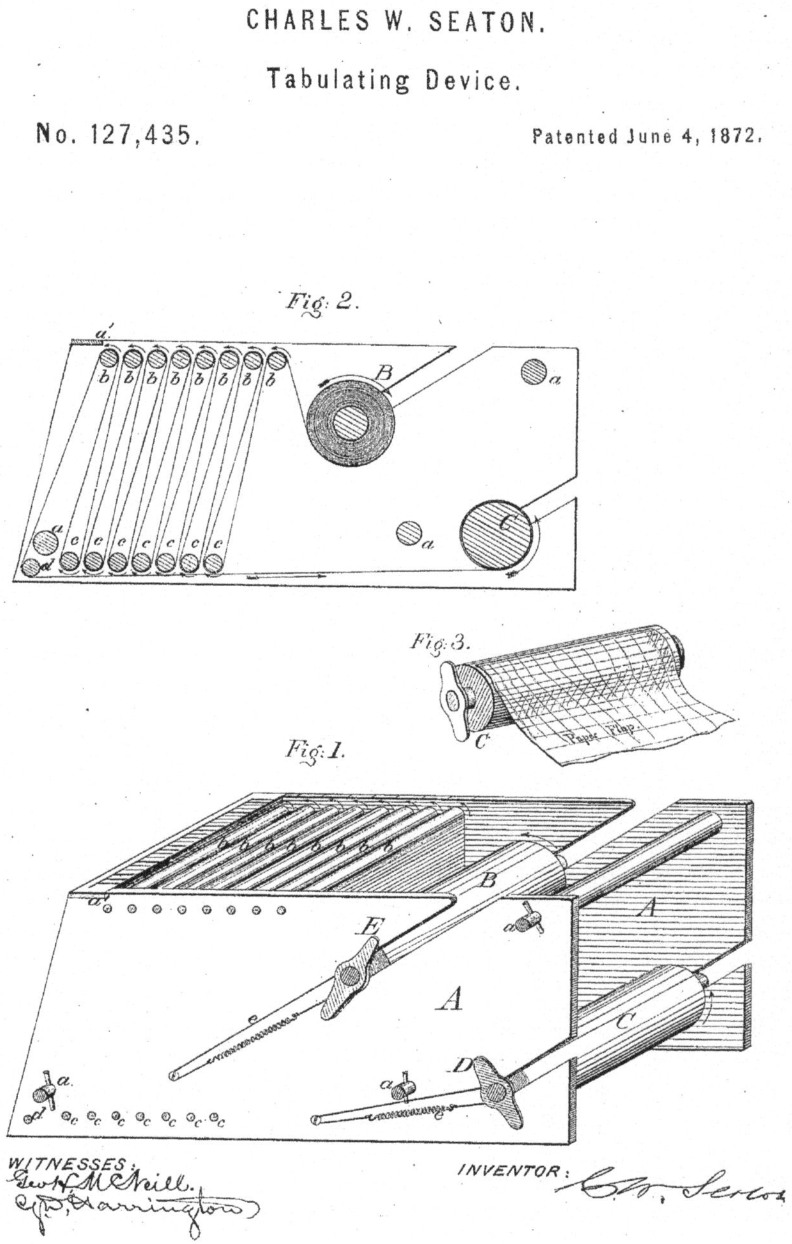

Fig. 3.1 Patent drawing of Charles W. Seaton's tabulating device (U.S. Patent No. 127435A), which in fact was merely a wooden box for making tables

approximately 3,000 pounds—about one-tenth of Fogg's fortune—and certainly a considerable sum. This success left a deep impression on Hollerith, and after personally operating Seaton's device, he quickly formed his own ideas.

Hollerith later left the Census Bureau and briefly held a teaching position at MIT. In the school's laboratory, he assembled a genuine tabulating machine. This machine was no longer the kind of wooden box used for formatting tables like Seaton's but rather a new statistical device that used 0 s and 1 s to represent data. It used holes punched into a paper tape to represent information, electric current to detect the presence or absence of holes, and a punching device to record data. Hollerith devoted tremendous passion to this machine, but showed little interest in teaching. In 1883, he resigned from MIT and, with a recommendation from Charles W. Seaton, took a position as an assistant examiner at the U.S. Patent Office.

The latter half of the nineteenth century was a time of rapid technological advancement, and Western countries placed great emphasis on protecting scientific and technological achievements. Many top-tier scientists served as patent examiners. They not only enjoyed high social status but also earned incomes far above the societal average. For example, in Hollerith's time, one of the most famous patent examiners in history—and one of the greatest scientists of the twentieth century—also worked in the patent field: Albert Einstein. After graduating from ETH Zurich, Einstein joined the Swiss Patent Office in 1902 with an annual salary of 3500 Swiss francs. As one of the most brilliant minds in human history, the routine work of patent examination was a breeze for him. In his spare time, he proposed the quantum theory of light and the principle of special relativity. Not only did this not interfere with his job performance, but his excellence also earned him a raise—his salary increased by 400 Swiss francs per year.[4]

Although Hollerith's salary at the U.S. Patent Office did not match that of Einstein, who held a doctoral degree, he still earned \$1200 annually.[5] Considering that the average annual income in the United States around 1900 was only \$449,[6] this was quite a respectable income. However, Hollerith did not take the position of patent examiner in order to earn a stable salary. His real goal was to become a great inventor and entrepreneur like Edison or Bell. To achieve that goal, Hollerith began preparing early on, with a clear plan to use the patent system to support his future endeavors.

In 1884, believing he had accumulated enough experience in the patent field, Hollerith resigned from the Patent Office and started his own business. He presented himself publicly as a patent expert and agent while continuing to develop his tabulating machine. At the time, the patent agency industry in the United States was already well developed, but there was still no standardized terminology. Some called themselves "patent solicitors," others "patent agents," and still others "patent counsel." The title we are familiar with today, "patent attorney," had not yet become

---

[4] Sparknotes. Albert Einstein. Retrieved March 30, 2024 from Sparknotes Website: https://www.sparknotes.com/biography/einstein/section4/

[5] Austrian, G. D. (1982). *Herman Hollerith: Forgotten giant of information processing*. Columbia University Press.

[6] Susan Eck. (2001). Salaries in 1901. Retrieved March 30, 2024, from Doing the Pan Website: https://panam1901.org/visiting/salaries.htm

widespread.[7] Thus, Hollerith simply referred to himself as an "expert" and "agent." In a private letter dated September 1884, Hollerith wrote:

> *As regards patent rights, as I am the first in the field of invention, I can secure what is technically known as a 'foundation patent' covering the ground broadly and, therefore, all subsequent improvements would be subject to my broad patent...I could also, of course, secure patents in foreign countries (England, Germany, France, etc., all of which countries compile elaborate statistics).*[8]

This passage shows that even though his invention was still in its infancy, Hollerith already had a clear understanding of patent strategy, including international planning. In the following decades, Hollerith and the early IBM indeed followed this basic approach to expand their patent portfolio. From its very inception, this century-old company possessed the three critical elements needed for a technology firm to build a strong intellectual property position: a great invention, high-quality patents, and a carefully thought-out patent strategy—laying a solid foundation for its later success.

On September 23, 1884, Hollerith submitted a patent application for his invention titled "Art of Compiling Statistics." On October 27, 1885, he filed a divisional application. On January 4, 1887, and again on September 8, 1888, he made amendments to the case through reissue procedures, extending the authorization timeline. On January 8, 1889, both the main case and the divisional were granted, with patent numbers US395783A and US395782A, respectively. These two patents primarily protected the basic concept and operational steps of using holes in paper tape or cards to record data.

As shown in Fig. 3.2, Hollerith's tabulating system mainly consisted of cards or paper tape, a punching device, a card reader, and a dial display. Each card represented a U.S. resident, and the punched holes at different positions on the card encoded information such as age, gender, occupation, income, and address. Clerks would punch the cards or tape according to census data, marking the corresponding positions. After punching, the cards or tape were inserted into the card reader. Inside the reader were metal probes, each aligned with a potential punch position on the card or tape, and each corresponding to a dial on the machine's top panel. Each dial represented one type of information—such as age, gender, occupation, income, or address. Once the card was inserted into the reader, the probes would drop into the holes. Where there was a hole, electric current would pass through; where there was none, it would not. If current passed through, the dial corresponding to that probe would move by one increment.

For example, suppose a census clerk receives a form for a white male farmer from the state of Florida. Based on the information on the form, the clerk would punch a new card accordingly and then insert it into the card reader. Once the card is in the reader, the four dials corresponding to "Florida," "male," "farmer," and "white" would each advance by one increment. If the data had to be read and recorded manually from the form, not only would it be time-consuming, but it would also be prone to error.

---

[7] Kara W. Swanson, (2009). The Emergence of the Professional Patent Practitioner. Technology and Culture, 50(3), 519–548. https://doi.org/10.1353/tech.0.0298

[8] Austrian, G. D. (1982). Herman Hollerith: Forgotten giant of information processing. Columbia University Press. p34:

Fig. 3.2 The Hollerith census tabulator. (Source: Marcin Wichary. (n.d.). *The Hollerith census tabulator* [Image]. Wikimedia Commons. https://commons.wikimedia.org/wiki/File:The_Hollerith_census_tabulator_(2,585,340,347).jpg. This image is shared under the Creative Commons CC BY 2.0 license.)

In 1886, Hollerith received a government assignment to compile statistics on mortality in the city of Baltimore. During this work, he borrowed a tool used by railroad conductors to punch tickets in order to complete the punching process. Comparing the cards to the paper tape he had previously used, Hollerith noticed the advantages of cards: they were easier to organize, more portable, could be re-sorted, and subsets could be extracted to compile new datasets. In consideration of future sorting and classification needs, Hollerith decided to add a mechanical device to his tabulating machine dedicated to sorting cards.

In June 1887, Hollerith submitted a new patent application for the tabulating machine equipped with a card-sorting box. This patent was also granted on January 8, 1889, under patent number US395781A. Together with US395782A and US395783A, these three consecutively numbered patents formed Hollerith's earliest patent portfolio. As he had planned, these patents covered the fundamental concepts used in all future improvements of the tabulating machine and laid a solid foundation for further patent portfolio development (see Table 3.1).

From today's perspective, the scope of patent protection in the late nineteenth and early twentieth centuries was broad, and Hollerith's patents were no exception. Patent US395783A covered the basic concept and procedure of recording data with punched holes. The term "perforated sheets of non-conducting material" referred to punched paper tapes or cards, "perforations representing statistical items" was the key feature of such cards or tapes, and "electro-magnetic mechanical counters"

Table 3.1 Hollerith's three consecutively numbered patents

	Date of Application	Number of Claims	Claim 1
US395783A	September 23, 1884	7	*The combination, with perforated sheets of electrically non-conducting materials, said perforations representing statistical items, of electro-magnets and mechanical counters in circuits controlled by said perforated sheets, substantially as and for the purposes specified*
US395782A (a division of patent US395783A)	October 27, 1885	2	*The herein-described improvement in the art of compiling statistics, which consists in, first, forming or arranging a standard or templet indicating the relative position in which each item or characteristic of the individual is to be recorded; secondly, forming a record of each individual or thing by locating index-points upon a strip or tablet, said index-points representing the characteristics of the individual and bearing a determinate relation to each other and to the standard, and, finally, submitting said separate records successively to the action of circuit-controlling devices for operating the registering devices representing the statistical items to be compiled, whereby each statistical item or combination of items when contained in the record of any individual is accurately registered.*
US395781A	June 7, 1887	21	*The improvement in the art of compiling statistics, which consists in first preparing a series of separate record-cards, each card representing an individual or subject; second, applying to each card at predetermined intervals circuit-controlling index-points arranged, according to a fixed plan of distribution, to represent each item or characteristic of the individual or subject; and, third, applying said separate record-cards successively to circuit-controlling devices acted upon by the index-points to designate each statistical item represented by one or more of said index-points, substantially as described.* [3]

referred to the card-reading devices. Claim 1 of US395783A protected the combination of perforated sheets and counters—simple and sweeping.

Patent US395781A described a more complete punched card statistical system, adding a "sorting box" to categorize and store cards. It also narrowed the protected medium from "perforated sheets" to "cards." The feature of the "sorting box" did not appear until claim 12 of this patent.

The fact that early IBM computers used holes in cards or paper tapes to represent data is widely discussed in many science popularization articles and computer science textbooks. Readers of this book are likely already familiar with it. However, the concept of the "sorting box" may be unfamiliar to most readers. Compared to the innovative and impactful idea of "recording data via punched holes," the sorting box seems unimpressive—it is mentioned only in an inconspicuous dependent claim. Yet history has proven that it possessed significant commercial value. Although Hollerith's first patents had broad enough coverage, by the time the tabulating machine had developed into an industry, their protection period of 17 years had already expired. During those 17 years, Hollerith faced no real competitors. For more than half a century afterward, there was almost no advancement in the technology of representing information via holes in paper cards within the tabulating machine field. In contrast, the technology of sorting, classifying cards, and controlling hardware continued to evolve. Starting with the sorting box, these improvements gradually formed IBM's earliest patent barriers.

Figures 3.3 and 3.4 show, respectively, the patent drawing of Hollerith's US395781A and a physical image of the sorting box. The sorting box was a wooden box slightly larger than the punched cards, fitted with a lid, and mounted on the surface of a low table, within easy reach of the tabulating machine operator. Each box was equipped with an electromagnetic device on one end, which was connected to the probes inside the card reader. When the probes detected a specific signal, the lid of the corresponding box would open. Each sorting box could be configured to correspond to various attributes represented by the dials, such as race, gender, or income.

For example, under the broader context of racial segregation at the time, the Census Bureau often archived cards for Black and white individuals separately by gender and income. Without the sorting box, after each card's data was processed, clerks had to manually file it—an extremely time-consuming task. With the sorting box, however, the counter would automatically trigger the lid of the appropriate sorting box to open. For instance, when a clerk inserted a card for an unemployed

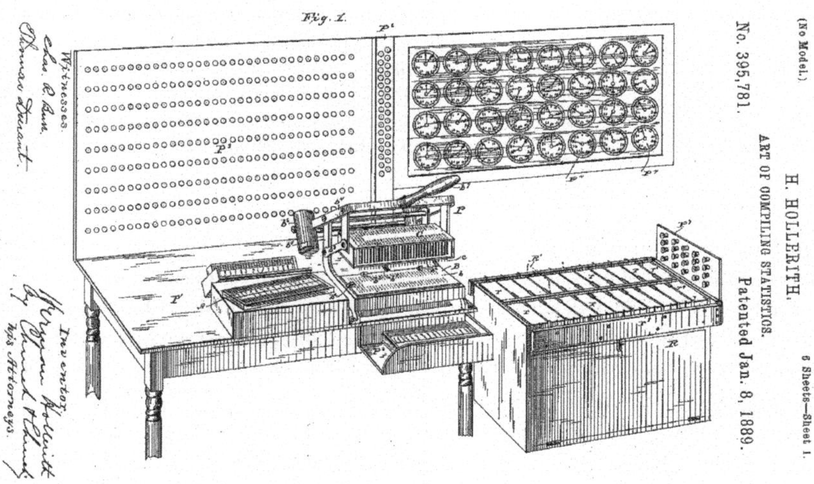

Fig. 3.3 Patent drawing of Hollerith's US395781A

Fig. 3.4 Sorting box of Hollerith's tabulator (photograph by the author at the Computer History Museum, USA)

Black male into the card reader, not only would the corresponding dials for "Black," "male," and "low income" each move one increment, but the lids of the associated sorting boxes would also be triggered by the electromagnetic devices to open immediately. The clerk then simply needed to pull the card out and drop it into the open box, and the filing was complete.

On the surface, the sorting box seemed to merely save clerks the effort of manually organizing cards. But its real significance extended far beyond that. Once sorted, the cards were no longer a chaotic pile but had become an organized and filtered dataset—the very definition of "data." Thanks to the sorting box, the tabulating machine could now offer not only numeric output on dials, but also categorized, sorted datasets. The neatly stacked cards in each compartment of the sorting box were, in fact, the original form of what we now see in spreadsheet software like Excel. From this perspective, Hollerith's tabulating machine can truly be considered a computer.

After leaving the Patent Office, Hollerith secured a number of government contracts for local censuses or similar statistical tasks. He continued improving the tabulating machine as he used it in practice. In 1889, the U.S. Army leased his machine to compile statistics on ill soldiers. This leasing contract laid the foundation for the business model that would later be adopted by IBM—and eventually by Xerox as well.

The leasing model offered two key advantages. First, it saved customers the initial capital investment. Hollerith's early clients were mostly government agencies, where cost-saving was seen as a political achievement. Second, under a leasing model, Hollerith was obligated to maintain and service the machine, eliminating post-sale disputes. Bell Telephone Company had also carried out extensive leasing operations in the 1870s. In the 1960s, Xerox used the same model for its high-speed but expensive and fragile photocopiers. This business model emphasized long-term relationships with major clients and government agencies and relied on elite sales teams who were well-versed in the technology and customers' needs. Both the rise and fall of IBM were closely tied to this model—a subject we'll explore further in later chapters.

At the time, Hollerith set the annual rental fee for his tabulating machine at around $1000. By comparison, the annual salary of a professional statistician was about $550.[9] But a single tabulating machine could process far more data than one or two statisticians.

The 1890 U.S. Census marked a milestone in Hollerith's entrepreneurial journey. Thanks to his tabulating machine, this census took only 2 years to complete. In contrast, the previous census—despite covering a smaller population and handling fewer data—took the Census Bureau 10 years to finish. The 1890 census saved the government an estimated $5 million in costs.[10] After that, Hollerith secured more contracts with the U.S. government, as well as some orders from Europe and Canada. He also received numerous personal honors and memberships in scientific societies across Europe, earning him widespread respect in political, business, and academic circles as a technical expert. However, the work was not easy: on one hand, Hollerith worked alone and had no permanent employees; on the other hand, since census projects in various countries were short-term, he had no stable long-term contracts and had to travel constantly, exhausting himself in the process. Seeking stability, Hollerith began exploring other business opportunities.

At the time, insurance companies and railway firms were also dealing with massive amounts of customer data. Once Hollerith gained fame, these companies approached him for collaboration. He continued improving his machines, adding new designs and features tailored to the needs of government agencies and large corporations—including various types of punching devices (such as those in patents US487737A and US676433A), and systems that used hydraulic power to transport and sort cards (e.g., patent US526130A).

In the 1890s, Hollerith also conceived the idea of adding a printing function to his tabulating machine. In a letter to his patent attorney, he described his design for this printing function and even drew a diagram. The system included an input dial, a roller for printing, paper, and an ink container—all designed to interface with the tabulating machine. He also mentioned having built a counter with similar functionality, which was unfortunately lost when the client company relocated. Regrettably,

---

[9] Heide, L. (2009). Punched-card systems and the early information explosion, 1880–1945. JHU Press. p34

[10] Cortada, J. W. (2000). Before the computer: IBM, NCR, Burroughs, and Remington Rand and the industry they created, 1865–1956. Princeton University Press. p48

whether due to Hollerith's own oversight or that of his patent attorney, this design was never patented.[11]

In 1900, Hollerith signed a contract with the Census Bureau to develop an automatic tabulating machine. The resulting patent, US685608A, which he filed on March 13, 1901, further advanced the machine's automation. It stated: "After the card is positioned and the device is set in motion, including feeding the card, editing or tabulating the data on it, and removing and classifying the card—all these operations are fully automatic and controlled by the data itself." This meant the machine could not only receive input but also perform computations and output results, organize and classify data, and even control the hardware. This technology marked a major step toward transforming the tabulating machine into a true computer.

Thanks to these technical improvements, the punched card machine's processing speed far outpaced that of human labor. In the 1890s, Hollerith was still struggling to secure contracts in the European and American markets. But after 1900, as automation increased, he finally began to achieve notable commercial success.

In 1896, after securing the major contract for the 1897 All-Russia Census and landing a key client in the New York Central Railroad, Hollerith finally had sufficient capital to formally register his "Tabulating Machine Company"—one of IBM's predecessor companies. He transferred all of his patents to this new firm. In today's business lingo, the Tabulating Machine Company was undoubtedly a unicorn: it possessed unique technology, enjoyed strong governmental ties, and counted high-credit clients—mainly government agencies and large corporations—among its customers.

This unicorn company soon encountered its first heavyweight competitor. After the turn of the twentieth century, the Census Bureau grew increasingly dissatisfied with the high rental fees for Hollerith's tabulating machines and sought to eliminate dependence on them by developing its own statistical equipment. At the time, Hollerith's earliest three patents were nearing expiration, but more than a dozen subsequent patents remained valid. To address this, the Census Bureau brought in an expert named Harry H. Allen from the Patent Office to study Hollerith's patents and oversee the development project. In 1907, shortly after Hollerith's initial three patents expired, a Census Bureau employee named James Powers completed the design for a new tabulating machine.

As mentioned earlier, Hollerith's machines detected holes in punched cards using electrical current—a feature protected by his still-valid patents. To avoid this feature, Powers went to great lengths to devise a mechanical method for detecting holes. In his machine, the punched card used to input data would be inserted into a "pin box," where a probe was installed for each potential punch position. If a hole was present, the corresponding pin would drop through it and push a rod on the other side of the card (labeled "push rods" at position 29 in Fig. 3.5). These rods then triggered a complex set of mechanical linkages that moved the dials, thereby

---

[11] Austrian, G. D. (1982). Herman Hollerith: Forgotten giant of information processing. Columbia University Press.

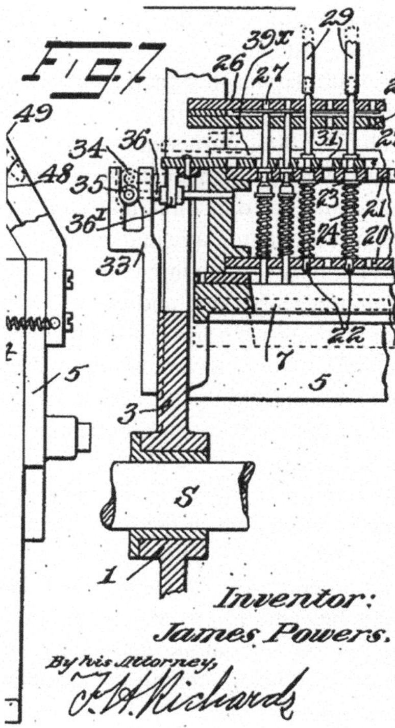

Fig. 3.5 Patent drawing of Powers' US1061118A

completing the input process. In the description of patent US1061118A, Powers criticized existing technology for relying on "complicated electrical apparatus," requiring "a large number of intricate adjustments between various conductors," and being prone to "poor contact caused by dust or small foreign matter." His explanation came off as mere stubborn justification. After all, the idea of using mechanical pins and rods to detect card perforations was reminiscent of the Jacquard loom from a hundred years prior. In an era when the electrical revolution was in full swing, reverting to purely mechanical structures because of "complexity" and "contact issues" could hardly be called innovative—it was merely a workaround to avoid infringing on patents.

Patent expert Harry H. Allen concluded that Powers's machine was essentially a scaled-up version of Hollerith's tabulating machine with a new sorting box tacked on. The problem was, Powers's sorting boxes were far inferior in performance: it took 30 of Powers's sorting boxes to match the capability of 10 of Hollerith's.[12] Hollerith's patent US526130A, granted in 1894, used hydraulic mechanisms to control the sorting boxes. Another patent, US685608A, implemented machine control

[12] Austrian, G. D. (1982). *Herman Hollerith: Forgotten giant of information processing*. Columbia University Press. p285

via the cards themselves. Hydraulics was the fundamental power source of the time, and the latter patent dealt with automation and core computing principles—obstacles Powers couldn't circumvent. Allen ultimately concluded that the Census Bureau had only three options: buy new sorting boxes from Hollerith's Tabulating Machine Company; purchase a license for Hollerith's sorting box patents; or retrofit the 20 sorting boxes Hollerith had sold to the Census Bureau in 1902. However, even the retrofit option risked infringing Hollerith's patents. After weighing the options, the Census Bureau chose to proceed with retrofitting.

In 1910, shortly before the start of a new census cycle, Hollerith filed a patent lawsuit against the Census Bureau and requested a temporary injunction to stop it from modifying his sorting boxes. According to limited historical records, the Census Bureau was in no position to mount a strong defense on patent grounds and could only argue that the sorting boxes delivered by Hollerith in 1902 were of poor quality and required repair and modification. However, this contradicted earlier public statements from Census Bureau officials, who had long praised Hollerith's machines.

The initial ruling was not in Hollerith's favor, but the appellate court overturned that decision. The two sides continued to clash in court, but due to the passage of time, few reliable records of these proceedings remain. All we know is that a settlement was reached in May 1912, but the Census Bureau did not pay Hollerith any settlement fee.

This case involved more than just patent issues. In the early twentieth century, the U.S. government still enjoyed a certain degree of sovereign immunity in patent litigation. Hollerith lobbied Congress intensively, pushing for legislation that would grant him the right to sue the Census Bureau. Even after gaining that right, Hollerith never received any compensation. In a letter to his lawyer, he explained that proving damages would be extremely difficult and resource-intensive, detracting from the company's daily operations. Moreover, the machines in question were over a decade old, and most people familiar with them had already left the company. Given these uncertainties, he doubted that any damages awarded would even cover the legal costs.[13] The challenges Hollerith faced over a century ago remain highly relevant to modern businesses seeking to enforce patent rights.

We believe that it often takes a long time for a technology to evolve from scientific principle to prototype and finally to industrial application—sometimes longer than the patent's 20-year protection period (and back then, U.S. patents lasted only 17 years). Only after a product is implemented and refined through market feedback and user needs can truly powerful, game-changing patents be forged. The tabulating machine is a perfect example: the first batch of patents was granted in 1890, but these patents expired without facing any real competition. Once those patents expired, the Census Bureau was free to design its own tabulating machines. Yet to avoid infringing on Hollerith's sorting box patents, the Bureau went to

[13] Austrian, G. D. (1982). Herman Hollerith: Forgotten giant of information processing. Columbia University Press.

extraordinary lengths, even bringing in patent office experts to assist. Still, the resulting sorting box performed poorly, and the Bureau had to retrofit decade-old machines. When Hollerith filed his lawsuit, only one machine had been modified, and the next census was about to begin. Without the sorting box, the Census Bureau would have been unable to complete its statistical work. Switching to manual sorting would have cost at least $500,000.[14] This clearly underscores the importance of Hollerith's sorting box patents (such as US526130A and US685608A).

We will further explore the relationship between technical principles, foundational patents, and subsequent technological improvements in the chapter titled "The Limits." After all, the Census Bureau was Hollerith's client, not his competitor, so there was no need for him to push them into a corner. Next, we will turn our attention to Hollerith's true competitors.

[14]Austrian, G. D. (1982). *Herman Hollerith: Forgotten giant of information processing*. Columbia University Press. p291

Chapter 4
From Unicorn to Monopoly

In 1911, Hollerith's Tabulating Machine Company merged with two other commercial machine companies to form the Computing-Tabulating-Recording Company (CTR). In the same year, Powers left the Census Bureau and founded the Powers Accounting Machine Company, which would become CTR—and later IBM—IRA's biggest competitor. Between 1911 and 1914, Powers's company won numerous clients, forcing CTR to scale back its research budget and temporarily shift machines to Europe for rental before Powers expanded there.

By this time, Hollerith was in his 50s and had become more conservative in product development. During this period, Thomas J. Watson Sr. entered CTR, strengthened marketing and key account management, and helped CTR reclaim its market position. In 1924, CTR was renamed International Business Machines Corporation—the IBM we recognize today.

Watson's rise within IBM is well-documented in numerous sources—this book will not retell the story. Interested readers can refer to *The Maverick and His Machine: Thomas Watson, Sr. and the Making of IBM* or IBM's "Heritage" section on its official site. Here, our focus remains on IBM's early patent history.

Although Watson came from a business background and had no patent experience, he was well aware of the power of patents. Before joining CTR, Watson had worked as an executive at National Cash Register (NCR), a leading office equipment firm. NCR, founded in 1884, created history's first mechanical cash register and by mid-twentieth century became one of the earliest entrants into electronic computing—remaining a Fortune 500 company today .

NCR's patent story is equally intriguing. In the 1870s, restaurant owner James Ritty and his brother invented and patented a cash register (US 221 360A) to prevent employee theft at checkout. Despite being licensed, their product sold only one unit—a testament to their lack of business savvy despite their inventive spirit.

That single machine was purchased by John Patterson, the so-called "father of modern sales." Patterson recognized its potential, acquired the patent, and led NCR to national success.

Q. Yue, F. Yang, *Governing Innovation: A Global History of Patent-Driven Technological Progress*, https://doi.org/10.1007/978-981-95-5570-3_4

Unlike tabulating machines, cash registers had simpler designs and were easier to replicate, with a broader market—from banks and insurance companies to street-level shops. As a result, NCR quickly faced stiff competition. Like those we covered previously, NCR cultivated an aggressive patent-enforcement approach—filing lawsuits every year from its founding into the 1910s, often forcing competitors to shut down.[1] However, NCR's unchecked dominance soon sparked backlash. The Sherman Antitrust Act of 1890, the world's first antitrust legislation, began to shape sentiment. By the 1910s, NCR's market share hovered at around 90%, drawing the attention of the U.S. Department of Justice.

One defeated competitor was Henry S. Hallwood, who, after going bankrupt, founded a new cash-register company and was again sued by NCR. Unlike Dr. Chalfant, Hallwood did not resort to violence—instead, he petitioned the Department of Justice. In 1911, the DOJ filed criminal charges against over 20 NCR executives for "restraint of trade," including Watson. The initial ruling was against NCR, and Watson faced possible jail time. Fortunately, a devastating flood in NCR's hometown in 1913 prompted NCR to offer aid—opening facilities to the public and mobilizing employees to assist recovery efforts. This philanthropic gesture improved NCR's public image and brought political favor, ultimately resolving the case via consent decree .

Learning from NCR's costly overreach, Watson avoided using patents to drive competitors out of business. Soon after joining CTR, he personally visited Powers's company to offer licensing terms. His proposal was generous: 25% of gross rental revenue plus 18% of card sales—equating to about 20% of Powers's total income. However, he withheld key patents: since Hollerith machines used electrical detection and Powers used mechanical detection, Watson did not license Hollerith's electrical detection patents to Powers.[2] As *The Maverick and His Machine* notes, some viewed his demands as "*magnanimous*"—a sign of his willingness to welcome competition Moreover, tabulating machines remained a niche, far more technically complex and small-scale than cash registers—meaning antitrust scrutiny had not yet caught up.

The licensing fee became a heavy burden hanging over Powers' company. In the following years, its cash flow remained tight, and by 1922, the company was on the brink of bankruptcy. At that point, the tabulating machine market had grown large enough to attract attention from antitrust authorities. Not wanting to drive his only competitor out of business, Watson generously agreed to cut the licensing fee in half. Even so, Powers' business remained unsustainable and was eventually acquired by Remington Rand in 1927. At the time of acquisition, Powers' company still owed IBM $350,000 in unpaid licensing fees, which Remington Rand later paid to IBM.

[1] Cortada, J. W. (2000). Before the computer: IBM, NCR, Burroughs, and Remington Rand and the industry they created, 1865–1956. Princeton University Press. p75

[2] Pugh, E. W. (2009). Building IBM: shaping an industry and its technology. MIT press. p250

Remington Rand would go on to become another key player in the early world of business machines and computing. Like NCR and IBM, it was a company built on proprietary, patent-protected mechanical technology.

In 1868, American inventors Christopher Sholes and Carlos Glidden made significant improvements to traditional typewriter technology, for which they received U.S. Patent No. 79,265. The improved machine became known as the "Sholes and Glidden typewriter." Sholes himself was not optimistic about the invention's future and sold the patent to the Remington Company. Remington, however, saw great value in the patent and refined the design before bringing it to market—with tremendous success. One technical adjustment, in particular, had a lasting impact that most readers will have experienced firsthand: because Sholes' original design allowed typing speeds so fast that the machine would jam, Remington altered the keyboard layout to the now-infamous "QWERTY" arrangement. This layout, illogical and hard to memorize, deliberately slowed typing speed to reduce jams, thereby making the machine more durable. And so "QWERTY" became the global standard. Between 1882 and 1887, Remington's typewriter sales multiplied several times over,[3] and the machine became standard equipment in offices across government and industry. The company emerged as a leading American technology enterprise. After acquiring Powers' company, Remington Rand became IBM's primary competitor.

After Watson took over, Hollerith gradually stepped away from company affairs, and all R&D decisions were handled by Watson alone. Though IBM maintained its technological lead, this didn't guarantee success in every area. Readers familiar with international patent practice may know that early U.S. patent law followed a "first to invent" system. A corresponding procedure called an "interference proceeding" was used by the U.S. Patent Office to determine patent ownership when multiple parties claimed the same invention. One of Hollerith's major patent applications was challenged by Powers' company and entered such a proceeding.

In Hollerith's early designs, the tabulating machine included a feature called a "stop card." When the machine detected a stop card, it would halt operations, compute the total, and reset the counters to zero.[4] Of course, this stop card had to be manually inserted by the operator in the appropriate position. However, in a patent application submitted in March 1914 (U.S. Patent No. 1,830,699), Hollerith invented a feature for "automatic control of the tabulating machine," which enabled the machine to automatically stop and execute calculations after processing a batch of cards. The patent faced a prolonged authorization process: first, Powers' company claimed it had invented the feature earlier, and later, inventors from two smaller firms—Charles A. Tripp and Royden Peirce—joined the interference proceeding with their own patent applications. Of the two, Tripp's patent posed the greater

[3] Megan Jenkins. (2019). The rise and fall of the female typist. Retrieved March 30, 2024 from The Center for Growth and Opportunity at Utah State University Website:https://www.thecgo.org/benchmark/the-rise-and-fall-of-the-female-typist/

[4] Pugh, E. W. (2009). Building IBM: shaping an industry and its technology. MIT press. p43

threat. CTR's patent attorney believed that whoever won this interference proceeding would be able to control the future of the automatic printing tabulator market.[5]

In the end, all three patents were granted: Hollerith's U.S. Patent No. 1,830,699, Tripp's U.S. Patent No. 1,824,581, and Peirce's U.S. Patent No. 1,930,283. Each covered distinct aspects of the technology. While CTR retained Hollerith's patent, it could not sidestep those held by Tripp and Peirce. Watson had no choice but to open the checkbook: in 1922, he purchased Peirce's company and patents for $261,000, and Peirce later became one of IBM's core research staff. Then, in 1924, Watson paid Tripp a lump sum of $212,500 to obtain a license for his patent. These two large expenditures, combined with the 1922 decision to halve the licensing fee for Powers, plunged CTR into financial difficulty, forcing it to implement company-wide pay cuts and layoffs to stay afloat.

From this episode, it is clear that although early IBM did earn revenue from patents, when faced with strategic business needs, patent income was often a secondary concern. In fact, patents could even function as a cost center when it came to supporting strategic product development—a situation quite different from IBM's approach today.

In the late 1920s, the United States was hit by an unprecedented Great Depression. While most companies tightened their belts and struggled to survive, Watson went against the tide. He not only avoided layoffs but increased hiring and production, believing the Depression was only a temporary downturn. In hindsight, Watson's judgment was mistaken, but he was saved by a stroke of luck. In 1933, Franklin D. Roosevelt assumed the presidency and launched the New Deal. Despite strong political opposition, Roosevelt pushed Congress to pass the Social Security Act, establishing systems for old-age and unemployment insurance. The law was formally enacted in 1935. To ensure the effective operation of the social security system, the U.S. government needed to immediately begin recording contributions and distributions for 27 million American citizens. At the time, IBM was the only company in the world capable of managing such a massive task.[6]

The Social Security Act brought IBM much more than a single contract. Beyond the federal agency, American corporations—especially large and mid-sized enterprises—also needed to maintain more detailed payroll and social insurance records. This made IBM's tabulating machines and punched cards indispensable to the business world. From 1935 to 1939, during a period of overall economic stagnation in the United States, IBM's revenue grew by 81%. Its U.S. workforce increased from 6268 employees in 1935 to over 10,000 by 1941.

During the Depression, while most American industries withered, only a few companies like IBM were profitable. With cash in hand, Watson took the opportunity to buy up patents in large quantities. Many companies were cash-strapped, allowing IBM to drive prices down. In 1952, during an antitrust investigation by the

[5] Pugh, E. W. (2009). Building IBM: shaping an industry and its technology. MIT press.

[6] Robert McMillan. (2012). How Social Security Rescued IBM From Death by Depression. Retrieved April 8, 2024 from Wired: https://www.wired.com/2012/06/how-social-security-saved-ibm/

U.S. Department of Justice, one official bluntly said to Watson: "The record shows and you cannot deny, I use a bad word to shorten a long discussion—you stole the French Bull patent."[7]

In the later antitrust lawsuit, the U.S. Department of Justice alleged that many of the patents Watson acquired during the Great Depression were not related to tabulating machines but instead belonged to other areas of business machines. The DOJ claimed that Watson didn't purchase these patents to develop new products, but rather to construct market barriers—using them to intimidate major companies that had yet to enter the tabulating machine market, thereby discouraging them from encroaching on IBM's territory.[8] Unfortunately, since IBM ultimately reached a settlement with the Department of Justice, we'll never know for sure whether Watson actually used patents to threaten other companies, what really transpired during that time, or what evidence the DOJ had in its possession.

Not all of the patents IBM acquired during the Depression were left sitting on a shelf, used solely for deterrence. One notable example was the exam scoring machine invented by Reynold Johnson (U.S. Patent No. 1,997,178; Fig. 4.1 shows the patent drawing, and Fig. 4.2 depicts the prototype). Though this technology is nearly a century old, it is surely familiar to every reader. The machine could automatically recognize answers that students had shaded in with a pencil on their answer sheets and would tally the scores accordingly.

Reynold Johnson was originally a high school teacher. In his spare time, he invented the scoring machine, which caught IBM's attention. Later, due to the economic devastation of the Great Depression, Johnson lost his teaching job and had to sell stockings on the street with his wife to make ends meet. Just a few days into street vending, IBM contacted him and invited him to New York, offering a weekly salary of $50 to join a project team headed to Columbia University to test the scoring machine. Although the results of the trial were mediocre, the project manager recommended that IBM proceed with the acquisition. IBM paid $4000 for Johnson's patent and hired him with an annual salary of $4000.

In 1937, IBM released the Model 805 International Test Scoring Machine. Within 5 years, its annual rental revenue reached $100,000, and accompanying equipment sales approached $200,000. Although this product accounted for only 0.3% of IBM's annual revenue at the time—a mere drop in the bucket—it had a lasting impact. After all, who hasn't used a No. 2 pencil to fill in standardized test answer sheets? This machine had a complex mechanical design that required long-term maintenance and ongoing development, and its primary customers were governments and schools. Such an endeavor was far beyond the reach of an individual high school teacher like Johnson. Only a large corporation like IBM could successfully commercialize the machine and integrate it into a mature after-sales service system. Reynold Johnson later focused on R&D at IBM and, 20 years later, led the team that

---

[7] Maney, K. (2003). *The maverick and his machine: Thomas Watson, Sr. and the making of IBM*. John Wiley & Sons. p389

[8] T Maney, K. (2003). *The maverick and his machine: Thomas Watson, Sr. and the making of IBM*. John Wiley & Sons. p391

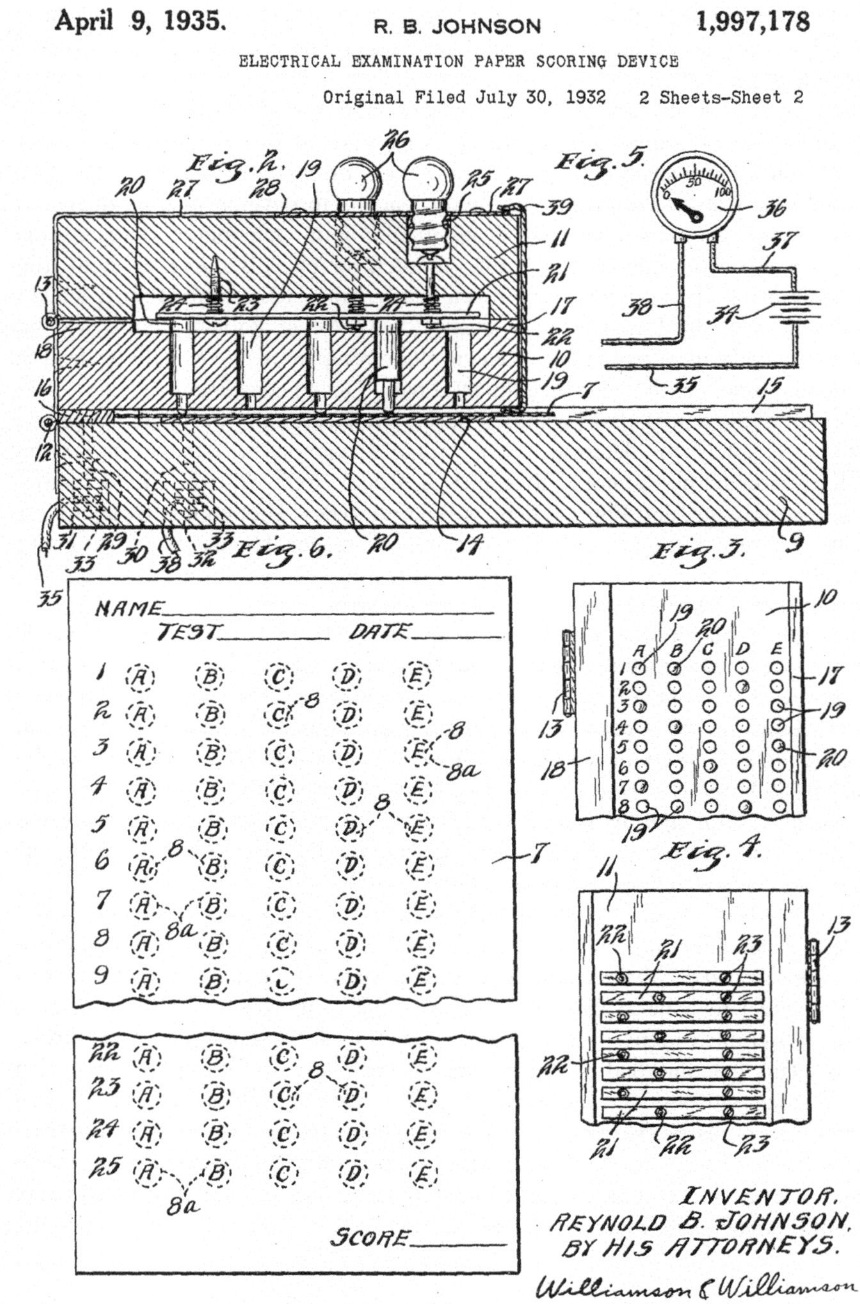

Fig. 4.1 Patent drawing of U.S. Patent No. 1997178A

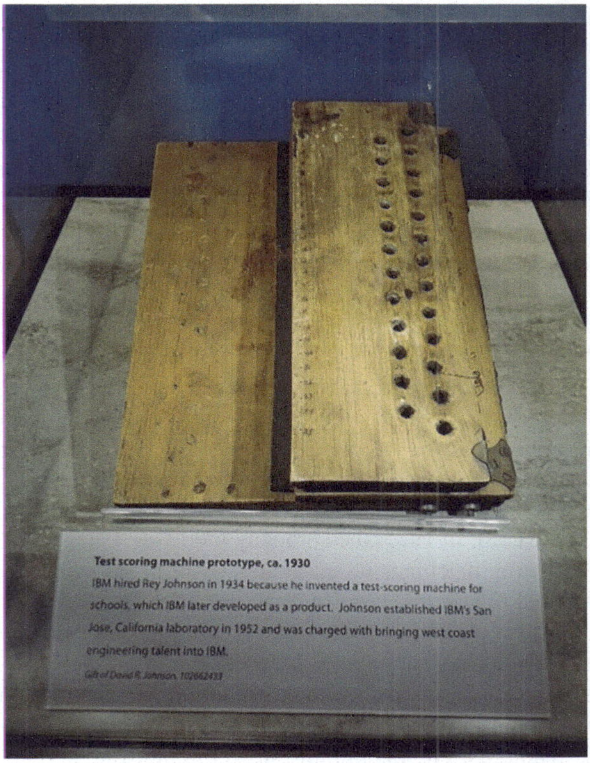

Fig. 4.2 Prototype of Johnson's test scoring machine (photograph by the author at the Computer History Museum, USA)

developed the first commercially viable hard disk drive. Some would go on to call him the "father of the disk drive."[9] This acquisition represents a perfect blend of corporate profit, individual achievement, and social contribution.

While acquiring patents, IBM also invested significant manpower and funding into improving its own products. As noted in *The Maverick and His Machine: Thomas Watson, Sr. and the Making of IBM*, the key to IBM's success was its ability to *keep customers happy.* In Watson Sr.'s view, every customer complaint represented a potential vulnerability that a competitor might exploit—and it was imperative to have the engineering team address it immediately.[10]

IBM wasn't alone in this endeavor. Powers and, later, Remington Rand (which acquired Powers's company) also invested heavily in developing new products and

[9] L.A Times. (1998). Reynold Johnson; 'Father' of Disk Drive. Retrieved from Los Angeles Times Website: https://www.latimes.com/archives/la-xpm-1998-sep-18-me-23976-story.html#:~:text=Reynold%20B.%20Johnson%2C%20known%20as%20the%20father%20of,metal%20disk%20instead%20of%20tape%20or%20a%20drum.

[10] T Maney, K. (2003). *The maverick and his machine: Thomas Watson, Sr. and the making of IBM*. John Wiley & Sons p251

features. The 1920s became a decade of rapid advancement in office automation, marked by a constant stream of new innovations.

Both Powers and Remington Rand made notable strides in input and output technology. In 1913, Powers was the first to file a patent for a tabulating machine with a printing function (U.S. Patent No. 1,245,502A, "Tabulating-Printer for Statistical Purposes"). For a long time, this was the only model on the market that supported printing. Although Hollerith had conceived of this idea before 1900, either he or his patent attorney failed to file a patent, allowing Powers to take the lead—and causing CTR to suffer as a result. In 1914, Powers filed U.S. Patent No. 1,271,614A (Fig. 4.3 shows the diagram), which integrated a typewriter with a punching mechanism. This allowed numeric keys on the typewriter to control the punches on a card, a feature that received positive feedback from the industry.

Remington Rand, having risen to prominence through its typewriters, complemented Powers's technological strengths. After acquiring Powers's company, Remington Rand filed U.S. Patent No. 1,682,451A in 1924 (Fig. 4.4 shows the diagram), integrating typewriter letter keys, punctuation keys, and the space bar into the punching system. Pressing a letter key would punch a hole in a specific location on the card. In 1925, the company filed U.S. Patent No. 1,727,471A, which enabled tabulating machines to automatically print statistical results directly onto cards for manual review and confirmation.

In this process, more and more office workers began to set aside pen and paper in favor of keyboards and spreadsheets. Their work habits began to resemble those of today's corporate white-collar employees. And so, bit by bit, these seemingly modest innovations reshaped the world.

Though CTR was initially slower to innovate in output technologies, it—by then renamed IBM in 1924—began filing several patents for tabulating machines with printing capabilities (such as U.S. Patent Nos. 693572A and 1780685A) and launched its own line of tabulating-printers. According to the patent specifications, many of these tabulating-printers were card-controlled, showing a high level of automation. U.S. Patent No. 1626871A protected an "accounting machine" that kept pace with the times by incorporating a numeric keypad, which was soon released as a product.

Over time, more and more IBM patents began to feature "computing" machines. These included James W. Bryce's "computing machine" (U.S. Patent No. 2271248A) and "multiplying machine" (U.S. Patent No. 2237335A), Gustav Tauschek's "computing machine" (U.S. Patent No. 1950527A), and Paris Robert Edward's "electrical computing machine" (U.S. Patent No. 2185681A) and "automatically controlled computing method" (U.S. Patent No. 2165298A). This shift illustrates how IBM was no longer content with simply "tabulating"; it had begun to pour substantial effort into processing, computation, and data analysis. Gustav Tauschek later also invented the magnetic drum, a bold and creative breakthrough in early magnetic storage history. Thanks to these accumulated technologies, IBM was able to build the Harvard Mark I electromechanical computer for Harvard University in the 1930s—an important milestone in the history of computing.

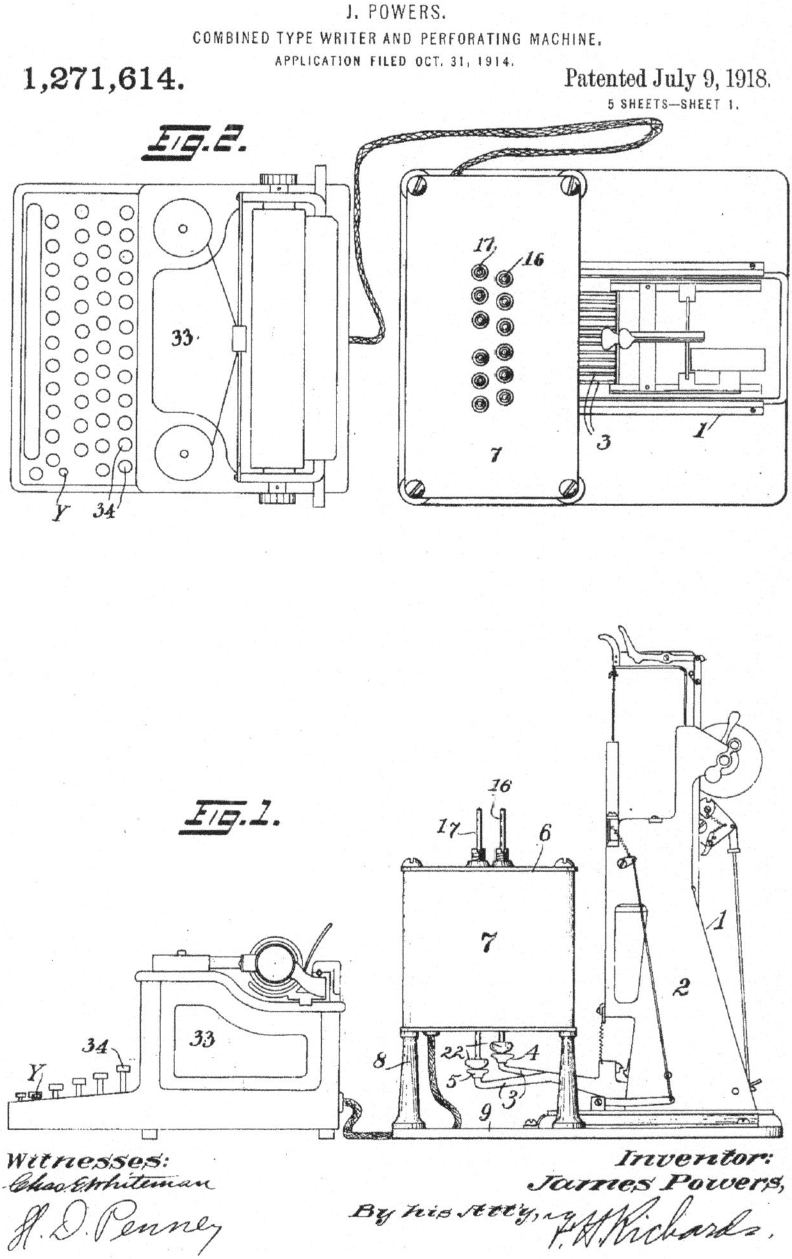

Fig. 4.3 Patent drawing of U.S. Patent No. 1271614A

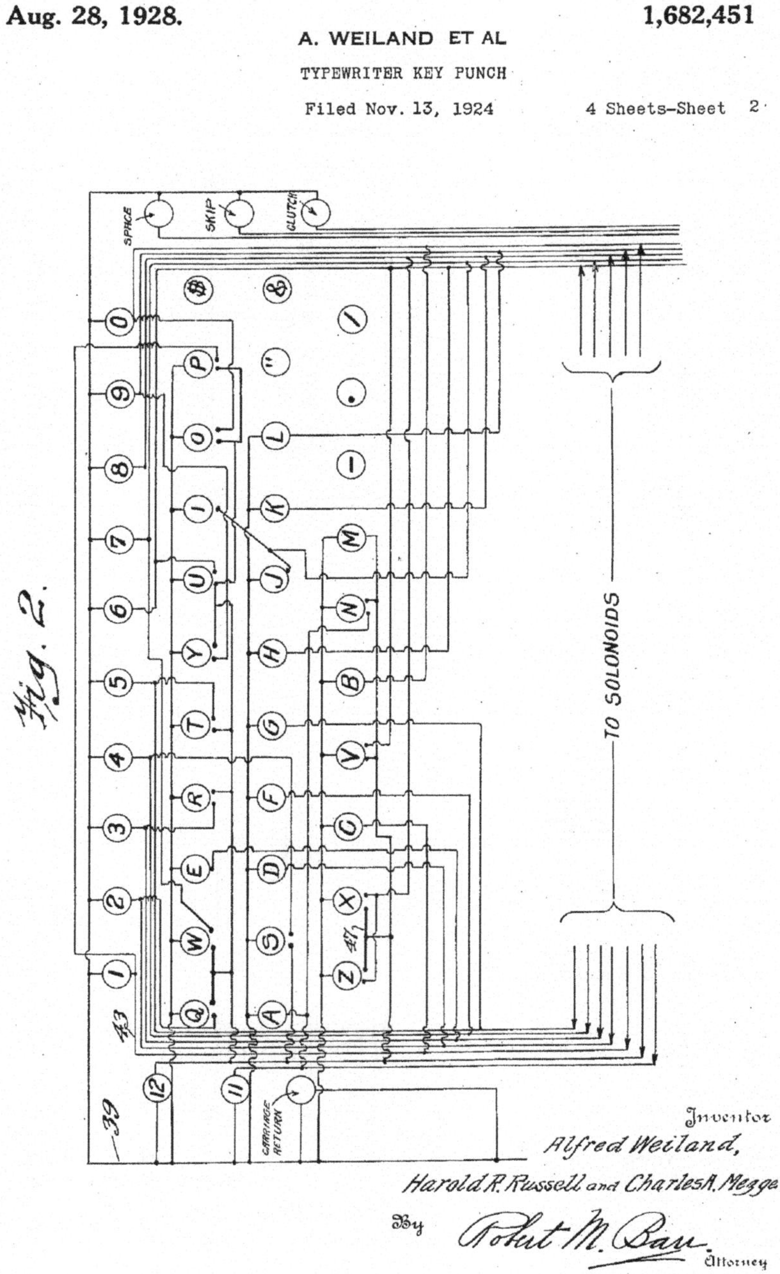

Fig. 4.4 Patent drawing of U.S. Patent No. 1682451A

At this juncture, as patent professionals, we offer a brief reflection. From Hollerith to Watson Sr., their numerous patent strategies were textbook examples of how to harness a well-constructed patent system. A mature patent portfolio can push competitors to innovate, develop new products, and meet a wide range of user needs. This in turn drives technological progress and advances society as a whole.

As shown in Fig. 4.5, beginning in the late nineteenth century, U.S. government and corporate demand for office equipment exploded, and the growth in the 1920s was particularly phenomenal. New technologies drastically boosted productivity and also fueled demand for a more skilled labor force (Table 4.1 illustrates the growth in office and white-collar jobs in the U.S. from 1900 to 1940). IBM, Powers, Remington Rand, and Watson's former employer NCR all made indelible contributions to this transformation.

There's a story from the first century AD: an inventor presented Emperor Vespasian with a device that could transport columns for construction, saving human labor. Vespasian rewarded him—but sealed the invention and forbade its manufacture. His reasoning: "Then what shall I do to feed my people?"

For a Roman emperor 2000 years ago, that concern may have made sense: every job saved by technology was a job lost. But in the rapidly advancing twentieth century, relieving people from repetitive labor meant they could devote themselves to more creative endeavors.

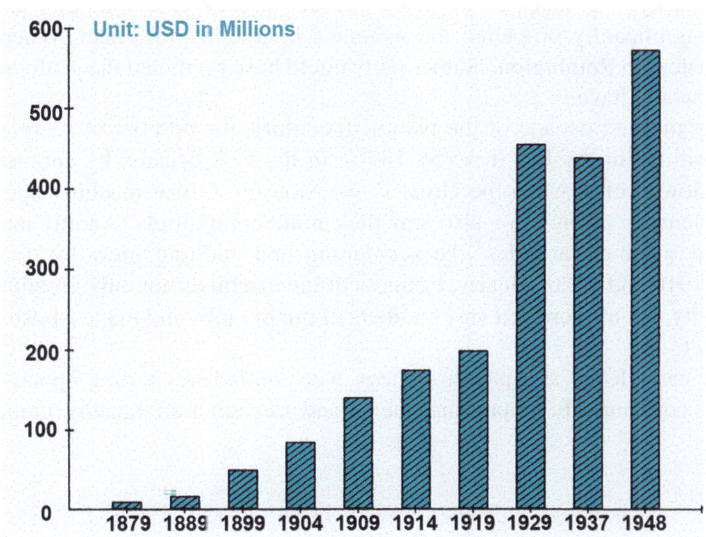

Fig. 4.5 Capital investment in the manufacture of office equipment in the United States, 1879–1948 (in million USD). (Source: Cortada, J. W. (2000). *Before the computer: IBM, NCR, Burroughs, and Remington Rand and the industry they created, 1865–1956*. Princeton University Press, p. 83)

Table 4.1 Growth of office jobs and white-collar positions

Occupation	1900–1910	1910–1920	1920–1930	1930–1940
Growth Rates of U.S. Office Jobs, 1900–1940 (unit: %)				
Labour force	26[a]	23	15	6
Managers	45[b]	14	29	4[c]
Clerks	127[b]	70	28	15[c]
Growth Rates of U.S. White-Collar Jobs, 1900–1940 (unit: %)				
Typists/Secretaries	189	103	40	11
Bookkeepers/Cashiers	93	38	20	–2
Office Machine Operators	178	102	30	31
Accountants/Auditors	70	61	63	24

Source: Cortada, J. W. (2000). *Before the computer: IBM, NCR, Burroughs, and Remington Rand and the industry they created, 1865–1956*. Princeton University Press
[a]Estimated value
[b]In 1900, managers and clerks accounted for 8.9% of all civilian employees
[c]In 1940, managers and clerks accounted for 16.9% of all civilian employees

The typewriter, for example, played a vital role in women's emancipation. Its design, particularly suitable for slender female fingers,[11] opened up a massive employment market for educated women. In the early twentieth century, the number of women working as typists grew dramatically, becoming a social phenomenon. The popular British drama *Downton Abbey* portrayed a housemaid who leaves service to study, passes the typing exam, and achieves financial independence and social mobility. Advances in women's education, economic standing, and social position significantly propelled the women's liberation movement. When Sholes sold his patent to Remington, neither party could have predicted the profound social impact it would have.

During the first decade of the twentieth century, the number of secretarial and typist positions in the U.S. rose by 189%; in the next decade, by another 103%. Growth slowed only when the Great Depression hit. Office machine operators—jobs emblematic of the era—also saw their numbers multiply. Additionally, more traditional white-collar roles like accounting and auditing grew by over 200% between 1910 and 1920. Clearly, business office machines not only greatly boosted productivity, but also created vast numbers of quality jobs and made a positive societal impact.

IBM's early R&D and patent strategy were rooted in creating epoch-defining products, continuously improving them, and engaging in healthy competition.

[11] BBC. (2013). The arrival of women in the office. Retrieved April 8, 2024 from BBC website: https://www.bbc.com/news/magazine-23432653 *"The typewriter is especially adapted to feminine fingers. They seem to be made for typewriting. The typewriting involves no hard labour and no more skill than playing the piano," wrote John Harrison, in his 1888 Manual of the Typewriter*

These efforts delivered tangible benefits to users, aligned with societal needs, paved the way for the advent of computing, and constrained competitors like Powers. While patents played a crucial role in IBM's rise, the company's core revenue—and its social contributions—always came from its products.

Chapter 5
Thrill of Achieving Monopoly via Patents

In Chap. 4, we discussed many patents, particularly those concerning automation and usability improvements in tabulating machines after Hollerith's foundational invention patents had expired. However, the patent we'll examine in this chapter is quite different: its application and use were less about protecting technical innovation and more about strategic business considerations.

We are all familiar with the famous "razor-and-blades" business model: first, sell the razor handle at a low price to ensure customer loyalty; then design blades that only fit that handle and use patents to protect the design, thereby locking out competitors entirely. This model is often attributed to the Gillette Company, whose founder, King Gillette, was a contemporary of Hollerith. In 1901, King Gillette filed a patent application for a "safety razor." In May 1904, he submitted a divisional application. That November, both the parent patent (US775134A) and the divisional patent (US775135A, see Fig. 5.1) were granted.

In US775134A, King Gillette noted that maintaining the sharpness of reusable blades required significant effort and expense. He improved upon this by introducing a disposable blade. These blades "*require but a small amount of material and can be ground very quickly and easily.*" This allowed Gillette to produce and sell blades at low cost, enabling consumers to discard them once dull and simply replace them with new ones. Both patents protected the combination of razor and blade. During the patent's term, the price for a razor with a dozen blades was $5, and a set of replacement blades cost around $1. Though not inexpensive, this was far cheaper than the then-popular razors, which sold for $12–20 and required professional sharpening. Gillette's disposable blades quickly gained favor, and the company's growth entered a rapid ascent.

In 1921, as King Gillette's patents neared expiration, competitors were eager to enter the disposable blade market. In response, Gillette slashed the price of the razor to below $1 and included three blades, while keeping the price of blade refills at

Q. Yue, F. Yang, *Governing Innovation: A Global History of Patent-Driven Technological Progress*, https://doi.org/10.1007/978-981-95-5570-3_5

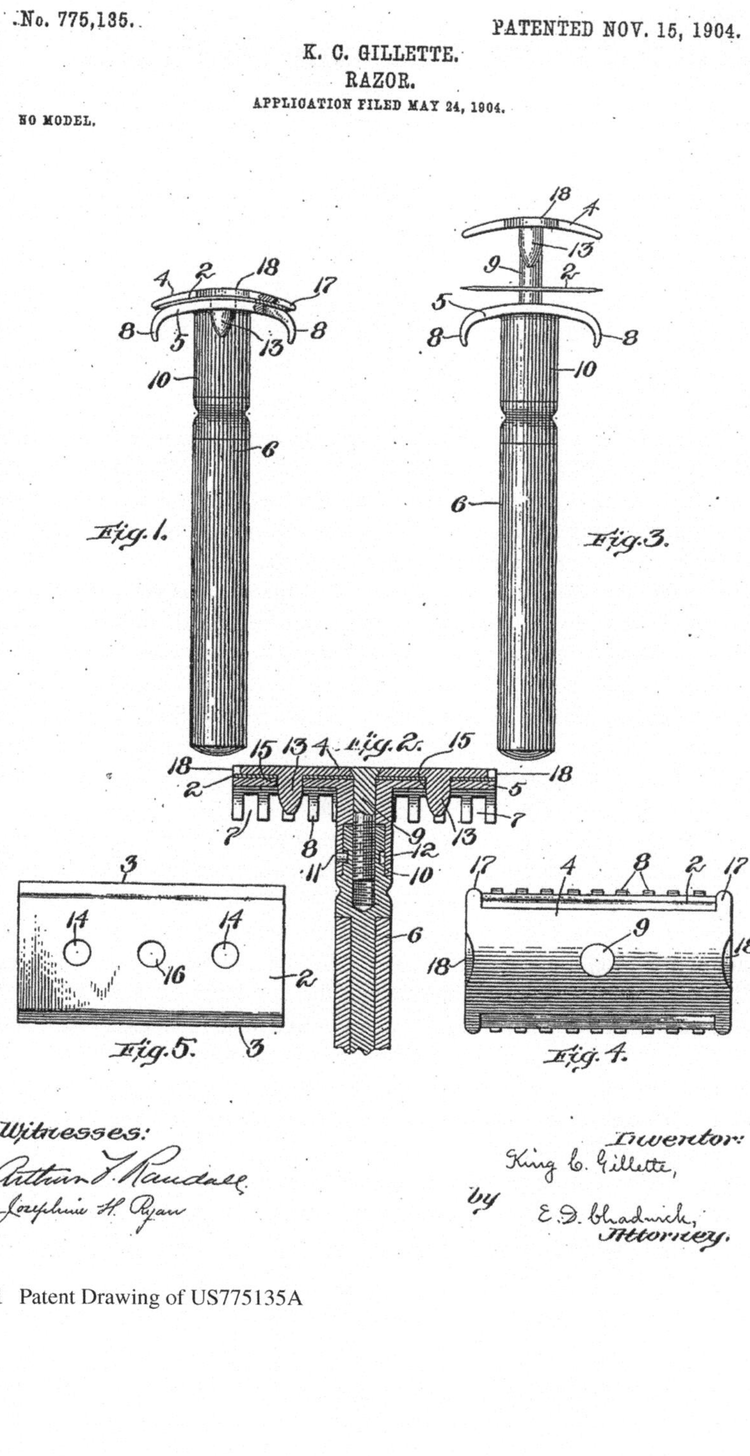

Fig. 5.1 Patent Drawing of US775135A

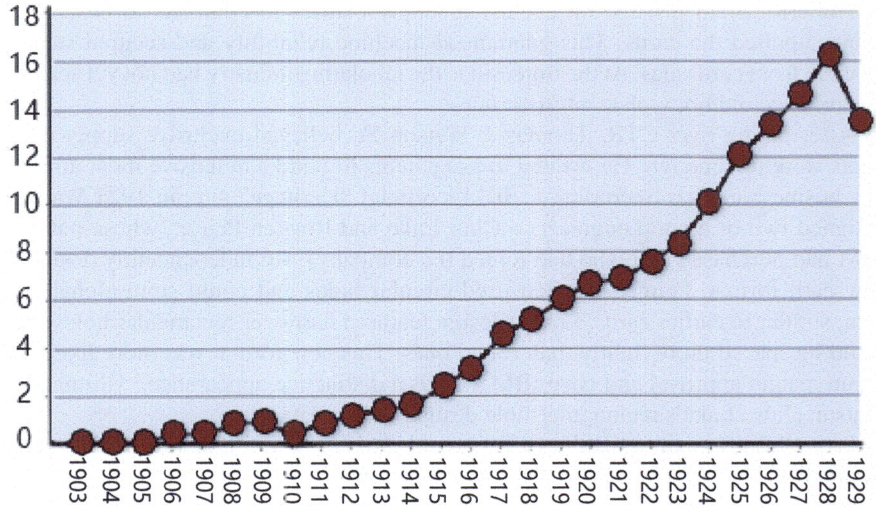

Fig. 5.2 Net profits of the Gillette Company, 1903–1929 (in millions of U.S. dollars)

about $1.[1] Despite losing patent protection, Gillette's razor sales doubled that year, and blade demand soared, pushing profits up even faster. The razor-and-blades model had shown its formidable power (see Fig. 5.2).

Following Gillette, IBM was among the earliest and most prominent companies to adopt this model. As previously mentioned, Hollerith initially used punched tape with his tabulating machine, later switching to cards. But Hollerith never patented the punched paper tape or card itself. Prior uses of punched cards—such as in Jacquard looms—and their matured application in telegraphy were already well known. It would have been easy for patent examiners at the time to find such prior art.

Still, even without a patent on punched cards, Hollerith made substantial profits from them. Early tabulating machines were purely mechanical, and with the manufacturing limitations of the time, machines were prone to failure. To avoid malfunctions caused by users using subpar cards, Hollerith chose to manufacture the cards himself and sell them directly to customers.

He priced the cards between $0.85 and $1 per thousand, while the production cost was only $0.30. Despite the low unit price, the high consumption volume ensured considerable revenue. For instance, government projects in Russia and the U.S. used millions of cards each. While enterprise consumption was smaller, the explosive growth of data due to advancing office automation and management practices kept demand robust. In 1907, one of Hollerith's major clients—the Southern Railway—used 200,000 cards in just the initial months of a project, which later ramped up to 550,000 cards per month.

[1] Picker, R. C. (2011). The razors-and-blades myth (s). *U. Chi. L. Rev.*, *78*, 225.

Hollerith often insisted on exclusive supply clauses in contracts to ensure he alone supplied the cards. This guaranteed machine reliability and secured steady revenue from card sales. At the time, since the tabulating industry had not yet scaled, this practice did not violate antitrust laws.

After taking over CTR, Thomas J. Watson Sr. believed exclusive supply contracts were not enough. He wanted to use patents to build a defensive moat around this business model. According to IBM's official "Heritage" site, in 1927 Watson assigned two of his top engineers—Clair Lake and Royden Peirce (whose patents IBM had purchased and who had joined the company)—to independently design a new card format. Peirce's version used circular holes and could store alphabetic data, similar to earlier cards. Lake's design featured narrower rectangular holes that could be spaced more tightly than round ones. This new format was more likely to secure patent approval and gave IBM's cards a distinctive appearance.[2] Ultimately, Watson chose Lake's rectangular-hole design.

From this effort came IBM's 1928 patent US1772492A, titled "*Record Sheet for Tabulating Machines*" (see Fig. 5.3), which had just two claims:

Claim 1: *A tabulating card having oblong control apertures arranged in vertical columns and horizontal lines with the greater dimension in the vertical direction whereby the card is stronger and contains more perforations than a similar card having similarly arranged and spaced circular perforations of a diameter equal to said vertical dimension.*

Claim 2: *A card as in claim 1 in which the apertures are rectangular.*

This patent ensured IBM's punched cards had a distinctive design while locking competitors into using round-hole formats. Though Remington Rand developed its own cards, IBM's exclusive role in handling the U.S. Social Security system's data made its 80-column punched card (see Fig. 5.4) the de facto industry standard. For decades—even as business machines evolved—the 80-column card persisted. At one point, punched cards accounted for over 30% of IBM's total profits.[3]

As described in the specification of US Patent No. 1772492A, rectangular-hole cards did have their advantages; for other industries, standardization generally brings more benefits than drawbacks. This patent became IBM's fundamental weapon for excluding competitors and monopolizing the market. IBM's purpose in applying for this patent was to sell cards. From a business perspective, this approach was understandable. However, IBM's actual practice was to make it so that companies everywhere could only buy cards from IBM. This behavior can be defined as a form of patent hegemony.

In 1930, Remington Rand and IBM signed a cross-licensing agreement and jointly adopted the 80-column card as the industry standard. In 1932, the U.S. Department of Justice launched an antitrust investigation into the two

[2] IBM. (n.d.). *The IBM punched card.* In *IBM 100: Icons of progress.* Retrieved March 31, 2024 from https://www.ibm.com/ibm/history/ibm100/us/en/icons/punchcard/

[3] Pugh, E. W. (2009). *Building IBM: shaping an industry and its technology.* MIT press. p254

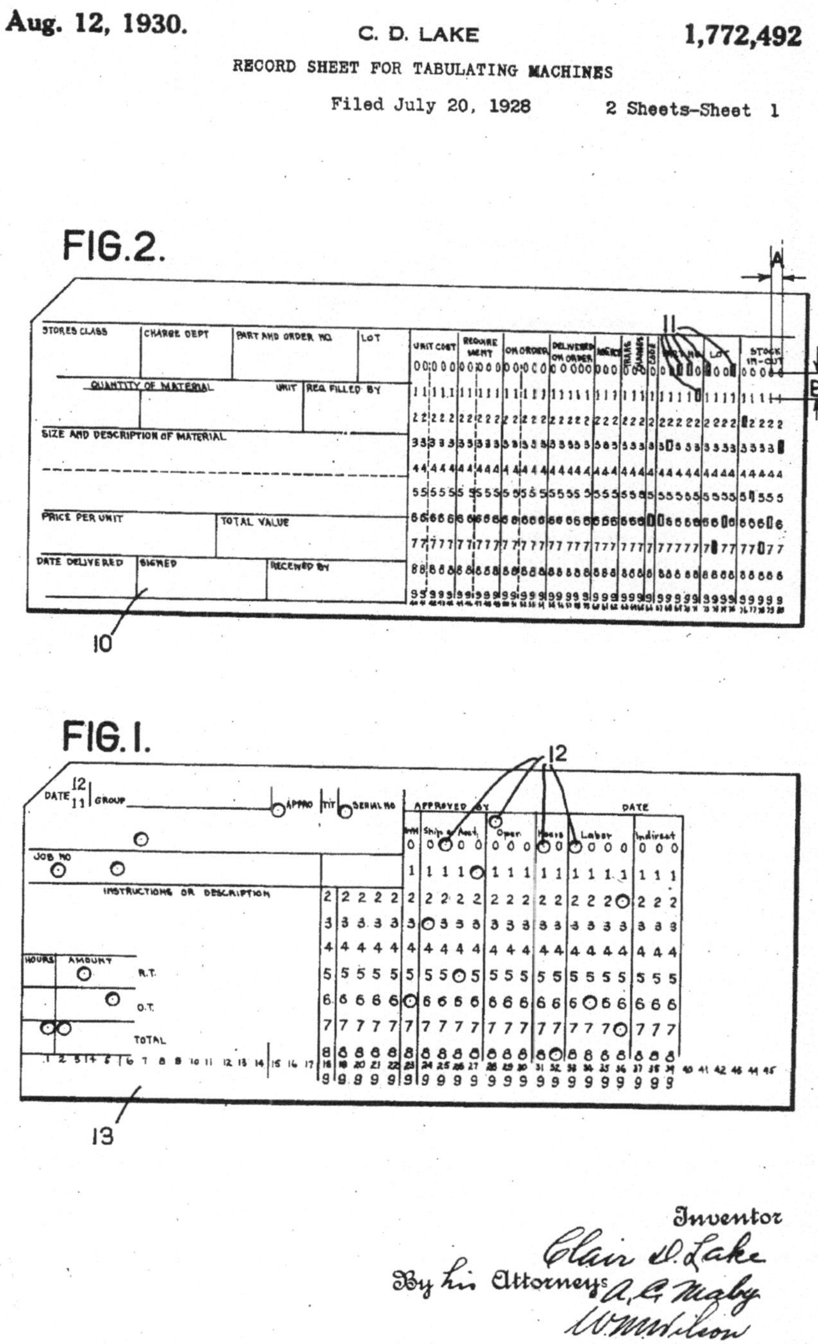

Fig. 5.3 Patent drawing of U.S. Patent No. 1772492A

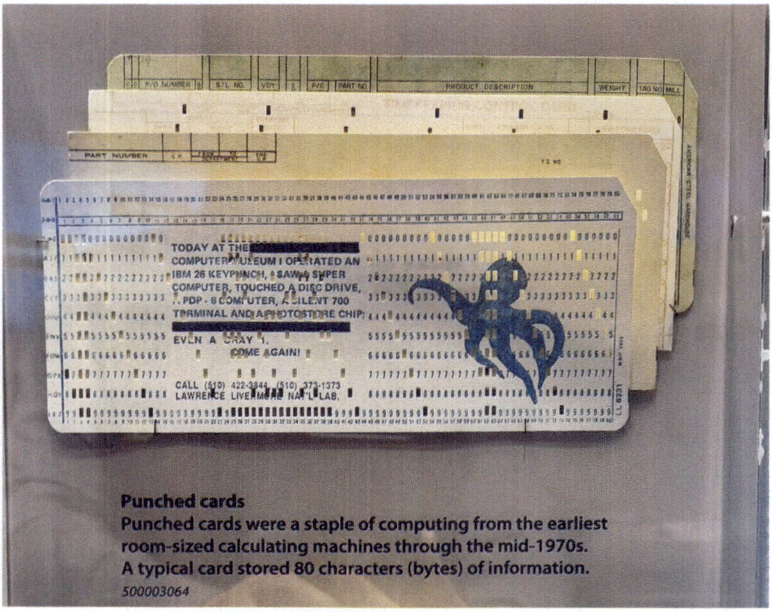

Fig. 5.4 Physical 80-column card. (Photograph by the author at the Computer History Museum, USA)

companies, examining behaviors such as renting but not selling, forcing users to buy their cards, and refusing to sell cards to non-customers.[4] In 1936, the U.S. Supreme Court ruled that IBM's card sales practices violated antitrust laws. According to the ruling, IBM produced 3 million punched cards annually, accounting for 81% of the market share; 10% of the company's total revenue came from punched cards.[5] Following the court's decision, IBM was no longer allowed to prohibit customers from using non-IBM cards. Before the ruling, Remington Rand and IBM had voluntarily rescinded their cross-licensing agreement, thereby nullifying two of the DOJ's charges. IBM also lowered the prices of the cards it sold to the government in exchange for a settlement.

From today's perspective, given IBM's size, using patents and contracts to force bundled card sales and exclude competitors was clearly monopolistic behavior. But it's important to remember that before IBM was found guilty, there were virtually no precedents for such a case. It was precisely because of this IBM case that such practices came to be recognized as common antitrust violations.

This case is a classic in antitrust law and is mentioned in nearly all textbooks on the subject. However, most textbooks fail to mention that even after this case, IBM's patent remained valid and its card business was not significantly affected. By the

[4] Cortada, J. W. (2019). *IBM: The rise and fall and reinvention of a global icon*. MIT Press.

[5] International Business Machines Corp. v. United States 298 U.S. 131 (1936).

late 1930s, revenue from card sales accounted for 15% of IBM's total income[6] — a noticeable increase from the 10% figure prior to the antitrust litigation.

This may suggest that using patents or similar methods to restrict the choices of third parties or even consumers is ultimately a minor tactic. In contrast, product competitiveness is what truly matters. As we will see in later chapters, many companies became overly reliant on the "razor-and-blades" model, prioritizing quick profits from consumables over the actual needs of users. When the product itself is outstanding, pulling a few tricks with the consumables might not cause much harm; but during critical periods of technological transformation, such missteps in R&D strategy could lead to disastrous consequences.

IBM's tabulator–card model lasted for quite a long time. But by the time Thomas Watson Jr. entered IBM's management, the limitations of punched cards had already begun to emerge. Whether in government or business, the volume of data to be processed was continually increasing, while the physical area and thickness of cards could not be further reduced. In the 1940s, some of IBM's major clients began to complain about the overwhelming number of punched cards piling up. In 1948, a vice president of Metropolitan Life Insurance Company, one of the world's largest insurers, complained to Watson Jr., saying: "*...we already have three floors of this building filled with punch cards and it's getting worse. We just can't afford to pay for that kind of storage space.*" He also expressed hope for new technologies: "*I'm told we can put our records on magnetic tape.*".[7] The president of Time magazine expressed similar frustration. For each magazine subscriber, the publisher used three cards to record contact details, subscription information, and so on. He directly threatened IBM: "We've dedicated an entire building to your equipment... if you can't come up with something new, we'll have to find another solution."

Although punched cards were read by machines, they were still physical objects—errors could be manually checked, and for an operator, replacing a card and punching a few holes was no big deal. But if data were digitally stored on magnetic tape, it would be a different story: no organism on Earth has evolved eyes capable of reading the information recorded on magnetic tape.

Readers who lived through the CD and DVD era will recall that in the early twenty-first century, many people preferred to burn important data to discs. As hard drive prices fell, many transferred their data to backup drives. In the age of the cloud, cloud storage became a go-to solution for many "lazy" users, but it also made those accustomed to burning discs or using hard drives uneasy. For the same reasons, Watson Sr. was skeptical of magnetic tape. According to Watson Jr., his father "*distrusted magnetic tape instinctively. On a punch card, you had a piece of information that was permanent. You could see it and hold it in your hand. Even the enormous files the insurance companies kept could always be sampled and*

[6] Cortada, J. W. (2019). *IBM: The rise and fall and reinvention of a global icon*. MIT Press. p89 "*By the late 1930s, card sales made up 15 percent of the company's revenue*"

[7] von Simson, E. (2010). *The Limits of Strategy: Lessons in Leadership from the Computer Industry*. iUniverse. P21

hand-checked by clerks. But with magnetic tape, your data were stored invisibly on a medium that was designed to be erased and reused."[8]

Some sources suggest that to address the storage capacity issue, Watson Sr. once asked the company to develop a larger 180-column card to increase the data stored per card by sevenfold. However, we found no corresponding patent records. Judging from subsequent technological developments, considering that the 80-column card had already been the industry standard for decades, the idea of replacing it with a 180-column version seems unrealistic.

In the 1950s, the U.S. Department of Justice once again launched an antitrust investigation into IBM. Thomas Watson Sr. was a man of backbone and resilience, growing only more resolute with age. When faced with this renewed threat from the DOJ, he refused to compromise, preparing to fight it out in court. Thomas Watson Jr., however, believed that the company's future lay in the field of computing and saw no need to continue battling over tabulators and punched cards.

Watson Sr. spared no expense in hiring a top-tier legal team to confront the DOJ. Every lawyer on the team was seasoned and well-connected. According to Watson Jr.'s autobiography, one of the lawyers, Joseph Keenan, had previously served as a federal judge. Whenever Watson Sr. consulted him, Keenan would confidently say, "*These things can always be handled, Mr. Watson,*" and then send a "fantastic" bill—without ever resolving anything. Another lawyer, Robert Patterson, who had served as Secretary of War under President Truman, joined IBM's board and persuaded Watson Sr. to dismiss Keenan. However, the day after the lawsuit was filed, Patterson tragically died in a plane crash and never got the chance to contribute. In addition, IBM's general counsel Burke Marshall, who had held a senior post in the DOJ handling civil rights and was personally close to the Attorney General, was also unable to stop the lawsuit.[9]

The legal battle and the company's strategic direction sparked repeated debates and fierce arguments between father and son, straining their relationship to the breaking point. In 1955, as Watson Jr. was on his way to court, he ran into his father. Watson Sr. asked where he was going. When Watson Jr. replied that he was going to meet with the judge, his father lost his temper. Ultimately, though, he agreed to let his son attend the meeting—but strictly forbade him from making any decisions. Bearing immense pressure, Watson Jr. arrived at the courthouse, where an IBM staff member slipped him a small note. It read:

100%
Confidence
Appreciation
Admiration
Love
Dad[10]

---

[8] von Simson, E. (2010). *The Limits of Strategy: Lessons in Leadership from the Computer Industry*. iUniverse. P21

[9] Hart, D. M. (1999). IBM in American politics, 1970–1999. *Business and Economic History, 28*(2), 49–59.

[10] Maney, K. (2003). *The maverick and his machine: Thomas Watson, Sr. and the making of IBM*. John Wiley & Sons.

That note marked the symbolic transfer of power between two legendary leaders of IBM. In January 1956, Watson Jr. signed a consent decree, formally reaching a settlement with the U.S. Department of Justice. IBM abandoned its rental-only policy, agreed to sell tabulators at reasonable prices, and permitted other companies to sell punched cards. Watson Sr. supported his son's decision and insisted that every IBM employee adhere to full compliance. On their first day, all new hires were required to study the terms of the decree.[11] Most importantly, IBM agreed to license all of its patents to others at "reasonable rates."

On that day, IBM left behind the defensive moat built from its tabulator and card patents. It stepped forward, unburdened, to lead humanity in the pursuit of new dreams with new tools.

[11] Maney, K. (2003). *The maverick and his machine: Thomas Watson, Sr. and the making of IBM*. John Wiley & Sons.

Chapter 6
The Fall of a Legendary Patent

As mentioned in previous chapters, by the late 19th and early 20th centuries, patent litigation and operations in the United States were already flourishing. However, with the advent of the electronic computer age, the patent industry fell into decline.

This downturn was largely due to the public and governmental reaction to the Great Depression. Having witnessed the destructive effects of monopolistic behavior, the U.S. became deeply averse to monopoly. As patents functioned as monopolistic tools in the realm of technology, they naturally became a prime target of judicial scrutiny. From the 1930s through the 1970s, antitrust investigations became the top issue facing large tech firms, and the patent industry entered nearly half a century of stagnation.[1]

From a legal perspective, this period marked the emergence and gradual refinement of the concept of "non-obviousness." Early U.S. patent law required only "novelty" and "utility." It wasn't until a 1941 court decision introduced the "flash of genius" standard that the law began demanding a higher threshold for inventiveness. In 1952, the U.S. enacted a new Patent Act—its first major revision since 1836—which formally codified the requirement of "non-obviousness" in Section 103. This new standard created significant legal uncertainty, as courts across jurisdictions interpreted Section 103 differently, making patent holders increasingly hesitant to litigate.

From 1951 to 1966, an independent inventor named William Graham battled a corporation in court for 14 years, eventually appealing all the way to the Supreme Court. The Court's decision in this case would define the interpretation of Section 103 and establish what are now known in patent textbooks as the "Graham factors." As Tom Morris, the defense attorney in the case, recalled, when the case was heard publicly by the Supreme Court, the courtroom was packed to capacity, with even the

[1] United States. Federal Trade Commission. (2003). *To promote Innovation: The proper balance of competition and patent law and policy*. DIANE Publishing. Retrieved April 8, 2024 from: https://www.ftc.gov/sites/default/files/documents/reports/promote-innovation-proper-balance-competition-and-patent-law-and-policy/innovationrpt.pdf

Q. Yue, F. Yang, *Governing Innovation: A Global History of Patent-Driven Technological Progress*, https://doi.org/10.1007/978-981-95-5570-3_6

hallways crowded with lawyers anxiously awaiting the outcome.[2] Although Graham's name became immortalized in patent history, his 14-year legal battle ended with both of his patents being invalidated.

Judges of the 1960s had mostly been appointed during the Roosevelt era. Having lived through the Great Depression and witnessed the monopolistic tricks played by early industrial giants, they were deeply wary of monopolies and harbored little sympathy for patents. One lawyer lamented, "*The patent law basically ceased to exist for about 15 years [in the 8thCircuit]*"[3] A 1971 report from the Second Circuit revealed that 80% of patents were invalidated on appeal.[4] This anti-patentee judicial climate persisted until the formation of the Federal Circuit Court of Appeals in the 1980s.

Despite this challenging environment, IBM remained a constant target of antitrust scrutiny by the Department of Justice. While the company maintained a high volume of patent filings, it was rarely aggressive in enforcement. Even so, IBM continued to thrive through its unmatched product competitiveness and emerged from this downturn as the undisputed leader in the tech industry, entering the 1980s with its dominance intact.

During this period, IBM demonstrated both the noble spirit of focusing on research and innovation, and the aggressive posture of leveraging its massive scale to pressure smaller players. These dual traits will be explored in greater detail below.

The defining theme of this new era was the electronic computer. Unlike the punched card era, IBM no longer held a clear first-mover advantage. In fact, one of the most fundamental patents of early electronic computers wasn't owned by IBM—it belonged to IBM's longtime rival, Remington Rand.

In 1945, engineers John Presper Eckert and John W. Mauchly at the University of Pennsylvania designed and built the world's first general-purpose electronic computer for the U.S. military. It was named the "Electronic Numerical Integrator and Computer," or ENIAC for short. In early 1946, ENIAC was delivered to the military, primarily for ballistic calculations. Weighing around 30 t and containing 18,000 vacuum tubes, ENIAC could perform 5000 calculations per second. It took 30 s to compute a single ballistic trajectory—while this speed is far behind that of modern smartphones, it was 2400 times faster than even the most experienced human

---

[2] Patricia Busa McConnico. (2016). 50 Years: Graham v. Deere. Retrieved April 8, 2024 from: https://www.texasbar.com/AM/Template.cfm?Section=Table_of_Contents&Template=/CM/ContentDisplay.cfm&ContentID=33532 also available at: https://blog.texasbar.com/2016/02/articles/people/50-years-graham-v-deere/

[3] Schnaidt, J. (2016). The Mad Med of IP. *Managing Intell. Prop., 261*, 21. Retrieved April 8, 2024 from: https://www.oblon.com/A11960/assets/files/News/MIP-Cover-story-Sept161.pdf *"The patent law basically ceased to exist for about 15 years [in the 8th Circuit]"*

[4] United States. Federal Trade Commission. (2003). *To promote Innovation: The proper balance of competition and patent law and policy*. DIANE Publishing. Retrieved April 8, 2024 from: https://www.ftc.gov/sites/default/files/documents/reports/promote-innovation-proper-balance-competition-and-patent-law-and-policy/innovationrpt.pdf

calculators in the military at the time. It marked the beginning of the electronic computing era.

After completing this groundbreaking invention, Eckert and Mauchly resigned from the University of Pennsylvania and founded the Eckert-Mauchly Computer Corporation. Since the technologies behind ENIAC were developed during their tenure at the university, they wrote to the university president, requesting the transfer of patent rights related to ENIAC. Fortunately, the president generously agreed. Unfortunately, the pair's business acumen didn't match their engineering brilliance, and by 1950, the company was on the verge of bankruptcy.

At that time, IBM's old rival, Remington Rand, was undergoing a post-war transformation and had identified the computer industry as a key development focus. It acquired the Eckert-Mauchly company. Eckert and Mauchly then abandoned business operations and dedicated themselves to R&D, jointly designing a new generation of computers—the UNIVAC (UNIVersal Automatic Computer). This enabled Remington Rand to rise swiftly in the computer market and pose a serious threat to IBM's dominance. Meanwhile, the patenting of ENIAC's technologies was taken over by Remington Rand's experienced patent team.

After a lengthy prosecution process, on February 4, 1964, ENIAC's core technology was finally granted patent protection as **US Patent No. 3,120,606**. The document spanned 207 pages, with 12 pages of claims totaling 148 individual claims. Due to the sheer number of claims, we cite only Claim 1 here:

> *Means for producing electric pulses in sequence, electronic means for alternately transmitting certain ones of said pulses as recurrent differentiated groups, electronic means for selecting particular pulses from one of said differentiated groups to represent quantitative values, electronic means for selecting particular pulses from another of said differentiated groups to represent certain qualitative values, reading means responsive to pulses representing both the qualitative and quantitative values for reading data to be processed upon command of at least one of said qualitative pulses, storing the data thus read, and making the data available in the form of data pulses in response to at least one other of said qualitative pulses, and electronic means for receiving said data pulses and responsive thereto for performing electrical switching operations of a nature determined by selected ones of said qualitative values and of a degree determined by selected ones of said quantitative values.*

Much like Herman Hollerith's tabulating machine patents, this patent protected the foundational concept of the electronic computer. By then, Remington Rand had merged with another large corporation to form Sperry Rand Corporation. Upon securing the patent, Sperry Rand immediately began negotiating with IBM, demanding licensing fees. According to Charles McTiernan, who managed patent licensing at Sperry Rand, IBM ultimately paid a total of $11.1 million in licensing fees for this patent.

This transaction recalls the earlier episode where Powers had to pay steep licensing fees to Thomas Watson Sr. However, unlike IBM, Sperry Rand did not become the dominant player in the computer era despite owning this groundbreaking patent.

After settling with IBM, Sperry Rand turned its attention to other data-processing companies, including major cross-industry giants like RCA (Radio Corporation of America), General Electric, and Honeywell. Most of these companies found themselves powerless against the sweeping scope of the ENIAC patent and had no choice but to concede. George Elgroth, the head of patents at General Electric and a former Sperry Rand employee, was intimately familiar with his old employer's patent portfolio. Yet even he found no effective counters to challenge the ENIAC patent. He could not identify any threatening patents in GE's massive portfolio, nor could he construct any compelling arguments for invalidity. All he could do was delay, repeatedly citing his inability to reach company leadership. Only AT&T's Bell Labs mounted a legal challenge, attempting to invalidate the ENIAC patent through judicial proceedings—but they ultimately failed.[5]

Among all the adversaries, only Honeywell posed a true threat to the ENIAC patent. On May 26, 1967, Sperry Rand and Honeywell filed lawsuits against each other on the very same day. Sperry Rand accused Honeywell of patent infringement and demanded $20 million in licensing fees and damages. Honeywell countered by accusing Sperry Rand of fraud and monopolistic practices and petitioned the court to declare **US Patent No. 3,120,606** invalid.

This lawsuit became one of the largest patent litigations in U.S. history. The trial alone lasted 135 days, with 77 witnesses taking the stand and over 20,000 pages of court transcripts. It was one of the most time-consuming cases in the federal court system. According to reports, when the lead partner at Honeywell's law firm handed the case to attorney Charles G. Call, he was asked if interested on taking a case which might take 10 years of his time.[6] The technical documentation in the trial was not only voluminous but also rigorous, including some of the earliest seminal works in computing history—such as the historic "First Draft of a Report on the EDVAC" by computing pioneer John von Neumann, which first articulated the von Neumann architecture. By the time the case concluded in 1973, the total legal costs had reached $8 million. It's hard to say how many attorneys would willingly trade 10 years of their career for such a fee.

In 1973, the court issued a landmark preliminary ruling declaring **US Patent No. 3,120,606** entirely invalid. Some claims were struck down for lack of novelty and other grounds, but the primary reason for invalidation was "public use." Honeywell presented evidence proving that ENIAC had been publicly demonstrated more than a year prior to the patent's priority date of June 26, 1947. Additional evidence showed that certain components of ENIAC had already been delivered to the military as early as July 1944. The judge determined that these components alone already satisfied the definition of an "electronic computer" as claimed in the patent, and that their delivery constituted public use.

The previously mentioned "First Draft of a Report on the EDVAC" by John von Neumann also disclosed several key technical elements protected by **US Patent No. 3,120,606**. This report had been made public in June 1945 by ENIAC project security officer Herman Goldstine, and it served as a crucial piece of prior art.

[5] *Atanasoff-Berry Computer Court Case*. (n.d.). JVA Initiative Committee and Iowa State University. Retrieved March 31, 2024 from https://jva.cs.iastate.edu/courtcase.php

[6] *Atanasoff-Berry Computer Court Case*. (n.d.). JVA Initiative Committee and Iowa State University. Retrieved March 31, 2024 from https://jva.cs.iastate.edu/courtcase.php

With the judge's gavel, the immense value of **US Patent No. 3,120,606** vanished in an instant. Sperry Rand came away empty-handed and was instead ordered to pay a settlement to Honeywell. From the University of Pennsylvania to the Eckert-Mauchly Computer Corporation and eventually to Sperry Rand, over 20 years of patent work by numerous professionals came to nothing. As one of the earliest and most important patents in electronic computer history, the technologies it protected now entered the public domain.

For the nascent electronic computer industry, the invalidation of the ENIAC patent may have been a blessing in disguise. Granted in 1964, **US Patent No. 3,120,606** was originally set to remain in force until 1981 (since U.S. patents at the time were valid for 17 years). It would have covered an extraordinarily critical phase in the development of electronic computing. Had it not been invalidated, an untold number of tech startups from the 1960s and 1970s might have had to pay Sperry Rand hefty licensing fees.

Looking back, history gave Sperry Rand a precious 9-year window. But historical opportunities are often fleeting, and Sperry Rand failed to seize the moment to grow stronger. Ultimately, it could only remain on the fringes of a computer market dominated by IBM. In 1986, Sperry Rand merged with another early computer company, Burroughs, to form Unisys. The name Remington Rand was thus consigned to history.

This story reminds us that holding foundational patents and having legendary inventors does not guarantee business success. For a product to become a market favorite, it takes an immense amount of meticulous, comprehensive effort. Herman Hollerith, for example, spent over a decade refining his tabulating machine, exploring application scenarios, and building a customer base before founding the company that laid the groundwork for IBM's rise. By the time IBM faced real competition, Hollerith's original patents had already expired. Therefore, among the many factors behind a product's commercial success, the filing date of a patent or the inventor's academic fame is not necessarily decisive.

There were many reasons behind Sperry Rand's failure in the early computer market. First, Sperry was significantly larger than Remington Rand before their merger, with multiple product lines. It lacked a leader like Thomas Watson Jr. who was willing to go all in on computing.[7] Second, scientists Eckert and Mauchly remained at the helm of Sperry Rand's computer division. Their team, scholarly and honest, wasn't afraid to admit product shortcomings—but such candor often dampened customer enthusiasm.[8] In contrast, many IBM employees were seasoned salespeople, well-versed in interpersonal dynamics and skilled in customer engagement.

Third, Sperry Rand's organizational structure was even more bloated and bureaucratic than IBM's at the time. In 1959, the company recruited IBM executive Dause

[7] Greg Satell. (2016). Take a Long Look at IBM and You'll Understand the Importance of Focus. Retrieved April 8, 2024 from https://www.forbes.com/sites/gregsatell/2016/01/10/take-a-long-look-at-ibm-and-youll-understand-the-importance-of-focus/?sh=745665652d0f

[8] Rosen, S. (1969). Electronic computers: A historical survey. *ACM Computing Surveys (CSUR)*, *1*(1), 7–36.

L. Bibby to overhaul the management of its computer division, but the effort failed. According to a senior Sperry Rand executive: "*At IBM…Bibby would press a button and a thousand guys in pin-striped suits would come out and salute; when he pressed [the same] [button] at UNIVAC, nothing happened. To mobilize an effective organization takes one kind of talent—and resuscitating a dying organization takes another.*"[9]

Though IBM did not possess a killer foundational patent like ENIAC, it had years of deep experience in business accounting, a clear understanding of clients' computing needs, and profound expertise in electromechanical systems. These assets quickly propelled IBM to the forefront of the burgeoning electronic computer market.

To meet the needs of clients across various industries, IBM launched multiple computer product lines. For example, the IBM 701 was primarily used for scientific computing, while the IBM 702 focused on financial and accounting applications. Although IBM secured a solid market position across most of these lines, the use of different hardware systems with incompatible architectures meant that each required its own programming language—resulting in significant resource waste.

As shown in Figs. 6.1 and 6.2, IBM's early computer systems required a range of adapters to connect to specific types of disk storage. Moreover, the software used on these systems varied widely. Apart from Assembly language, there was no single programming language that could run across all IBM systems.

To solve the compatibility issue, Thomas Watson Jr. made a $5 billion bet—one of the boldest in business history. The result was the revolutionary System/360 mainframe, launched in 1964 (a physical photo of which is shown in Fig. 6.3). In his book *Good to Great*, management guru James Collins ranked the System/360 alongside the Boeing 707 and the Ford Model T as one of the three greatest commercial successes in human history. IBM's official website proudly quotes this praise when introducing the System/360.[10] Many other experts have also ranked it as one of the finest products ever created.[11] The release of the System/360 marked a watershed moment in the history of the Information Age and laid the groundwork for IBM's dominance over the next two decades.

The System/360 mainframe came in various models with differing performance levels and memory capacities. Customers could customize configurations to a certain degree, but all models used essentially the same instruction set. This allowed IBM to avoid rewriting software for every industry-specific system. Additionally, the System/360 featured standardized interfaces, enabling customers to mix and match a wide variety of peripheral devices freely.

---

[9] *7.7 UNIVAC Falls to IBM*. (n.d.). In *Bit by Bit: History of Computing in the Twentieth Century*. Retrieved March 31, 2024, from https://ds-wordpress.haverford.edu/bitbybit/bit-by-bit-contents/chapter-seven/7-7-univac-falls-to-ibm/

[10] IBM. (n.d.). *System/360*. In *IBM 100: Icons of progress*. Retrieved March 31, 2024 from https://www.ibm.com/ibm/history/ibm100/us/en/icons/system360/

[11] Cortada, J. W. (2019). *IBM: The rise and fall and reinvention of a global icon*. MIT Press.

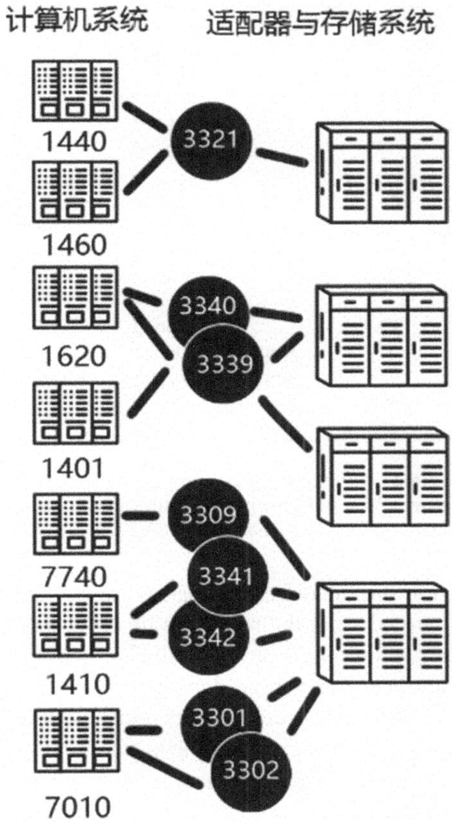

Fig. 6.1 Different storage devices used in early IBM computer systems (Evans, B.O. (1986). System/360: A Retrospective View. IEEE Annals of the History of Computing, 8(2), 155–179. https://doi.org/10.1109/mahc.1986.10033)

This feature of the System/360 gave rise to a new business model. First, customers could pair the mainframe with various peripherals and services—including magnetic tapes, disks, and drums for data storage, as well as printers and displays for output, plus industry-specific software. This became IBM's new "razor-and-blade" combination: the mainframe replaced the tabulating machine, and accessories and software took the place of punch cards as higher-margin revenue sources. Second, software development no longer needed to address cross-platform compatibility, significantly reducing programmer workloads. This also gave IBM more pricing flexibility in its software services, making it a key competitive advantage.

According to some scholars, in the 1960s, the value of software in pre-installed computer systems accounted for over 40% of the total value.[12] Yet under IBM's pricing strategy at the time, hardware and software were bundled, and software was

[12] Morris, C. R., & Ferguson, C. H. (1993). How architecture wins technology wars. *Harvard Business Review*, *71*(2), 86–96. *"By one estimate, the contribution of software to the value of a computer system had grown from eight per cent in the early days to something like 40 per cent by the 1960s".*

IBM Computer systems	Programming Languages						
	COBOL	FORTRAN	Auto-coder	Assembler	Sort	Utilities	RPG
305				Support		Support	
650		Support		Support	Support	Support	
704		Support		Support			
705	Support			Support	Support	Support	
709		Support		Support	Support	Support	
1401	Support	Support	Support	Support	Support	Support	Support
1410	Support	Support	Support	Support	Support	Support	Support
1620		Support		Support		Support	
1710				Support		Support	
7030		Support		Support	Support		
7040	Support	Support		Support	Support	Support	Support
7070	Support	Support	Support	Support	Support	Support	Support
7080	Support			Support	Support	Support	
7090	Support	Support		Support	Support	Support	

Fig. 6.2 Different programming languages used in early IBM computer systems (Evans, B.O. (1986). System/360: A Retrospective View. IEEE Annals of the History of Computing, 8(2), 155–179. https://doi.org/10.1109/mahc.1986.10033)

often provided free of charge. As some sources recount,[13] after the U.S. economy emerged from the recession in 1972, there was a nationwide shortage of programmers—marking the beginning of a golden era for the profession. Even 50 years later, it seems that golden age has yet to end.

We found several patents related to the System/360 mainframe, such as two 1964 filings for "data processing systems" (US Patent No. 3,400,371 and US Patent No. 3,315,235). The documentation for these patents is quite lengthy: US Patent No. 3,315,235, for instance, runs 102 pages and contains 27 claims—longer than the average master's thesis. Due to space limitations, this book does not delve into the details of these patents. Furthermore, due to the restrictions imposed by IBM's consent decree, these patents could only serve as bargaining chips to exchange for "reasonable" licensing fees, or occasionally be used as counterclaims in lawsuits. This situation persisted into the 1980s.

Additionally, computers were extremely expensive at the time and required dedicated teams for maintenance and software development—well beyond the reach of most small businesses. Before the System/360, people often spoke of "IBM and the Seven Dwarfs," referring to IBM and its seven smaller competitors. After the System/360 debuted, two of those competitors dropped out: RCA sold its computer division to Sperry Rand in 1971, and General Electric sold its computer business to Honeywell in 1970, effectively exiting the industry. For IBM at the time, using patents to crush the few remaining competitors would not have been a wise move.

[13] Grad, B. (2002). A personal recollection: IBM's unbundling of software and services. *IEEE Annals of the History of Computing*, *24*(1), 64–71.

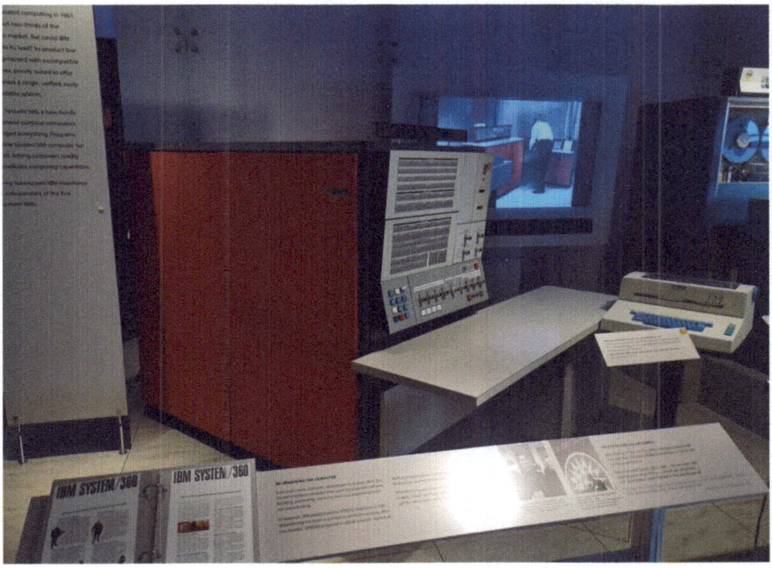

Fig. 6.3 Photograph of the System/360 mainframe. (Photograph by the author at the Computer History Museum, USA)

Moreover, to ensure a closed ecosystem, IBM protected many of its component interface technologies as trade secrets.[14] For these reasons, discussing the role of patents in the mainframe market during the System/360 era is of limited relevance.

In the razor-and-blade business model, the truly interesting part is often not the "razor" but the "blade." Therefore, in the following sections, we will focus on IBM's two most profitable segments during its peak: disk storage and software.

Thanks to System/360's compatibility, third-party vendors no longer needed to design hardware and software for each platform individually. They quickly began producing peripherals compatible with the System/360 and developing software tailored for it, thriving within IBM's ecosystem. Although IBM dominated the mainframe market, it continually faced fierce competition in peripheral hardware and software. In responding to these challengers, IBM demonstrated both the majestic poise of the world's top tech company and the overbearing tactics often associated with monopolistic giants. A fascinating tale indeed—one we shall now explore.

[14]Farrell, J., & Weiser, P. J. (2003). Modularity, vertical integration, and open access policies: Towards a convergence of antitrust and regulation in the internt age. *Harv. JL & Tech.*, *17*, 85. "IBM kept the interfaces between the different parts of its system secret and proprietary"

Chapter 7
From Cards to Disks

In his debut story *Rescue Party*, Arthur C. Clarke tells a tale set in the near future: one day, an interstellar alliance of alien civilizations detects a catastrophic disaster brewing in the solar system. Out of humanitarian concern, the alliance dispatches a rescue team to Earth, hoping to save a few human lives. At the time, they know little about humans—only that an intelligent species exists on Earth with access to radio technology, as evidenced by radio signals received 200 years earlier. In the aliens' understanding, it takes thousands of years for a civilization to progress from discovering radio waves to mastering space travel. However, upon landing on Earth, the rescue team finds the entire planet eerily silent and devoid of life. As it turns out, humankind had already developed spacefaring technology before the disaster and departed the solar system en masse, destination unknown.

This story reflects the post–World War II American optimism about technology, and a confidence—even pride—in the rapid progress of human innovation. The same mindset appears in *The Three-Body Problem*: it is this fear that backward civilizations might rapidly catch up technologically that leads advanced civilizations to carry out brutal "Dark Forest" strikes, snuffing out potential competitors in their cradles.

Clarke, an author with a solid background in engineering theory, foresaw many technologies that would later become reality. However, in *Rescue Party*, the seemingly "hacked" human civilization had not yet fully abandoned punched cards. In the story, although the alien rescue team does not find a single living soul on Earth, they do come across a facility once used to store human population data—and the punched cards leave a deep impression. In a silent room rows of machines and endless filing cabinets stand motionless, and inside each drawer lie thousands of thin, hole-punched cards. Clarke lingers on this image, marveling that the fate of billions of individuals—every man, woman, and child—could be condensed into such fragile slips of paper. He remarked that it had once been a miracle of information technology now lay forgotten, a poignant reminder of both human ingenuity and its transience.

Q. Yue, F. Yang, *Governing Innovation: A Global History of Patent-Driven Technological Progress*, https://doi.org/10.1007/978-981-95-5570-3_7

This storage technology, which Clarke imagined would last into the space age, was ultimately replaced by magnetic storage. As we all know, the most common application of magnetic storage—the hard disk drive—was invented by IBM. By tracing the history of hard disk development, we may uncover the true reasons behind IBM's success in the age of electronic computers.

In the 1950s, magnetic storage technology was already fairly common, but the invention of the hard disk was not the result of incremental improvements to magnetic storage. Rather, it arose from efforts to overhaul the punched card system.

This brings us to a curious bit of trivia: the origin of the concept of the "file." We're all intimately familiar with the term—whether in Word or WPS for drafting patents, AutoCAD for design, or Windows File Explorer, "File" is always the first item in the menu bar, usually with the shortcut key "F." But few know that the idea of a "file" in this context dates back to the 1950s, the era of punched cards, and came long after terms like "document" or "dossier."

A single punched card, with its mere 80 columns, could not hold much information. Often, multiple cards—dozens, hundreds, or more—had to be grouped together to represent a meaningful dataset, similar in scale to a modern Excel file. These card sets were usually stored in drawers, cabinets, or boxes, and a full box or drawer of such data would be called a "file." In English, the word "tub" is used both for an open container and for a bathtub, so a box of such data cards was also known as a "tub file."[1]

"Tub files" were widely used in the 1950s. Take a financial workflow, for instance: company clerks processing invoices would often work with repetitive data, as the same products were frequently sent to the same clients. Since no visual word processing software existed—and certainly no Word or Excel templates—the idea of pre-filled templates was not an option. To simplify input of recurring invoice details, staff would prepare customer info cards and product info cards. The former encoded names, addresses, and phone numbers; the latter included product specifications and categories. All of these were stored in tub files.

When issuing an invoice, the operator would pull the relevant customer and product cards from the tub file, feed them into a specialized accounting machine, and then input variable information such as quantity, discount, and date via punched card or keyboard. The machine would then automatically print the completed invoice.

By the early 1950s, some clients complained that tub files were cumbersome to operate and manage. IBM swiftly launched a project to streamline tub file workflows. Beginning in April 1952, the original idea was to install a mechanical conveyor system in offices, allowing entire boxes of tub files to be delivered to operators at their desks—saving them the trouble of walking back and forth. The plan resembled something from *The Three-Body Problem*, like Qin Shi Huang's human-powered computer.

Using this crude concept as a starting point, IBM invented the hard disk drive within a few short years, completely phasing out punched cards. The rotating disk

[1] Benj Edward. (2021). What Are Computer Files and Folders? Retrieved April 8, 2024 from https://www.howtogeek.com/757092/what-are-computer-files-and-folders/

replaced the conveyor belt, and the magnetic head replaced the human operator. The "human computer" described in *The Three-Body Problem*—where ancient Trisolarans develop manpower-based computing to calculate orbital chaos—suggests that their civilization repeatedly failed to survive catastrophes and only much later developed electricity and machinery. If the Trisolarans had seen how humanity evolved from semi-manual data systems to fully automated magnetic storage within mere years, they might have felt a shiver down their spines.

At the same time as they explored conveyor systems, IBM also studied the potential of magnetic storage. Two mainstream options existed then: magnetic tape and magnetic drum. Magnetic tape had an inherent flaw: the physical movement of the tape was required to locate specific data, and even 30 years later, this problem remained unresolved in music cassette players.

We mentioned magnetic drums earlier—an invention by Gustav Tauschek, patented in 1937 (US Patent No. 1,880,523), and a technology IBM could claim as its own. IBM's 650 computer used magnetic drums as memory. Though magnetic drums allowed random access, the data was stored on the curved surface of a large cylinder—offering relatively little usable surface area compared to its bulk. IBM used magnetic drums in some products, and Remington Rand (by then renamed Sperry Rand Corporation) extended the technology. But their experience confirmed that it was exceedingly difficult to overcome the space limitation of magnetic drums (Fig. 7.1).

Sperry Rand Corporation had an opportunity to develop its own disk storage system as part of the UNIVAC project. However, due to internal corporate politics, the initiative was ultimately abandoned. The R&D teams were split between Saint Paul and Philadelphia. The team in Saint Paul designed a disk-based storage system, while the Philadelphia team proposed using magnetic drums. Because Philadelphia was home to the University of Pennsylvania and the stronghold of Eckert and Mauchly, it wielded greater influence within the company. As a result, Sperry Rand discarded the disk plan and opted for magnetic drums with a diameter of 18 inches.[2]

Ultimately, Sperry Rand released a product called the FASTRAND drum memory unit for use with UNIVAC computers (see Fig. 7.2). At the time, FASTRAND was the fastest storage device available, showcasing impressive technological sophistication. Yet its shortcomings were just as glaring: the unit weighed about 2.27 metric tons (approximately 5000 pounds), far exceeding what standard flooring could support, and required specially reinforced structural support.

The drum itself was cast iron—an oversized iron column more than 1.8 m (roughly 6 feet) long. When spinning at high speeds, it generated an enormous amount of angular momentum—enough to make the entire machine bounce off the ground. To prevent this, the unit had to be firmly anchored to the floor. It's said that some units suffered catastrophic failures during operation, with the drum tearing through walls as it broke apart. One documented case involved the U.S. military attempting to conduct field tests using UNIVAC equipment, loading the entire setup—including the FASTRAND unit—onto a truck. During transit, the spinning drum's momentum caused the truck to flip over while turning a corner.

[2] Rostky, G. (1998). Disk drives take eventful spin. *EE Times*, *8*. Retrieved April 8, 2024 from https://www.disktrend.com/disk3.htm

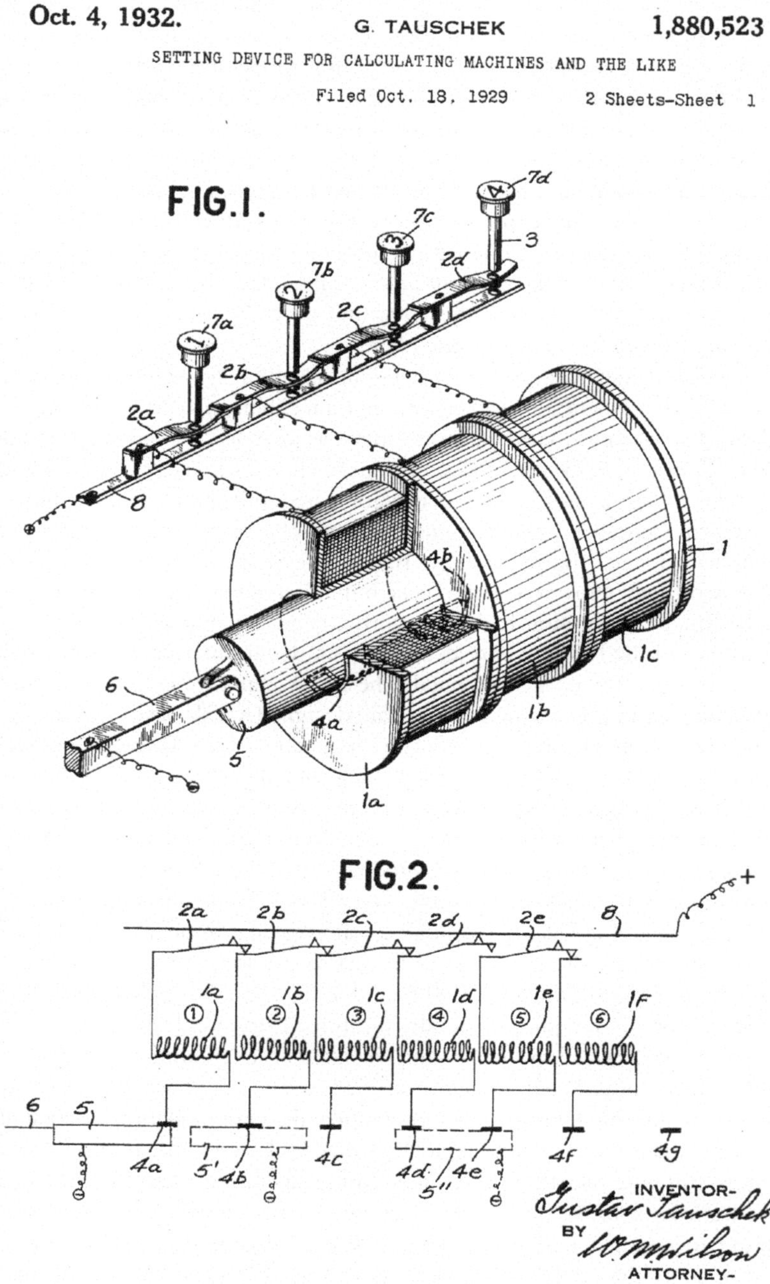

Fig. 7.1 Patent drawing of Gustav Tauschek's magnetic drum

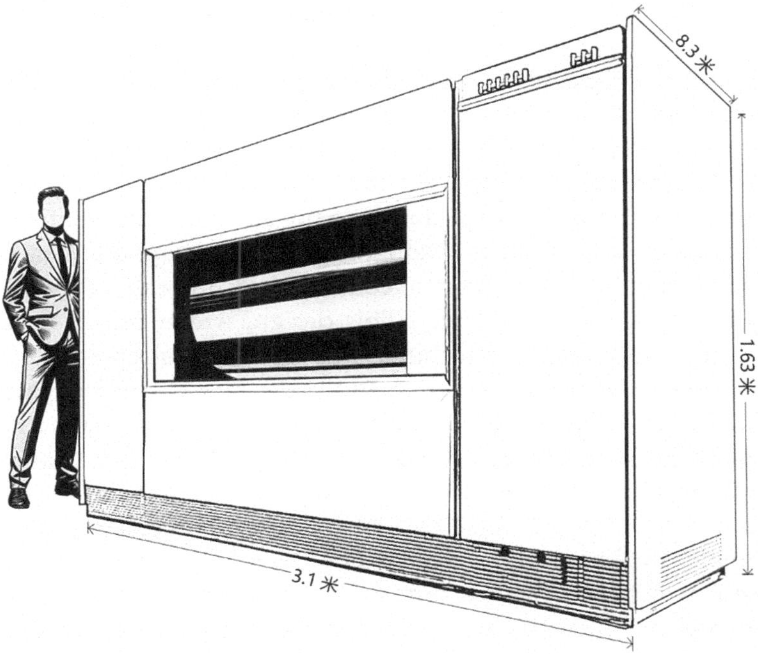

Fig. 7.2 Comparison between the FASTRAND magnetic drum memory and a human figure. (Illustration by the author, dimensions in meters)

To address the angular momentum issue, the second-generation FASTRAND used two drums rotating in opposite directions to cancel out each other's torque.

American tech companies often harbor biases against technologies not invented within their own organizations—a phenomenon some researchers call the "not invented here" problem. This bias can sometimes override rational judgment and lead companies astray in their R&D efforts. However, IBM, despite being the inventor of drum memory technology, did not become infatuated with it to the point of stagnation. During its exploration of magnetic storage technology, IBM's researchers found inspiration in an academic paper that led them to consider the possibility of disk-based storage.

The author of the paper was Jacob Rabinow, an engineer at the U.S. National Bureau of Standards. Rabinow made significant contributions to electromechanical engineering and held more than 200 patents in his lifetime. He was one of the early pioneers in the computer industry. In his paper titled "The Notched-Disk Memory," he discussed the concept of magnetic disk storage and applied for a patent in 1951. Figure 7.3 shows a screenshot of Rabinow's paper, and Fig. 7.4 shows his patent diagram.

common shaft, and the notches are aligned so as to permit the magnetic heads that are to scan the disks to pass through the channel formed by the notches. This makes it possible to move the heads rapidly to the desired disk.

It soon became apparent that a more effective arrangement of the disks is to mount them on a circular shaft in the form of a toroid or "doughnut." The magnetic heads then can be rotated on a central shaft instead of being moved in a straight line. This design, which results in a simple and rugged machine, was therefore adopted.

When the heads are in position, the disk is rotated past them while information, in the form of coded magnetic pulses, is recorded or read out. A large number of heads, spaced along a radius of the disk, are used so that the whole useful area of each disk can be read or written upon in a single revolution of the disk. The bank of heads is divided into two halves, with a narrow slit between them through which the disk passes. The general arrangement of the

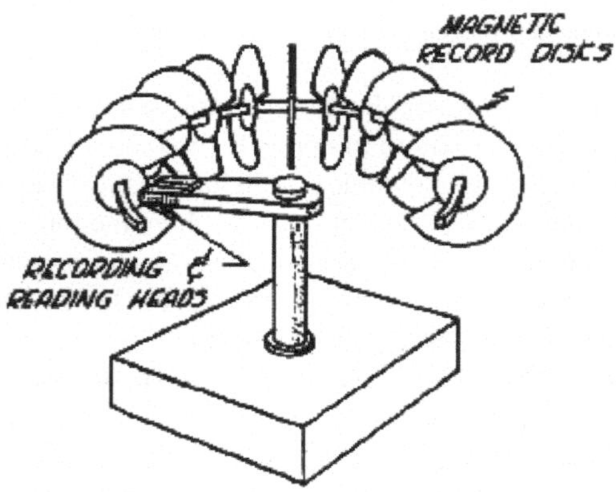

Figure 1. Schematic diagram of the notched-disk memory

machine is shown in Figure 1. Thin aluminum disks, coated on both sides with magnetic material, are being used in the experimental models at NBS.

The notched-disk memory "doughnut" can be thought

Fig. 7.3 Screenshot of Rabinow's paper *"The Notched-Disk Memory"* (Rabinow, J. (1952). The notched-disk memory. *Electrical Engineering*, *71*(8), 745–749. Retrieved April 8, 2024 from http://www.ed-thelen.org/RAMAC/NotchedDiskMem_EE_1952-1.pdf)

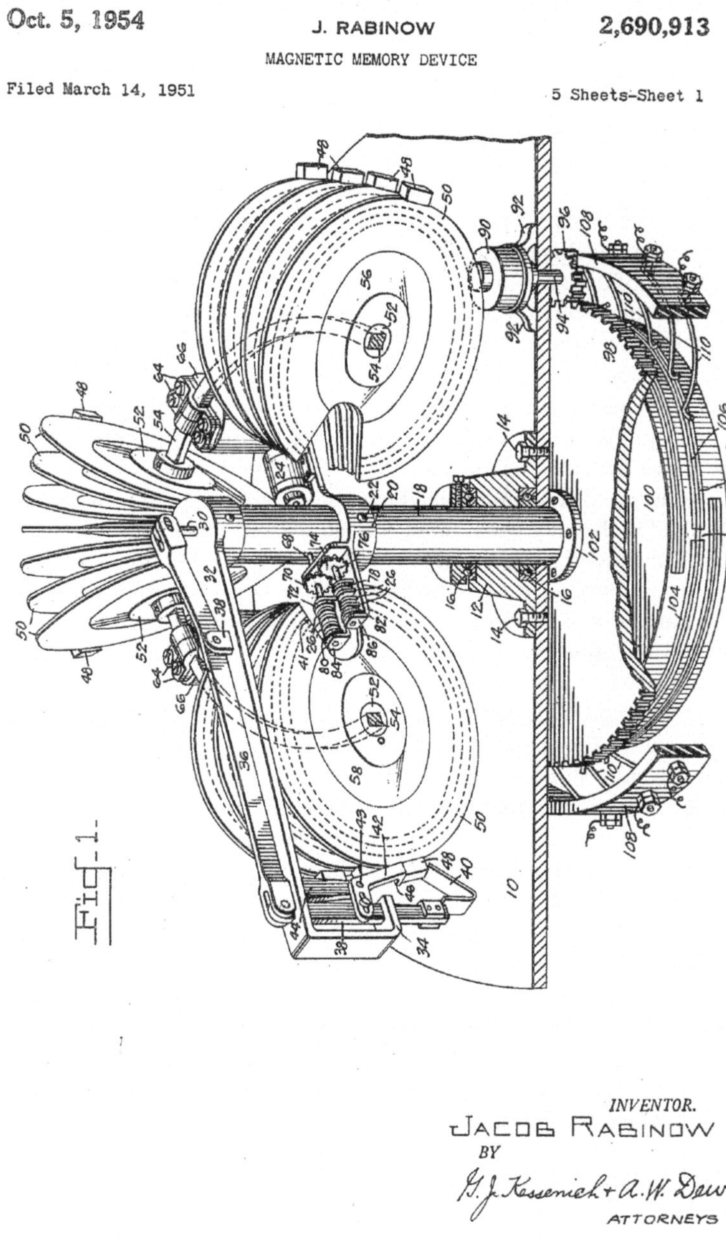

Fig. 7.4 Patent drawing of U.S. Patent No. 2690913A

Based on Rabinow's paper, a development team at IBM's San Jose laboratory drafted an initial design proposing to replace tub files with rotating circular magnetic disks.

We previously mentioned Reynold Johnson, who invented the test scoring machine using marked cards. This invention secured him a job at IBM during the Great Depression, saving him from poverty. Johnson later rose to prominence and became the head of the San Jose lab. When he saw the new disk-based proposal, he defied opposition and threw his full support behind the project.

In 1953, Johnson traveled to IBM's headquarters to present the year's R&D proposals to the board of directors, with the disk project being a primary focus. When he arrived at the office, senior Watson and other board members were present, but junior Watson was not. Watson Sr. was aware of the disk project and jokingly told the board that Johnson's team was out to "kill the punch card," which raised some concerns among the directors. Not wanting to appear too radical, Johnson quickly clarified that the application of disk storage would actually boost punch card sales. However, the board remained unconvinced and voted to reject the project.

Johnson was deeply disappointed, but he wasn't ready to give up. He decided to proceed covertly. On the way back, he instructed the two team members traveling with him to keep the board's rejection a secret. Upon returning to San Jose, Johnson ordered the team to continue development under a new codename—RAMAC—to avoid attracting attention. Fortunately, junior Watson visited San Jose shortly thereafter and gave formal approval for the RAMAC project.

Because Rabinow had already publicly disclosed the basic concept of magnetic disk storage, IBM could not patent it as they had done with tabulating machines. IBM's goal was to turn Rabinow's idea into reality, transforming the steampunk-esque mechanical rings into a robust and commercially viable product.

For example, ensuring that the read/write head could swiftly position itself on a rapidly spinning disk without damaging it was a real engineering challenge. To save space, disk thickness had to be minimized and the platters closely stacked. Initially, the disks were too thin and wobbled during high-speed rotation, interfering with data reads. As a result, thicker disks were used, which made the spacing tighter and increased the risk of the read/write head damaging the surface. Eventually, engineers fitted the heads with three pneumatic pistons, creating an air cushion between the heads and the disk surface—this became known as an air bearing.

This air bearing required extremely smooth disk surfaces, leading to challenges in material selection and coating processes. Rabinow's paper recommended "thin aluminum disks coated with magnetic material on both sides," but the team experimented with copper, magnesium, glass, and plastic before ultimately returning to aluminum. The coating process had to ensure a flat surface without irregularities. The team adopted a spin-coating method, where a liquid coating was poured onto the center of a spinning disk, and centrifugal force distributed it evenly. In the early days, they measured the liquid with paper cups and filtered it through nylon stockings—a process that lasted a year before automation took over.

In February 1954, IBM's development team succeeded in transferring information from a card to a disk and back again, signaling that the technology was nearly mature. By year's end, IBM had filed at least three patents for the RAMAC disk system:

- US3134097A: "Data Storage Machine," filed December 24, 1954, granted May 19, 1964.
- US3503060A: "Direct Access Magnetic Storage Device," a continuation of the above, filed December 24, 1954, granted March 24, 1970.
- US3037205A: "Magnetic Record Disc with Gas-supported Transducer," filed October 9, 1956, granted May 29, 1962.

The first two patents covered the fundamental structure of the hard disk drive, while the third solved a key technical problem. Since Rabinow's paper and patent had already disclosed the basic concept of circular magnetic disk storage, IBM couldn't patent that idea itself. Instead, the first two patents protected the complex mechanical system: the spindle at the disk's center, concentric data tracks, belt-pulley rotation controls, and a sophisticated read/write head assembly. Though the head assembly was nearly the length of a human arm, it could move precisely through the narrow gaps between disks.

Some researchers argue that despite the dawn of the electronic computer era, traditional electrical and mechanical technologies remained vital. IBM's manufacturing capabilities in these areas provided critical support for its core product development, contributing significantly to the company's commercial success in the 1960s and 1970s.[3] These three patents—US Patent No. 3,134,097, US Patent No. 3,503,060, and US Patent No. 3,037,205—fully demonstrate IBM's deep technical foundation in traditional fields. They validate the view that IBM's success was driven not just by innovation in new domains, but also by its mastery of established engineering disciplines. The fusion of conventional electromechanical expertise with swift responsiveness to user needs and an openness to academic research produced these landmark inventions.

The specification of US Patent No. 3,134,097 notes that "the transducer is of the air bearing type, with an air gap between the head and the recording surface," although this feature was not included in the patent's claims. However, in its continuation, US Patent No. 3,503,060, claims 1 and 2 specifically mention the air bearing: "moving the transducer (i.e., the read/write head) in the opposite direction and positioning it upon an air bearing." Both of these patent documents exceed 60 pages in length, while the third patent, US Patent No. 3,037,205, is much shorter—only three pages including drawings. This third patent protects the air bearing head solution: "The support means is adapted to retain the transducer (i.e., the head) upon a rim until spring bias is overcome by the establishment of an air film induced by rotation of the disk, said support means further being adapted to support said transducer for movement from a position on said rim to a position on the magnetic surface without interfering with the air film" (Fig. 7.5)

The IBM 305, which used the RAMAC disk system, had a rental cost of $650 per month, weighed 785 kilograms, and could store 3.75 MB of data.[4] Today, 3.75 MB barely holds a single medium-quality MP3 file or one medium-high resolution

[3] Flamm, Kenneth 1988. *Creating the Computer*. Washington: Brookings Institution

[4] *RAMAC 350 myths*. (n.d.). *RAMAC 350 Restoration Web Site*. Retrieved March 31, 2024 from http://www.ed-thelen.org/RAMAC/Myths.html

Fig. 7.5 Hard disk drive of the RAMAC disk system. (Photograph by the author at the Computer History Museum, USA)

smartphone photo. Yet at the time, that amount of data required 52 disks, each 24 inches (about 61 cm) in diameter. Even so, 24-inch disks occupied less physical space than punch cards, and the speed of access was dramatically improved. From the advent of the RAMAC disk system, punch cards and tabulating machine operators began to fade into history. Decades later, the space age anticipated by Arthur C. Clarke never quite materialized, while the punch card storage technology he had seen as a symbol of human achievement had vanished without a trace. Punch cards and tabulating machines can now only be glimpsed in museums, history books, and patent archives.

In the 1950s, disk storage stood at the frontier of storage technology alongside magnetic tape and magnetic drums, but its market remained relatively small. That changed in April 1965, when IBM launched a new disk product—the 2314 Direct Access Storage Facility (DASF), abbreviated as "2314," which could be paired with the System/360 mainframe released the previous year. This marked a major leap forward for the disk storage market.

The 2314 introduced innovation in read/write head technology by using ceramic materials to reduce the distance between the head and the high-speed rotating disk to just 2.159 microns. This technology was documented in US Patent No. 3,631,425. The patent specification identified a key issue in head design: the farther the head is from the disk surface, the weaker the signal. To increase data density, the head needed to be closer to the disk to achieve a high signal-to-noise ratio. US Patent No. 3,631,425 employed elastic elements to press a floating block toward the disk, allowing the head to penetrate the air bearing's cushion and draw nearer to the disk surface. Holes in the floating block created a pressure differential that maintained the minimal separation from the disk surface, while the elastic elements ensured that the head could follow minute surface variations and sweep away fine contaminants.

The 2314 also improved the coating process to create smoother disk surfaces. A nozzle would dispense coating material beginning at the outer edge, then move inward to the inner circle, and back out again—avoiding the radial streaks caused by traditional spin-coating. This technique was recorded in US Patent No. 3,198,657.

These innovations in head and coating technology, coupled with the widespread adoption of the System/360, helped make the 2314 a commercial success. As the standard external storage device for both the System/360 and its successor, the System/370, the 2314 opened up a multi-hundred-million-dollar market, becoming a major source of revenue for IBM beyond mainframe leasing. In February 1970, IBM declared external devices to be a "core strategic issue" for the company.[5] Mainframe systems and their peripherals had officially replaced tabulating machines and punch cards as the new "razor-and-blade" combo.

IBM applied its usual high-margin pricing strategy to the 2314. The largest-capacity version rented for $5675 per month or sold for $256,400. Despite its high brand premium, IBM's sales volume was modest and its product quality was not the best—creating opportunities for smaller vendors.[6] In 1966, Telex Corporation began producing magnetic tape drives compatible with IBM's System/360, which were well received by the market. Buoyed by success, Telex expanded into disk drives, printers, and memory.

In April 1970, Telex released the Model 5314 disk drive and controller, and by year's end, had captured 5.3% of the market from IBM's 2314.

Given the market conditions at the time, IBM could not simply avoid competition by refusing to license its patents. Instead, it responded with other strategies.

First, in September 1970, IBM launched a new product—the 2319A disk drive (hereafter referred to as "2319A")—designed to work with the latest System/370 mainframe. From a technical standpoint, this new product offered little innovation. The main change was reducing the number of spindles from four in the 2314 to three, thereby cutting storage capacity by a quarter and reducing both cost and price. Additionally, the drive's control functions were split: one part remained in the drive itself, while the other was integrated into the System/370 Model 145 CPU. Clearly, this was intended to complicate cloning or compatibility efforts by competitors. The

---

[5] Christopher Norgaard, Monopolization of Computer Peripheral Equipment: Telex Corp. v. International Business Machines Corp., 510 F.2d 894 (tenth Cir. 1975), 53 Denv. L.J. 295 (1976).

[6] Sara Baase. (1974). IBM: Producer or Predator. Retrieved from: https://reason.com/1974/04/01/ibm/

monthly rental for the 2319A was $300 cheaper than Telex's comparable offering, yet IBM still maintained a profit margin above 20%.

By the end of 1970, IBM released the 2319B disk drive, with prices further reduced—more than $1000 less per month compared to the 2314 with similar storage capacity. Unlike the 2319A, the 2319B did not separate control functions. Apart from that, it was nearly identical. In subsequent antitrust litigation between IBM and Telex, the district court found that the 2319B was a product launched solely for the purpose of price-cutting, offering no performance improvements over the 2314.

Nonetheless, these two new models failed to halt the rise of smaller competitors. Companies like Telex withstood the price war and employed more flexible tiered pricing strategies to expand their market share. This approach, common today in the consumer electronics industry, worked much like modern-day players such as Huawei, OPPO, Vivo, and Xiaomi offering a range of product options—"medium cup," "small cup," "half cup," even "one-third cup"—to meet different consumer price points. IBM, by contrast, was like Apple, offering only "large" and "extra-large" options. When dominant firms act arrogantly, this kind of pricing tactic can be quite effective. By the end of 1972, third-party peripheral manufacturers, led by Telex, had captured 21.6% of the disk drive market that competed directly with the 2314, snatching a sizable bite from IBM's business.

In response, IBM's marketing team conducted research and proposed an even more aggressive price reduction strategy—"kill a thousand enemies, lose eight hundred of our own"—cutting prices by 15% to 50% to drive out smaller competitors entirely. This was clearly at odds with IBM's usual business practices, and senior management rejected the proposal. Instead, under the directive of new CEO Frank Cary, IBM unveiled a killer weapon in 1971: the "Fixed-Term Lease Plan."

Prior to this, IBM's leases were typically month-to-month, offering flexibility. The "Fixed-Term Lease Plan" shifted to annual contracts. Customers signing a 1-year lease would receive an 8% discount, while 2-year leases earned a 16% discount.

On the surface, this appeared to be a simple volume-discount incentive, but it carried hidden implications.

First, the plan applied only to peripherals such as disk drives, tape drives, and printers, while deliberately excluding IBM's core products: CPUs and memory. At the same time the fixed lease plan launched, IBM raised prices on its entire CPU product line by 4% to 8%—an increase roughly equivalent to the leasing discount. IBM claimed this was merely a routine price adjustment in response to rising costs. Given that the United States was in the midst of the Great Inflation of the 1970s, the explanation seemed plausible, and the increase was approved by the federal Price Commission.

In 1972, IBM introduced its next-generation mainframes: System 370/Model 158 and System 370/Model 168. These new models did indeed offer performance improvements over their predecessors, but their rental prices soared disproportionately. The monthly rental for the previous models—System 370/Model 155 and 165—had been $20,600 and $36,400, respectively. The new models jumped to $30,700 and $48,600, an increase of over $10,000. Meanwhile, IBM significantly reduced the rental prices of its new memory products that supported the new systems. In later antitrust proceedings, the district court concluded that the purpose of the higher system prices was twofold: first, to offset the discounts offered by the Fixed-Term Lease Plan, and second, to absorb the losses from memory price reductions. Third-party vendors—Telex among them—had started offering IBM-compatible memory products. IBM hoped to kill off these competitors early by slashing memory prices. But, as the saying goes, "the wool comes from the sheep's back": though IBM seemed to be lowering prices, it ultimately shifted the cost burden to customers via high-priced mainframe systems.

Moreover, IBM imposed steep penalties for breach of the Fixed-Term Lease Plan. If a customer terminated a 2-year lease within the first year, they would owe a penalty equivalent to 5 months' rent. Even if the contract was terminated during the second year, the penalty would still be 2.5 months' rent.

The Fixed-Term Lease Plan dealt a heavy blow to small peripheral manufacturers. With System/360 and System/370 mainframes standing out in the market, most customers had no real alternatives and limited bargaining power, leaving them with little choice but to lease IBM's peripherals long-term to cut costs. On July 22, 1971, 40% of IBM's disk, tape, and printer business customers had opted for fixed-term leasing, effectively tethering them to IBM's platform. This move curbed the momentum of the rising peripheral makers.

In 1972, Telex, whose growth had been seriously hampered, filed an antitrust lawsuit against IBM, demanding the company be broken up. Telex won the first round of litigation, but in IBM's countersuit, the court ruled that Telex had stolen trade secrets and ordered it to pay IBM $18.5 million in damages. Telex was financially incapable of covering such a hefty penalty[7] and ultimately had to settle with IBM.

In the ruling issued on September 17, 1973,[8] the judge acknowledged that IBM had engaged in exclusionary conduct, but also highly praised the company's contributions to the computer industry:

> *IBM's growth and success in the industry have been due in substantial measure to its skill, industry and foresight. It has tended to set the standard for quality in the EDP industry for products and services. It has met notably favorable response in the market, and has been deeply involved in the phenomenal growth of the industry since almost its beginning. Each succeeding generation of IBM products has represented some technological improvement over the preceding generation and has involved development of new processes, storage devices, input/output devices and software. In the approximately twenty years that the EDP industry has been in existence IBM has introduced more than 600 products. Some of these products include major technological innovations. By virtue of its own research and development, IBM has obtained more than 10,000 patents which are freely licensed. I therefore cannot fully agree with Telex's contention that "IBM did not gain, nor has it maintained its position in the industry through skill, industry and foresight".*

---

[7] Robert Dorr, Donald M. Duft & Cheryl Hodgson, Patents, Trademarks, Copyrights, and Unfair Competition, 53 Denv. L.J. 247 (1976).

[8] Telex Corp. v. International Business MacHines Corp., 367 F. Supp. 258 (N.D. Okla. 1973) https://law.justia.com/cases/federal/district-courts/FSupp/367/258/1425543/

The ruling also revealed some fascinating facts: IBM conducted exit interviews for key employees, attended not only by the direct supervisor but also by a patent attorney, to determine what trade secrets, confidential information, or proprietary knowledge the employee had been exposed to. After the interview, this information was documented and archived. This suggests that even in the 1970s, the U.S. already had a mature system for trade secret protection. The American legal process, particularly its discovery phase, is notoriously intricate—so much so that even Europeans find it daunting. But American companies had long developed sophisticated responses to it.

Of course, given IBM's size and market share in the mainframe business at the time, it couldn't act entirely without constraint. Besides Telex, several other peripheral manufacturers also filed antitrust suits against IBM around 1970. In terms of core products, Control Data Corporation (CDC) petitioned the U.S. Department of Justice in 1967 to launch an antitrust investigation against IBM. CDC alleged that IBM would strategically time product announcements to sow doubt at critical decision points, nudging customers to delay purchases in anticipation of new IBM offerings instead of buying from its competitors. IBM also mastered the art of "drip-feeding" its technology roadmap—releasing planned improvements incrementally only when its market share was threatened. Additionally, IBM sometimes made lofty public claims about future products that were unlikely to be realized—"paper machines" and "phantom computers," as CDC called them. According to CDC, they were burned by such tactics at least twice. Once in the early 1960s, when trying to enter the scientific computing market, IBM announced a high-performance scientific computer whose specs looked impressive but never materialized. Delivery was delayed four times, the design repeatedly altered, and in the end, the machine was halved in computing power and shipped in only a few units. Another instance was when CDC launched its 6600 computer. IBM responded by loudly announcing an even more powerful machine, drawing away many of CDC's prospective customers. But again, IBM's product barely sold a few units.[9]

From 1968 onward, IBM became embroiled in a series of antitrust lawsuits, generating an enormous volume of discovery material—40 million pages in total. These included meeting minutes from IBM's senior management, discussions on new computer systems, detailed data on product profitability and costs, R&D expenses, pricing strategies, customer research, and more. The court record for the Telex case alone ran over 40,000 pages. After the case concluded, the court allowed the Computer Industry Association to obtain copies of IBM's discovery documents. The association then sold this trove of material for $5000 per set[10] — turning it into a textbook for a new generation of American tech companies.

When Emperor Xuan of Han once instructed the crown prince, he famously said: "The Han dynasty has its own institutions, which blend the ways of kings and

---

[9] Sara Baase. (1974). IBM: Producer or Predator. Retrieved from: https://reason.com/1974/04/01/ibm/

[10] Sara Baase. (1974). IBM: Producer or Predator. Retrieved from: https://reason.com/1974/04/01/ibm/

hegemons." IBM, too, did not rely solely on domineering tactics. While its business teams, empowered by IBM's massive scale, engaged in elaborate maneuvers, the disk development team at the San Jose lab remained equally active. In 1975, IBM released two groundbreaking disk drive products that once again left competitors in the dust. One was the IBM 3330 Model 11, codenamed "Iceberg." The other, even more famous, was externally branded as the IBM 3340 and internally codenamed "Winchester"—a name that would later become synonymous with a whole class of hard drives.

Compared to earlier models, the hallmark of the Winchester drive was that the disk, spindle, actuator arm, and read/write head were all sealed together inside a dustproof cartridge. This cartridge was sold as a single, non-serviceable unit. This key feature was captured in US Patent No. 3,786,454. Claim 1 of the patent listed standard disk components—frame, center-perforated rotating disk, central spindle, and the data read/write head—all of which were consistent with previous designs. The novelty appeared in the seventh element:

> ...a sealed, interchangeable cartridge integrally connected with said frame assembly for enclosing said disk apparatus, spindle, actuator arm, and head assembly...

Before the Winchester, hard drives had removable disks—much like how optical discs can be removed from their drives. Users would detach single or multi-platter disk packs for maintenance or archival storage, much like handling punched cards, vinyl records, or magnetic tapes. This was considered entirely normal at the time. The Winchester project leader worried that making disks non-removable would frustrate customers and repeatedly consulted the marketing department about the advantages of removable disks. The marketers had no clear answers and simply replied, "It's always been done this way."

But removable, open-architecture drives had one major issue: contamination and surface damage. The clearance between the read/write head and the disk surface had already reached a technical limit. Although US Patent No. 3,631,425 (for the IBM 2314 drive) claimed that the head could remove some contaminants, in practice, on-site engineers still had to wipe down disk surfaces with alcohol regularly. Any delay in cleaning could result in "bad sectors" and data loss—an unacceptable risk for clients paying tens of thousands of dollars per month to lease disk systems.

The sealed cartridge of the Winchester created a dust-free environment, solving this problem. But sealing the disk also meant the read/write head now had to be sold as part of the cartridge. Therefore, its cost had to be minimized. The Winchester project's core innovation was a newly designed magnetic head, jointly developed by the manufacturing and R&D teams, leveraging production expertise to control costs. The final head featured three guide rails and rested on a specific part of the disk surface when inactive. During read/write operations, it "took flight" and sought the correct track—an approach recorded in US Patent No. 3,823,416.

Because the disks were sealed in a cartridge and couldn't be swapped out when bad sectors developed, the team created another crucial technology: a method for circumventing defects in disk media, protected under US Patent No. 3,997,876. The

Winchester's innovations redefined the hard disk drive and set the industry standard for the next 20 years.

In October 1979, IBM filed for a patent titled "Thin Film Inductive Transducer" (US Patent No. 4,295,173), inventing the thin-film head—a breakthrough that further reduced the gap between the head and the disk surface, establishing a true technological moat for IBM. Harvard professor Clayton Christensen, a renowned scholar on technological innovation, called this a "competence-destroying technological innovation." According to his research, between 1981 and 1986, more than 60 new companies entered the disk drive market. Only five were capable of implementing thin-film head technology, and none succeeded commercially.

Even during the crisis-ridden 1980s, IBM continued to make significant advances in head technology—a testament to its enduring commitment to innovation.

Although this chapter focuses primarily on disk technology, it must be noted that IBM, at the time, was also responsible for many other groundbreaking inventions that changed the way people lived and worked. While the company's commercial conduct could be heavy-handed, its R&D efforts consistently pushed boundaries and remained true to the spirit of a genuine technology enterprise. Two examples are particularly noteworthy.

The first is the invention of Dynamic Random Access Memory (DRAM). In 1966, IBM's renowned computer scientist Robert Dennard designed a memory cell using just a single MOS transistor—an invention of enormous impact. According to IBM's official website:

In 1966, while relaxing at his home in Westchester County, New York, gazing out over the Croton River Gorge, Dennard was struck by inspiration on his living room couch. He began to consider the properties of MOS technology—which could fabricate capacitors capable of holding or not holding a charge, thus representing a binary 1 or 0. A transistor could control whether the capacitor was charged. The more he thought about it, the more convinced he became that he could design a simple memory.

In April 1967, Dennard presented his elegant design—where a transistor both wrote charge to and read it from a capacitor—at an internal meeting. His audience did not immediately grasp the significance. As Dennard concluded, the room fell silent—no questions, no discussion. His concept was simply too advanced. At the time, further industry progress in integrated circuit design and MOS processing technology was needed before DRAM could be practically implemented.

Even IBM's patent attorneys were puzzled by the idea. One lawyer reportedly said the invention was too simple—just a transistor and a capacitor—and thus lacked inventiveness. With help from colleagues, Dennard eventually convinced them, and the patent (US3387286A) was granted in 1968. IEEE would later describe Dennard's invention as "the most widely produced man-made artifact on Earth" (Fig. 7.6)

Though Dennard filed his patent in a timely manner, IBM did not immediately follow up with product development and failed to bring the first commercial DRAM to market. Three years after his invention, newcomer Intel released the 1103

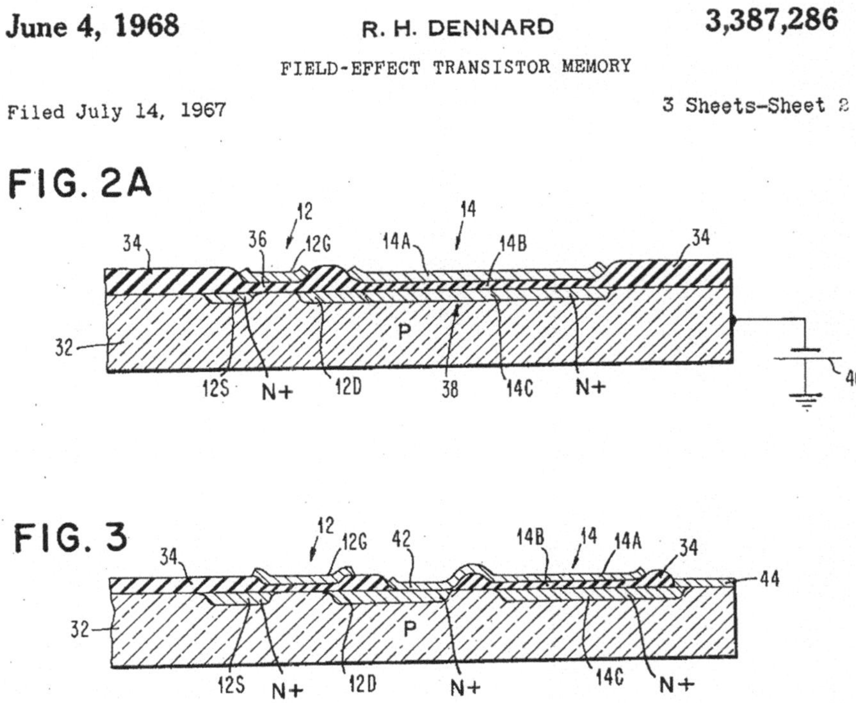

Fig. 7.6 Patent drawing of U.S. Patent No. 3387286A

memory chip—the first commercially successful DRAM product. The 1103 was so influential that many still mistakenly believe Intel invented DRAM.

Years later, despite being universally acknowledged as the "father of DRAM," Dennard remained indignant about this misconception. In a 2009 interview, IEEE Spectrum reporters noted that Dennard always carried his original DRAM design notes and never missed a chance to point out the popular myth of Intel's role.[11]

The second example is the ubiquitous barcode. In 1949, Norman Woodland and Bernard Silver co-filed a patent for a "Classifying Apparatus and Method," which was granted in 1952 (US2612994A). As students, they were inspired by a supermarket manager's complaint about time-consuming inventory management. One day, sunbathing on a beach, Woodland drew inspiration from Morse code—specifically its use of dots and dashes to convey information. He wondered, why not use thick and thin bars to represent more complex numeric data? He dragged four fingers through the sand to draw a circle, sketching what would become known as the "bull's-eye" circular code—the precursor to modern barcodes (Fig. 7.7).

[11] Adee, S. (2009). Thanks for the memories. *IEEE Spectrum*, *46*(5).

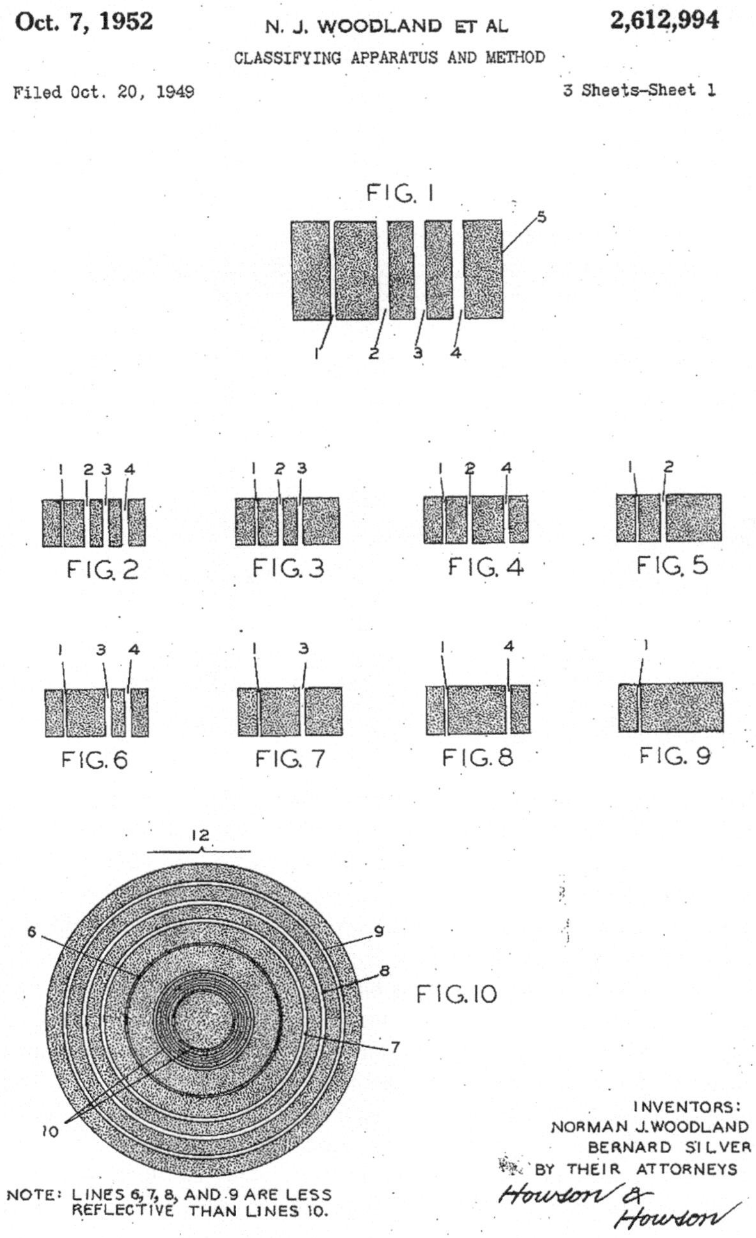

Fig. 7.7 Patent drawing of U.S. Patent No. 2612994A

IBM's senior Watson was deeply impressed by Woodland's idea and hired him. However, due to technological limitations, barcodes could not yet be input automatically into computer systems, and Woodland was unable to advance his invention further. Records show that the barcode patent changed hands several times: first sold by Woodland to Philco for $15,000, then resold to Ford Motor Company, and eventually passed to RCA.

In the late 1960s, thanks to the development of laser scanning technology, RCA finally found practical use for the bull's-eye barcode. During a trade show, an IBM representative saw RCA's demo and proposed launching a similar project. IBM initiated development through standard procedures, including patent searches—only to discover, to their astonishment, that the inventor of the foundational patent had been working at IBM for 15 years. Woodland was brought back into the project and made significant contributions.[12]

Barcodes revolutionized retail by automating the entry of product and price information, enabling streamlined checkout. IBM launched the IBM 3650 and 3660 storage systems in 1973, driving transformation in the retail industry. Today, barcodes are omnipresent and essential to modern life.

For many years, Woodland's contributions remained obscure—until a serendipitous event in 1992. That year, President George H. W. Bush visited a supermarket in Orlando and was shown a barcode scanner by an NCR executive. Intrigued, Bush asked several questions and praised the technology. A reporter captured his puzzled expression, and The New York Times ran a story titled "Bush Encountered Supermarket, Amazed," accusing the president of being out of touch with middle-class life.[13]

Bush's PR team defended him, claiming his surprise was at the scanner's ability to read damaged barcodes—then a new feature. To bolster the president's image, the Republican Party organized a public event where Bush presented Woodland with a medal, underscoring the president's understanding of barcode technology and his recognition of its inventor.[14]

---

[12] Lissack, M., & Roos, J. (1999). *The next common sense: Mastering corporate complexity through coherence*. Nicholas Brealey Publishing.

[13] Andrew Rosenthal. (1992). Bush Encounters the Supermarket, Amazed. Retrieved April 8, 2024 from https://www.nytimes.com/1992/02/05/us/bush-encounters-the-supermarket-amazed.html

[14] Weightman, G. (2015). *Eureka: how invention happens*. Yale University Press. p128

Chapter 8
From Opponent to Champion of Software Patents

In the 1970s, IBM was involved in patent litigation with companies like Xerox and Ampex, both of which sued IBM for patent infringement and alleged monopolistic practices. Ultimately, the cases ended in cross-licensing agreements and settlements.[1] These purely defensive maneuvers cannot be regarded as proactive patent strategies. If IBM can be said to have had any active strategic posture in the realm of patents during this period, it was primarily in the domain of software. As one of IBM's patent attorneys later admitted:

> *IBM addressed this problem by lobbying for software to be unpatentable. When the President set up a commission to study the U.S. patent system, a senior IBM manager was appointed head of this commission. Not surprisingly, the commission concluded that software should not be patentable because the patent examiners were not prepared to examine software patents.*[2]

Between the 1960s and 1970s, IBM played an active role in several key cases and policy discussions opposing the patentability of software.

In 1963, two researchers from Bell Labs—Gary Benson and Arthur Tabbot—submitted a patent application for a method that converted BCD code to pure binary. This was a purely algorithmic patent. The U.S. Patent Office unsurprisingly rejected the application. Bell Labs appealed to the Patent Trial and Appeal Board (PTAB), which again rejected it. The case was then brought to the United States Court of Customs and Patent Appeals, the predecessor to the Federal Circuit, where Bell Labs secured a reversal. IBM and Honeywell then filed opposition briefs, and the Patent Office petitioned the Supreme Court to review the decision.

During the Supreme Court proceedings, IBM participated as amicus curiae, arguing that "a computer program is simply a mode of expressing ideas." The Court

[1] Clare M. Reckert. (1974). Ampex and I.B.M. Reach a Settlement in Patents Dispute. Retrieved April 8, 2024 from https://www.nytimes.com/1974/07/29/archives/ampex-and-ibm-reach-a-settlement-in-patents-dispute-dispute-is.html

[2] Leith, P. (2007). *Software and patents in Europe*. Cambridge University Press,. P25

Q. Yue, F. Yang, *Governing Innovation: A Global History of Patent-Driven Technological Progress*, https://doi.org/10.1007/978-981-95-5570-3_8

agreed with IBM's position and ruled in favor of the Patent Office, rejecting Benson's application.

Why was IBM opposed to software patents?

In 1965, Martin A. Goetz, founder of the software company Applied Data Research (ADR), filed a patent application for a program he developed called Autoflow. The patent was granted on April 28, 1968 (US Patent No. 3,380,029). Autoflow was a flowchart generation software. At the time, nearly all programmers were expected to create flowcharts for their programs, which was the final step in the development process. However, manually drafting these diagrams was tedious, leading many to skip the task—causing major maintenance issues later. Autoflow solved this by automatically generating flowcharts from source code. The software reportedly cost around $10,000 to develop and sold several thousand copies.[3]

Like many software patents today, Goetz's patent cleverly framed the invention as a system involving computers, tape units, and control systems, rather than software per se, enabling it to pass examination. As the first-ever "software patent," the application's attorney—uncertain of the Patent Office's standards—included a near-complete explanation of the source code within the specification. However, despite several competing products emerging by 1972, none achieved commercial success, and ADR never used the patent to sue. Thus, the actual power of US Patent No. 3,380,029 remains unclear.

As ADR promoted Autoflow, IBM released a rival product called Flowcharter, which quickly became Autoflow's strongest competitor. Flowcharter's automation was inferior and required user input,[4] but it had a major advantage: it was free—or at least appeared so. As previously discussed, IBM bundled software with mainframe rental fees, where customers had little pricing power. Renting an IBM mainframe effectively meant accepting Flowcharter, making additional software purchases unnecessary. Although Autoflow's automation was better, it wasn't enough to justify its cost. IBM also deployed a "blank check" strategy, frequently promising customers future upgrades and new features, discouraging them from switching.[5]

According to Goetz, ADR approached IBM regarding infringement of US Patent No. 3,380,029 and hoped to dissuade them from developing similar software. Though there was no litigation, Goetz maintained that Flowcharter remained technically inferior to Autoflow.[6]

---

[3] Campbell-Kelly, M. (2004). Not all bad: An historical perspective on software patents. *Mich. Telecomm. & Tech. L. Rev.*, *11*, 191.

[4] Goetz, M. A. (1998). How ADR Got Into the Software Products Business and Found Itself Competing Against IBM. Retrieved April 8, 2024 from: https://archive.computerhistory.org/resources/access/text/2012/10/102746516-05-01-acc.pdf

[5] Timothy Lee. (2014). Will the Supreme Court save us from software patents? Retrieved April 8, 2024 from https://www.disputesoft.com/wp-content/uploads/2014/03/Will_the_Supreme_Court_save_us_from_software_patents.pdf

[6] Goetz, M. A. (1998). How ADR Got Into the Software Products Business and Found Itself Competing Against IBM. Retrieved April 8, 2024 from: https://archive.computerhistory.org/resources/access/text/2012/10/102746516-05-01-acc.pdf

As one of the earliest independent software companies, ADR continuously avoided direct competition with IBM. However, IBM's unpredictable moves often blindsided small firms. For example, while ADR was developing a flagship product called ROSCOE, IBM introduced a similar product named CRBE. In 1970, IBM contacted all of ROSCOE's potential customers to announce a free alternative called TSO.

To combat IBM's dominance, Goetz joined forces with other developers to form the Association of Data Processing Service Organizations (ADPSO), advocating for software patentability. At the time, small software companies focused on unmet customer needs within the IBM ecosystem. However, because their products relied on IBM platforms, the tech giant had a structural advantage. Once these smaller players established a viable product, IBM could easily add similar features to its free software offerings—or just promise to do so—effectively suppressing competition.

ADPSO argued that "software is actually a form of machine device which structures the general-purpose hardware." IBM, however, insisted that software consisted merely of "mental steps," and was therefore unpatentable.

In 1968, the U.S. Department of Justice initiated a new antitrust case against IBM, accusing it of bundling software with hardware, thereby stifling competition in the emerging software market. In 1969, ADR filed an antitrust suit as well. Soon after, IBM issued a major policy shift: it publicly renounced software bundling. ADR then reached an out-of-court settlement with IBM, which paid $2 million in damages.

This era of legal and academic debate about software patentability—beginning with Gottschalk v. Benson and culminating in Diamond v. Diehr (1981)—slowly brought about legal clarity.[7]

By the 1980s, several developed nations began drafting copyright legislation specifically for computer software. During this period, IBM's lobbying apparatus became highly active on a global scale, advocating for copyright protection for software.[8] As is well known, copyright primarily protects against direct copying, whereas patents protect technical methods. If software were protected solely under copyright law, it would be legal to bypass protection by replicating functionality without copying the code. By this time, IBM had accumulated a substantial portfolio of software assets and was mainly interested in preventing reverse engineering.

In the 1980s, Japan's Ministry of International Trade and Industry proposed a sui generis legal framework for software protection, suggesting a protection term of 15 years—far shorter than the standard copyright term (usually life of the author plus 50 years). Around the same time, the World Intellectual Property Organization

---

[7] Smith, G. (2019). Unsung innovators: Marty Goetz, holder of first software patent. Retrieved April 8, 2024 from https://www.computerworld.com/article/2540020/unsung-innovators%2D%2Dmarty-goetz%2D%2Dholder-of-first-software-patent.html

[8] Drahos, P., & Braithwaite, J. (2017). *Information feudalism: Who owns the knowledge economy*. Routledge. P185

(WIPO) began drafting a treaty on software protection. In 1984, under IBM's strong lobbying, Western nations jointly pressured Japan to abandon the 15-year proposal. Seeing the shift in international policy, WIPO also dropped its draft treaty.[9]

In 1989, the European Commission attempted to harmonize software copyright laws across member states by proposing a directive on the legal protection of computer programs (commonly known as the Software Directive). Due to lobbying by firms like Bull (France), Fujitsu (Japan), Olivetti (Italy), NCR (U.S.), and Sun Microsystems, the draft avoided restricting reverse engineering. These companies also formed the European Committee for Interoperable Systems (ECIS) in 1989 to advance their lobbying efforts. In response, IBM established the Software Action Group for Europe (SAGE), whose members included Microsoft, Apple, and Lotus.

According to the book *Information Feudalism*, in the following years, copyright and computer experts flooded the corridors of the European Commission—one of the largest lobbying efforts in its history. ECIS received some support from developing countries, but several, including Argentina, reversed their stance under U.S. pressure. As a result, ECIS's efforts ultimately failed. On May 14, 1991, the European Community adopted the Software Directive. While it made some concessions to ECIS, such as allowing limited reverse engineering, it mostly favored the position of IBM-led SAGE.

From the 1970s to the 1980s, IBM's influence on global intellectual property legislation—particularly software protection—was unmatched. However, ironically, the primary beneficiaries of the new system turned out to be the Wintel alliance (Microsoft and Intel), particularly the then up-and-coming Microsoft. IBM's own OS/2 operating system failed spectacularly, and it struggled to compete with the emerging wave of nimble software companies. After circling back to the very policies it once opposed, IBM discovered its software products weren't even worth copying—ultimately pushing it to embrace software patents.

In 1994, the U.S. Patent and Trademark Office held a public hearing on software patents, inviting all major tech companies to weigh in. Companies like Oracle and Adobe opposed software patentability. Oracle argued that software innovation occurred rapidly, without the need for massive capital investment, and typically involved novel combinations of known techniques. IBM, once a staunch opponent of software patents, had now completely reversed its stance. It argued that inventions related to computer programs should not be treated differently from hardware or microprocessor inventions. It also appealed to patriotic sentiment, claiming that "as the industry matures and faces increasing foreign competition, patents will be the key to protecting the most important innovations produced in the United States."

Prior to the hearing, even Bill Gates had expressed concerns about software patents. In a 1991 memo to Microsoft executives, he wrote:

> *If people had understood how patents would be granted when most of today's ideas were invented, and had taken out patents, the industry would be at a complete standstill today...*

[9] Drahos, P., & Braithwaite, J. (2017). *Information feudalism: Who owns the knowledge economy.* Routledge. P185

I feel certain that some large company will patent some obvious thing related to interface, object orientation, algorithm, application extension or other crucial technique… they have a 17-year right to take as much of our profits as they want.[10]

Nonetheless, Microsoft actively pursued software patents and supported IBM's position during the hearings.

At the beginning of this volume, we referenced Huntington's theory on "values and ideology." Now is an opportune moment to revisit that discussion. In the 1960s, IBM's opposition to software patents was justified on the grounds of protecting innovation and industrial development. In the 1990s, its support for software patents was again framed as serving the cause of innovation and industry. So, are software patents ultimately beneficial or detrimental to innovation?

This question remains hotly debated. Both sides have compelling arguments and internally consistent theories. The American patent system, too, has evolved amid such debates. But does this evolution reflect a deeply rooted set of superior values and ideologies? That remains an open question.

According to expert research, under Thomas Watson Jr.'s leadership, IBM's involvement in politics was relatively limited. The company had only two or three personnel stationed in Washington D.C. whose main responsibility was to monitor publicly available congressional information. Watson Jr. reportedly forbade them from attending congressional hearings.[11] He preferred a subtler method of influence, once telling his successor, Frank Cary: "The most elegant and effective form of political influence is face-to-face. The worst way is to open a Washington office full of professional lobbyists."

Despite this, in 1975, Frank Cary established IBM's Government Programs Office—initially small but steadily growing. Eventually, it became one of the largest of its kind in Washington, with over a hundred staff members. These were top experts in taxation, trade, compliance, R&D policy, and export controls. They became regular participants in key industry forums and were known internally as "issue managers." Their job wasn't mere lobbying—it was high-level issue management. As managers of policy issues, their mission was not to passively respond to government actions, but to proactively research political issues in detail, formulate IBM's positions, track policy discussions, and insert IBM's demands into the discourse to exert influence.

In the long run, how effective was this influence? That's a difficult question. In the 1980s, IBM both manufactured DRAM chips for its own use and procured DRAM from external suppliers. Its Japanese subsidiary exported DRAMs to the U.S., and IBM also held shares in Intel, a leading DRAM producer. IBM also had IP disputes with various Japanese firms. This complex web of interests made it unclear even to IBM which policy direction would most benefit the company.

[10] Timothy B. Lee. (2014). Will the Supreme Court save us from software patents? Retrieved March 31, 2024 from The Washington Post Website: https://www.washingtonpost.com/news/the-switch/wp/2014/02/26/will-the-supreme-court-save-us-from-software-patents/

[11] Hart, D. M. (1999). IBM in American politics, 1970–1999. Business and Economic History, 28(2), 49–59.

According to Harvard professor David M. Hart, regardless of whether the chips were cheap or expensive, American or Japanese, IBM had a stake in all of them.[12]

As a member of the Semiconductor Industry Association (SIA), IBM played a key role in negotiating the Japan–U.S. semiconductor agreement. But after the agreement was signed, not only did Japanese firms suffer, IBM's own Japanese subsidiary was also caught in the fallout. Meanwhile, rising chip prices in the U.S. severely impacted IBM's domestic computer business.

After reassessing its interests, IBM led the formation of the Computer Systems Policy Project (CSPP), a coalition of more than ten companies. CSPP negotiated with the SIA to promote the second Japan–U.S. semiconductor agreement, signed in 1991. This episode illustrates how even a company as vast as IBM could struggle to discern which policy would best serve its interests.

Yet, up until at least 1997, IBM maintained a strong political presence. Its public lobbying expenditure reached $5 million that year—dwarfing Hewlett-Packard's $500,000, Intel's $600,000, and Microsoft's $1.9 million, despite Microsoft being under antitrust investigation at the time.

What, then, did all this high-cost lobbying achieve? IBM had once successfully opposed software patents, winning the Benson case. Its push for copyright-based software protection was embraced in 1980s judicial and legislative reforms. But throughout that decade, IBM failed to launch any software product that could rival Oracle, Microsoft, or Adobe. No matter how successful it was on the lobbying front, those wins could not be translated into strategic gains without world-class products.

In 1995, IBM acquired Lotus, gaining patents for Lotus 1-2-3 and Lotus Notes. Some sources[13] suggest that IBM's leadership at the time had little understanding of Lotus and pursued the acquisition out of an urgent desire to break into software. The deal was finalized in what many saw as a near-hostile takeover. After the acquisition, the entire Lotus executive team, including the CEO, resigned. The core development team behind Lotus Notes also left within 2 years. With the departure of these key players, Lotus quickly lost market share and was eventually overtaken by Microsoft Office.

In business, pragmatism reigns. Within the legal frameworks of the U.S. and Europe, IBM actively engaged in legislation involving cutting-edge technologies, advocating for positions aligned with its interests. This behavior should not be moralized. However, while studying and emulating the strengths of American institutions, one should also beware of non-practicing entities (NPEs) that masquerade under the banner of innovation protection while engaging in legal extortion. As the old saying goes, "Their mouths are full of principles, but their bellies are full of business."

In conclusion, prior to the 1980s, IBM's patent strategy was relatively unremarkable due to the restrictions of the consent decree. But in the realm of intellectual

[12] Hart, D. M. (1999). IBM in American politics, 1970–1999. Business and Economic History, 28(2), 49–59.

[13] Matsumoto, S. (松本茂) (2018). Japanese Outbound Acquisitions: Explaining What Works. Springer.

property legislation, its influence was unrivaled. This indirect method of shaping national law may be the most "elevated" form of IP strategy—perhaps even more sophisticated than dominating technical standards. Regrettably, by that point, IBM no longer had the world-class products to match, and its victories in policy circles merely paved the way for others.

Chapter 9
From Paper Mill to Patent Factory

The 1980s marked a pivotal turning point in the U.S. patent system. While IBM had been locked in prolonged antitrust battles with government regulators for decades, another tech giant, AT&T, was similarly entangled in antitrust investigations. Under the intense pressure of these probes on leading corporations, even in the high-tech fields of electronics and semiconductors, patents were not regarded as a strategic priority.

In the 1970s, facing mounting domestic economic crises, President Jimmy Carter advocated for measures to stimulate technological innovation and commercialize the outcomes of invention—signaling the beginning of America's transition into the information age[1]. In 1982, to address the long delays and inconsistent standards in patent litigation, Congress passed the Federal Courts Improvement Act. This act merged the former Court of Customs and Patent Appeals with the appellate division of the U.S. Claims Court to establish the Court of Appeals for the Federal Circuit (CAFC), a pioneering court specialized in patent cases.[2] Following its formation, the CAFC unified the standards for patent adjudication and issued a series of rulings favorable to patent holders. Notably, it raised the evidentiary threshold for invalidating a patent from a "preponderance of the evidence" to "clear and convincing evidence."[3] Against this backdrop, the volume of patent filings across the U.S. began to surge rapidly.

Between 1983 and 1988, Hewlett-Packard's granted patents rose by 275%, Intel's by 213%, Kodak's by 187%, and Texas Instruments' by 168%. Even the slower-moving AT&T saw a 32% increase. In contrast, IBM appeared largely

[1] 孔译珞. (2014). 专利专门性法院的先驱者——美国联邦巡回上诉法院的发展 (Doctoral dissertation).

[2] 孔译珞. (2014). 专利专门性法院的先驱者——美国联邦巡回上诉法院的发展 (Doctoral dissertation).

[3] Henry, M. D., & Turner, J. L. (2006). The Court of Appeals for the Federal Circuit's Impact on Patent Litigation. The Journal of Legal Studies, 35(1), 85–117.https://doi.org/10.1086/498834

Q. Yue, F. Yang, *Governing Innovation: A Global History of Patent-Driven Technological Progress*, https://doi.org/10.1007/978-981-95-5570-3_9

oblivious to these shifting tides, registering only about a 10% increase in patent grants during the same period.

Researchers analyzing data from the United States Patent and Trademark Office (USPTO) compiled statistics on IBM's patent registrations from 1969 to 2014, as illustrated in Fig. 9.1.[4] In this chart, the gray shading represents the number of IBM-registered patents, while the black squares indicate the proportion of IBM patents relative to all granted patents in the U.S. As shown, between the 1970s and the late 1980s, IBM's annual granted patents fluctuated between 1500 and 2000, with only modest changes in quantity. From 1984 to 1990, while the number of registered IBM patents saw a slight uptick, their share of total U.S. granted patents noticeably declined.

Most studies on patent management agree that the 1980s represented a major transformation period both for the U.S. patent legal system and for the intellectual property management models in the American high-tech sector. Yet the driving force behind this transformation was not IBM, but Texas Instruments (TI)—a company on the verge of bankruptcy, fighting desperately amid U.S.-Japan semiconductor trade tensions. In 1985, the same year as the signing of the Plaza Accord between Japan and the United States, TI filed lawsuits against several Japanese firms, seeking damages for infringement on a series of patents related to integrated circuits and manufacturing processes. The signing of the 1986 U.S.-Japan Semiconductor

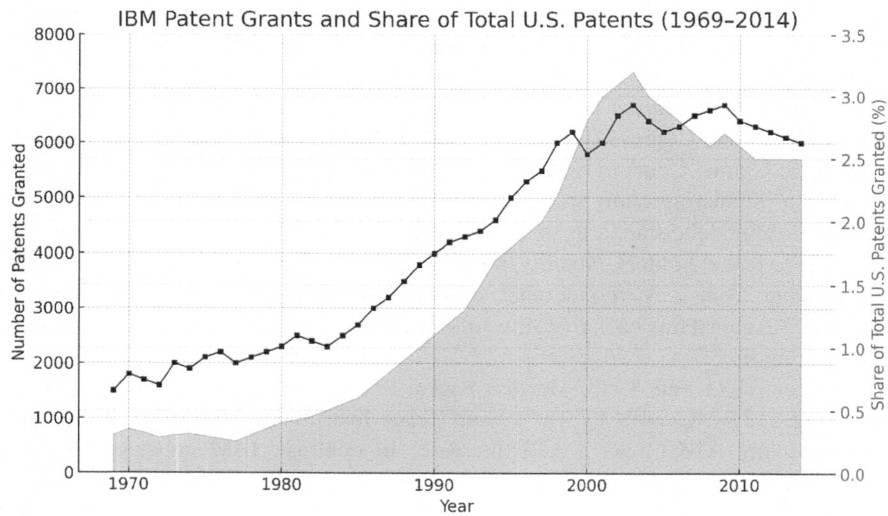

Fig. 9.1 IBM Patent Grants and Share of Total U.S. Patents (1969–2014)

[4] De Paiva Britto, J. N., Costa Ribeiro, L., Araújo, L. T., da Matta Machado, G. T., & da Motta e Albuquerque, E. (2018). Knowledge flows, changing firms' competences and patent citations: an analysis of the trajectory of IBM. *Economics of Innovation and New Technology*, 1–31.

Agreement marked Japan's definitive defeat in the semiconductor war. That same year, TI won an initial victory in its lawsuits against Japanese companies.

Having tasted success, TI swiftly turned its attention to domestic competitors. Between 1986 and 1993, the company won case after case, ultimately securing nearly $2 billion in damages and licensing fees over a 7-year span. In 1992, TI's patent licensing revenue reached $391 million—exceeding the company's total operating income for that year, which stood at $274 million. TI's triumph became a landmark example of next-generation IP management for all of Silicon Valley. Companies such as Motorola, Hewlett-Packard, AT&T, and IBM quickly followed suit, ushering in a new era of aggressive patent operations—among them, IBM's performance would prove most striking. In management studies of patent strategy in the twenty-first century, TI and IBM are frequently cited as having established two dominant models of intellectual property management.[5]

Though both were pioneers of this new era, IBM's response lagged behind TI's. As previously mentioned, IBM had been bound by consent decrees since 1956, which made it exceedingly cautious—almost timid—in asserting its patent rights. Moreover, IBM's immense scale as one of the world's few mega-multinational corporations made it particularly difficult to pivot or enact institutional reform. Even after Lou Gerstner's sweeping reforms, scholars into the 2010s still wrestled with the complexities of IBM's organizational structure. Jay R. Galbraith, the prolific management theorist from Switzerland's IMD business school, once remarked that IBM's organizational structure was the most complex he had ever seen.[6] Within such a system, reforming IP management practices proved extraordinarily challenging.

Prior to 1989, IBM's approach to intellectual property management was still relatively rudimentary. The entire management model was decentralized.[7] Once researchers achieved technological breakthroughs, an internal committee made up of R&D managers and corporate lawyers would evaluate the innovation and choose between two options:

- File a patent application;
- Publish the innovation in the *Technical Disclosure Bulletin* (TDB).

Publication in the TDB served a defensive function: by placing innovations in the public domain, it blocked competitors from securing patent rights on similar ideas. According to IBM's internal performance evaluation system at the time, a successful patent filing earned 3 points, while a TDB publication was worth only 1 point. However, publishing in the TDB was far easier—typically just a two- or three-page

[5] For example, Colleen Chien, From Arms Race to Marketplace: The Complex Patent Ecosystem and Its Implications for the Patent System (2010), Available at: http://digitalcommons.law.scu.edu/facpubs/69

[6] Jay R. Galbraith, Designing Matrix Organizations That Actually Work: How IBM, Procter & Gamble, and Others Design for Success (San Francisco, 2009), 129.

[7] Ajay Bhaskarabhatla, Deepak Hegde (2014) An Organizational Perspective on Patenting and Open Innovation. Organization Science 25(6):1744–1763. https://doi.org/10.1287/orsc.2014.0911

project report—so many inventors opted for this path instead of navigating the complex patent application process. As of 1988, only 18% of IBM's innovations were submitted for patents; the rest were published openly and, effectively, given away for free.

IBM's scientists also frequently published in academic journals. While this did not count toward performance evaluations, it held weight in the selection of IBM Fellows—an elite internal recognition. Researchers have noted that during the 1980s, IBM's research division was steeped in an academic atmosphere. Scientists prided themselves on scholarly accomplishments and the distinction of becoming IBM Fellows.[8] At the time, IBM Research was led by renowned mathematician Ralph Gomory and physicist John Armstrong, both of whom were highly accomplished academically but not known as corporate executives. James McGroody, who would later lead patent reforms, was himself a prominent physicist, a former president of the American Physical Society, and a member of the National Academy of Engineering. According to Gomory, prior to 1989, IBM's scientists were no different from university researchers—they were under no obligation to create commercial value for the company. The research division exuded the aura of an ivory tower, devoted to the pursuit of open science and generally resistant to aggressive patent strategies. Ironically, even as IBM was declining from 1981 to 1987, its R&D budget doubled—from $3 billion to $6 billion.

In both 1986 and 1987, IBM scientists won the Nobel Prize—poetic highlights of that academic era.

IT journalist Robert Cringely, in his book *The Decline and Fall of IBM: End of an American Icon,*[9] echoed these sentiments. Cringely often likened IBM's organizational structure to Japan's feudal caste system (though Japan lacked a true caste system): the CEO was the monarch of the IBM kingdom; division heads were the nobility; the sales department made up the samurai class—since only samurai could ascend to nobility, and a particularly accomplished noble could even be crowned king. Programmers and engineers were akin to artisans, possessing a certain income and social standing. At the bottom were customer service, manufacturing, and systems maintenance staff—the equivalent of serfs (despite the lack of a formal serf class in Japan). In this rigidly stratified hierarchy, researchers occupied a lofty priestly status. They were respected across ranks, even though "no one really understood what they were doing."

Yet as the saying goes, "what prospers must decline." Eventually, the axe fell even on the scientists. In his managerial memoir *Who Says Elephants Can't Dance?*, Lou Gerstner repeatedly praised his own decisions to slash R&D spending and downsize research departments. In late 1994, Gerstner appointed John Thompson to overhaul the software division. Thompson enacted a brutal streamlining, reducing 30 labs to just 8, and consolidating 60 brands into 6. Gerstner was lavish in his

[8] Ajay Bhaskarabhatla, Deepak Hegde (2014) An Organizational Perspective on Patenting and Open Innovation. Organization Science 25(6):1744–1763. https://doi.org/10.1287/orsc.2014.0911

[9] Cringely, R. X. (2014). The decline and fall of IBM: end of an American icon?. BookBaby.

praise: "His team cut hundreds of millions from the research budget, redirecting funds into new markets and sales efforts." Since the 1990s, IBM has had its ups and downs, but its R&D spending never again reached the heights of the 1980s Nobel Prize era. In the European Commission's *2022 EU Industrial R&D Investment Scoreboard*, IBM ranked 36th—trailing not only Intel and Microsoft, but also Huawei. From a global top-10 position in 2004, to outside the top 20 in 2012, to 35th in 2022, IBM's decline has been dramatic.[10]

Marc Levenson's story is a textbook example of how IBM Research had become disconnected from industry and burdened by a convoluted management system. In 1981, Levenson invented phase-shift mask (PSM) technology and submitted a report to IBM's patent office. While the team recognized the invention's novelty, they declined to file a patent because IBM had already met its yearly quota for optical patents. Levenson subsequently published a paper on the invention in an IEEE journal, putting the technology into the public domain. For Levenson, this merely meant forfeiting a bonus; for IBM, it was a potentially game-changing misstep in future industry positioning.

In 1982, Levenson improved his invention by using polymethyl methacrylate (PMMA) as the phase-shifting material, thereby classifying the innovation under polymer technology—a field in which IBM aimed to file more patents that year. This time, IBM submitted patent applications in the U.S., Japan, and Europe (U.S. application number US36567282A).

Unfortunately, the invention still failed to secure a U.S. patent. Soon after filing, the U.S. Patent Office identified a prior MIT patent (US4360586A) that potentially threatened the novelty of Levenson's claim and requested a response. While Berkeley professor Andrew Neureuther, a key advisor to IBM's chip division, argued that the prior art posed no threat, IBM's patent team decided it wasn't worth pursuing and withdrew the application.

In Japan, Levenson's patent was granted without issue. But when it later faced a validity challenge, IBM Japan dropped the defense after discovering that headquarters had already abandoned the U.S. application—leading to the patent's invalidation.[11]

By the mid-1980s, Japanese semiconductor firms began paying close attention to Levenson's work. IBM, however, was fixated on X-ray lithography, assigning Levenson to that field instead. Though Levenson remained confident in his own approach, he failed to gain managerial support. At one point, he sought budget approval to visit IBM's chip plant and evaluate his invention in a production setting. He eventually secured the funding from a high-level chip division executive, but his

[10] Grassano, N., Hernandez Guevara, H., Fako, P., Nindl, E., Georgakaki, A., Ince, E., Napolitano, L., Rentocchini, F., Tubke, A. (2022). The 2022 EU Industrial R&D Investment Scoreboard, Publications Office of the European Union, Luxembourg, 2022, :10.2760/485748, JRC132035. Retrieved March 31, 2024, from https://iri.jrc.ec.europa.eu/sites/default/files/contentype/scoreboard/2022-12/EU%20RD%20Scoreboard%202022%20FINAL%20online_0.pdf

[11] Carroll, P. (1994). *Big blues: The unmaking of IBM*. Crown Publishing Group.

direct supervisor diverted the money elsewhere, preventing Levenson from going—a missed opportunity for vital industrial engagement.

In 1984, Levenson turned to squeezed light research, a topic rooted in fundamental science with no immediate application to IBM's product line. He wasn't concerned—at least it meant no further battles with the patent office. But in 1993, amid major downsizing, Levenson returned from a business trip to find his office stripped bare. All his equipment was gone. This incident prompted his departure from IBM.

Ironically, IBM's chip division later concluded that X-ray lithography was a dead end, while Levenson's technology had become mainstream. But by then, IBM had lost both the inventor and the patent.

When management scholars reflect on IBM's history, they invariably focus on Lou Gerstner's dramatic turnaround after he became CEO in 1993. In these accounts, his predecessor—John F. Akers, who led IBM from 1985 to 1993—is often cast as a cautionary tale. Steve Jobs once described Akers as *"smart, eloquent, fantastic salesperson, but he didn't know anything about product*."[12] Yet in truth, the reform of IBM's intellectual property management system did not begin with Gerstner; it was initiated under Akers' tenure.

Marshall Phelps—an influential figure in the U.S. patent world who oversaw IBM's patent operations for over 20 years and earned the nickname "the godfather" of American IP—once remarked that Gerstner was the archetypal manufacturing CEO. When Gerstner first took the helm, he was unfamiliar with the intricacies of patent licensing in the high-tech sector and even questioned why anyone would license technology to a competitor. Only after Phelps explained the necessity of cross-licensing in the tech world did Gerstner fully support the strategy and give Phelps a free hand.

Because of Gerstner's prominence in business management, many academic studies mistakenly peg 1993 as the starting point of IBM's transformation in IP strategy, with much of the related statistical data beginning in that year. However, research by Bhaskarabhatla and Hegde (2014) shows that Gerstner's reforms only lightly touched on IP matters; the real changes began in 1989 and were largely complete by around 1993. In Gerstner's own *Who Says Elephants Can't Dance?*, the overhaul of IBM's IP system is mentioned only in passing.

By 1989, IBM was deeply mired in crisis. Its PC business was floundering, and its R&D budget had ballooned to an all-time high of $7.2 billion. As previously mentioned, IBM's R&D spending never again reached that level, remaining in the $5–6.5 billion range well into the twenty-first century. Faced with mounting pressure, then-CEO John F. Akers initiated sweeping personnel changes and launched the first mass layoff in the company's 70-year history. One-quarter of the workforce was cut, morale plummeted, and many departments descended into disarray. While widely criticized, this wave of reform also marked the beginning of IBM's transformation in IP management: John Armstrong, former head of the Research Division,

[12] Robert McMillan. (2011). "Steve Jobs: HP Implosion Was an iTragedy | Enterprise". WIRED. Retrieved April 8, 2024 from https://www.wired.com/2011/10/steve-jobs-itragedy/

was promoted to Chief Scientific Officer (CSO) and no longer oversaw patents; renowned physicist James McGroody—formerly of the ThinkPad division—took over the Research Division and assumed full responsibility for intellectual property.

Upon taking office, McGroody swiftly implemented bold reforms to the company's patent management system. First, he declared that patent registration—not trade secrets or free disclosure—would be IBM's primary method of protecting technical innovation. In 1990, McGroody founded the "Patent Academy" to train technical staff on writing, evaluating, and processing patent applications, as well as on the criteria for determining patent quality. In 1991, John Cronin assembled the "Patent Factory" team, which began scouring IBM's global R&D operations to identify and patent hundreds of previously overlooked inventions.

McGroody also revamped IBM's inventor incentive system. According to interviews with researchers,[13] prior to 1989, IBM rewarded inventors for either "writing + publishing" or "writing + filing a patent application." Because the former was easier, many inventors colluded with in-house patent attorneys to classify their work as "publishable" rather than "patentable," opting for simpler, lower-reward routes. McGroody abolished this loophole, instituting a policy that only patent applications would earn incentives. Moreover, if a patent generated licensing revenue, the inventors would receive bonuses ranging from $20,000 to $50,000.

He also tasked Marshall Phelps with leading a team focused on licensing IBM patents to third parties. McGroody set a clear goal for them: maximize licensing income to help fund the company's R&D.

In 2003, Phelps, then semi-retired, was personally recruited by Bill Gates to join Microsoft and restructure its patent department—continuing his influence in the patent world. On the "Patent Hall of Fame" website hosted by IAM, Phelps is credited with "convincing senior management (and Wall Street) to stop viewing IP as a legal cost and start seeing it as a profit center."

In addition, IBM took creative steps to reduce its legal expenses in the patent domain. One innovative move was the creation of a "Virtual Law Firm"—a team composed of retired or semi-retired attorneys and patent agents who worked remotely. These professionals provided patent-related legal services at fixed rates, which significantly lowered costs while maintaining quality.

Thanks to this series of sweeping reforms, IBM's patent grants surged dramatically: from 637 in 1988 to 3777 in 1998—an almost fivefold increase. This explosive growth stands in stark contrast to the mere 10% uptick during the 1982–1988 period. By 1998, a full 85% of IBM's inventions were being patented, and the long-running *Technical Disclosure Bulletin* (TDB) was officially discontinued that same year. Under Phelps' leadership, IBM's outbound licensing revenue experienced exponential growth. Before 1991, the company earned less than $20 million annually from licenses. By 1992, that number had climbed to $30 million; in 1993, it skyrocketed to $345 million, and by 1998, it had reached approximately $1.1

[13] Ajay Bhaskarabhatla, Deepak Hegde (2014) An Organizational Perspective on Patenting and Open Innovation. Organization Science 25(6):1744–1763. https://doi.org/10.1287/orsc.2014.0911

billion—equivalent to 25% of IBM's annual R&D budget. In the early 2000s, virtually every scholarly study on intellectual property management cited IBM's "$1 billion per year in licensing revenue" as a benchmark of success.

It is during this period that IBM's patenting strategy took on a new character—one distinctly oriented toward patent monetization.

One study[14] indicate that before 1989, IBM's patent filing activity across various technical domains showed relatively stable trends, generally aligned with the company's business focus. For example, patent filings decreased in sectors like semiconductor manufacturing, which IBM was gradually exiting, while they increased in areas IBM was investing in more heavily—such as software fields like file management, databases, and data transmission. Notably, in the early 1990s, most software companies had very few software patents. Paradoxically, since software itself could not be patented directly and had to be tied to hardware, IBM—a traditional hardware company—ended up amassing a large portfolio of software patents.[15] Between 1978 and 1988, IBM accounted for over a quarter of all software patents granted by the U.S. Patent and Trademark Office (USPTO). As previously discussed, IBM had also become a vocal proponent of software patentability.

Starting in 1989, however, the citation rates of IBM patents and technical publications declined sharply. Figures provided by various researchers illustrate this shift. In one chart, the line tracks how often IBM patents were cited in other companies' patent applications from 1980 to 1998. Another chart shows the decline in citations to IBM's technical disclosures, such as those published in the TDB. Since citation frequency is often seen as a proxy for patent quality, the drop might seem concerning. But some scholars argue that this decrease doesn't necessarily reflect lower quality. Instead, it may indicate a shift in IBM's patent strategy—away from the ideals of "open science" and toward a more guarded, proprietary model. As a result, citing IBM patents became legally and financially riskier for other companies (Figs. 9.2 and 9.3).[16]

---

[14] Ajay Bhaskarabhatla, Deepak Hegde (2014) An Organizational Perspective on Patenting and Open Innovation. Organization Science 25(6):1744–1763. https://doi.org/10.1287/orsc.2014.0911

[15] Chien, C. V. (2010). From arms race to marketplace: the complex patent ecosystem and its implications for the patent system. Hastings Lj, 62, 297.

[16] Ajay Bhaskarabhatla, Deepak Hegde (2014) An Organizational Perspective on Patenting and Open Innovation. Organization Science 25(6):1744–1763. https://doi.org/10.1287/orsc.2014.0911

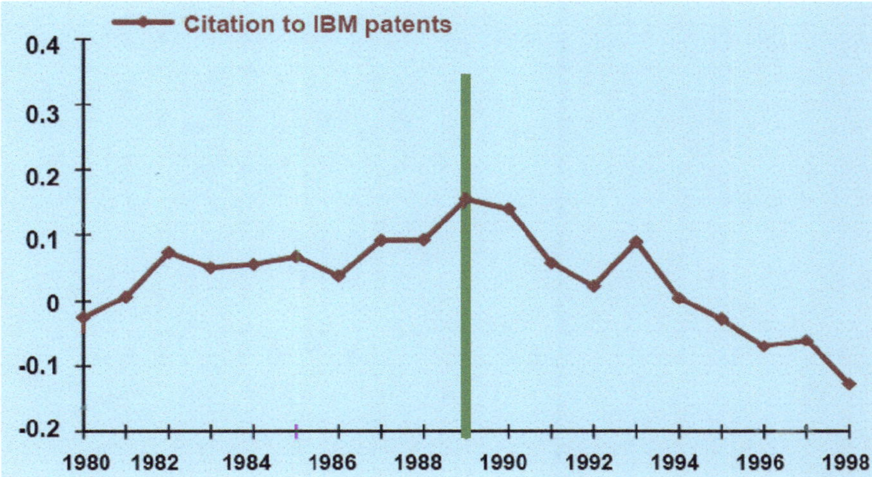

Fig. 9.2 Citations of IBM patents in other companies' patent applications, 1980–1998

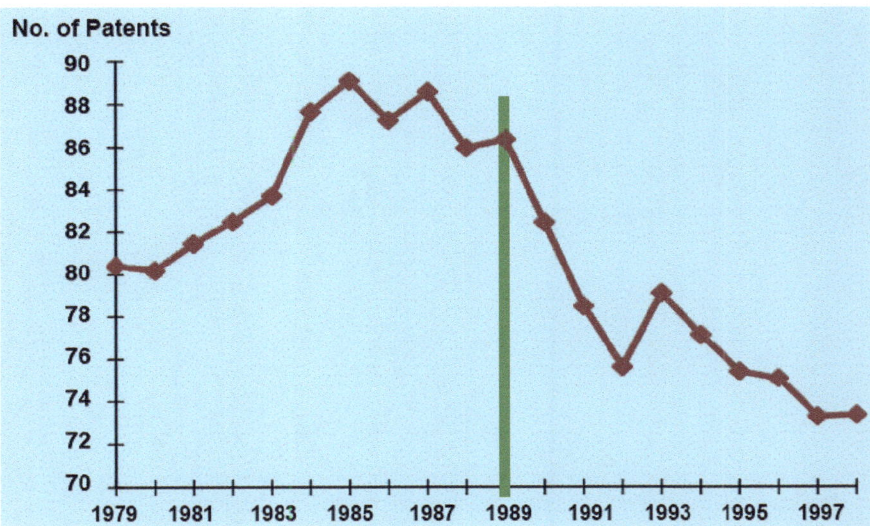

Fig. 9.3 Citations of IBM's technical disclosure publications (e.g., the TDB) in other companies' patent applications, 1979–1998

Chapter 10
From Mainframes to Microcomputers

When discussing the rise and fall of IBM, the history of the personal computer (PC) is absolutely indispensable. This period has been extensively chronicled, and any technology enthusiast is likely familiar with it. For those interested, references such as *The Rise and Fall of High-Tech Empires* offer more detail. This chapter will not delve into that narrative, but instead focuses on patents and copyrights.

In brief, during the 1970s, personal computers were still niche products—playthings for tech geeks and small businesses. Apple's launch of the Apple II (see Fig. 10.1) marked a major turning point. This machine was excellent in terms of quality, performance, and design, and came with a suite of desktop-level office software. IBM followed with its own desktop computer product—the IBM PC. IBM and Apple took divergent approaches to their products: IBM was "open", while Apple remained "closed". However, under the assault of a pack of aggressive small and mid-sized firms, both companies were struggling by the late 1980s. This led to the legendary stories of Lou Gerstner saving IBM and Steve Jobs' triumphant return to Apple.

One particularly noteworthy feature of the Apple II was a killer app—VisiCalc, a spreadsheet software (Fig. 10.2 shows a screenshot of its interface). As the progenitor of spreadsheet programs, VisiCalc marked a key moment in the entrance of microcomputers into the office environment. Recall that when Hollerith worked for the Census Bureau, clerks used Seaton tabulating equipment, which essentially generated paper-based tables. This simple tool was Hollerith's inspiration. In this sense, VisiCalc's success carries great significance—IBM was born from "tabulating", but with VisiCalc, that tabulating work began shifting to lightweight personal computers.

VisiCalc was a blockbuster, selling 700,000 copies at $250 apiece. Due to the era, this heavyweight software received no patent protection. VisiCalc's lead

Q. Yue, F. Yang, *Governing Innovation: A Global History of Patent-Driven Technological Progress*, https://doi.org/10.1007/978-981-95-5570-3_10

Fig. 10.1 Apple II computer released in 1977. (Photograph by the author at the Computer History Museum, USA)

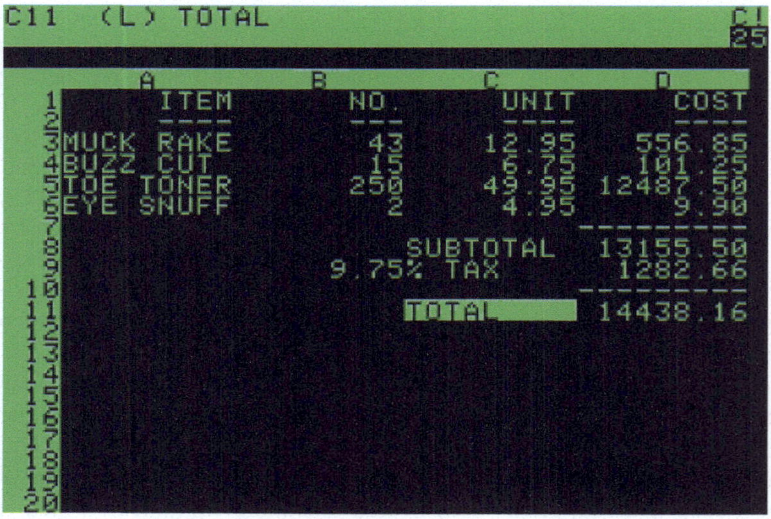

Fig. 10.2 VisiCalc running on the Apple II. (Source: Gortu. (2005, September). *VisiCalc* [Image]. Wikimedia Commons. https://commons.wikimedia.org/wiki/File:Visicalc.png. This image is in the public domain)

developer Dan Bricklin graciously stated, *"I'm not rich because I invented VisiCalc," Bricklin said, "but I feel that I've made a change in the world. That's a satisfaction money can't buy."*[1] However, he also admitted elsewhere that he would patent it if

[1] Mary Bellis. (2018). The First Computerized Spreadsheet. Retrieved March 31, 2024, from ThoughtCo. Website: https://www.thoughtco.com/dan-bricklin-bob-frankston-spreadsheet-visicalc-4078060

Fig. 10.3 Apple II computer and IBM PC on display at the Computer History Museum. (Photograph by the author at the Computer History Museum, USA)

he invented the spreadsheet today.[2] Incidentally, another key productivity software of the time—WordStar, a word processor—sold 650,000 copies at $495 each,[3] and it too had no patent protection.

With spreadsheets and word processors—the precursors of Excel and Word—personal computers evolved from a nerd's toy into genuine productivity tools. Besides Apple, more and more companies began developing personal computer products. IBM released the IBM PC in 1981 (see Fig. 10.3), mostly assembled from third-party components and running Microsoft's MS-DOS operating system. IBM's permissiveness allowed a swarm of smaller firms—HP, Compaq, Dell, and others—to produce IBM-compatible PCs, dubbed "IBM clones." These swiftly captured the emerging PC market and fueled a wave of software company growth. PCs, with their low cost, ease of use, and rich software ecosystem, came to dominate offices large and small, delivering a heavy blow to IBM's traditional mainframe business.

To understand IBM's patent strategy regarding personal computers, it helps to examine its main competitor at the time—Apple. Apple held relatively few patents in its early years. When the Apple II launched in 1977, the company's sole patent

[2] Dennis Crouch. (2018). Patenting a Spreadsheet. Retrieved March 31, 2024 from PatentlyO: https://patentlyo.com/patent/2018/10/25074.html

[3] Campbell-Kelly, M. (2004). Not all bad: An historical perspective on software patents. *Mich. Telecomm. & Tech. L. Rev.*, *11*, 191.

was titled "Microcomputer for Use with Video Display" (U.S. Patent No. US4136359A). This patent was involved in several lawsuits between 1983 and 1984.

At the time, computers were Apple's only core product. But fundamentally, Apple was a classic startup, assembling hardware primarily from off-the-shelf components. While Apple was always aggressive in protecting its intellectual property, its early patent portfolio was thin, and patent litigation rare. However, Apple was extremely assertive in copyright protection. Two major copyright lawsuits—the 1982 Apple v. Franklin Computer Corp. case (hereafter "Franklin") and the 1988 Apple v. Microsoft case—were legal milestones. Apple won the former and lost the latter.

Apple co-founder Steve Wozniak was a technically minded individual with an open philosophy. Out of a spirit of sharing, he included detailed technical documentation with early Apple computers to help third-party developers write software. This inadvertently lowered the barrier for copycat manufacturers. At the time, U.S. law had no clear provisions regarding machine-readable code. In 1982, Franklin introduced a personal computer modeled on the Apple II, named the ACE 100—the first compatible machine to mimic Apple's design.

In May 1982, just 2 months after the ACE 100 went on sale, Apple sued Franklin for copyright infringement. At the time, the U.S. Copyright Act didn't explicitly cover machine-readable code, and by a literal reading, such code couldn't be considered a "work" under the law. The trial court sided with Franklin.

In June 1982, Franklin released an updated model, the ACE 1000, priced at $1595. Anticipating Apple's lawsuit, Franklin modified the design to avoid infringing Apple's hardware patents. For instance, the new ACE 1000 motherboard lacked the chip responsible for color video output, possibly as a way to avoid "divided infringement," since buyers could add the chip for $50 at retail.[4] The ACE 1000's ROM contained only boot code for starting the disk; the rest of the software loaded from disk into memory after booting, sidestepping Apple's patent claims related to loading code directly from ROM into memory. We don't know whether these patent workarounds were legally effective, as there is no record of further litigation.

Because Franklin's software copying was blatant, Apple focused its legal attack on copyright. The ACE 1000's operating system was nearly identical to Apple's, with only superficial changes—like renaming "COPYA" to "COPY." Even the ROM contained strings like "Applesoft" and "James Huston"—the latter being the name of an Apple programmer who had no counterpart at Franklin.

Faced with such clear evidence, Franklin could hardly deny its actions. Its defense rested on the argument that Apple's ROM code was machine-readable and thus outside the scope of copyright law. But in August 1983, the U.S. Court of Appeals for the Third Circuit ruled that both ROM and floppy disk code were eligible for copyright protection. The court famously stated:

[4] Steven Weyhrich. (2023). The Apple II Abroad / Clones. *Apple II History*. Retrieved March 31, 2024 from Apple II History Website: http://apple2history.org/history/ah12/#07

The category of 'literary works'... is not confined to literature in the nature of Hemingway's For Whom the Bell Tolls.

This ruling was a watershed in the history of copyright law and had a profound global impact on software legislation. From that point on, machine-readable code became protected under copyright. In January 1984, Franklin agreed to pay Apple $2.5 million to settle the case and filed for bankruptcy later that year.[5] Though the company survived, Apple's relentless legal pressure drove it out of the desktop computer market, and it pivoted to PDA and e-book products, remaining in business until 2009.

In the face of its weak patent protection, Apple aggressively pushed for judicial change in the realm of copyright—and succeeded in suppressing a major competitor. In 1988, Apple again sued, this time targeting Microsoft for allegedly copying its graphical user interface (GUI). This case is even more well-known in copyright history and will be explored in more detail in a later chapter. Had Apple won, copyright protection for GUIs might have reached a new legal height. But the court ruled against Apple, noting that Apple's own GUI had been "borrowed" from Xerox.

These two landmark lawsuits reflect Apple's firm commitment to protecting its core products. Both cases were brought in the absence of clear statutory guidance or legal precedent, involving significant uncertainty, yet Apple still chose to litigate. As a result, both cases had far-reaching influence on intellectual property law.

In stark contrast to Apple's relentless pursuit of legal protection, IBM adopted a laissez-faire attitude toward the cloning of its IBM PC by smaller manufacturers throughout the 1980s—a fact well known in the industry. However, exactly what patent strategy IBM pursued to protect the IBM PC remains unclear. According to Wikipedia, most components of the IBM PC were sourced from third parties, and thus IBM obtained "only a handful of patents on the PC, covering such features as the bytecoding for color monitors, DMA access operation, and the keyboard interface." In more recent years, many media sources have cited the existence of "nine foundational patents" for the IBM PC, a claim invariably traced back to interviews or biographies of former IBM engineer Mark Dean.[6]

Mark Dean, one of the rare Black engineers in 1980s high-tech firms, has received considerable media attention in recent years. Nearly every profile of him mentions that he holds three of the "nine foundational patents" for the IBM PC. However, these reports never specify what the nine patents actually are.

Computer magazine, published by IEEE, once referenced this "urban legend" and directly asked Dean about it.[7] He responded: "*Two of them had to do with the*

[5] New York Times. (1984). Franklin and Apple Settle Suit. Retrieved March 31, 2024 from New York Times Website: https://www.nytimes.com/1984/01/05/business/franklin-and-apple-settle-suit.html

[6] For example, Mark Dean and Dennis Moeller (https://lemelson.mit.edu/resources/mark-dean-and-dennis-moeller), The Black Digital Elite: African American Leaders of the Information Revolution; Continuing innovation in information technology: workshop report

[7] Goth, G. (2011). Ibm pc retrospective: There was enough right to make it work. *Computer*, *44*(8), 26–33.

way you put pixels on the screen – the graphic interface and color and pixel creation and control on the monitor. The third had to do with refreshing the memory interface… Those nine patents provided a definition of what became known as IBM PC-compatible".[8] Based on these clues, the three patents Dean referred to likely include:

- US4442428A: *Composite Video Color Signal Generation From Digital Color Signals*
- US4437092A: *Color Video Display System Having Programmable Border Color*
- US4575826A: *Refresh Generator System for a Dynamic Memory*

Around the same time, IBM also filed several patent applications from its Boca Raton facility (a city in southeastern Florida), including four patents by David J. Bradley:

- US4437093A: *Apparatus and Method for Scrolling Text and Graphic Data in Selected Portions of a Graphic Display*
- US4408200A: *Apparatus and Method for Reading and Writing Text Characters in a Graphics Display*
- USRE32201E: *Apparatus and Method for Reading and Writing Text Characters in a Graphics Display*
- USRE33894E: *Apparatus and Method for Reading and Writing Text Characters in a Graphics Display*

Bradley is credited with a notable contribution to personal computing history: he was the original designer of the Ctrl-Alt-Del key combination—intended as the least likely combination to be pressed accidentally—for rebooting the IBM PC. This key combination has endured to this day. Without a doubt, Bradley was one of the key figures on the original IBM PC project team.

The three patents held by Mark Dean and the four patents by David Bradley all share a priority date of August 11, 1981—just one day before the IBM PC was officially released on August 12, 1981.[9] The applications were filed in Boca Raton, matching the IBM PC's development site. Technically, these patents all relate to display and manipulation of graphical content on screens, placing them within the same domain. Considering the inventors, filing dates, and locations, it's highly likely that these patents form part of the so-called "nine foundational patents" of the IBM PC. Notably, none of these patents were involved in litigation during the 1980s.

According to the book *Big Blues: The Unmaking of IBM*, IBM's legal team later identified several critical patents related to the IBM PC that could pose a significant threat to mainstream personal computer manufacturers.[10] However, IBM decided to

[8] Goth, G. (2011). Ibm pc retrospective: There was enough right to make it work. *Computer*, *44*(8), 26–33.

[9] James. Cortada. (2021). How the IBM PC Won, Then Lost, the Personal Computer Market. IEEE Spectrum. Retrieved March 31, 2024 from IEEE.org.: https://spectrum.ieee.org/how-the-ibm-pc-won-then-lost-the-personal-computer-market#toggle-gdpr "*The big day came on 12 August 1981.*"

[10] Carroll, P. (1994). *Big blues: The unmaking of IBM*. Crown Publishing Group.

take a passive approach, opting to focus its energy on promoting the new PS/2 bus standard.

This evidence suggests that if IBM had wanted to, it could have used its patents to shut down all IBM PC clone manufacturers, or at the very least, impose heavy licensing fees—just as Thomas Watson Sr. had once done to Powers Accounting Machine Company. But IBM chose not to. While Apple spared no effort in using copyright law to protect its products, IBM took a markedly open stance toward the personal computer. At the heart of this difference was IBM's own positioning of the personal computer: it was never seen as a true core product.

As documented in *Big Blues: The Unmaking of IBM*, then-CEO John Akers once remarked in a meeting that if personal computers ever became a "commodity" business, IBM would shut down the operation immediately.[11] IBM's later sale of its PC business to Lenovo further validated this viewpoint.

The development of follow-up products to the IBM PC did not go smoothly. The IBM PC itself had been a top-secret project led by an engineer named Don Estridge and was carried out in Boca Raton, away from IBM's main headquarters. Most departments within the company were unaware of it. This secrecy was largely intended to bypass IBM's notoriously bureaucratic internal processes.[12] As a result, the PC project was able to break free from IBM's traditional "vertical integration" approach—most components were sourced externally, and software like BASIC and MS-DOS was procured from Microsoft. However, once the IBM PC had achieved a measure of success, the project team found it increasingly difficult to maintain its original agility.

On August 1, 1983, the IBM PC team was reorganized into the Entry Systems Division (ESD). An article in *IEEE Spectrum* noted[13] that this marked the team's full reintegration into IBM's bureaucratic system. The article also pointed out that the rebranded ESD swelled from 4000 to 10,000 employees, which deeply frustrated Estridge. What bothered him most was that the newly added programmers knew very little about the IBM PC. The PC's software had to run within extremely limited memory—similar to the software constraints faced by mainframes in the 1950s. The new programmers, having grown accustomed to the more generous environments of 1970s mainframe systems, lacked experience with such tight constraints. Meanwhile, Estridge himself became overwhelmed with administrative and managerial responsibilities. IBM's foray into the home computer market with the IBM PCjr also ended in failure.

Tragically, Estridge's story ended abruptly. On August 2, 1985, he was killed in a plane crash. He was posthumously honored and widely hailed in the tech community as the "father of the PC." After his death, there was even less interest among IBM's top brass to treat the IBM PC as a core business.

---

[11] Carroll, P. (1994). *Big blues: The unmaking of IBM*. Crown Publishing Group.

[12] Chapman, M. R. (2003). *In Search of Stupidity*. New York: Springer-Verlag.

[13] James. Cortada. (2021). How the IBM PC Won, Then Lost, the Personal Computer Market. IEEE Spectrum. Retrieved March 31, 2024 from IEEE.org.: https://spectrum.ieee.org/how-the-ibm-pc-won-then-lost-the-personal-computer-market#toggle-gdpr

As mentioned in earlier chapters, IBM had been firmly opposed to software patents during the 1970s. After the launch of the IBM PC, however, its prestige attracted virtually every software company to develop for its platform, and every hardware company wanted to make products compatible with it. This dual ecosystem gave IBM PC a commanding position in the personal computing market, leaving competitors like Apple and Compaq gasping for air.

Compaq had originally developed its own line of personal computers. But with software developers flocking to the IBM PC, Compaq had no choice but to imitate its architecture in order to ensure compatibility. Rod Canion, the founder of Compaq, stated that if software companies were unwilling to adapt to Compaq's computers, then Compaq's machines would instead adapt to the software.[14] Thus, the PC-compatible market was born. As Compaq's sales soared in 1983, many small computer manufacturers joined the trend of making IBM-compatible machines. Five years after the IBM PC's release, IBM PCs and compatibles accounted for 75% of the U.S. market, leaving Apple with only 15%.

In 1984, IBM PC revenue reached $4 billion—more than double Apple's. Yet this windfall had a downside: with every new company adopting the IBM PC, the market for IBM's mainframes shrank. For IBM, a giant built on systems, software, and services for mainframes, losing even a fraction of this market meant tremendous losses. Compared to that, the revenue from PC sales was a mere drop in the bucket.

A few years later, IBM attempted to reclaim control by introducing a new bus standard: the **Personal System/2 (PS/2)**. The project was led by senior manager **Bill Lowe**. IBM believed that competitors would eagerly adopt this new architecture, just as they had previously copied the original IBM PC, and would seek licensing.

As previously mentioned, IBM's legal team had already identified several key patents related to the original IBM PC. These patents could pose a serious threat to most mainstream personal computer manufacturers if enforced. However, wary of litigation uncertainty, Bill Lowe chose not to launch a legal offensive immediately. Instead, his strategy was to wait until the new system was launched and then settle both old and new scores in one go.

To extract higher royalties and license fees, IBM set the licensing standard at **5% of each machine's selling price**. At that time, a typical personal computer sold for about **$3000**, meaning each unit could net IBM around **$150** in licensing revenue.[15]

This model resembled Microsoft's: in the PC era, every time a manufacturer sold a machine, it paid Microsoft for the operating system. Had IBM succeeded in implementing the same model, it might have made a fortune from licensing. However, the situation wasn't so simple. The new generation of PC manufacturers—nurtured by IBM's prior openness—were not passive. Each had its own R&D roadmap, peripheral products, and matching business strategies. By the late 1980s, these companies

[14] Canion, R. (2013). Open: How Compaq Ended IBM's PC Domination and Helped Invent Modern Computing. BenBella Books, Inc.

[15] Carroll, P. (1994). *Big blues: The unmaking of IBM*. Crown Publishing Group. p139

had accumulated their own patent portfolios—not as vast as IBM's, but strong enough to provide bargaining power for reverse engineering and design-around tactics.

In his book *Open: How Compaq Ended IBM's PC Domination and Helped Invent Modern Computing*, Compaq co-founder **Rod Canion** documented the patent standoff between IBM and Compaq over PS/2. Shortly after PS/2's release, IBM approached Compaq for licensing negotiations. Talks began in **1987** and lasted until **September 1988**, with little substantive progress. Frustrated, IBM began to apply pressure through patent infringement claims. Canion recalled: "*At first I thought it was planning to use the tactic TI had used on us in 1982 and try to slow us down with a lawsuit. Instead, it was more like part of a larger program to extract some value for its intellectual property that competitors were using.*"[16]

Canion's assessment proved accurate. Although IBM sought royalties, its motive this time differed from the era of Thomas Watson Sr. licensing punch card tech to hinder competitors. Back then, the goal was market domination through high licensing costs. Now, IBM wasn't trying to dominate the PC market—it wanted to collect rent as a landlord.

Despite being smaller, **Compaq** was not defenseless. It quickly identified several patents that IBM had potentially infringed. One key patent was invented in **1982** by **Ken Roberts**, titled "Video Display System Having Multiple Selectable Screen Formats" (**US4574279A**). After nearly **9 months** of negotiations, the two companies reached a **cross-licensing agreement** on **July 14, 1989**, with Compaq paying IBM a **one-time fee of $130 million**.[17] According to Canion, this amount was significantly less than IBM's initial demand—thanks to Compaq's own patent leverage.

Compaq also spearheaded the creation of a **royalty-free alternative standard**, the **Extended Industry Standard Architecture (EISA)**, further marginalizing IBM's PS/2.

On the surface, this might appear to be a textbook case of "first-rate companies sell standards, third-rate companies sell products." IBM dispatched a single negotiation team and squeezed $130 million in real cash out of Compaq—at the cost of a few travel expenses. But from a strategic viewpoint, this $130 million was a meager consolation prize following a major strategic failure: IBM's grand vision to control the personal computer industry through patents had not materialized. Meanwhile, the disruption to IBM's core business caused by the rise of the PC industry was a long-term process—so long as companies like Compaq, HP, Sun Microsystems, and Dell remained active and continued to release new products, IBM's market share would continue to erode.

Another takeaway from this story is that although IBM possessed world-class technology, it was detached from end users. Compaq, by contrast, focused on user needs in its product development. Though it had fewer patents, their high quality

---

[16] Canion, R. (2013). Open: How Compaq Ended IBM's PC Domination and Helped Invent Modern Computing. BenBella Books, Inc. p64.

[17] Canion, R. (2013). Open: How Compaq Ended IBM's PC Domination and Helped Invent Modern Computing. BenBella Books, Inc.

posed obstacles that even IBM struggled to circumvent. This gave Compaq valuable strategic room to operate.

Other PC manufacturers were equally unimpressed by Bill Lowe's prized new standard. The book *Big Blues: The Unmaking of IBM* describes the dramatic shift that unfolded:

> Bill Lowe spent all of 1987 and 1988 waiting for Dell, Tandy, Toshiba, or some other company to develop clones. At first, he was very confident. He said that anyone attempting to copy the new PS/2 Micro Channel architecture would need a team of very smart engineers and an army of good lawyers. A few months later, as competitors bypassed the Micro Channel technology one by one, Lowe changed his tune, saying he might be willing to license it under the right terms. Then he said he'd license it under any terms. Finally, he was practically begging other companies to adopt it.

According to the book's authors, IBM's biggest blunder was placing its hopes for the PC business on the bus architecture—something that had minimal impact on PC performance at the time. Consumers were interested in visible performance improvements, such as faster software startup and execution speeds. IBM's new bus architecture offered no discernible advantage in this area. Even **Ed Belove**, a senior executive at software giant Lotus, admitted: "If IBM says Micro Channel is useful, then I believe it must be—even though I have no idea what it actually does."

Bill Lowe remained stubbornly convinced of his strategy. For over a year, he crisscrossed the United States promoting the PS/2, spending only 3 days in his own office. But no one was buying.

Earlier chapters noted that in the tabulating machine era, IBM's main profits came from selling inexpensive punch cards. In the mainframe era, its most important assets were software. But by the 1980s, IBM—beset by internal and external difficulties—could no longer afford to re-engineer this software for the IBM PC and its clones. As personal computers became widespread, the real value of IBM's once-prized software assets declined day by day. The massive investments in lobbying and global legislative efforts became nothing more than sunk costs.

Within IBM's vast product portfolio, the IBM PC was neither technologically advanced nor prioritized by senior management. Yet it was this seemingly insignificant product that rendered IBM's long-standing intellectual property strategy obsolete. All the well-dressed executives lobbying in Washington, all the high-level politicking, and all the millions spent securing legislative gains—were suddenly rendered meaningless by one humble consumer product.

Chapter 11
Holding Patents for Quick Profits

The success of IBM's patent strategy had a profound impact on the industry. Most studies on intellectual property management recognize IBM and Texas Instruments as pioneers of the new era of patent monetization. However, due to the disparity in corporate scale, IBM's influence was far more significant.

In 1995, the United States Patent and Trademark Office (USPTO) granted a total of 113,834 patents. By 2004, the number had risen to 181,229—a growth of approximately 59%.[1] During this period, IBM consistently topped the patent grant rankings, increasing its annual patent count from 1383 in 1995[2] to 3248 in 2004[3]—a surge of 135%. Ian Ayres of Yale Law School and Gideon Parchomovsky of the University of Pennsylvania pointed out that IBM's patent count surged even as its R&D spending declined, while its patent licensing revenue skyrocketed. As a high-profile industry leader, IBM set a benchmark emulated by many tech companies.[4] Chien (2010) also observed that the patent "arms race" launched by American firms in the 1990s was, to a large extent, a chain reaction triggered by IBM's aggressive pursuits in the high-tech field.

In the 1990s, when IBM's patent strategy first began to shine, "intellectual property management" was still a novel term. David Kline, co-author of several books with Marshall Phelps, wrote in his 2000 book *Rembrandts in the Attic: Unlocking the Hidden Value of Patents*[5] that even around the year 2000, very few CEOs would

[1] U.S. Patent and Trademark Office. (2021, May). *U.S. patent statistics chart: Calendar years 1963–2020*. Retrieved March 31, 2024 from https://www.uspto.gov/web/offices/ac/ido/oeip/taf/us_stat.htm

[2] U.S. Patent and Trademark Office. (1995). *Patenting trends by organization, 1995* [PDF]. Retrieved March 31, 2024 from https://www.uspto.gov/web/offices/ac/ido/oeip/taf/topo_95.pdf

[3] U.S. Patent and Trademark Office. (2004). *Patenting trends by organization, 2004*. Retrieved March 31, 2024 from https://www.uspto.gov/web/offices/ac/ido/oeip/taf/topo_04.htm

[4] Ayres, I., & Parchomovsky, G. (2007). Tradable patent rights. *Stan. L. Rev.*, *60*, 863.

[5] Rivette, K. G., & Kline, D. (2000). *Rembrandts in the attic: Unlocking the hidden value of patents*. Harvard Business Press.

Q. Yue, F. Yang, *Governing Innovation: A Global History of Patent-Driven Technological Progress*, https://doi.org/10.1007/978-981-95-5570-3_11

ever use the words "patent" and "strategy" in the same sentence. Only a handful of American companies had an IP strategy—and they kept it highly secret, afraid competitors would learn their "secret weapon."

Rembrandts in the Attic remains one of the seminal early twenty-first-century works on intellectual property management. While researching the book, Kline interviewed many heavyweights in the American tech sector, including G. Richard Thoman, who served as CEO of Xerox around 2000. Kline noted that in the late 1990s, most CEOs still defaulted to conventional approaches like cost-cutting and M&A when talking about value creation. But Thoman explicitly stated that Xerox's future lay in intellectual property. "My focus is on IP… I believe IP management is the key to enhancing Xerox's value… Companies that manage IP well will win; others will lose." He also claimed Xerox was "the first Fortune 500 company to hire a CEO specifically to maximize the value of IP assets." Notably, before joining Xerox, Thoman had served as CFO (Chief Financial Officer) at IBM.

Thoman wasn't the only former IBM executive known for aggressively pushing IP strategy. Dan McCurdy of Lucent Technologies also spent many years at IBM. From 2000 to 2001, he led Lucent's IP department, licensing out 26,000 patents from Bell Labs and Lucent worldwide—earning $500 million in licensing revenue. Brian Hinman, IBM's Vice President of Patent Licensing from 1996 to 2007, later became Chief Intellectual Property Officer (CIPO) at Philips from 2013 to 2017.

Among prominent managers who emerged from IBM's IP division, the most influential might be David Kappos. In 2009, newly elected President Obama appointed Kappos as Director of the USPTO. After stepping down in 2013, Kappos went into legal practice, shuttling between Silicon Valley and Washington, remaining highly active in the IP world. Some of the more radical opponents of software patents in the U.S. staunchly believe that he is a lobbyist hired by IBM and Microsoft, tirelessly portraying him as public enemy number one in the software world.[6]

Kappos was a strong proponent of software patents and believed the proliferation of patent litigation was a positive sign. Around 2012, during a period of intense patent litigation between Apple and Samsung, he stated: "*The explosion of litigation we are seeing is a reflection of how the patent system wires us for innovation …It's natural and reasonable that innovators would seek to protect their breakthroughs using the patent system…Our patent system is the envy of the world.*"[7]

In contrast to Kappos' stance, many in the industry oppose litigation abuse. Jon Ellenthal, a prominent investor in tech companies, once remarked that any industry in which most deals are settled in court is an industry in trouble. Marshall Phelps also lamented that while the word "patent" once evoked "invention," today it mostly brings to mind "litigation"—a clear regression. James Bessen, a professor at Boston

---

[6] *David Kappos*. (n.d.). TechRights. Retrieved March 31, 2024 from http://techrights.org/wiki/index.php/David_Kappos

[7] Timothy Lee. (2012). US patent chief to software patent critics: "Give it a rest already". Retrieved March 31, 2024 from ArsTechnica: https://arstechnica.com/tech-policy/2012/11/us-patent-chief-to-software-patent-critics-give-it-a-rest-already/

University School of Law, noted that from the late 1980s to the 2010s, litigation costs in non-chemical and non-pharmaceutical fields (which he viewed as functioning well under the patent system) had soared nearly threefold. He estimated the burden of unnecessary patent litigation as a "tax" of 10–20% on innovation.[8]

To be fair, prior to the 2010s, IBM, though already notorious for extracting licensing fees, rarely initiated lawsuits. Lindsay Moore and other American scholars pointed out that in the early twenty-first century, IBM was still more inclined to settle than to litigate in cases of patent infringement.[9] The surge in litigation can't be entirely blamed on IBM; what most companies envied was IBM's enormous licensing revenue. However, the quality of IBM's patents has long been under scrutiny.

Criticism of IBM's patent quality dates back to the 1990s. Many in the software industry accused IBM and other patent giants—such as Hitachi, AT&T, DEC, Toshiba, and Sharp— "*to be totally incapable of delivering innovative software products in the marketplace*" despite holding the most software patents.[10]

Simon Phipps, former president of the Open Source Initiative who had worked at both IBM and Sun, recalled his experience at IBM: "*When I worked at IBM, I asked a patent lawyer at the company what was needed to file a patent. I was told 'a rough idea — we can fill in the details for you — and then all the ways you can think of by which we could tell if someone else is using that idea'.*"[11]

While IBM's leadership clearly took pride in its massive patent portfolio—publicizing its totals every year—this attitude increasingly seemed out of step with the times.

In 2008, Mike Dillon, General Counsel at Sun Microsystems, announced on his blog that the company would reduce its annual patent filings from around 1000–700, shifting focus to a smaller number of high-quality patents. He emphasized that patent filing and maintenance were costly, and that given Sun's business model, a portfolio sufficient for customer support and self-defense was enough.[12]

At the end of 2009, Cisco's General Counsel Mark Chandler declared the company would abandon "patent quantity" in favor of "patent quality," aiming to reduce filings by 30%.[13] That same year, Kevin Light, Deputy Director of HP's IP

---

[8] Annie Lowery. (2011). Die, Patent Trolls! Will IBM's new meta-patent help fix a broken system or just make things worse? Retrieved March 31, 2024 from Slate Website: https://slate.com/business/2011/01/patent-trolls-and-ibm-will-big-blue-s-meta-patent-fix-a-broken-system.html

[9] Moore, L., & Craig, L. (2008). Intellectual capital in enterprise success: strategy revisited. John Wiley & Sons.

[10] Gordon Irlam. (1994). Negative Correlation of Innovation and Software Patents. Retrieved March 31, 2024 from Gordon Irlam's personal website: https://www.gordoni.com/software-patents/scoreboard.html

[11] Simon Phipps. (2012). Why software patents are evil. Inforworld. Retrieved March 31, 2024 from Infoworld Website: https://www.infoworld.com/article/2078547/why-software-patents-are-evil.html

[12] Mike Dillon. (2008). The Patent Arms Race, Legal Thing (May 26, 2008), Retrieved June 21 2019 from Sun's Website: http://blogs.sun.com/dillon/entry/the_patent_arms_race

[13] Michael Arndt, Cisco's Patent Strategy: It's More Than Numbers, Bus. Wk. (Dec. 21, 2009), http://www.businessweek.com/innovate/next/archives/2009/12/has_the_recessi.html

department, echoed the sentiment, stating HP would prioritize "quality over quantity." Reportedly, after Mark Hurd became HP CEO in 2005, he pushed for patent efficiency—cutting money and time spent on patenting, minimizing waste, and focusing on high-quality patents relevant to the core business. A 2009 article on tech site CNET headlined this trend with: "*HP focuses on patent quality, IBM on quantity*".[14]

But behind IBM's enormous patent numbers, had its patent quality and innovation truly declined?

Gideon Parchomovsky and R. Polk Wagner, in a 2005 study,[15] noted that although IBM's patent count had steadily increased since the 1990s, its patent quality had indeed shown a downward trend. The key metric they used was "patent intensity"—that is, the R&D expenditure per patent. Their research showed that since Lou Gerstner took over as IBM's CEO in 1993, the R&D expenditure per patent at IBM dropped sharply, reaching only one-sixth of what it was a decade earlier by 2003 (see Table 11.1).

Clearly, "patent intensity" is a rather crude metric—R&D expenditures usually correlate with the number of patents granted years later, not those granted in the same year. Nevertheless, in the absence of more precise data, this indicator may still offer a rough assessment of IBM's patent quality. As a supplement to Table 11.1, we compiled IBM's patent intensity data from 2004 to 2022 (see Table 11.2).

As shown in Fig. 11.1, after Gerstner took the helm at IBM, the company's R&D spending was slashed drastically. For IBM, which in 1993 was teetering on the edge of collapse, this was a completely rational decision. Although R&D expenditures fluctuated in subsequent years, they remained largely stable, hovering between $5 billion and $6 billion, with noticeable dips only in particularly poor-performing

Table 11.1 Changes in IBM's patent numbers and R&D expenditures, 1992–2003

Year	Number of patents	R&D expenditures (Million USD)
1992	842	6522
1993	1107	5558
1994	1298	4363
1995	1383	4170
1996	1867	4654
1997	1724	4877
1998	2657	5046
1999	2756	5723
2000	2866	5084
2001	3411	4986
2002	3288	4750
2003	3415	5077

[14] Matt Asay. (2009). HP focuses on patent quality, IBM on quantity. Retrieved March 31, 2024 from CNET: https://www.cnet.com/news/hp-focuses-on-patent-quality-ibm-on-quantity/

[15] Parchomovsky, G., & Wagner, R. P. (2005). Patent portfolios. *U. Pa. L. Rev.*, *154*, 1.

Table 11.2 Changes in IBM's patent numbers and R&D expenditures, 2004–2022

Year	Number of Patents	R&D Expenditures (Million USD)
2004	3248	5673
2005	2203	5842
2006	3621	6107
2007	3125	6337
2008	4169	6153
2009	4887	5820
2010	5866	6026
2011	6148	6258
2012	6478	6302
2013	6788	5743
2014	6969	5437
2015	7355	5247
2016	8088	5751
2017	9043	5787
2018	9100	5379
2019	9262	5989
2020	9130	6333
2021	8682	6488
2022	4743	6567

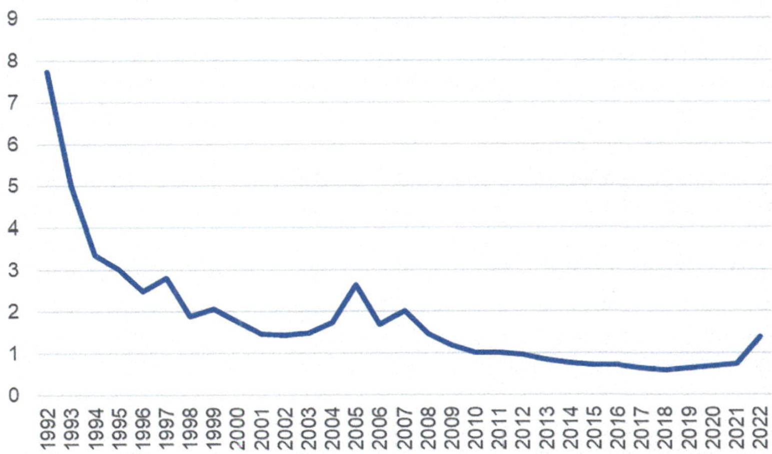

Fig. 11.1 Trends in IBM's patent intensity, 1992–2022. Vertical axis unit: Million USD

years. In stark contrast to this relatively steady R&D spending, IBM's patent output soared, with a noticeable drop only after the 2002 dot-com crash. Taken together, the data shows a sharp decline in IBM's patent intensity—by 2017, R&D spending per patent was less than one-tenth of what it had been in 1992. If inflation is factored in, the per-patent R&D expenditure dropped even more dramatically.

In today's high-tech industry, where R&D investment is booming, many observers have begun to question IBM's anomalously high patent output. Among top-ranking companies at the USPTO, Microsoft and Intel consistently trailed IBM by over 50% in patent count. Companies like Samsung, Toshiba, and Canon—East Asian firms with defensive patent strategies and a focus on mass-market consumer electronics—barely kept pace with IBM in patent volume. Moreover, IBM's R&D spending was significantly lower than that of these peers. For instance, Samsung's R&D budget in 2018 exceeded $13 billion—more than twice that of IBM. Clearly, IBM's massive patent filings raised eyebrows.

One industry expert noted: "*IBM's 'more is more' strategy when it comes to obtaining rights has become a dubious distinction of sorts. In fact, the company allows many more patents to lapse than it maintains. It also makes broad use of defensive publications to ensure that certain inventive ideas remain unpatentable and stay out of the wrong hands. With what amounts to a volume strategy, IBM is strangely cynical about patent quality. Originally a hardware company that evolved into a software and services provider, IBM's main business is no longer computing. While it still sells a lot of mainframes and peripherals, how many patents does a consulting firm need to compete? Few of the most innovative businesses maintain tens of thousands of patents, and none of the most successful consulting firms do.*"[16]

This expert observed that one-third to half of IBM's patents were ultimately abandoned, with many lasting no more than 24 months. "Over the past 20 years, IBM's patent strategy has been driven by the mantra 'more is better.' Whether IBM has fallen into the trap of chasing its image as the intellectual property leader—sacrificing quality and real returns for inflated patent numbers—remains to be seen. … Patents should not be the goal; they should be a tool to achieve the goal."

We strongly agree with some industry colleagues' view:[17] IBM's massive patent arsenal has come at the cost of declining quality. By around 2005, IBM's vast portfolio had already fallen behind that of Microsoft, Cisco, and Sun in terms of quality. A study by intellectual property consultancy Ocean Tomo found that between 2005 and 2010, although IBM ranked first in patent grants, it ranked only eighth in overall patent value.[18] Based on metrics such as citation counts, maintenance fee payments, and litigation costs, Ocean Tomo concluded that Microsoft's smaller patent pool was more than three times as valuable as IBM's. The head of Ocean Tomo's patent valuation unit pointed out that many of IBM's patents were service-related and paled in value compared to Microsoft's gaming and software patents.

---

[16] Bruce Berman. (2013). The need to lead: IBM under the microscope. Intellectual Asset Management January/February 2013. Retrieved March 31, 2024, from: http://www.brodyberman.com/pdf/IAM_57%20Intangible%20investor%2004.pdf

[17] Parchomovsky, G., & Wagner, R. P. (2005). Patent portfolios. *U. Pa. L. Rev.*, *154*, 1.

[18] Steve LeVine. (2010). IBM May Not Be the Patent King After All. Retrieved March 31, 2024 from Bloomberg: https://www.bloomberg.com/news/articles/2010-01-13/ibm-may-not-be-the-patent-king-after-all

The tide of history is vast and irresistible; those who follow it will flourish, and those who defy it will perish. From the 1980s to the 1990s, IBM, though slow to transform, managed to emerge as a dominant force in the patent world thanks to its deep technical reserves and massive patent count, becoming an admired model for others. But times have changed. As more and more companies publicly commit to prioritizing patent quality, is IBM once again falling behind?

Colleen Chien, a professor at the University of California, Berkeley, argues that the U.S. patent landscape is shifting from a frenzied "arms race" phase to a more peaceful "trading market" stage.[19] She noted that the cold war of patents among major corporations had ended: from 2000 to 2008, most patent litigation involved mismatched opponents—either big versus small or vice versa—while only about one-third took place between equally matched rivals. By around 2010, lawsuits initiated by non-practicing entities (NPEs) already accounted for 20% of all U.S. patent litigation. In contrast, relations among major competitors had stabilized, with fewer lawsuits, resulting in a delicate equilibrium.

Chien further noted that since IBM's patent war victory over Texas Instruments, corporations were compelled to hoard patents in a mad scramble. The outcome of this "cold war-style" arms race was that major firms amassed huge patent arsenals. While these were of limited use against NPEs, among themselves the firms reached a kind of ceasefire. By around 2010, lawsuits among high-tech competitors dropped sharply. After 20 years of escalation, tit-for-tat litigation, bad press for patent trolls, and changing corporate cultures, large firms began thinking twice before initiating lawsuits.

Chien also pointed out that if the purpose is purely defensive, patent holders don't need every patent to be of high quality. In a corporate patent arms race, the vast majority of patents will never be challenged or licensed. In this cold war of intimidation, quantity matters more than quality. Thus, even without any intent to extort, companies widely adopted a strategy of filing many low-quality patents. Studies suggest that companies investing limited time and resources during this phase often skipped prior art searches altogether. A 2008 study revealed that such projects typically incurred patent prosecution costs 25% below average.[20]

Although this surplus of patents helped maintain a cold-war-style balance among corporations, it also fueled the rise of NPEs. Since NPEs have no operating business, patent giants found it difficult to retaliate effectively—aside from selectively invalidating patents, there was little they could do. In this context, patent quantity became less relevant for operating firms.

Earlier, we noted that IBM's massive patent count was its ace in extracting licensing fees. Admittedly, quantity matters—especially in the U.S., where high legal costs make evaluating numerous patents burdensome for small firms. But as leading IP experts observe, it's NPEs that place the most emphasis on quantity. For

[19] Chien, C. V. (2010). From arms race to marketplace: the complex patent ecosystem and its implications for the patent system. Hastings Lj, 62, 297.

[20] Craig Opperman & Carina Tan, Getting Less for More, Intell. Asset Mgmt., Dec./Jan. 2008, at 8, 8–9, (discussing low-cost, volume-based patenting).

instance, the notorious NPE Intellectual Ventures is estimated to hold 30,000–60,000 patents, scattered across more than 1300 shell companies.[21]

Why do NPEs need so many patents? Terry Ludlow, CEO of chip analysis firm Chipworks and a former employee of prominent NPE MOSAID, explained in a 2014 article[22] that today in the U.S., "*asserting a single patent or even a small portfolio of patents is almost suicidal.*"

Ludlow explained that if a patent troll asserts only a few patents, most tech firms will ignore it—these letters are so common that a mid-sized firm might receive several per week. Since the 2007 *SanDisk* decision greatly lowered the threshold for declaratory judgment actions, such letters tend to be extremely vague to avoid giving the target company grounds to sue first or to file for inter partes review (IPR). Vague letters are often tossed aside. However, if specifics are provided, the target may initiate litigation or IPRs in retaliation. Thus, when the number of patents is small, most companies are bold enough to fight. But when the patent portfolio is massive, the cost of reviewing them all or filing numerous IPRs becomes prohibitively expensive—and even then, there's no guarantee of invalidating all claims.

A 2013 study from UC Berkeley echoed this view:[23] While NPEs do rely on some high-quality patents, they benefit more from large pools of low-quality ones. This trend has two consequences. First, as aggressive tactics become popular, more patents are filed for minor tweaks or incremental improvements that can't survive scrutiny. This wastes USPTO resources and leads to a vicious cycle—more bad patents create even more bad patents. Second, large portfolios enable "super-patent" effects, conferring broader de facto exclusivity and effectively taxing entire industries.

The study further noted that while NPEs don't attack with just a handful of patents, they also don't show their full hand at once. They begin by presenting a curated batch and demand a license covering not just those patents but others not yet disclosed. If the target refuses, they sue using the initial batch and prepare a second round. During this process, NPEs reveal their tenacity—once known for being relentless, future targets are more likely to capitulate. "Even if each patent has only a 10% chance of surviving validity challenges… it may still make sense to license the whole portfolio."

From these studies, it becomes clear why IBM, despite spending far less on R&D than Samsung, Huawei, or Google, still maintains a massive patent arsenal: this obsession with quantity likely stems from a one-sided pursuit of licensing revenue.

The aforementioned studies also point out that when operating companies engage in discussions about patent infringement, their approach varies depending on the opponent. In encounters between large firms, both sides are aware that a conclusive victory is unlikely, and matters typically end in cross-licensing. But when facing

[21] Feldman, R., & Ewing, T. (2012). The giants among us. Stanford Technology Law Review, 1.

[22] Ludlow, T. (2014). Trends in Technology IP Licensing. *IPO Law Journal*.

[23] Orr, J. R. (2013). Patent Aggregation: Models, Harms, and the Limited Role of Antitrust. *Berkeley Tech. LJ*, *28*, 525.

smaller companies, large firms often aim to intimidate their opponents into backing down, thereby reducing competition. NPEs, by contrast, are not interested in cross-licensing or locking up key technologies. The more players in the market, the better it is for NPEs. Once a certain technology gains market acceptance and related products are established, NPEs will lie in wait, patiently stalking their "prey" until it's fattened up—then they strike.

In a 2014 research paper,[24] American scholar Kristen Osenga introduced the concept of "Formerly Manufacturing Entities" (FMEs), as a counterpart to "Non-Practicing Entities" (NPEs), referring to companies that once had operational businesses but later shifted to patent licensing as their main revenue stream. IBM, she argued, is a quintessential FME. In this chapter, we will look at several case studies to examine whether IBM has indeed grown more and more like a typical NPE. We will also explore, from the perspective of IBM's corporate culture and overall strategy, how its management gradually transformed a company once destined to explore the stars into a pirate ship set on plunder.

In 2002, American attorney Gary L. Reback published a commentary[25] on the Forbes website recounting an experience representing Sun Microsystems in a clash with IBM's patent team. According to him, in the mid-1990s, IBM asserted seven patents against Sun Microsystems in licensing negotiations. Among them was the so-called "fat line" patent, which effectively claimed a basic geometric method for thickening a line on a computer screen. Sun's counsel, many of whom were trained both as engineers and as lawyers, systematically challenged each claim and concluded that none posed a credible infringement risk.

Despite this, IBM's representatives emphasized the breadth of their portfolio—approximately 10,000 U.S. patents—and suggested that, even if the asserted set proved weak, additional patents could easily be substituted. Ultimately, Sun agreed to a settlement payment, illustrating IBM's strategy of leveraging portfolio size rather than the intrinsic strength of individual patents.

This story took place in the 1980s. At the time, IBM mostly relied on discussions and meetings to collect royalties, maintaining the dignity of a blue-blooded aristocrat. Today's IBM, however, increasingly resorts to litigation. At least from the perspective of patents, it is obvious that IBM has undergone a massive transformation in its business priorities.

In 2016, IBM sued Groupon for patent infringement. Unlike other IBM opponents, Groupon fought back aggressively and publicly expressed its hostility toward IBM in the media. A Groupon spokesperson derisively referred to IBM as a dial-up-era dinosaur,[26] and one of their attorneys described IBM's behavior as "extortion." Another lawyer, David Hadden, told Bloomberg that Amazon had only paid IBM

[24] Osenga, K. (2014). Formerly manufacturing entities: Piercing the patent troll rhetoric. *Conn. L. Rev.*, *47*, 435.

[25] Gary L. Reback. (2002). Patently Absurd. Retrieved March 31, 2024 from Forbes Website: https://www.forbes.com/asap/2002/0624/044_print.html

[26] De Leon, I., Donoso, J. F., & de León, I. (2017). *Innovation, Startups and Intellectual Property Management*. Springer.

licensing fees out of fear, whereas Groupon wasn't afraid—someone had to stand up to IBM's bullying.[27]

The media broadly sympathized with Groupon and found it incomprehensible that IBM was suing twenty-first-century internet companies based on patents from the 1980s.

Prodigy was once an IBM subsidiary, co-founded in the 1980s by IBM and the legacy retailer Sears, with IBM executives serving as CEOs.[28] It was one of the early regional network service providers before the rise of the internet, alongside CompuServe, Genie, and others—platforms that held the earliest internet memories for millions of American users.

Beginning in the late 1980s, Prodigy built a nationwide network in the U.S. and launched a paid network service called "Prodigy Service." For $12.95 a month, users could access the Prodigy network through their personal computers. From the Prodigy main interface, users could browse ads, news, weather, and sports—much like modern portals—and they could also communicate via email and BBS forums.

Among the internet's early pioneers, Prodigy was the first to offer a graphical user interface. Competing services like CompuServe and Genie were text-only. With its visual-rich design, Prodigy was among the earliest to provide online banking, stock trading, advertising, and shopping, generating considerable advertising revenue. To overcome the limitations of weak PC performance at the time, Prodigy made a series of technical innovations, developing an early form of a CDN (Content Delivery Network) as early as 1993. The achievements of Prodigy did not escape the attention of IBM's patent team; even today, IBM still holds the intellectual property rights to some of these key technologies.

Prodigy could be considered a truly innovative enterprise, but it made repeated missteps on the business front and was ultimately abandoned by IBM due to massive financial losses. The Prodigy development team had assumed that users would spend most of their time viewing ads, news, and other content rather than engaging in interactive activities. Consequently, their approach to remote access was quite conservative. Prodigy deployed its content at Points of Presence (POPs) located in various cities, which were connected to Prodigy's main servers via leased lines to refresh local content caches. The developers believed users would simply connect to their nearest POP via dial-up when needed, making long-term network connections unnecessary.

However, Prodigy grossly underestimated early internet users' enthusiasm for email and BBS-style interaction. In 1990, the most active 5% of users sent 35 million emails, forcing POPs to make frequent dial-up calls and generating unexpectedly high costs. To reduce traffic, Prodigy began moderating content in 1991,

---

[27] Christopher Yasiejko. (2018). IBM and Groupon Trade Final Blows as Patent Case Goes to Jury. Retrieved March 31, 2024 from Bloomberg Website: https://www.bloomberg.com/news/articles/2018-07-26/ibm-and-groupon-trade-final-blows-as-patent-case-goes-to-jury

[28] Michael Banks. (2008). Prodigy: The pre-Internet online service that didn't live up to its name. Retrieved March 31, 2024 from TechRepublic: https://www.techrepublic.com/blog/classics-rock/prodigy-the-pre-internet-online-service-that-didnt-live-up-to-its-name/

removing offensive information and comments. This sparked a strong backlash, with angry users accusing Prodigy of reading their emails or even attacking their personal computers. Such rumors led to significant user attrition. In 1994, Prodigy launched a chatroom feature that became hugely popular. Thousands of users stayed online 8–10 hours a day, again driving up operational costs. Prodigy was forced to shut down the chatrooms, provoking yet another wave of user dissatisfaction. By 1996, Prodigy had racked up $1.3 billion in losses, and its user base had dropped below one million. IBM finally let it go. That same year, Prodigy was acquired by International Wireless Inc. and connected to the emerging Internet,[29] becoming one of the world's earliest Internet Service Providers (ISPs) offering dial-up access. The original Prodigy network survived for a time under the name "Prodigy Classic," but was shut down around the turn of the millennium due to the Y2K issue. In the twenty-first century, Prodigy was acquired by SBC and eventually merged into AT&T, also acquired by SBC, disappearing from public view.

Though Prodigy is now long gone, its pioneering role in online advertising and e-commerce deserves lasting recognition. To make the most of personal computers at the time, Prodigy's development team came up with a number of valuable innovations. Although IBM sold off Prodigy and did not pursue online shopping or portal services, it retained a large portfolio of Prodigy-related patents. Years later, as the internet economy flourished, these patents became a Damoclean sword hanging over American tech firms.

In recent years, many of IBM's patent lawsuits have involved key Prodigy-era patents. In November 2013, just before Twitter's IPO, IBM sent the company an official letter demanding licensing fees for three patents (US6957224B1, US7072849B1, and US7099862B2). In March 2014, Twitter and IBM reached an agreement: Twitter paid $36 million for 900 IBM patents, a sum equal to 5% of its 2013 revenue ($665 million).

In February 2015, IBM filed a patent infringement lawsuit in Delaware against Priceline and its subsidiaries Kayak and OpenTable. In 2016, IBM sued Groupon in the same court, claiming infringement of four patents (US5796967A, US7072849B1, US5961601A, and US7631346B2), demanding $167 million in damages. In January 2018, IBM filed suit against Expedia for infringing the same four patents. In September 2019, IBM sued real estate data firm Zillow over patent US7072849B1. In 2020, IBM sued Airbnb using the same patent and demanded licensing fees from pet supply retailer Chewy, citing four patents (US7072849B1, US9569414B2, US7076443B1, and US6704034B1). Chewy responded with a declaratory judgment action for non-infringement in the Southern District of New York.

Among these lawsuits, the two most frequently cited patents were US7072849B1 (hereafter the "'849 patent") and US5796967A (the "'967 patent"). These patents originated from inventions by IBM software engineer Robert Filepp in the late

[29] *Prodigy Communications Corporation history.* (n.d.). *Funding Universe: Company histories.* Retrieved March 31, 2024 from http://www.fundinguniverse.com/company-histories/prodigy-communications-corporation-history/

1980s. The '849 patent focuses primarily on online advertising. Its claim 1 reads as follows:

> *A method for presenting advertising obtained from a computer network, wherein the network includes multiple user receiving systems, each enabling users to request applications comprising interactive services. Each receiving system includes a display, and at least a visual portion of the application is displayed over one or more screens on the display. The method includes:*
>
> (a) *Constructing the application such that it can be transmitted over the network and presented as a first portion of said screen(s) display content.*
> (b) *Constructing the advertisement content to match the application such that it can be transmitted over the network and presented as a second portion of the screen(s) display content simultaneously with the application. The advertisement is configured as an object comprising ad data.*
> (c) *Selectively storing the advertisement object in the memory of the receiving system.*

The '967 patent focuses on the display of network applications, with content very similar to that of the '849 patent. Its claim 1 outlines a method for presenting interactive applications over a computer network, using structured data objects that can be selectively stored in the receiving system, with fallback retrieval from the network if local storage is unavailable.

The '849 patent protects network technologies that were in their infancy over thirty years ago. At the time, there was not yet a clear concept of "internet technology." David Eastman, an expert witness for Priceline, classified the technology in the '849 patent as belonging to the domain of "electronic publishing." In terms of practical application, the technology covered by the '849 and related '967 patents was primarily used in the Prodigy system of the 1990s. Other contemporaneous regional networks and earlier commercialized videotex systems may also constitute prior art. However, individuals with knowledge of these networks and legacy systems—mostly technical experts or enthusiasts from thirty years ago—are now aging, and much of that early technology is difficult to document or verify.

In the lawsuits, Priceline and Zillow both retained David Eastman as an expert witness. Eastman had held management positions at CompuServe since 1974. Like Prodigy, CompuServe was a commercial network that predated the widespread popularity of the internet. Eastman's expert testimony referenced earlier systems such as videotex platforms and Apple's HyperCard system from the 1980s—technologies that were either conceptual experiments or outright commercial failures. Like Prodigy, CompuServe, and Videotex, these systems now lie forgotten in the dusty corners of tech history. Due to their minimal impact, only specialist historians of technology tend to examine them, and it is difficult to argue that these outdated systems hold much relevance today.

A 2015 report by the American Intellectual Property Law Association (AIPLA) revealed that the median cost of an Inter Partes Review (IPR) hearing before the Patent Trial and Appeal Board (PTAB) was $275,000, with appeals bringing the total to $350,000. For patent infringement lawsuits where damages ranged from $10 million to $25 million, the median cost at the close of discovery was $1.9 million, and $3 million by trial. In cases where damages exceeded $25 million, median litigation costs rose to $3 million at discovery and $5 million at trial. These sums, which could otherwise have funded research and innovation, instead flowed into the hands of patent attorneys, expert witnesses, bureaucrats, and non-practicing entities (NPEs).

As for the '849 patent, IBM's opponents filed at least three challenges asserting lack of patentable subject matter—all of which failed. Priceline was the first to file a motion in court claiming that the '849 patent covered unpatentable subject matter, but the Delaware court denied the motion. Priceline also petitioned the PTAB to institute a Covered Business Method (CBM) review based on lack of patentability, but the PTAB denied that request after preliminary review. In the Groupon case, lacking compelling prior art, Groupon also challenged the patent solely on grounds of patent ineligibility—again unsuccessfully.

Challenges based on novelty and non-obviousness likewise failed. Except for Groupon, the defendants managed to find some relevant prior art and brought invalidity challenges in court or before the PTAB. Priceline and Zillow's prior art mainly referenced videotex systems from the 1980s, supported by academic articles and technical literature; Expedia's cited prior art came from the "Boston Community Information System," a regional network from the same era. Yet none of these attempts succeeded. To date, among all lawsuits involving the '849 patent, only Priceline (later renamed Booking Holdings) has succeeded, using a non-infringement defense based on divided infringement principles established in the *Akamai* case. Most other companies lost their cases against IBM.

IBM claimed that both the Groupon website and its mobile app infringed on the '967 and '849 patents. Regarding the '849 patent, IBM alleged in its complaint:

> Groupon has infringed at least claim 1 of the '849 patent, for example:
>
> Presenting information from a computer network (e.g., the Internet), wherein the network includes multiple user receiving systems (e.g., computers or mobile phones used by Groupon users). Users request applications (e.g., Groupon's homepage, local deals, or product pages), which offer interactive services (e.g., goods or services at discounted prices). These receiving systems include a display (e.g., a monitor or smartphone screen), and at least a visual portion of the application is presented over one or more screens. The method includes the following steps:
>
> (a) Constructing the application (e.g., Groupon homepage, local, or product page) to be transmitted over the network and to constitute the first portion of the display content on one or more screens (e.g., Groupon's website or content).
> (b) Constructing advertisement content (e.g., promotional coupon ads) to match the application and be transmitted over the network, forming a second portion of the display content on one or more screens (e.g., banners), displayed simultaneously with the application (e.g., Groupon homepage, local, or product pages). The advertisement is configured as an object containing ad data (e.g., woff or jpeg files).

(c) Selectively storing the advertisement object (e.g., embedding a cache-control parameter) in the memory of the receiving system (e.g., browser cache).

Groupon filed a motion arguing that the '967 and '849 patents claimed abstract ideas—namely, the local storage and segmented display of information on a user's device—and lacked any "inventive concept," making them unpatentable under U.S. Patent Law Section 101.

IBM cited the 2016 *Enfish LLC v. Microsoft* case, arguing that the '967 and '849 patents improved computer functionality and were thus patent-eligible under the *Alice* test. Groupon countered that the patents merely described desired outcomes using conventional computing methods, offering functional goals without teaching how to achieve them. IBM responded that, prior to these patents, interactive services over networks were delivered via "dumb terminals," requiring central processing and storage of all user applications. This approach limited the number of users each host could serve and resulted in slow response times. The patents at issue introduced an architecture that offloaded application and ad content into discrete objects stored locally, improving performance and scalability.

The court, citing *Enfish* and similar precedents, stated that claims representing "improvements to computer functionality" are not abstract ideas. Filepp's patents enhanced computer network performance, making them patent-eligible. For example, claim 1 of the '967 patent improved memory efficiency and system speed, allowing more users to be served. The court also found that the patents did not rely on vague functional language but described specific architectures for generating screen content, including methods of data retrieval and storage.

Thus, the court concluded that under the *Alice* framework, the patents represented concrete solutions to computing problems, not abstract ideas, and therefore did not require further analysis under the second step of the *Alice* test.

Groupon's invalidity motion was denied. A jury awarded IBM $83 million in damages. Groupon announced plans to appeal, but IBM, seeking to end the litigation, negotiated a lower settlement: $57 million. Groupon's challenge to IBM's dominance ultimately failed.[30]

IBM's Prodigy introduced online shopping before Amazon, pioneered the portal website model before Yahoo, and monetized online advertising before Google. Yet despite its technological prowess, IBM never used these patents to build its own e-commerce platform or internet products. Instead, as internet companies surged ahead, IBM cashed in by selling its soon-to-expire "vintage" patents.

When Prodigy began clashing with its user base over mounting costs from prolonged online activity, it was already a sign of its—and IBM's—impending failure. User demand drives commercial development, and the much-lauded concept of "traffic" ultimately boils down to meeting user needs. When IBM had a flood of user

---

[30] Jan Wolfe. (2018). Groupon to pay IBM $57 million to settle U.S. patent dispute. Retrieved March 31, 2024 from Reuteurs Website: https://channels.theinnovationenterprise.com/articles/groupon-settles-patent-suit-with-ibm-for-57m

attention opening the door to a new era, it slammed that door shut on itself, clutching a few patent needles and waiting in the shadows for a chance to pounce.

In *The Decline and Fall of IBM: End of an American Icon*, the author argues that "IBM is a sales company run by salespeople"—a structural flaw rooted in its chronic preference for sales over product. Aside from founders Thomas Watson Sr. and Jr., and outsider Louis Gerstner, nearly every IBM CEO has risen from the sales ranks, giving the sales department disproportionate influence over company decisions. Steve Jobs once commented on IBM:

> If you're in the product division at IBM or Xerox and you create a better printer or computer, what does it matter? When your company already has a market monopoly, there's nowhere to go but down. The people who can still move the company forward are the sales and marketing folks. So, they take over. Product people get kicked out of the boardroom. The company forgets what makes great products. The instincts and brilliance that propelled the company to dominance now rot in a corner, while those in charge have no idea what makes a good product or a bad one. Deep down, they don't care about helping the customer at all.

The Decline and Fall of IBM also notes that IBM imposes a strict incentive-and-punishment system on its sales staff. Top performers rise quickly and may be promoted to management or even CEO. Those who fail to close deals get poor performance reviews and may be placed on "improvement plans." To avoid blame, sales teams often scapegoat other departments. Furthermore, like many large corporations, IBM builds its teams mainly through campus recruiting and internal promotions. From the bottom to the top, many employees spend their entire careers at IBM, with limited exposure to the "outside world."

With a management culture rooted in sales and limited external perspective, IBM tends to solve problems through discipline rather than improving internal processes, optimizing products, or enhancing operations. According to the book, even under CEO Gerstner, IBM "never truly improved the quality, efficiency, or productivity of its services." Instead, it focused relentlessly on cost-cutting—especially in HR—slashing R&D staff and offshoring manufacturing to replace one American worker with eight Indian workers. Between 1997 and 1998, IBM's Global Services Division even developed the mindset that paying penalties for service failures was cheaper than actually providing after-sales service. Since some clients didn't know they were entitled to penalties or didn't bother to claim them, IBM concluded it might as well eliminate support altogether.

Over time, IBM's Global Services Division overtook its hardware operations to become the company's main cash cow. It's well known that IT services are more profitable than hardware. But as more companies entered the IT outsourcing market, IBM faced fierce competition. At first, it tried to maintain premium pricing on the strength of its brand. However, in the highly competitive B2B market, price rules, and IBM's brand advantage faded. So, as a "sales company," IBM had no choice but to lower prices. Yet it did so without improving service—offering cheaper but also lower-quality service, rather than delivering equal or better service at a better price.

The book uses an interesting metaphor: Early IBM was like Switzerland—precise craftsmanship, premium prices, strong brand identity. Over time, it turned into feudal Japan: rigid hierarchy, entrenched factions. The "serf" departments were crushed and voiceless, while the salespeople were the samurai, willing to do anything to climb the ranks. Gerstner played the role of MacArthur, leading the reforms.

Interestingly, some Japanese scholars see strong parallels between IBM and Japanese companies. Scholar Shigeru Matsumoto points out[31] that while Japanese corporations have long modeled their M&A practices after GE, in reality, IBM is a better and more "realistic" model. In Japanese diversified firms, business unit leaders are also board members, representing their unit's interests—just like IBM, not GE. Given these similarities, Matsumoto argues that IBM offers more relevant lessons for Japanese firms.

A 2016 study from Johns Hopkins University expressed cautious optimism about IBM's long-term stock prospects but warned of major risks: "Failure to integrate acquired businesses with Watson," and that "the company's transformation is highly dependent on technical talent. The loss of top engineers to competitors would severely impair IBM's project delivery capabilities." From 1995 to 2016, IBM has faced the same problems—evidence of just how difficult it is to turn around a behemoth.

As a giant corporation, IBM has recently undertaken an extreme "demonic disassembly" of its own business. In 2002, it sold its hard drive business to Hitachi; in 2005, its PC division to Lenovo; in 2007, its digital printer business to Ricoh; and in 2012, its POS division to Toshiba Tec. Matsumoto praised IBM's downsizing strategy, contrasting it with the delayed retreats typical of Japanese firms. IBM exits businesses while they are still profitable and hold strong market positions—moves Japanese companies could learn from. Between 2002 and 2011, despite only a 30% increase in total sales, IBM's profit margin jumped from 9% to 20%, and free cash flow increased by $10 billion. Matsumoto attributes this to IBM shedding low-margin commodity businesses and acquiring high-margin companies.

Shigeru Matsumoto noted, "During the tenure of Samuel Palmisano as CEO, as soon as a business was deemed a commodity, he would sell it off without hesitation. Even business units he had once overseen himself—such as the PC division—were not exempt." Matsumoto emphasized that throughout its century-long history, IBM had always excelled at pioneering markets with original products, rather than copying competitors and engaging in price wars in already-established markets. Starting in 2002, IBM began a comprehensive retreat from what it deemed "commodity" businesses—because commodities rely on volume to win, a model incompatible with IBM's long-term strategy of pursuing high profit margins. This relentless pursuit of profit aligns perfectly with Colin Greeley's observations.

The Decline and Fall of IBM: End of an American Icon states that IBM never truly wanted to produce commodity goods or invest in large-scale manufacturing of

[31] Matsumoto, S. (松本茂) (2018). Japanese Outbound Acquisitions: Explaining What Works. Springer.

any product. "The idea of investing millions of dollars in manufacturing consumer-grade digital phones would probably terrify IBM. The company culture simply doesn't support creating highly integrated devices that can be produced at low cost and on a large scale." Unlike Matsumoto's enthusiastic praise, the author of *The Decline and Fall of IBM* holds a degree of hostility toward IBM's salespeople, interpreting the company's continual divestment as a sign that its sales-trained executives lack competitive grit: "As a sales-driven company, IBM tends to abandon a business as soon as sales become difficult. IBM doesn't understand that it should continuously improve its products and services so that even in the face of fierce competition, it doesn't become harder to sell."

From a Japanese perspective, Matsumoto's admiration for IBM's asset shedding is understandable. In 2002, IBM sold its hard disk business to Hitachi for $2 billion. Not long after, storage device prices plummeted, leading to five consecutive years of losses for Hitachi, which was forced to inject large sums to keep the business afloat, eventually offloading it to Western Digital. Hitachi felt relieved and unburdened after the sale. Similarly, IBM sold its POS business to Toshiba Tec just as mobile payments were on the rise. Shortly after the deal closed, PayPal and Rakuten launched mobile payment apps that severely impacted the POS market. In 2016, the POS division acquired by Toshiba Tec posted a loss of 69.6 billion yen. Likewise, when Lenovo bought IBM's PC division for $1.25 billion in 2004, it acquired a 51% stake in NEC's PC division for just 1.8 billion yen in 2011—prompting Matsumoto to remark on IBM's uncanny ability to sell at just the right time.

Through this analysis, it becomes clear that IBM's corporate culture places heavy emphasis on sales and profit. So then, is the patent licensing team a high-performing sales team? Does the patent licensing business generate high profits?

Undoubtedly, revenue from patent licensing cannot be considered pure profit. R&D, legal fees, travel, and administrative costs are all part of the equation. However, it's difficult to find such cost breakdowns in publicly available documents, and few studies have examined the issue. In *From Assets to Profits: Competing for IP Value & Return*, author Bruce Berman recounted a 2006 conversation with Robert Phelps, who estimated that 95% of IBM's patent licensing revenue could be considered pure profit.[32]

Given this, it's no wonder that a company like IBM, so obsessed with profitability, would see patent licensing as critically important. After all, what other business can deliver over 95% profit margins? As patent licensing evolved into a form of non-practicing entity (NPE) activity and the number of patents began to outweigh their quality, management's attitude toward R&D became increasingly self-evident.

Is IBM the only company trapped in such a conundrum due to its sales-first tradition, profit-driven DNA, and massive institutional structure?

[32] Berman, B. (2008). From assets to profits: competing for IP value and return (Vol. 36). John Wiley & Sons.

The answer is likely no. IBM's path toward becoming an NPE may also stem from the very nature of capital itself.

As we know from Silicon Valley lore, the history of startups is filled with dramatic highs and lows. Among the most attention-grabbing are the stories of founders being ousted from their own companies: Steve Jobs of Apple, Mike Lazaridis of BlackBerry, and Jack Dorsey of Twitter are all well-known examples. Some founders were removed due to poor managerial ability; others were ousted for sticking to their entrepreneurial vision and refusing to cater to investors' insatiable demand for capital gains. As the old saying goes, "If the profit is sufficient, capitalists will sell the rope to hang themselves." After all, capital's sole nature is to pursue profit. We often assume that founders and managers are the true entrepreneurs, seeking to change the world for the better, while investors are rent-seeking freeloaders, willing to use any means to chase short-term gains.

Even setting aside emotional factors, the conflict between managers and investors is a well-established reality in academia. Around 1976, Michael Jensen and others from the University of Rochester published the landmark paper *Theory of the Firm: Managerial Behavior, Agency Costs and Ownership Structure*, introducing the now-famous "agency theory," which identifies inherent conflicts between a firm's owners (shareholders) and its managers. By the 1980s, agency theory had gained wide acceptance, sparking a wave of research into resolving these conflicts. Many scholars focused on the relationship between CEO compensation and corporate performance. Some studies suggested that stock-based pay and incentive structures helped improve company results.[33] Over time, academics and business leaders reached a consensus that tying executive pay to shareholder value through stock-based incentives offered a seemingly win-win solution.

One study summarized changes in CEO compensation for S&P 500 companies from the 1970s to the mid-1990s (see Fig. 11.2).[34] Thin lines in the graph represent salaries and bonuses, while thick lines reflect total income including stock options and grants. By 1996, stock-related income had become an indispensable part of executive pay. Interestingly, the bar chart in the graph reflects the volume of academic papers published on the topic, illustrating the field's heightened interest.

By around 1996, the median annual compensation for CEOs of S&P 500 companies was approximately $2.5 million. Subsequent studies showed that this figure rose to $seven million by 2009 and exceeded $10 million by 2015.[35]

An increasing number of CEOs had their compensation packages linked to stock prices—IBM was no exception. Both Sam Palmisano and Lou Gerstner experienced

[33] Murphy, K. J. (1999). Executive compensation. *Handbook of labor economics*, *3*, 2485–2563. Retrieved March 31, 2024 from SSRN: https://ssrn.com/abstract=163914 or https://doi.org/10.2139/ssrn.163914

[34] Murphy, K. J. (1999). Executive compensation. *Handbook of labor economics*, *3*, 2485–2563. Retrieved March 31, 2024 from SSRN: https://ssrn.com/abstract=163914 or https://doi.org/10.2139/ssrn.163914

[35] Bennett, B., Garvey, G., Milbourn, T., & Wang, Z. (2017). Managerial compensation and stock price informativeness. *WUSTL Working paper*.

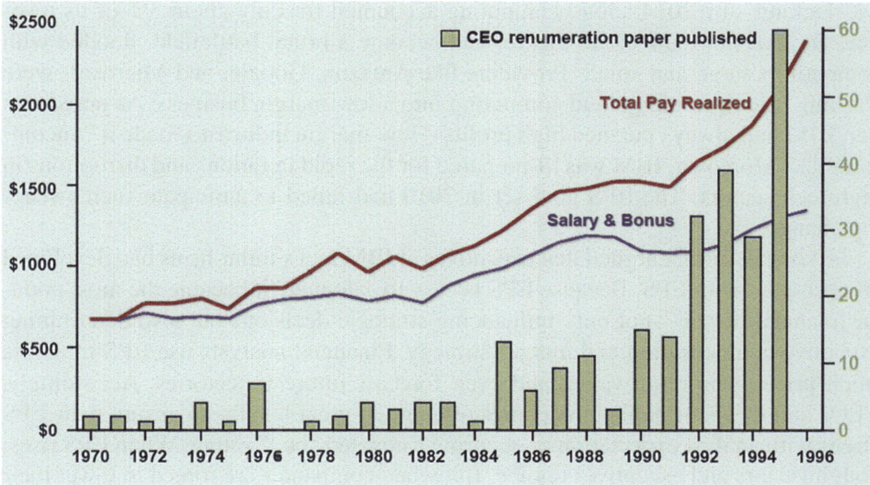

Fig. 11.2 Changes in CEO compensation at S&P 500 companies, from the 1970s to the mid-1990s

this trend. Palmisano, upon retirement, walked away with $271 million, surpassing even Gerstner, who had saved the company from near-collapse. The author of *The Decline and Fall of IBM: End of an American Icon* argues that it was precisely because compensation was tied to stock price that Palmisano, in 2005, made IBM's corporate goal the enhancement of earnings per share (EPS). In 2010, just before stepping down, Palmisano set a second EPS goal: to increase IBM's EPS to $20 by 2015.

Jack Welch, the former CEO of General Electric, expressed strong disapproval of such practices in a 2009 interview with the *Financial Times*: "Shareholder value is the dumbest idea in the world. Shareholder value is a result, not a strategy. Your main constituencies are your employees, your customers, and your products." Welch insisted that neither managers nor investors should treat stock price gains as their primary goal.

In October 2014, IBM was forced to abandon the $20 EPS target. Then-CEO Ginni Rometty stated, "We are disappointed in our performance… The results demonstrate that our industry is facing unprecedented change."

Aleksander Jaskowiak, a former senior management consultant at IBM, noted[36] that in order to meet EPS targets, Rometty resorted to cost-cutting, divesting business units, layoffs, tax avoidance, and aggressive stock buybacks, while also entering into numerous high-stakes agreements. Facing an "unprecedented transformation," Rometty pivoted IBM toward cloud computing as a strategic focus. Although IBM had made some preparations for this shift, its investment enthusiasm

[36] Aleksander Jaskowiak. (2014). Why IBM CEO should listen to vox populi. Retrieved March 31, 2024 from Linkedin: https://www.linkedin.com/pulse/20141125090104-7490986-why-ibm-ceo-should-listen-to-vox-populi

was lacking—by 2014, cloud computing accounted for only about 3% of its revenue. By that time, the cloud market had become a brutal battlefield, flooded with competitors large and small. Providers like Amazon, Google, and Microsoft were slashing prices, turning cloud computing into a low-margin business. As noted earlier, IBM had always pursued high profits—low-margin industries made it "uncomfortable." Moreover, IBM was ill-prepared for the rapid iterations and disruptions in high-tech sectors. The EPS goal set in 2010 had failed to anticipate such sweeping change.

Jaskowiak further argued that executives at IBM and similar firms had developed an obsession with EPS. Because EPS is easy to calculate, it became the most popular financial metric—not only influencing strategic decisions but also determining executive compensation and merger strategy. Financial analysts use EPS to assess stock prices, corporate value, and even forecast future trajectories. Accordingly, CEOs and CFOs of public companies often go to great lengths to ensure their EPS aligns with analyst expectations—securing a coveted "buy" rating. When EPS rises, both investors and executives rejoice. But when companies are forced to lower their EPS due to pressure, markets often react brutally—with double-digit stock declines.

In the short term, stock buybacks have a near-magical effect on EPS. Reducing the number of outstanding shares increases demand and drives up stock prices—boosting earnings per share and helping hit quarterly EPS targets. If executives hold stock options, they benefit directly and thus have even greater incentive to pursue buybacks. Jaskowiak stated bluntly that IBM's executives routinely exercised their stock options and then immediately sold their shares.

With management effectively captured by capital and incapable of formulating meaningful long-term strategic plans, the future becomes opaque and unpredictable. To impress Wall Street investors, management initiates a string of strategic partnerships—projects that appear innovative on the surface but either lack monetization potential or fail to offset deterioration in core business lines. According to Jaskowiak, IBM's partnerships with Twitter, Tencent, and Apple are all examples of this phenomenon. *The Decline and Fall of IBM* also notes that to impress Wall Street with its resolve and strength, IBM constantly sought $10 billion-scale acquisitions. Because the ultimate goal was to grow EPS, smaller deals became irrelevant—IBM's true "customers" had become its investors on Wall Street.

In 2015, IBM's board approved another $5 billion stock repurchase plan. In the first nine months of 2014 alone, IBM spent $13.6 billion on stock buybacks—more than double its net income for 2013 and more than twice its R&D expenditure that same year.

Jaskowiak believed that after acquiring SoftLayer, IBM should have focused on internal integration and business stabilization instead of chasing further cloud acquisitions. Cloud is a capital-intensive industry; companies must invest heavily in load capacity to keep pace with market leaders and meet compliance standards.

Jaskowiak advocated reforming executive compensation—from stock- and option-based packages to KPI systems that prioritize long-term corporate value. Stock-based incentives encourage executives to manipulate stock prices via buybacks. Moreover, because stock option transactions are recorded as cash movements and not on income statements, they're difficult for outsiders to monitor.

Mark Cuban, one of America's most well-known investors, stated in 2014: "They have no vision. What they've evolved into is a company that does [arbitrage] on acquisitions…Who is IBM anymore?"[37] A commentary on the finance site *Fool* echoed this sentiment: "*Share buybacks are easy to do, and are often a sign of management's lack of ideas.*".[38]

Around 2016, Warren Buffett and his firm Berkshire Hathaway held 81 million shares of IBM. In the first two quarters of 2017, Buffett sold a third of those holdings. Speaking to CNBC, Buffett said: "*I don't value IBM the same way that I did 6 years ago when I started buying… I've revalued it somewhat downward… I think if you look back at what they were projecting and how they thought the business would develop I would say what they've run into is some pretty tough competitors.*"[39] By the first half of 2018, Buffett and Berkshire Hathaway had sold 94.5% of their IBM shares. Buffett admitted: investing in IBM was a mistake.[40]

The renowned economist David Stockman, often referred to as the "father of Reaganomics," wrote in *The Great Deformation* that:

> By 2013, IBM had earned about $100 billion in net profits over seven years—and spent every penny of it on stock buybacks. The former global titan of high technology had nothing better to do with its cash than shrink its own equity base. IBM's total share count had dropped by 20%, and 45% of its EPS growth was attributable solely to buybacks. Furthermore, over those seven years, IBM had also spent $20 billion on dividends. In total, IBM funneled cash equivalent to 120% of its 2007–2013 net profits into the insatiable maw of quick-money traders and hedge funds. Naturally, algorithmic traders welcomed such "shareholder-friendly" stocks with open arms—even if their ultimate fate was bankruptcy liquidation.

When corporate executives chase personal gain by manipulating stock prices with massive amounts of cash—abandoning strategic initiatives and pandering to Wall Street investors—how can they set meaningful goals for their R&D teams, let alone guide the mission of their patent licensing departments?

In Chap. 8 of *From Assets to Profits: Competing for IP Value & Return*, Harvard Business School professor Willy Shih discusses common attitudes of executives

---

[37] Bort, J. (2014, October 22). *Mark Cuban slams IBM: It's 'no longer a tech company. They have no vision.' Business Insider*. Retrieved March 31, 2024 from https://www.businessinsider.com/mark-cuban-ibm-is-not-a-tech-company-2014-10

[38] Jeremy Bowman. (2018). The 3 Biggest Mistakes Warren Buffett Made With IBM. Retrieved March 31, 2024 from The Moltley Fool: https://www.fool.com/investing/2018/02/18/the-3-biggest-mistakes-warren-buffett-made-with-ib.aspx

[39] Becky Quick. (2017). Warren Buffett has sold IBM shares, and 'revalued' tech icon downward, cites 'big strong competitors'. Retrieved March 31, 2024 from CNBC: https://www.cnbc.com/2017/05/04/warren-buffett-has-revalued-ibm-downward-cites-big-strong-competitors.html

[40] Colleen McKown. (2018). Over time, Buffett softens his stance on tech stocks. Retrieved March 31, 2024 from CNBC: https://buffett.cnbc.com/2018/04/02/over-time-buffett-softens-stance-on-tech-stocks.html

toward patent licensing teams.[41] He notes that although academia and industry have explored numerous methods to objectively quantify patent value (such as Ocean Tomo's valuation or the relatively crude "patent strength" metric), in reality, the only truly reliable indicators of a patent's value are commercial transactions and litigation. However, even business transactions may fail to reflect a patent's true commercial worth. Consequently, patents typically do not appear as assets on balance sheets—unless they were acquired externally through transactions. Likewise, if a specific patent generates licensing revenue, it is usually difficult to attribute corresponding R&D costs to that single patent. Thus, for many executives, patent licensing revenue is often seen as "found money."

If licensing revenues reach a significant and stable level, management may expect patent teams to forecast upcoming income accurately. This helps balance R&D budgets, polish financial statements, and impress investors. Yet under financial reporting pressure, patent teams frequently make price concessions to accelerate deal closure and count licensing fees as current-period revenue. As a result, many companies outside the U.S. have discovered that the best time to resolve patent disputes with American firms is at the end of a fiscal quarter or year.

Professor Shih warns that treating patents purely as a cash source is "extremely shortsighted—if not dangerous." He argues that pursuing patent licensing as the sole objective of R&D is not a sustainable business model, for the following reasons:

1. **The best patents often arise from practice.** Real-world testing and iterative refinement of technology yield effective solutions to practical problems, which in turn lead to new inventions—creating a virtuous cycle.
2. **The monetization window for patents is long**, often out of sync with investors' time horizons. Investors seek quick returns; patent monetization, including enforcement, usually lies far outside their planning cycles.
3. **High-value patents often stem from fundamental research in emerging fields.** Such foundational patents may take a decade or more to reach mainstream markets—well beyond the typical investor's timeline for returns.
4. **In highly visible, fast-moving R&D domains**, it's difficult to design patents with broad, clear, and defensible claims. To build a strong portfolio, a company must begin R&D early—before a field becomes crowded. This is becoming increasingly rare today.

We strongly agree with Professor Shih's analysis. For this reason, we reject the notion that "first-rate companies sell standards, second-rate companies sell technology, and third-rate companies sell products." Without excellent products, standards and technology are rootless—water without a source, trees without roots. When a "former manufacturing entity" turns into a Non-Practicing Entity (NPE), the path to failure is already laid bare.

[41] Berman, B. (2008). From assets to profits: competing for IP value and return (Vol. 36). John Wiley & Sons.

In *Dream of the Red Chamber*, characters like Qingwen and Xiren serve as shadow counterparts to Lin Daiyu and Xue Baochai, foreshadowing their fates. Inspired by this literary device, we present a "shadow enterprise" for IBM in this chapter—a company with a similar trajectory. Once a titan of innovation, it found itself struggling in a new era, eventually pinning its hopes on patents. Whether IBM can mount a comeback through patents remains to be seen, but this other company failed in its attempt and was ultimately cornered by fate. It is none other than Kodak—the "Yellow Giant" of yesteryear.

Chapter 12
A Supplementary Note: The Yellow Giant

Like IBM, Kodak also had a legendary founder—George Eastman. Similar to Herman Hollerith, Eastman began his journey by inventing a unique patented technology during his spare time, evolving from a lone inventor into a business magnate.

George Eastman was 6 years older than Hollerith and began his entrepreneurial journey slightly earlier. Before founding his company, Eastman worked as a bank clerk and was also an avid photography enthusiast. At the time, photography technology was still in its infancy, with many enthusiasts tinkering with rudimentary and improvised methods to improve this emerging craft. As we'll see in this chapter, many significant technological improvements came from amateur hobbyists.

As is well known, photography works by the principle of pinhole imaging—light passes through a small aperture and forms an inverted image on a photosensitive plate, which is then developed in a darkroom to produce a photograph. Early cameras used glass plates as their photosensitive medium, with two main processes: the "wet plate" and the "dry plate." The wet plate, also known as the collodion process, used collodion (a solution of nitrocellulose in ether and alcohol) as a solvent for silver nitrate, which was then coated on the glass plate. The plate had to be used while still wet and developed immediately after exposure—an inconvenient process. A decade later, technological advancements led to the emergence of the dry plate, which allowed exposure after drying but at the cost of reduced sensitivity and image quality compared to wet plates. Hence, the wet plate remained indispensable.

While experimenting with dry-plate photography in his spare time, Eastman invented a device for coating dry photographic emulsion on glass plates, enabling mass production of dry plates. The principle was straightforward: a roller applied a new type of photosensitive material evenly across the glass. In 1879, he filed a patent for the invention and received approval in 1880 (patent number US226503A; see Fig. 12.1 for the illustration). Subsequently, he founded the Eastman Dry Plate Company, dedicated to producing emulsion-coated dry glass plates. Although Eastman was formally the company's financial officer due to his banking background, and a partner held the title of president, Eastman retained full control over

Q. Yue, F. Yang, *Governing Innovation: A Global History of Patent-Driven Technological Progress*, https://doi.org/10.1007/978-981-95-5570-3_12

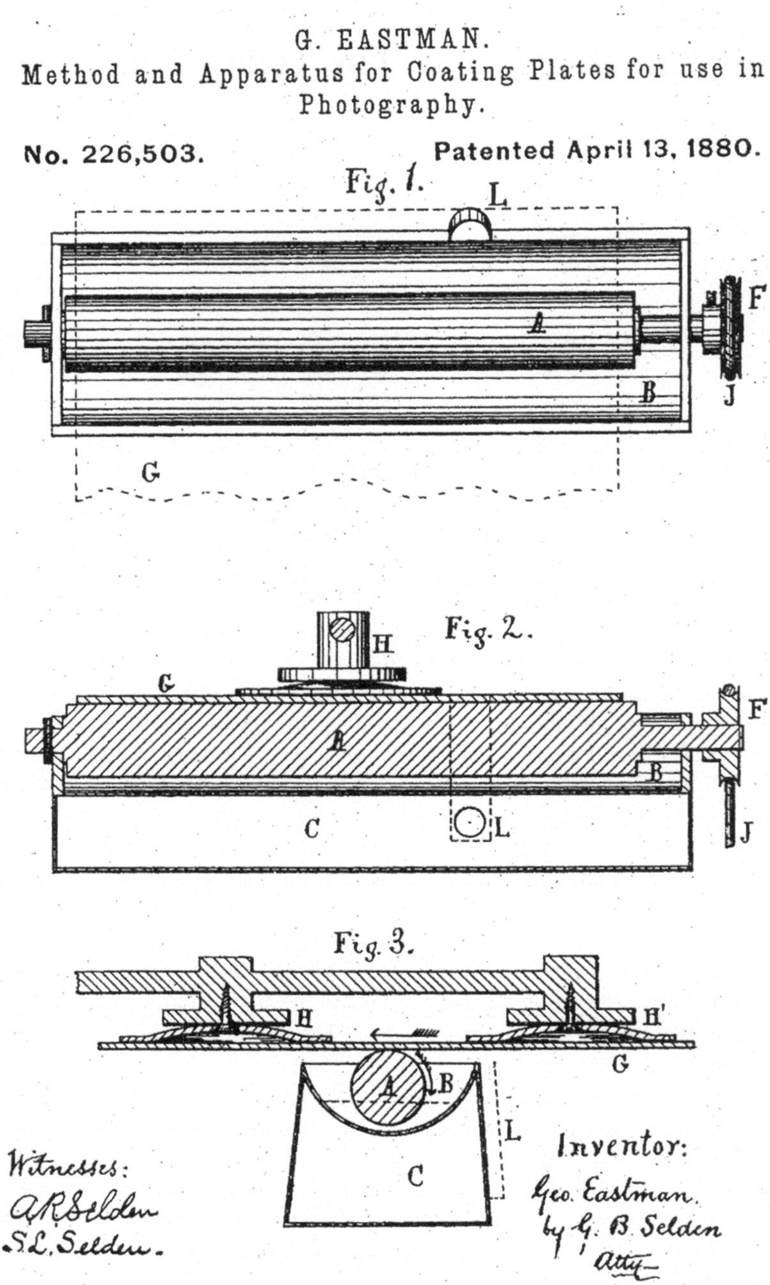

Fig. 12.1 Patent drawing from patent US226503A. In this drawing, "G" represents the glass plate used as film, "B" is a trough containing the emulsion. Roller "A" dips into the emulsion in trough B and rolls across the glass plate to evenly coat it

R&D. Two subsequent innovations led by him would forever transform the photography industry.

Despite the company name containing the word "dry plate," Eastman was not content with the bulky and fragile glass medium. He set his sights on flexible materials. While glass had advantages like transparency, hardness, and scratch resistance, it was heavy, fragile, and difficult to carry. Paper was an alternative emulsion carrier at the time. In 1881, an amateur photographer named Peter Houston pioneered the use of paper film in cameras. His brother, David Houston, saw its commercial potential and filed several patents (e.g., US526446A), which he sold to Eastman's company for $5000.[1] Eastman believed that flexible film was far superior to rigid glass for use as photographic negatives. In 1884, he restructured his company into the Eastman Dry Plate and Film Company.

Eastman once declared that his goal was to "make the camera as easy to use as a pencil." Between 1884 and 1885, he designed various "roll holders for photographic film," resulting in patents like US317049A, US317050A, and US316933A. In March 1888, he packed these roll holders into a wooden box, added a redesigned shutter and lens, and applied for a "camera" patent. It was granted in September that year, with patent number US388850A. Figures 12.2 and 12.3 show the patent illustration and the earliest Kodak camera.

That same year, Kodak launched the iconic slogan, "You press the button, we do the rest" (see Fig. 12.4). Eastman's simple wooden box camera was wrapped in sleek black leather and preloaded with a roll of film capable of taking 100 photos, selling for $25. Once all 100 pictures were taken, customers would send the camera and $10 to Eastman's company for film development and photo printing, after which the camera was returned.

Just like IBM's punched card business, Kodak was among the earliest practitioners of the "razor-and-blades" model—offering affordable cameras to attract users, while making film and developing services the true profit center. From this rudimentary 1888 camera, Eastman's company instinctively evolved into this business model, ultimately building a film empire.

The paper film and portable Kodak cameras greatly simplified photography and attracted many amateurs. However, paper film development was cumbersome and of inferior quality compared to glass plates, making it unpopular with professional photographers. The industry sought a material that combined the flexibility and lightness of paper with the durability and transparency of glass.

In 1869, inventor and entrepreneur John Hyatt developed a miraculous material that was hard but not brittle and could be molded into any shape. He named it "celluloid" and opened a factory producing celluloid billiard balls—ushering in the age of plastics. Photography enthusiasts, including Eastman, soon noticed that celluloid held potential as a replacement for both paper and glass. Some began slicing thin sheets from celluloid blocks for use as film bases, but these sheets were not flexible

[1] David Henderson. Online Artist Archive. Retrieved March 31, 2024 from Internet Archive: https://web.archive.org/web/20131029204317/http://www.nd.gov/arts/online_artist_archive/images-pdfs/H/Houston_DavidHenderson.htm

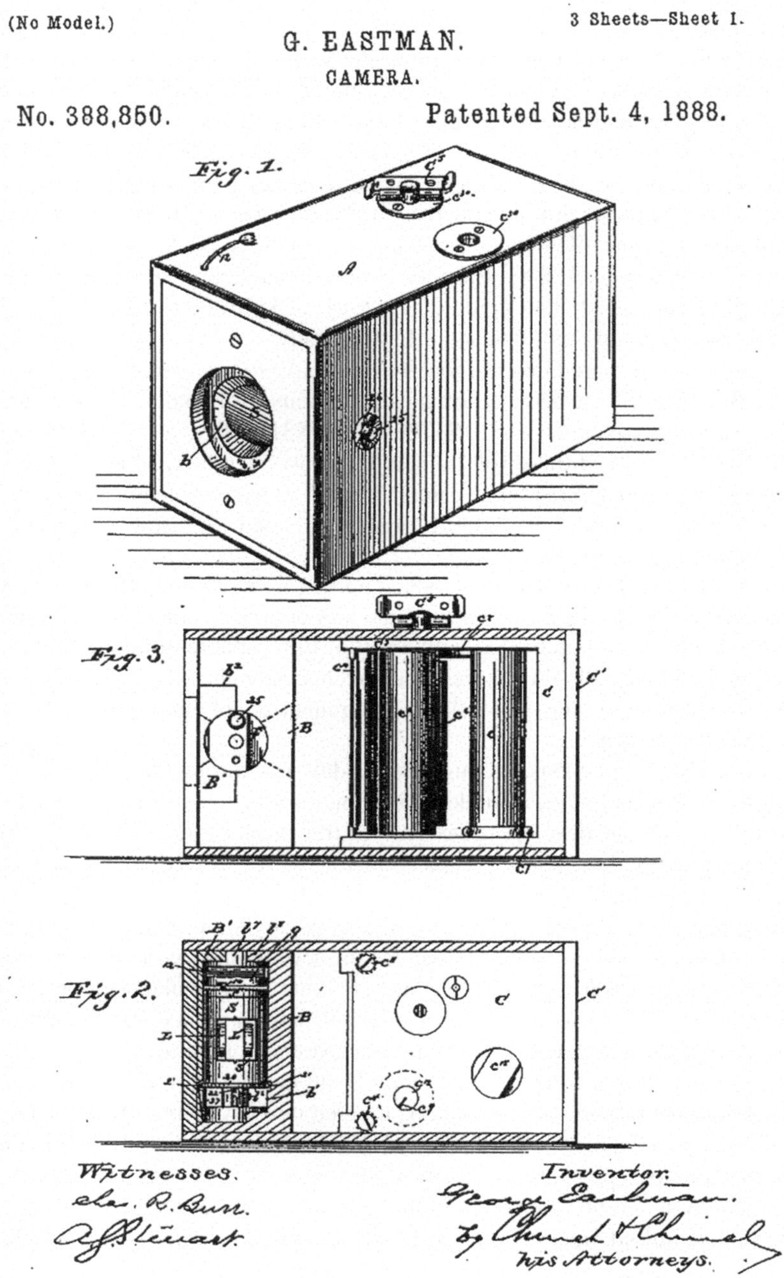

Fig. 12.2 Patent drawing of U.S. Patent No. 388850A

Fig. 12.3 Kodak No.1 camera sold starting in 1889. (Source: Bronger. (2012, July 29). *Kodak 1 camera* [Image]. Wikimedia Commons. https://commons.wikimedia.org/wiki/File:Kodak_nr_1.jpg. This image is in the public domain (CC0 1.0 Universal))

Fig. 12.4 Kodak camera advertisement from 1888

enough to roll like paper, lacked adequate transparency, and scratched more easily than glass.

By the 1880s, chemical advancements led to new formulations of celluloid to suit various industrial needs. Each time a new material hit the market, Eastman had his chemical engineer Henry Reichenbach study it. None met Eastman's standards. Finally, Reichenbach reversed his approach and began experimenting with raw celluloid ingredients. He discovered a formula yielding a transparent, flexible material ideally suited for photography. On April 9, 1889, Eastman's company filed a patent for the new formula.

However, the application process hit a snag when the examiner found a conflicting document and initiated an interference proceeding. The prior art came from an application submitted 2 years earlier by Hannibal Goodwin, a clergyman and amateur photographer. Though Goodwin had limited finances and no formal chemistry training, his passion and persistence led him to independently develop a formula almost identical to Reichenbach's. Unfortunately, Goodwin's attorney lacked expertise, drafting a long-winded, overbroad, and poorly written application. Goodwin himself didn't fully understand why his method worked and struggled to clarify the details, leading to repeated rejections. The examiner, despite frustration, remembered the application well. As soon as Eastman and Reichenbach's patent arrived, the examiner retrieved Goodwin's file.

In the end, Eastman and Reichenbach prevailed in the interference proceeding and received their patent on December 10, 1889. Eastman was confident in the market value of this patent, once writing, "Even if you gave me the telephone patent in exchange, I wouldn't trade it." At the time, Bell Telephone was thriving and its patents were famous, highlighting Eastman's faith in his technology.

While IBM faced little trouble with patents early on, Kodak encountered plenty of patent hurdles Kodak's explosive growth from its film business infuriated Goodwin. He spent his savings hiring a more capable patent attorney, continuously amending his application while diligently studying chemistry, gradually mastering the science behind his invention. In 1892, the patent office reopened the interference proceedings. This time, Goodwin achieved a partial victory: the office acknowledged that he had invented the formula first, but allowed Reichenbach's patent to remain valid. While Reichenbach's early film was smooth and transparent, it tended to crease. He improved it by adding camphor to the formula for strength. The patent office considered this camphor-enhanced version a legitimate improvement, while Goodwin's camphor-free formula was deemed the original invention. Eastman was unbothered—Goodwin's patent hadn't been granted yet, and even if it was, he figured he could simply buy it when needed.

Finally, on September 13, 1898, Goodwin's patent was granted (US610861A). By that time, Kodak had been selling film for nearly a decade, and Goodwin, now elderly, still harbored his American dream. In 1900, at age 78, he founded the Goodwin Film & Camera Company. Tragically, Goodwin died later that year in an accident, never witnessing his patent shake the foundations of the industry.

In just a decade, film had replaced glass plates as the preferred photographic medium. Moreover, the burgeoning movie industry had adopted film as its primary recording format, with Kodak claiming 80% of that market—a near-monopoly. At this point, Eastman was more concerned about patent risks from John Hyatt's Celluloid Manufacturing Company. As mentioned earlier, celluloid's recipe was Hyatt's patented innovation, and Reichenbach's formula closely resembled it, differing mainly in camphor content. To avoid infringing Hyatt's patents, Kodak reduced the camphor content in their formula. While this strategy dodged Hyatt's patent claims, it also made Kodak's film formula more similar to Goodwin's.

After Goodwin's death, his widow sold the Goodwin Film & Camera Company and the associated patent to Ansco, another early photography pioneer with far more resources than an independent inventor like Goodwin. Ansco demanded $1 million from Eastman, who refused. In response, Ansco, using Goodwin's company name, sued Kodak for patent infringement.

The lawsuit dragged on for over a decade—longer than Goodwin's original patent prosecution. Kodak submitted multiple prior art references to try invalidating Goodwin's patent, and their legal defense bitterly complained:

> *...7 years after the expiration of the Reichenbach patent 417,202, - 24 years after the Eastman Company began the manufacture of nitrocellulose film—and 26 years after the filing of the application on which the Goodwin patent issued. And this industry is enjoined on a patent the application for which was uniformly rejected by the five different Examiners who successively had it in charge during its eleven years' pendency in the Patent Office.*

To this, the judge remarked:

> *Truly an extraordinary and deplorable condition of affairs! But who was to blame for it—Goodwin or the five examiners who improperly deprived him of his rights during these eleven years?*[2]

In March 1914, the Second Circuit Court of Appeals ruled that Goodwin's patent was valid and that Kodak had infringed. Eastman was forced to settle with Goodwin's company and pay a substantial sum.[3] Though the exact amount remains unknown, Kodak clearly weathered the blow—it continued to dominate the film market. Before the Goodwin case concluded, the U.S. government had already launched an antitrust investigation against Kodak, officially suing the company in 1913. In a 1915 ruling from the Western District Court of New York, it was stated:

> ...In cameras, films, photographic plates, and photographic paper... Kodak held approximately 75% to 80% of the market share at the time this case began, effectively establishing a monopoly.[4]

This antitrust case primarily targeted Kodak's photographic paper business and its practices—acquiring competitors, signing exclusive supply contracts with foreign vendors, and exclusive distribution agreements with retailers. In 1921, Kodak agreed to a consent decree, ending the litigation. Under the terms, Kodak divested several acquired firms and agreed not to restrict retailers from selling rival products.

---

[2] Goodwin Film & Camera Co. v. Eastman Kodak Co., 213 F. 231 (1914). Retrieved March 31, 2024 from: https://cite.case.law/f/213/231/

[3] New York Times. (1914). EASTMAN CO. SETTLES CASE; Kodak Concern to Make Big Payment to Goodwin Company. Retrieved March 31, 2024 from New York Times Website: https://www.nytimes.com/1914/03/27/archives/eastman-co-settles-case-kodak-concern-to-make-big-payment-to.html

[4] United States v. Eastman Kodak Co., 226 F. 62 (1915) Retrieved March 31, 2024 from: https://cite.case.law/f/226/62/

Seventy-three years later, Kodak argued it no longer held market dominance and requested the court dissolve the decree.[5]

Following its antitrust settlement, IBM pivoted aggressively into computing, abandoning its punched card business and forging a new empire through innovations like the System/360 mainframe. Similarly, after signing its 1921 consent decree, Kodak invested heavily in color film R&D, opening a new market and maintaining its dominance in imaging.

Kodak's crown jewel in color film was the "Kodachrome" series. Known for its vivid and true-to-life colors, Kodachrome remained the medium of choice for professional photographers and enthusiasts throughout most of the twentieth century. Interestingly, the technology behind it wasn't developed by chemists, but by two musically gifted inventors: Leopold Mannes and Leopold Godowsky.

Mannes and Godowsky came from musical families and had been close friends since childhood. Disappointed by the poor visual quality of early color films, they resolved to develop a better color film process. Though both pursued professional careers in music—Mannes studying piano at Harvard and Godowsky majoring in violin at the University of California—they never abandoned their scientific curiosity, taking physics courses alongside their musical studies. After graduation, they juggled performance careers with experiments in photography and imaging.

In 1920, the duo filed a patent for "a method and apparatus for producing color cinematography" (US1619949A). From 1921 onward, they applied for several more patents under the title "color photography" (including US1538996A, US1997493A, US1516824A, US1980941A). Their core innovation involved layering multiple light-sensitive emulsions on film, each sensitive to a different spectral range. When exposed to light, each layer captured an image, which would then be developed and combined to reproduce true color. For instance, in patent US1516824A (see Fig. 12.5), two layers were used: the upper layer, sensitive to blue and green light, responded slowly; the lower layer, sensitive to red light, responded quickly. After development, the upper image would be bleached into a blue-green, while the lower image became orange-red—yielding vivid composite colors.

During their research, the two musicians connected with Kenneth Mees, head of Kodak's research labs, who took keen interest in their work.[6]

Though Mannes and Godowsky's method produced good results, it was notoriously complex. Additionally, the dyes used to differentiate color layers would bleed into adjacent layers—red-sensitive layers would turn green, and green-sensitive layers red. Kodak's labs, however, had just developed a new line of color-stable dyes that could solve this very problem.[7]

---

[5] United States v. Eastman Kodak Co., 853 F. Supp. 1454 (W.D.N.Y. 1994) Retrieved March 31, 2024 from: https://law.justia.com/cases/federal/district-courts/FSupp/853/1454/1517964/

[6] Daukantas, P. (2022). The Lingering Legacy of Kodachrome. *Optics and Photonics News*, *33*(6), 36–43. Retrieved March 31, 2024 from https://opnmedia.blob.core.windows.net/$web/opn/media/images/pdf/2022/0622/036-043_opn_06_22.pdf

[7] Barbara Flueckiger. (2012). Timeline of Historical Film Colors. Retrieved March 31, 2024 from Filmcolors.org: https://filmcolors.org/timeline-entry/1277/

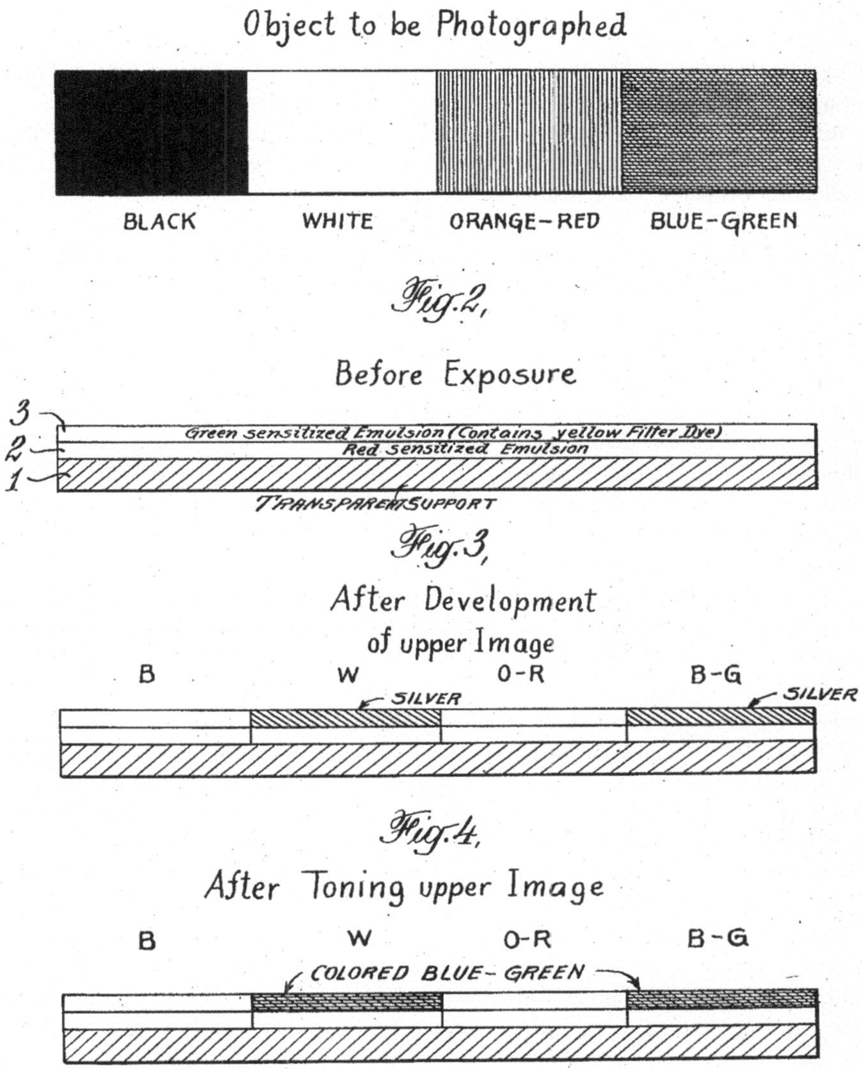

Fig. 12.5 Patent drawing of U.S. Patent No. 1516824A

In 1930, amidst the Great Depression, Kodak sought new growth avenues, identifying color film as a key strategic initiative. By then, the musician-inventors had burned through investor funds and still lacked a market-ready product. Mees invited them to Kodak's headquarters, cleared their $30,000 debt, paid them an annual salary of $7500, and provided them with laboratory space to continue their research, while Kodak engineers focused on production feasibility.

Once inside Kodak, the duo could no longer afford perfectionism; Mees pushed them to deliver a commercial product. Under his guidance, Kodak launched its 16 mm Kodachrome motion picture film in 1935, followed by the 35 mm still photography version a year later—ushering in a new era of imaging.

Interestingly, the two inventors retained ownership of their earlier patents when they joined Kodak, meaning that Kodachrome remained within the scope of their intellectual property. As a result, they received substantial annual royalties from Kodak, and their story became a cherished legend in the photographic world.

Kodachrome's development process remained complex even after improvements, making it unsuitable for amateur photographers to develop at home. Kodak incorporated the processing fee into the price of Kodachrome film—another example of the classic "razor-and-blade" business model.

In 1954, Kodak again faced an antitrust investigation. At the time, it controlled 90% of the U.S. color film and film processing market. Kodak quickly settled, signing a second consent decree agreeing to stop bundling processing services with color film and to license its processing technology to competitors. Despite these concessions, Kodak's market position remained unshaken. Forty years later, when Kodak petitioned the court to dissolve the decrees, it still held 67% of U.S. film sales by volume and 75% by revenue.

Just as IBM's PC team was forming, a Kodak engineer was quietly working on a small digital imaging project. Like the PC, it foreshadowed new modes of working and living—yet it ultimately brought Kodak to the brink of collapse.

This story is widely known: Kodak invented the digital camera, yet it was digital photography that nearly bankrupted the company. However, from a patent perspective, the story is more nuanced. First, it's important to clarify that Kodak did not invent the fundamental concept of the digital camera. In 1972, Texas Instruments filed a patent application for an "electronic photography system." After several amendments, a continuation of this application was granted in October 1976 as US4057830A, which is now considered the earliest digital camera patent (see Fig. 12.6 for its diagram).

This patent covered a system comprising an electronic camera, a storage device, a television receiver for displaying the images, and a playback mechanism to select and retrieve stored images. As illustrated in the figure, the electronic camera used a newly introduced CCD (charge-coupled device) image sensor to replace traditional film and was placed behind a mirror (106). Captured images were transmitted to an electromagnetic read-write head (110) and stored via a magnetic drum (102). Viewing the images required a conventional television set. The invention broadly encompassed key elements of digital photography, particularly the substitution of film with an image sensor and magnetic storage. However, Texas Instruments never commercialized the idea—not even building a prototype. Perhaps for that reason, neither the company nor the inventor is hailed as the "father of digital photography."

That title belongs instead to Kodak engineer Steven Sasson, who was inducted into the National Inventors Hall of Fame in 2011. In a 2007 blog post, Sasson

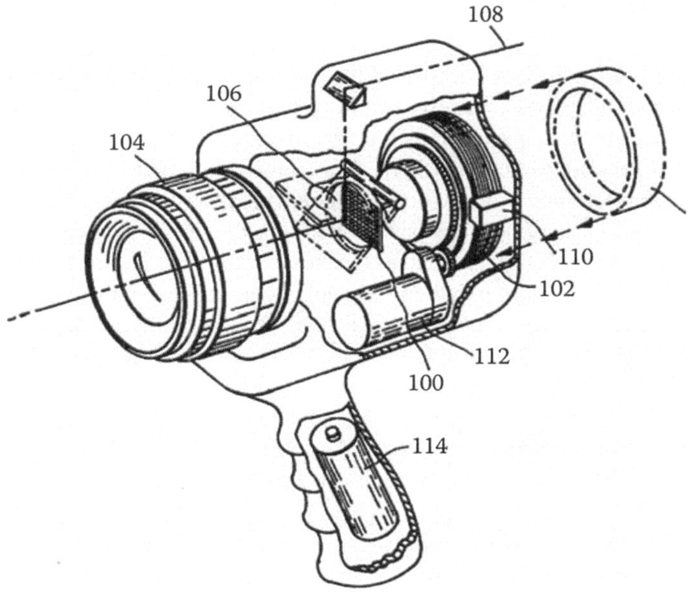

Fig. 12.6 Patent drawing of Texas Instruments U.S. Patent No. 4057830A

recalled the development of the first digital camera and the reactions of Kodak executives:

> "*In December of 1975, after a year of piecing together a bunch of new technology in a back lab at the Elmgrove Plant in Rochester, we were ready to try it. "It" being a rather odd-looking collection of digital circuits that we desperately tried to convince ourselves was a portable camera. It had a lens that we took from a used parts bin from the Super 8 movie camera production line downstairs from our little lab on the second floor in Bldg 4. On the side of our portable contraption, we shoehorned in a portable digital cassette instrumentation recorder. Add to that 16 nickel cadmium batteries, a highly temperamental new type of CCD imaging area array, an a/d converter implementation stolen from a digital voltmeter application, several dozen digital and analog circuits all wired together on approximately half a dozen circuit boards, and you have our interpretation of what a portable all electronic still camera might look like.*
>
> *...After taking a few pictures of the attendees at the meeting and displaying them on the TV set in the room, the questions started coming. Why would anyone ever want to view his or her pictures on a TV? How would you store these images? What does an electronic photo album look like? When would this type of approach be available to the consumer? Although we attempted to address the last question by applying Moore's law to our architecture (15 to 20 years to reach the consumer), we had no idea how to answer these or the many other challenges that were suggested by this approach. An internal report was written and a patent was granted on this concept in 1978 (US 4,131,919). I kept the prototype camera with me as I moved throughout the company over the last 30 years, mostly as a personal reminder*

> *of this most fun project. Outside of the patent, there was no public disclosure of our work until 2001".*[8]

Eventually, Sasson's project was allowed to continue by the head of Kodak's equipment division—though the executive candidly expressed that he hoped it would fail.[9]

Notably, Sasson was not listed as the first inventor on US4131919A; that honor went to his supervisor Gareth A. Lloyd. In a conversation that lasted less than 30 s, Lloyd had asked Sasson: "Can we build a camera using a solid-state imager?"[10] That simple question set Sasson on his path to building the first digital camera. Perhaps that's why Lloyd was listed as the lead inventor. But it was Sasson who carried the idea through experimentation and engineering—earning the rightful recognition as the father of digital photography. While flashes of inspiration are important, transformative inventions are built on tedious trial and error. The honor rightly went to the person who turned the idea into reality.

Sasson's US4131919A patent did not shy away from acknowledging Texas Instruments' earlier "electronic photography system" patent. In fact, Sasson explicitly discussed it in the background section of his patent, highlighting a key limitation:

> The disclosed camera has a drawback in that thermally generated minority carriers adversely affect the distribution of charge related to the incident light. From a one-second storage interval, a substantial amount of dark current can be expected, and these 'dark currents' may also be non-uniform.

To address this, Sasson proposed a solution:

> The total time for 'capturing' a single scene image and storing it in the buffer memory is approximately 75 milliseconds. The scene information then takes about 23 seconds to be transferred from the buffer to the recording device. In this way, the CCD dark current is maintained at a low level, because the CCD is not used for information storage, unlike in the previously disclosed British Patent 1,440,792 (i.e., Texas Instruments' US4057830A).

Thus, Sasson's digital camera took 23 s to capture and store a 10,000-pixel image on magnetic tape. A special television monitor was needed to view the image, and even that required another 23 s of data transfer.

Kodak shelved Sasson's prototype for many years, and it's hard to argue that it had any immediate industry impact. Throughout the late 1970s and 1980s, Kodak's contributions to the field went far beyond this single invention. Perhaps the most

---

[8] Steve Sasson. The First Digital Camera (via Kodak: Plugged In). Retrieved March 31, 2024 from ScanMyPhotos.: https://www.scanmyphotos.com/blog/2010/12/the-first-digital-camera-via-kodak-plugged-in.html

[9] Shamiyeh, Michael (2014). Driving Desired Futures (Turning Design Thinking into Real Innovation) || Disruptive Innovation at the Eastman Kodak Company (Part II)., 10.1515/9783038212843(), https://doi.org/10.1515/9783038212843.112

[10] The Irish Times. (2009). Portrait of inventor who saw the big picture - As a young Kodak employee, Steve Sasson invented the digital camera. Retrieved March 31, 2024 from The Irish Times: https://www.irishtimes.com/business/portrait-of-inventor-who-saw-the-big-picture-1.766671; Tampa Bay Times. (2005). The man who turned pictures into megapixels. Retrieved March 31, 2024 from Tampa Bay Times Website: https://www.tampabay.com/archive/2005/09/09/the-man-who-turned-pictures-into-megapixels/

foundational technology for digital cameras was the invention of the Bayer filter array.

In March 1975, Kodak engineer Bryce Bayer filed a patent application for a "Color Imaging Array," which was granted in July 1976 as US3971065A (see Fig. 12.7). Even today, this technology remains widely used in digital cameras and smartphones and is known as the "Bayer filter." The Bayer filter mimics the human eye's perception of color, using a pattern of 25% red, 50% green, and 25% blue (RGB) filters to direct light to the image sensor. This configuration allows a single sensor to capture color information.

Sasson once remarked in an interview that it was unimaginable to think of digital cameras before Bayer's invention. He explained that Bryce had been working on how to build a two-dimensional array of sensors capable of capturing color even though the sensors themselves could not distinguish it. This, he noted, became one of the foundations of digital imaging, and he regarded Bryce as one of his heroes.

While Bayer was solving the color capture challenge, Chinese-American physicist Ching W. Tang, who had earned a Ph.D. in physical chemistry from Cornell University, joined Kodak. In 1979, while retrieving a forgotten item from his lab, Tang noticed that an organic battery glowed faintly in the dark. He didn't

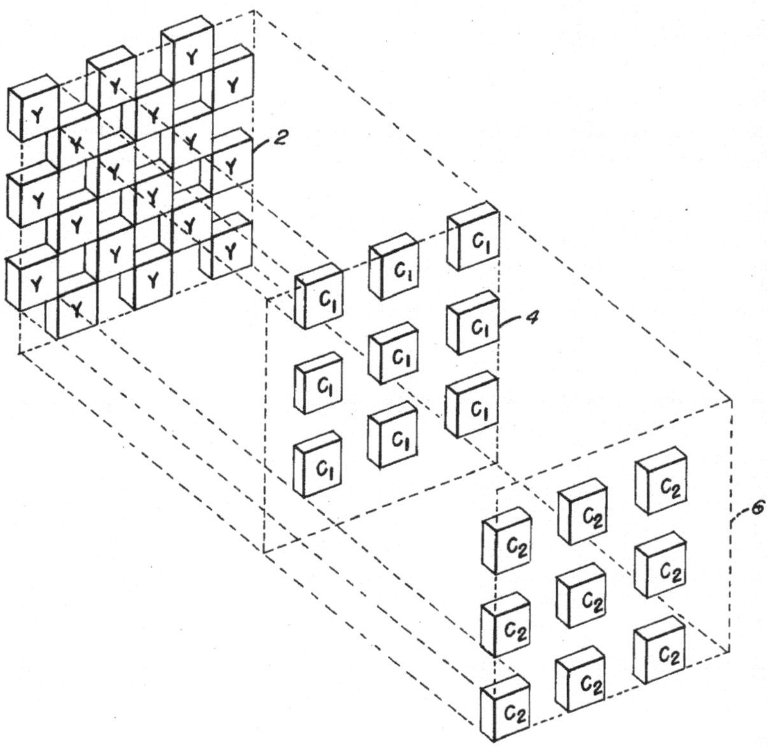

Fig. 12.7 Patent drawing of U.S. Patent No. 3971065A

dismiss the anomaly. After extensive research, he co-authored a landmark 1987 paper with Steven Van Slyke titled "Organic Electroluminescent Diodes," which introduced OLED (Organic Light-Emitting Diode) technology.

So when Kodak executives asked Sasson what an "electronic photo album" might look like, they could never have imagined that 40 years later, such an album would likely be a smartphone app displayed on an OLED screen—developed by a scientist working just down the hall from Sasson's lab.

Both CCD and Bayer filter technologies were already available, and magnetic storage kept improving—if Kodak didn't explore them deeply, other companies certainly did. In the years following Sasson's prototype, numerous companies began developing electronic cameras based on digital sensors (see Table 12.1).[11]

In August 1981, Sony unveiled an electronic camera called the "MAVICA," which caused a major stir in the industry. It was an SLR camera that could store analog video signals on magnetic disks and display images instantly on standard television screens. Sony aggressively promoted the product as a revolutionary innovation and heralded the dawn of a filmless era.

In hindsight, the MAVICA wasn't necessarily more practical than Steve Sasson's earlier invention. It relied on proprietary floppy disks known as "Mavipaks" and required a dedicated viewer, the Mavipak Viewer, to read them. Nevertheless, as one of the first cameras to function without film, MAVICA's release sent shockwaves through Kodak's leadership. Some employees were reportedly so distraught that they exclaimed, "*Oh, my goodness, photography is dead.*"[12]

Colby Chandler, Kodak's CEO at the time, remarked that a digital sensor with one million individual elements (i.e., pixels) would be required to produce an acceptable 3R print (approximately 5 × 3.5 inches or 12.7 × 8.9 cm). That pixel

Table 12.1 Electronic camera patents from the late 1970s to the early 1980s

Company	Patent No.	Title (English, official when available)	Year of Application
Polaroid	US4262301A	*Electronic Imaging Camera*	1978
Polaroid	US4541010A	*Electronic Imaging Camera*	1983
Nikon	US4420773A	*Electronic photographic camera*	1980
Nikon	US4456931A	*Electronic camera*	**1980**
Nikon	US4758883A	*Electronic picture camera with reduced memory capacity*	1982
Nikon	US4896226A	*Multi-motor electronic still camera having recording disk, focal plane shutter, and optical filters*	1982
Olympus	US4614966A	*Electronic still camera for generating long time exposure by adding results of multiple short time exposures*	1982

[11] Digitalkamera Museum. History of the Digital Camera and Digital Imaging. Retrieved March 31, 2024 from Digitalkamera Museum Website: https://www.digitalkameramuseum.de/en/history

[12] Gavetti, G., Henderson, R., & Giorgi, S. (2005). *Kodak and the digital revolution (A)*. Cambridge, MA: Harvard Business School Publishing.

count was four times greater than what sensors could achieve at the time—and still fell far short of the standard set by film. "Kodak's 35 mm Color Film II," he noted, "could deliver the equivalent of a 10-megapixel sensor." Chandler concluded that digital cameras were still a long way from being able to replace traditional film-based cameras.

During the early 1980s, Kodak was at the height of its power. In 1980, Kodak's revenue reached $9.7 billion—more than double Sony's $4.2 billion and nearly five times that of Fujifilm's $2 billion. That same year, Kodak invested $521 million in research and development, which increased to $615 million the following year.[13] A significant portion of this funding was funneled into the exploration of digital technologies.

After the release of MAVICA, Kodak established a Consumer Electronics Division and announced plans to enter the consumer electronics market by 1983. Not long after, the division launched the Kodavision 2000 series of 8 mm video cameras, which used Kodak-branded 8 mm video tapes. The cameras were manufactured by Panasonic, and the tapes were produced by TDK—both in Japan—putting Kodak in direct competition with Sony. However, the product failed commercially: shipment volumes were too small, and Panasonic's high manufacturing fees left little room for profit.

As previously discussed, Kodak's business model resembled the "razor and blades" strategy, where the camera—the "razor"—generated minimal profit compared to the film—the "blade." As such, Kodak had long outsourced its camera production and possessed limited experience and resources in camera design and manufacturing. Panasonic once proposed a partnership to help Kodak establish its own manufacturing capabilities in the U.S., but after weighing the costs, Kodak's management deemed the proposed $1 billion investment unjustifiable and ultimately decided to exit the consumer electronics business.[14]

However, Kodak did not completely abandon the development of electronic imaging technologies. Instead, it continued such work in secret and at a lower level of investment. Throughout the 1980s, Kodak was granted 271 U.S. patents related to CCD (charge-coupled device) technology—more than Olympus (247), Sony (152), Nikon (96), or Polaroid (47). Yet, in Japan, Olympus and Sony secured 710 and 524 CCD-related patents respectively, far outpacing Kodak. Canon, meanwhile, held 940 CCD-related patents in the U.S. and a staggering 2614 in Japan—demonstrating a level of commitment to new technology that surpassed all of its competitors.

In terms of CCD technology alone, Kodak remained a leader. It was the first to develop a megapixel-grade CCD sensor, marketed as the KAF 1400 or M1. In 1986, Kodak's subsidiary Videk used this CCD in the Videk Megaplus, a digital camera

---

[13] Vitton, J. J., Schultz, P. L., & Butz, N. T. (2014). Eastman Kodak: Facing disruptive technological change. *Journal of Critical Incidents*, *7*, 83–86.

[14] Gavetti, G., Henderson, R., & Giorgi, S. (2005). *Kodak and the digital revolution (A)*. Cambridge, MA: Harvard Business School Publishing.

priced between $10,000 and $40,000. It was the first digital camera capable of producing prints at 5 × 7 inches with sufficient resolution.[15]

In July of the same year, Canon released the RC-701 digital camera, which used a 380,000-pixel CCD made by Texas Instruments and sold for $2725. The model achieved moderate commercial success. Three months later, Kodak filed U.S. patent US4739409A, identifying a key flaw in the RC-701: its image sensor was susceptible to internal noise, such as "dark current"—a phenomenon where CCD "buckets" accumulate charge even in complete darkness. Kodak's patent proposed a solution known as "intelligent exposure control."

These facts demonstrate that Kodak never truly stopped developing digital technologies in the 1980s and kept a close eye on its competitors. However, many of Kodak's innovations remained in the lab and were never brought to market. Around 1990, after exploring various routes for commercializing digital imaging, Kodak believed it had finally discovered an electronic product that could best complement its traditional film business: the Photo CD.

From today's perspective, the Photo CD may seem like a simple storage medium for photographs, but at the time, it represented a significant innovation. Consumers could use their traditional film cameras, then have their negatives digitized and stored onto a CD at a photo lab. This allowed users to view and manage their images digitally—an appealing proposition before digital cameras became mainstream. The image quality stored on these discs surpassed that of all other digital photography products available at the time, which pleased Kodak's management, who had always prioritized image quality.

Kodak had high hopes for the Photo CD, envisioning it as a ubiquitous household product, much like its film, and as a cornerstone for establishing industry standards in electronic image display, transmission, and printing. Management believed the product combined the best of both analog and digital worlds and would extend the life of film in the upcoming digital age.

To realize this grand vision, Kodak attempted to collaborate with Microsoft to integrate Photo CD into early versions of the Windows operating system. The plan failed for unknown reasons, but a now-infamous meeting between Kodak's CEO Kay Whitmore and Bill Gates gained notoriety: Whitmore reportedly fell asleep in front of Gates during a high-level discussion. Due to this embarrassing incident, both CEO Magazine[16] and Yahoo Finance[17] later named Whitmore among the worst CEOs in history.

The Photo CD ultimately failed to meet expectations. At the time, CD drives were not yet widespread, and Kodak's CD reader device cost $400. Additionally,

---

[15] https://www.digitalkameramuseum.de/en/cameras/item/videk-megaplus

[16] Emma Woollacott. (2018). Lessons from history's worst CEOs. Retrieved April 8, 2024 from https://www.theceomagazine.com/business/management-leadership/lessons-from-historys-worst-ceos/

[17] Daily Ticker. (2014). The Seven Best Tips from the Seven Worst CEOs of All-time. Retrieved April 8, 2024 from https://finance.yahoo.com/blogs/daily-ticker/the-seven-best-tips-from-the-seven-worst-ceos-of-all-time-155124407.html

converting a single roll of film to CD cost $20—an unaffordable expense for average consumers. Within a year of its release, the Photo CD had become a niche tool for professional photographers, never achieving the mass-market penetration Kodak had hoped for.

Despite Kodak's strong R&D capabilities and historically healthy profit margins throughout the 1980s and 1990s, a combination of a devastating patent lawsuit and repeated strategic missteps plunged the company into crisis, leaving it ill-equipped to handle the challenges posed by digital photography and storage.

As mentioned earlier, while George Eastman was a patent novice, Henry Reichenbach, Kodak's co-founder, was a patent expert. This distinction may explain why IBM often fared better in patent litigation, while Kodak encountered repeated setbacks. One such setback was the Goodwin case, which, although troublesome, did not inflict lasting damage. However, Kodak's real downfall began with the disastrous patent infringement lawsuit filed by Polaroid over instant camera technology.

In 1947, American inventor Edwin Land unveiled the Polaroid instant camera, capable of developing a photograph within 1 min of being taken—an innovation that caused a sensation. The product launched in 1948 and quickly generated $5 million in sales, creating an entirely new market. Figures 12.8 and 12.9 (referenced in the original) illustrate the patent diagram for Land's US2520641A and the Model 95 instant camera released by Polaroid in 1948.

For several years, Kodak and Polaroid maintained a cooperative relationship, with Kodak supplying film to Polaroid. But as instant cameras evolved and gained popularity, Kodak decided to enter the market itself. In 1976, Kodak released its own instant camera and film system, putting it in direct competition with Polaroid.

That same year, in April, Polaroid filed a lawsuit in the District Court of Massachusetts, accusing Kodak of infringing ten of its patents. In 1986, after a decade of litigation, the court ruled in favor of Polaroid and issued a permanent injunction against Kodak. The verdict forced Kodak to exit the instant photography business, lay off 800 workers, and shutter over $200 million worth of facilities and equipment. Kodak also had to cease providing film and supplies for the 16 million instant cameras it had already sold, and offer compensation to affected consumers. The final blow came in the form of a staggering $900 million in damages—a historic penalty.

Records show that Kodak had made considerable efforts to avoid infringing on Polaroid's patents. By the standards of the 1970s, Kodak's actions might not have clearly constituted infringement. However, Kodak was up against a formidable opponent. Edwin Land, Steve Jobs's idol, shared Jobs's perfectionist and artistic spirit and was relentless in defending his inventions. He had an unusually strong awareness of patent law, far surpassing the typical entrepreneur or inventor.

At age 19, Land befriended patent attorney Donald Brown and filed his first patent. Over his lifetime, Land secured more than 500 patents. Brown later joined Polaroid's board, highlighting how seriously the company regarded intellectual property.

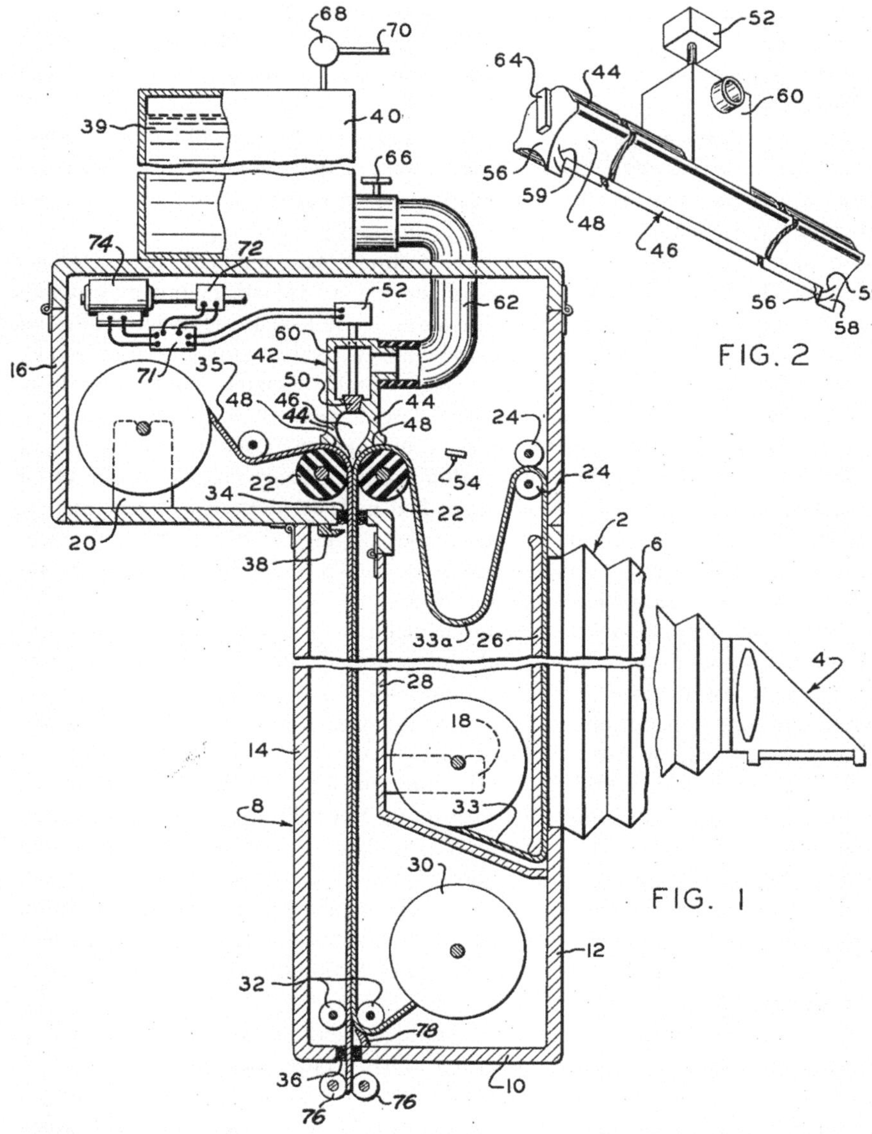

Fig. 12.8 Patent drawing of U.S. Patent No. 2520641A

Polaroid's patent attorney, Stanley Mervis, described Land as being deeply involved in patent filings. Land would scrutinize the scope and wording of claims before signing off, often demanding rewrites for insufficient originality. He even monitored the full course of patent prosecutions. Mervis recounted how Land, after losing a lawsuit over a rejected patent, confronted the judge at an event and

Fig. 12.9 Polaroid Land Camera Model 95 (1948). (Source: OppidumNissenae. (n.d.). *Polaroid Land Camera Model 95 (1948)* [Image]. Wikimedia Commons. https://commons.wikimedia.org/wiki/File:Coll._Marc%C3%A8_CL_-_Polaroid_land_camera_Mod_95_1948.jpg. This image is shared under the Creative Commons CC BY-SA 4.0 license)

passionately argued for his invention's originality, demonstrating Land's near-obsessive commitment to patents.[18]

Land's passion for patents might have seemed ill-timed—he came of age during a trough in the U.S. patent system. When Polaroid initiated its lawsuit in 1976, the judicial climate was still largely unfavorable to patent holders. There was little indication that the U.S. would soon swing back toward strong patent enforcement. But by the fourth year of the litigation, the tide had turned with the establishment of the Court of Appeals for the Federal Circuit. Land's unyielding defense of his patents finally paid off. Though Kodak was not the most heavily penalized company in those early Federal Circuit decisions, for its intellectual property team, the defeat felt no less punishing.

Polaroid had filed many patents as early as the 1940s, mostly related to black-and-white film. By the 1960s, the company began to submit a flood of patents for color instant film technology. Kodak believed many of these newer patents were mere extensions of earlier black-and-white technologies and lacked non-obviousness.[19] As evidence, Kodak's legal team challenged Polaroid's US3362821A patent with an earlier patent, US2584030A, both filed by Edwin Land. Both patents dealt with acid-reactive agents used in the photosensitive layer. While the latter applied to black-and-white photography, the former was for color. Kodak argued

[18] Mervis, S. H. (1994). Edwin Land—Champion of Patents. *Optics and Photonics News*, *5*(10), 50–53.

[19] Chisum, Donald and Mueller, Janice M., Book Review: Ronald K. Fierstein's A Triumph of Genius: Edwin Land, Polaroid, and the Kodak Patent War (American Bar Ass'n 2015) (June 26, 2015). Retrieved April 8, 2024 from https://ssrn.com/abstract=2623980

that the newer patent merely swapped "black-and-white" with "color," but the judge disagreed.

The court found that Kodak's interpretation ignored key differences in the development processes between sepia-toned and color images. Color diffusion transfer, unlike silver transfer used for sepia images, required longer processing times and the complete removal of alkali metal ions from the image environment. These unique technical demands were addressed in the newer patent but not in the earlier one.[20]

What's more, Kodak's legal team had been preparing for this fight for years. Seven years before entering the instant camera market, Kodak had already begun analyzing Polaroid's patents, ultimately commissioning 67 legal opinions on non-infringement or patent stability. But in the eyes of 1980s judges, these documents amounted to little.

Still, those legal opinions did have one use: when it came time to assess damages, the central question was whether Kodak's infringement had been willful. After reviewing Kodak's extensive legal preparations, the Massachusetts District Court concluded that Kodak had acted in good faith. This spared Kodak from treble damages, which could have pushed the payout far beyond $900 million.

Beyond the crushing lawsuit, Kodak was also hampered by strategic missteps. The company was searching for growth outside of its traditional film business. Though hesitant in electronics, Kodak, as a long-established chemical company, took a keen interest in pharmaceuticals. It had developed several promising chemical compounds for treating cancer and cardiovascular diseases, but lacked experience in drug registration and distribution.

In an ambitious move, Kodak borrowed heavily to acquire Sterling Drug, Inc., with dreams of becoming one of the world's top 20 pharmaceutical companies by the year 2000.[21] But this strategy marked a major turning point: for the first time, Kodak began relying on long-term debt to fund its operations, a shift that would prove disastrous.

By 1990, Kodak was burdened with $900 million in interest payments, which significantly eroded its profit margins. The same year, the Polaroid lawsuit concluded, and Kodak was ordered to pay an additional $900 million in damages, deepening the company's financial crisis.[22]

That year also marked the appointment of Kay Whitmore as Kodak's CEO. Faced with the twin burdens of debt from the Sterling acquisition and the Polaroid judgment, Whitmore decided to abandon his predecessor's diversification strategy. Following a path similar to IBM's Lou Gerstner 3 years later, he implemented

---

[20] Polaroid Corp. v. Eastman Kodak Co., 641 F. Supp. 828 (D. Mass. 1986) https://law.justia.com/cases/federal/district-courts/FSupp/641/828/1482978/

[21] Ansberry, C. 1988. Kodak Agrees to Acquire Sterling Drug — Offer of $5.1 Billion Ends Roche Bid, Would Give Buyer Major Drug Role. *Wall Street Journal*, 1988 Jan 25: 1

[22] Shamiyeh, Michael (2014). Driving Desired Futures (Turning Design Thinking into Real Innovation) || Disruptive Innovation at the Eastman Kodak Company (Part II)., 10.1515/9783038212843(), –. https://doi.org/10.1515/9783038212843.112

sweeping cost-cutting measures, including major staff reductions and R&D cutbacks. Whitmore also began divesting many of Kodak's non-core businesses to refocus on film and photographic chemicals.

Over the next few years, Kodak sold off 20 business units and their associated patents—many of them related to consumer digital technologies. These included the inkjet printing division, floppy disk maker Verbatim (sold to Panasonic), LCD maker Sayett Technology, electronics company Estek, Aquidneck Data Corporation, and Ultra-life Battery Company. In 1993, Whitmore targeted one of Kodak's oldest and most profitable divisions: Eastman Chemical was spun off as an independent company. The following year, Kodak sold Sterling Drug at half its acquisition price, officially ending its failed foray into pharmaceuticals.

In hindsight, Whitmore's divestment strategy appears shortsighted, depriving Kodak of potential profit sources and key opportunities in digital innovation. By 1997, companies like HP, Epson, and Canon were introducing low-cost home photo printers that quickly gained traction. These devices, sold under the classic razor-and-blade model—cheap printers, expensive ink—drew Kodak's attention. But Kodak's own inkjet printing division had been sold off 5 years earlier. Though Kodak later partnered with Epson to release home inkjet printers in 1999, they failed to gain market traction. By the time Kodak re-entered the space in earnest in 2003, HP had already established an unshakable market lead, and Kodak's late push floundered again.

Ironically, the spun-off Eastman Chemical thrived and remains a leading firm in the chemical industry—outlasting the original Kodak, which would later file for bankruptcy. The chemical offshoot now bears the founder's name as its sole legacy.

Meanwhile, throughout the 1980s and 1990s, Fuji Film emerged as Kodak's most formidable competitor in the traditional film market. Despite ongoing trade tensions between the U.S. and Japan, particularly in semiconductors, Kodak's core business—photographic film—didn't benefit from American political protection.

In 1984, Fuji introduced a 400-speed color film that was 20% cheaper than Kodak's comparable product. Initially, Kodak's executives refused to believe that American consumers would choose a foreign brand for something as iconic as film.[23] They dismissed Fuji's threat.

That same year, Peter Ueberroth, chairman of the U.S. Olympic Committee, revolutionized Olympic sponsorship by selling exclusive rights, turning the Los Angeles Olympics into the first profitable games in history. He approached Kodak, offering exclusive film sponsorship for $4 million. Kodak, however, only offered $1 million, and negotiations fell through. Fuji seized the opportunity, spending $7 million for exclusive sponsorship and significantly boosting its profile in the U.S. market.

This moment marked the beginning of a multi-decade global battle between Kodak and Fuji. Outside of their home markets, the two companies fought fiercely

[23] Michael Hiltzik. (2011). Kodak's Long Fade to Black. *Los Angeles Times*. 2011. December 11

for market share, with each holding about 70% in their respective domestic territories.[24]

Despite Fuji's aggressive expansion, Kodak had few patents it could use to fend off the competition. On May 18, 1995, Kodak filed a complaint with the Office of the U.S. Trade Representative, urging a Section 301 investigation against what it claimed were anti-competitive practices by the Japanese government favoring Fuji. Kodak CEO George Fisher argued that Fuji was using profits from its protected domestic market to fund below-cost pricing overseas, thereby undermining fair competition.

In response, Fuji issued a 588-page white paper titled "Rewriting History: Kodak's Revisionist View of the Japanese Consumer Photographic Market." It asserted that Kodak's weak performance in Japan was due more to poor management than to unfair trade. In fact, Kodak's Japanese subsidiary earned $1.2 billion in sales in 1994, making it the 43rd largest foreign firm in Japan at the time.

Ironically, Kodak's own CEO Kay Whitmore had publicly stated in 1990: "I don't believe there are many barriers to Kodak doing business in Japan. If there are problems, they must be due to Kodak's own lack of effort."

On January 30, 1998, the World Trade Organization's Dispute Settlement Body dismissed Kodak's claims. Many observers later concluded that technology—not trade policy—had already changed the playing field. As one analysis noted, Kodak and Fuji fought for 8 years over film and photo paper just as digital imaging was rapidly making those products obsolete. The conclusion: "Technology, not politics, proved to be the catalyst for change."[25]

By 2011, shortly before it filed for bankruptcy protection, Kodak's annual revenue had dwindled to $6 billion with a net loss of $760 million. In stark contrast, Fuji's revenue had grown to $27.4 billion, yielding a profit of $140 million. Both companies had once thrived on film, but Kodak, the larger and more dominant of the two, was the first to fall.

Kodak's management was frequently criticized for lacking vision, especially in failing to capitalize on the digital revolution. While this judgment may seem harsh, internal evidence shows that Kodak's leadership had been optimistic about digital technologies and had indeed tried to pivot. However, they consistently underestimated the speed and scale of the digital transformation. Rick Braddock, a Kodak board member since 1987, insisted that the company had long prepared for change—just not fast enough. Kodak's leadership believed that professionals would be the first to adopt digital photography, while mainstream consumers would continue to prefer printed photos for much longer.[26]

[24] Finnerty, T. C. (2000). Kodak vs. Fuji: The battle for global market share. Retrieved April 8, 2024 from https://digitalcommons.pace.edu/cgi/viewcontent.cgi?article=1018&context=business_cases

[25] Graham, E. M., & Richardson, J. D. (1997). *Snapshot: Kodak v. Fuji*. In J. J. Schott (Ed.), *Case studies in international trade policy* (pp. 153–167). Washington, DC: Institute for International Economics.

[26] Lucas, H. C. (2012). The search for survival: Lessons from disruptive technologies. ABC-CLIO

In 1993, George Fisher, the former CEO of Motorola, became Kodak's first externally appointed chief executive in its 117-year history. Known as the "digital man," Fisher had successfully guided Motorola into the digital era, and Kodak's board hoped he could lead a similar transition. He adopted a digital-first strategy that remained Kodak's core focus long after his tenure. However, Fisher struggled to gain the support of middle management, and his efforts to transform the company ultimately failed.

During Fisher's leadership, Kodak did release a series of digital cameras and sought to recreate its razor-and-blade model in the digital age. In 1996, it joined forces with Fuji, Canon, Nikon, and Minolta to launch the Advanced Photo System (APS), a type of film that could digitally record information. APS offered the flexibility to print photos or transfer them into digital form via special equipment. The capital markets responded enthusiastically. But in the looming digital age, any film-based solution—even hybrid ones—was destined to become obsolete, appealing only to niche or artistic users.

Like IBM, Kodak also revamped its executive compensation system in the 1990s. Under Fisher, bonuses were tied 50% to shareholder satisfaction, incorporating metrics like revenue growth, cash flow, and return on equity. Stock options became a key incentive. As a result, Wall Street's reactions played a growing role in shaping Kodak's strategies.

In the 1990s, even as Kodak's board and its "digital man" CEO Fisher pursued a transition to digital, the capital markets emerged as a major obstacle. A *Financial Times* commentary observed that beyond technical challenges, "Wall Street put the brakes on Kodak's digital transformation."[27]

Wharton professor Mary Benner conducted an illuminating study on how capital markets responded to disruptive technologies.[28] Her research showed that analysts tended to reward companies for incremental innovations to existing technologies, while penalizing them for pursuing strategies based on entirely new technologies. In the case of digital photography, Benner reviewed all Kodak-related English-language media reports from 1990 to 1996 in the LexisNexis database. Kodak's digital products garnered an average of 29 articles each, and Apple's rebranded QuickTake digital camera drew hundreds. Articles mentioning "Kodak" and "digital camera" totaled over 1400—clearly, there was significant media and industry attention to the technology.

But Wall Street told a different story. Among analyst reports, Kodak's 13 digital products were barely mentioned. In contrast, hybrid products that combined film with digital features received far more interest: Photo CD was cited 38 times, while APS (Advanced Photo System) was mentioned 144 times.

Morgan Stanley's 1991 report expressed optimism about Photo CD's long-term prospects. That same year, Salomon Smith Barney praised Kodak and other film

[27] Hill, A. (2012). Snapshot of a humbled giant. *Financial Times, Apr, 2*

[28] Benner MJ (2010) Securities analysts and incumbent response to radical technological change: evidence from digital photography and internet telephony. Organ Sci 21:42–62. https://doi.org/10.1287/orsc.1080.0395

makers for developing hybrid systems that merged chemical and digital imaging, calling Photo CD a prime example and forecasting it would "greatly extend the life of 35 mm film." By 1996, Salomon analysts hailed APS as "the most important development in photography in 20 years," while Credit Suisse First Boston predicted that APS would trigger an unprecedented consumer boom in the developed world's photography market.

These analyst opinions strongly correlated with Kodak's stock performance: whenever Kodak emphasized hybrid technologies, its stock price rose. In contrast, some firms, like Prudential Securities, were openly hostile to Kodak's digital efforts. A 1994 report dismissed digital imaging as offering "trivial (and remote) revenue potential," predicting shareholder backlash once it became clear how much money had been "wasted on 'digital nonsense.'" In 1995, Prudential declared, "Kodak is not a digital imaging player… It won't find much luck turning to digital for growth." Even in 1996, it stated, "Long-term, Kodak should focus on leveraging its strong cash flow through stock buybacks, launching APS, and cutting costs. We're not excited about its digital ventures."

Meanwhile, Morgan Stanley and Salomon, though less critical of digital strategy, consistently applauded Kodak whenever it reduced investment in digital.

After 1996, the trend toward digital cameras became increasingly pronounced, but Wall Street remained enamored with film-based hybrid products incorporating digital elements. Between 1997 and 2001, analyst reports mentioned APS 1917 times, whereas Kodak's entire portfolio of 32 digital cameras received just 156 mentions. Some investment banks like Salomon Smith Barney began to change their stance, calling Kodak's digital transition the "right direction," but skepticism lingered. For example, Morgan Stanley's 1999 report stated: "Digital imaging is eating into Kodak's high-margin film business and will continue to do so… We don't believe profits from digital imaging will ever compare with those from Kodak's film products."

Kodak's longtime rival, Polaroid, was also struggling through its own painful transition—and received similar treatment from capital markets. Whenever a film-digital hybrid was launched, analyst praise and share price increases followed. But purely digital products, throughout the 1990s, rarely earned Wall Street's favor.

Even Kodak's own annual reports hesitated to fully embrace digital. It wasn't until 1997 that Kodak first acknowledged in its annual filing that it had "developed (consumer) digital camera systems that do not use silver halide film."

Aggressive capital expectations often strip companies of their long-term vision. Under the weight of shareholder demands, both the "Yellow Giant" (Kodak) and the "Blue Giant" (IBM) found themselves dwarfed. These stories stand as cautionary tales.

Regardless of market skepticism, public enthusiasm for compact, convenient digital cameras grew rapidly. During Kodak's indecisive transition, the company finally faced a decisive consumer backlash. In a 2016 paper,[29] former Kodak executive Willy Shih noted that Kodak entered a phase of "scaling down." In growth periods, companies invest to increase scale and improve efficiency. This "scaling up" reduces unit costs and enhances capital returns. Scaling down, by contrast, is painful—shrinking demand for traditional cameras and film forced Kodak to reduce production, which in turn raised costs, further accelerating the decline in a vicious downward spiral.

Kodak had to amputate its traditional product lines to survive. But with every product line eliminated, it pushed remaining film users closer to digital—especially professional photographers who had clung to film. For a time, Kodak absorbed excess capacity by producing motion picture film. But when Hollywood fully embraced digital technology, Kodak lost that fallback as well.

At the same time, Kodak's retail distribution network also faced the challenges of downsizing. With fewer products, it became increasingly difficult to maintain visibility on store shelves. Film counters that once occupied prominent positions in retail stores and supermarkets were now disappearing altogether. Willy Shih noted that Kodak's executives were well aware of these issues, but had long remained silent—an attitude that outsiders often misinterpreted as ignorance of the coming disruption.

Contrary to popular belief, Kodak wasn't completely abandoned by the market during the digital camera era. In 2001, Kodak ranked second in U.S. digital camera sales, just behind Sony. However, this market position came at a steep cost: Kodak lost $60 on each camera sold.[30] In fact, until 2010, Kodak remained among the top four digital camera brands, though the entire market was already crumbling under the impact of smartphones.

During this time, Kodak carefully scaled back its expenses and downsized operations to cope with market shrinkage. Between 2004 and 2007, the company spent $3.4 billion on restructuring—only to be hit by the global financial crisis in 2008. With generous employee benefits and high pension liabilities, the post-crisis low interest rates dealt Kodak a devastating blow.

Desperate for a new growth engine, Kodak turned toward patent monetization.

In 2001, Kodak sued Sanyo, alleging that its digital cameras infringed Kodak's patents. After a brief legal skirmish, the two companies settled, with Sanyo agreeing to pay licensing fees. Soon after, Kodak struck a similar agreement with Olympus and began negotiations with Sony. After 3 years, talks with Sony collapsed, and on March 9, 2004, Kodak filed a lawsuit against the market leader in U.S. digital cameras.

In the amended complaint submitted on July 16, 2004, Kodak listed 11 patents—such as US5016107A, US5164831A, and US5493335A—most covering fundamental digital imaging technologies like image compression and storage. Leading

---

[29] Shih, W. (2016). The real lessons from Kodak's decline. *MIT Sloan Management Review*, 57(4), 11–13.

[30] MICHAEL HILTZIK. (2011). Kodak's long fade to black. Retrieved April 8, 2024 from https://www.latimes.com/business/la-xpm-2011-dec-04-la-fi-hiltzik-20111204-story.html

the list was Steve Sasson's 1989 patent for an "electronic still camera using image compression and digital storage" (US5016707A).

This lawsuit attracted significant attention. The *New York Times* noted that Kodak, once mostly a defendant in patent suits, had now become the plaintiff.[31] Bloomberg later remarked that Kodak's actions resembled Texas Instruments' 1980s strategy—using patent litigation as a lifeline in the face of bankruptcy.[32]

Sony struck back swiftly. Just weeks later, Sony filed a patent infringement lawsuit against Kodak in the U.S. District Court for New Jersey. While many details of this patent war remain undisclosed, public records show that much of the 2006 litigation centered on discovery rather than infringement or validity. Ultimately, in 2007, Kodak and Sony reached a settlement. They signed a cross-licensing agreement, which also extended to Sony Ericsson's mobile phone division. Under the deal, Sony agreed to pay Kodak licensing fees—marking a rare defeat for Sony in the courtroom.

Buoyed by this success, Kodak grew more ambitious. In 2008, after failing to negotiate licensing deals, Kodak sued Samsung and LG in federal court. Simultaneously, it filed a Section 337 complaint with the U.S. International Trade Commission (ITC), alleging that dozens of Samsung and LG mobile phones infringed two key Kodak patents: US5493335A and US6292218B1. The '335 patent, filed in 1993, described a single-sensor digital camera with selectable image resolution—aimed at reducing image noise in low-resolution modes. The '218 patent, filed in 1997, allowed users to capture still images while previewing motion video, effectively enabling preview modes in digital cameras.

Just before the ITC's initial ruling, LG capitulated on December 16, 2009, and reached a settlement with Kodak. The next day, the ITC administrative law judge ruled that the patents were valid and that Samsung had indeed infringed. In January 2010, Samsung also settled, agreeing to a cross-licensing deal and paying Kodak a licensing fee. Media reports[33] indicated that Kodak's largest revenue source that year was $838 million in patent licensing fees—most of which likely came from the Samsung and LG settlements.

Emboldened by its victories in Asia, Kodak set its sights on domestic tech giants. In January 2010, Kodak filed a Section 337 complaint against Apple and Research In Motion (RIM), the makers of BlackBerry, hoping to extract a $1 billion settlement.[34] The move drew sharp commentary from financial media, with Fool.com

---

[31] Claudia Deutsch. (2004). Kodak Accuses Sony of Patent Violations on Digital Imaging. New York Times. March 10, 2004. https://www.nytimes.com/2004/03/10/business/kodak-accuses-sony-of-patent-violations-on-digital-imaging.html

[32] Susan Decker. (2012). Kodak in Crisis Mines Patents for Cash Copying Texas Instruments. Retrieved April 8, 2024 from https://www.bloomberg.com/news/articles/2012-01-13/kodak-mines-patents-for-cash-copying-texas-instruments

[33] MICHAEL HILTZIK. (2011). Kodak's long fade to black. Retrieved April 8, 2024 from https://www.latimes.com/business/la-xpm-2011-dec-04-la-fi-hiltzik-20111204-story.html

[34] Susan Decker. (2012). Kodak in Crisis Mines Patents for Cash Copying Texas Instruments. Retrieved April 8, 2024 from https://www.bloomberg.com/news/articles/2012-01-13/kodak-mines-patents-for-cash-copying-texas-instruments

remarking, "it looks like **Eastman Kodak** (NYSE:EK) sees its future as being little more than a patent troll."[35]

This lawsuit had an air of tragedy to it. Kodak, struggling in the shrinking digital camera market, now found itself targeting the very companies—particularly Apple—whose smartphones had accelerated that decline. In the year before Apple and Samsung began their own smartphone wars, Kodak, once a titan of photography, stepped into the ring to challenge the new kings, adding a melancholic edge to the unfolding patent drama.

In this lawsuit, Kodak relied solely on US Patent No. 6292218B1 (the "218 patent"). Against RIM, Kodak asserted claims 15, 23, 24, 25, 26, and 27; against Apple, claims 15, 23, 25, 26, and 27. But in the lead-up to trial, Kodak voluntarily dropped all claims except for claim 15.

This was puzzling. IBM's legal strategy had always been to bring five patents to court—hoping that even if one stuck, the others could serve as legal padding. Kodak, however, put all its eggs in one basket. Whether this stemmed from overconfidence, dwindling financial resources, or simply a calculated bet, the result was risky. The '218 patent focused on capturing a low-quality preview image while recording video—a concept now embedded in nearly every smartphone.

Kodak's case may have required extensive discovery into Apple's internal systems, which would have meant sky-high legal fees and expert witness costs. That financial burden might explain why Kodak chose to pursue just one claim—the one it believed had the strongest chance.

On January 24, 2011, the ITC administrative law judge issued an initial determination: the accused products did not infringe, and claim 15 of the '218 patent was invalid. This was a devastating blow. In March, the ITC Commission decided to review the ruling, and on June 30, it remanded the case to the administrative judge after redefining two key terms—"still processor" and "motion processor." For Kodak, it was a flicker of hope.

But the reprieve was short-lived. On January 19, 2012, Kodak filed for bankruptcy protection, rendering the lawsuit's outcome effectively moot. Nonetheless, in May 2012, the ITC judge again found claim 15 invalid. The ITC Commission affirmed this final ruling on August 9, sealing Kodak's loss.

A 2014 report by the Intellectual Property Strategy Research Task Force of China's High-Tech Industry Development Promotion Association analyzed the Kodak bankruptcy case. Their paper, titled *"Kodak's Experience Suggests China's Patent Giants Should Moonlight as Patent Trolls Sooner Rather Than Later,"*[36] proposed a controversial thesis: that Kodak had failed because it didn't aggressively monetize its patents early enough. The report argued that stable, recurring patent

[35] Rich Duprey. (2017). The Worst Stocks for 2010: Eastman Kodak. Retrieved April 8, 2024 from https://www.fool.com/investing/general/2010/01/21/the-worst-stocks-for-2010-eastman-kodak.aspx

[36] 中国高技术产业发展促进会知识产权战略研究课题组. (2014). 柯达公司的经验说明,我国专利巨头应当尽早"兼职"当"专利楚奥". 科技促进发展(4), 75–77.

income could significantly improve a company's financial health. It praised companies like IBM, Philips, Qualcomm, Microsoft, and Oracle for generating substantial revenue by "moonlighting" as patent enforcers and claimed Kodak's downfall stemmed from nurturing patents that never turned a profit.

We must firmly reject this view.

Kodak failed not because it didn't transform into a "patent troll," but because it couldn't execute a self-renewing transformation—because it failed to produce the kind of products that consumers of a new era wanted. It became shackled by the successes of a bygone era, trapped in what Willy Shih—former Kodak executive—described as a "death spiral."

For further discussion on corporate self-renewal, please refer to the next chapter Disruption and Revolution.

Part II
Disruption and Revolution

Chapter 13
Introduction

Revolutions imposed from outside reduce their subjects to history; those born within make them legend.

The Author

Many have heard the oft-repeated tale: in 1803, American engineer Robert Fulton approached Napoleon Bonaparte to present his invention of the steamship and his forward-looking proposal to build a fleet powered by steam. Napoleon, reportedly thinking Fulton a fraud, responded:

> *You would make a ship sail against the wind and currents by lighting a bonfire under her decks? I don't have time for such nonsense.*

After Napoleon's rejection, Fulton turned to Britain and sold the patent rights to the Royal Navy. Two years later, in 1805, the combined French and Spanish fleets suffered a crushing defeat at the hands of the British Navy in the Battle of Trafalgar—the largest naval battle of the nineteenth century—and Napoleon abandoned his plans to invade Britain for good.

This story is widely circulated, appearing in various forms across both Chinese and English media. Some versions go further, claiming Napoleon's reliance on wind-powered sailing ships led to his defeat by Britain's steam-powered navy. It's a dramatically appealing anecdote—perfect for introducing a chapter on technological succession—but unfortunately, it's not true. While Napoleon did indeed reject Fulton's visionary proposal, the technological limitations of the time meant that every ship at Trafalgar, British or otherwise, was still a wind-powered wooden warship. The world's first true steam-powered warship wouldn't be built until 1849—and that ship, ironically, was French. Its name? *Le Napoléon*.

Military leaders like Napoleon were not usually slow to recognize the potential of new technology. As discussed earlier in Part 1, Napoleon personally reviewed the royalty fees for the Jacquard loom patents. According to *Unofficial Histories of the Qing Dynasty*, Hu Linyi—one of the four most esteemed ministers of the late Qing era—once witnessed a foreign steamship powering upstream with paddle wheels, "*racing like a galloping horse, swift as a whirlwind.*" He turned pale, said nothing,

Q. Yue, F. Yang, *Governing Innovation: A Global History of Patent-Driven Technological Progress*, https://doi.org/10.1007/978-981-95-5570-3_13

and, on the way back to camp, reportedly vomited blood from stress and nearly fell off his horse. He died within months in 1861.

By then, Western military technology was advancing so rapidly that each new breakthrough seemed to plunge Chinese patriots into despair. That same year, American inventor Richard Jordan Gatling created a hand-cranked multi-barrel machine gun. He received a patent in 1862 (US36836A; see Fig. 13.1), and within a year, the new weapon was deployed in the American Civil War. Historical records show that between August 1, 1873, and October 8, 1874, the Gatling Gun Company sold 245 units: 52 to the U.S. Army, 26 to the Navy, 10 to Brazil, and 4 to Spanish Cuba. Notably, the Qing government purchased 51 units, becoming one of Gatling's largest customers during that period[1] (Fig. 13.2 depicts the gun as illustrated in a late Qing military manual). In 1874, William Gardner invented the Gardner gun, which was patented in 1876 (US174130A). By 1880, the British Royal Navy, already equipped with Gatling guns, took interest in the Gardner design and invited Gardner to set up production in the UK, making him Gatling's strongest competitor.

Around this same period, another American inventor, Hiram Maxim, completed the first design for a fully automatic machine gun: the Maxim gun. Before turning to weapons, Maxim had been focused on civilian technologies. At the time, investors were heavily focused on electrification and devices like incandescent lamps, sparking intense competition. Edison ultimately emerged as the victor and became synonymous with invention. However, Maxim had also developed an incandescent bulb with impressive brightness and longevity even before Edison. But, as he recounts in his autobiography, an individual learned a critical detail of Maxim's design, filed a patent first, and enlisted relatives to provide false testimony in a patent interference proceeding. Maxim, who filed second, lost the rights and was unable to compete with Edison's well-oiled patent machine.[2] Afterward, Maxim was sent to Europe by his company to study German and French patents. While attending an exhibition in Vienna, he met a fellow American who gave him a fateful piece of advice:

> *Hang your chemistry and electricity! If you want to make a pile of money, invent something that will enable these Europeans to cut each other's throats with greater facility.*

Maxim took the advice to heart, and inspiration struck. Gatling and Gardner's guns were hand-cranked and required human effort to maintain continuous fire. They were mechanically complex, prone to jamming, and quite bulky—especially the Gatling gun, which resembled a small cannon. That's why, when imported to China, the Gatling gun was referred to as a "Cannon" rather than a firearm. Maxim's ingenious idea was to harness the recoil energy from each shot to cycle the weapon automatically. In 1884, he secured investment and founded the Maxim Gun Company. Early that year, his prototype successfully fired 600 rounds in 1 min. The demonstration caused a sensation, drawing visits from military leaders across Europe. Prince George, commander of the British Army, personally attended a trial.

[1] Chivers, C. J. (2011). *The gun*. Simon and Schuster.

[2] Sir Hiram S. Maxim http://www.lamptech.co.uk/Documents/People%20-%20Maxim%20HS.htm

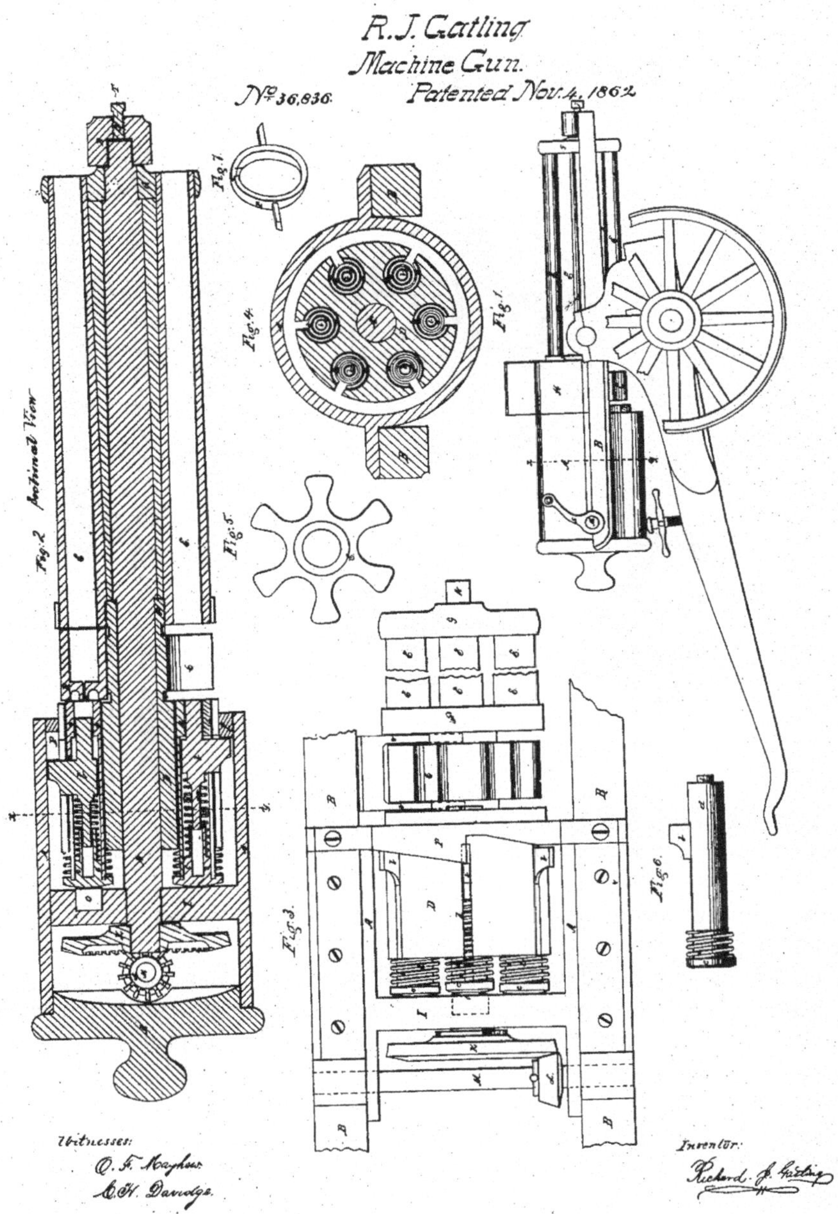

Fig. 13.1 Patent drawing of Richard J. Gatling's U.S. Patent No. 36836A, filed in 1861

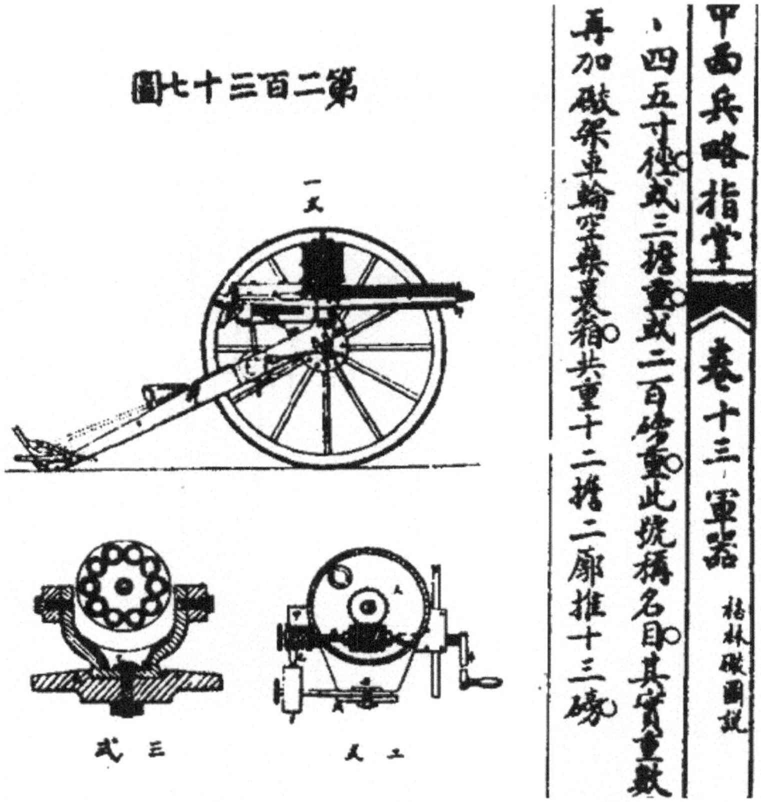

Fig. 13.2 Illustration of the Gatling multi-barrel machine gun (referred to at the time as the "Gelin Rapid-fire Cannon") from the late Qing military manual *Zhongxi Binglue Zhizhang*

The excitement became so overwhelming that Maxim couldn't get work done during the day.

In 1887, another British prince introduced the Maxim gun to German Emperor Wilhelm II. Upon seeing the weapon fire 333 rounds in just 30 s, the Kaiser immediately placed an order.[3] The Maxim gun was soon adopted by militaries across Europe.

On October 25, 1893, 700 men from the British South Africa Company were ambushed by 5000 Ndebele warriors. The Ndebele, inheritors of the only major sub-Saharan African military legacy founded by Shaka Zulu, were well-trained and better equipped than most regional forces. But in the face of British troops armed with four Maxim guns—making their battlefield debut—neither courage nor discipline could make a difference. The guns mowed down 1500 Ndebele fighters like

[3] Chivers, C. J. (2011). *The gun*. Simon and Schuster.

wheat before a scythe. British casualties? Four. This was the infamous Battle of Shangani.

Maxim gained wealth, fame, and recognition. He became a British citizen and industrialist, received a knighthood, and filed a number of patents in the UK (Fig. 13.3 shows one such machine gun patent). At the same time, the U.S. incandescent lamp market had turned into a brutal battlefield. Edison's top-tier patent team had gone into full offensive mode. In 1885 alone, Edison's company initiated nearly 100 lawsuits and administrative actions.[4] By 1890, Edison's market share in the incandescent bulb sector had dipped below 50%, prompting him to launch an unprecedented patent war. Under the shadow of Edison's reputation, every lamp manufacturer trembled. In 1895, a court ruling provided a broad interpretation of Edison's U.S. Patent No. 223898, effectively handing him victory.[5]

By 1901, Edison and his allies had spent $2 million on more than 200 lawsuits but achieved tremendous results: nearly every bulb manufacturer in the U.S. had to pay him royalties. By 1910, General Electric—Edison's company—controlled 97% of the incandescent lamp market.[6]

Edison also attempted to defend his patents overseas by registering them internationally and collecting royalties through licensing companies, often recruiting familiar business partners from earlier ventures in Europe.[7] However, U.S. patent law at the time had some now-unusual provisions: if a U.S. patent was also filed and granted earlier in a foreign country, then when the foreign patent expired, the U.S. patent would expire as well. Because U.S. patent approval often took years, while some European countries granted patents without substantive examination and at a faster pace, this led to Edison's patents expiring earlier in the U.S. than elsewhere.

Consequently, Edison filed relatively few patents in Europe, leaving room for local manufacturers to thrive. One of these was Philips—a company that would go on to become one of the central figures in this chapter of disruption and revolution.

In 1817, the Kingdom of the Netherlands—which included what is now the Netherlands and Belgium—enacted its first patent law. After more than half a century of development, the Netherlands still had not established a robust patent system. In fact, due to opposition from industry, that initial law was repealed in 1869 and not replaced until 1910.[8] It was in this legal vacuum that Philips—the company

[4]Arapostathis, Stathis; Dutfield, Graham (2013). Knowledge Management and Intellectual Property || ., 10.4337/9780857934390(), –. https://doi.org/10.4337/9780857934390

[5]Consol. Elec. Light Co. v. McKeesport Light Co., 40 F. 21 (C.C.W.D. Pa. 1889), aff'd, 159 U.S. 465 (1895)

[6]Shaver, L. (2012). Illuminating innovation: From patent racing to patent war. *Wash. & Lee L. Rev.*, *69*, 1891.

[7]Arapostathis, S., & Dutfield, G. (Eds.). (2013). *Knowledge Management and Intellectual Property*. Edward Elgar Publishing.

[8]Rules, discretion and industrialisation: The patent system of the Netherlands, 1817–1869 Wagenaar, H. (Author). Jul 2022 Student thesis: Doctoral Thesis › Doctor of Philosophy

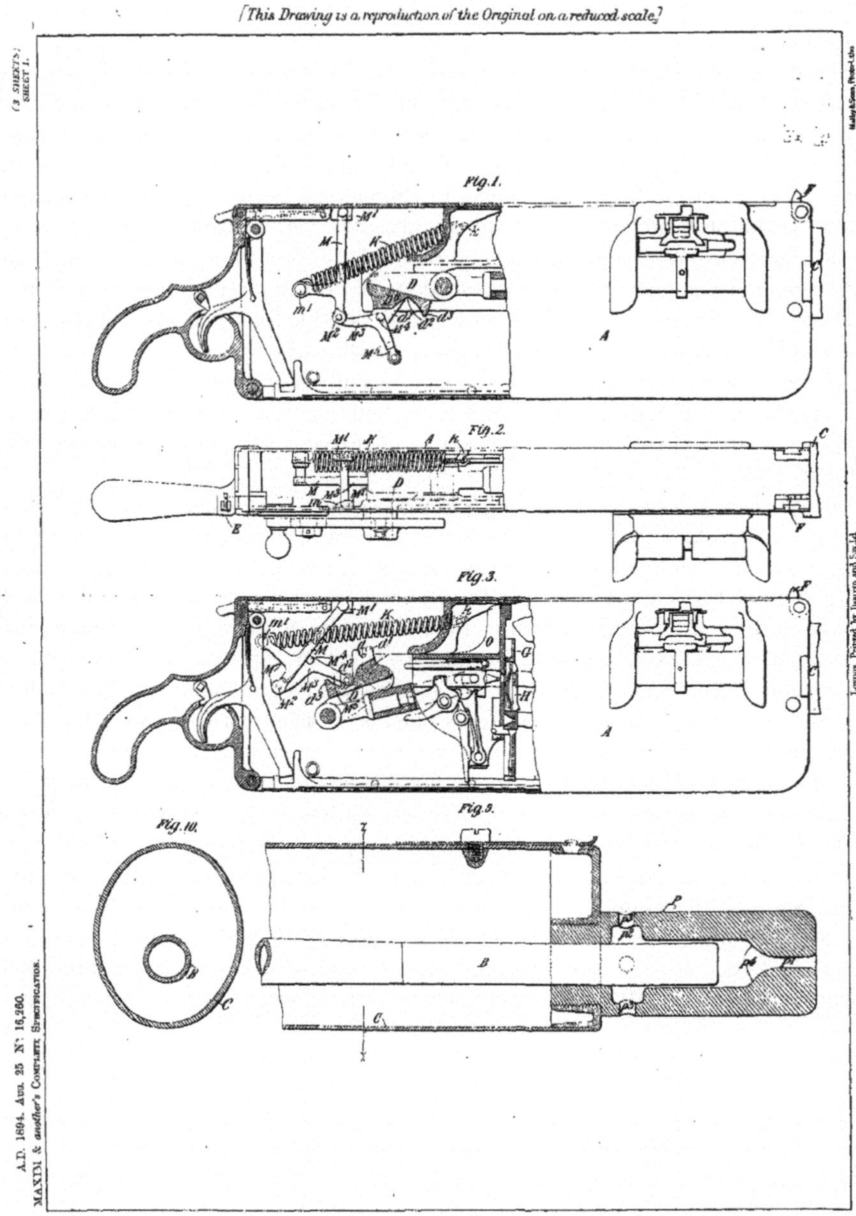

Fig. 13.3 Patent drawing of Hiram Maxim's British patent GB189416260A

we now associate with aggressive patent strategies and enforcement—was born. Reflecting on this history, Ad Teulings of the University of Amsterdam wrote:

> *...Gerard Philips, who founded the company in 1891, has been given a place in the Dutch industrial history as a brilliant inventor. He never was, actually; he was a good craftsman tlhough and used the absence of patent laws in the Netherlands upto 1911 as an excuse to copy the products and processes of his competitors. After 1912 his role has been waning. He remained a general director and was more eager to be praised and honoured as the Dutch equivalent of a von Siemens or Edison.*
>
> *Anton Philips, his brother and commercial director, was, for a time, successful in acquiring access to foreign markets. After 1930, when the market for incandescent lamps was choked, he seemed to have lost his commerial assertiveness and had no idea of how to overcome the economic crisis. He was actually set aside by a "coup" in his senior management and research staff. They were responsible for the diversification into the radio business, the most successful decision in the company's history, resulting in a leading position for the company in the world market. But even when his personal contribution to the firm was declining he was heralded as the native Emil Rathienau, the builder of the AEG industrial and financial empire, and finally, even the radio success was attributed to him.*[9]

At a time when the Netherlands had no patent protection, it embarked on industrialization. Philips took full advantage of this, boldly copying products and manufacturing techniques from General Electric and Siemens to achieve its initial growth. Additionally, wages in the Netherlands were lower than in Germany, and child labor was still legal—whereas Germany, during the Second Empire, had implemented numerous labor protections and banned child labor. This gave Philips a cost advantage over German competitors like Siemens and AEG. Edison, with limited patent presence in Europe, was unable to prevent Philips from marketing incandescent bulbs across the continent.

Coincidentally, Switzerland also lacked a patent law during this period. It wasn't until 1888, under pressure from the Paris Convention and the League of Nations, that Switzerland passed its first patent law. Yet, due to lobbying by the Swiss chemical industry, the law included a unique clause: only inventions that could be demonstrated with a mechanical model were eligible for protection. This allowed chemical companies to freely copy foreign patents, especially German ones. It wasn't until 1907, under pressure from Germany, that this restriction was removed.[10]

After settling in Britain, Maxim remained bitter over losing the electric lamp race, but at least he no longer had to deal with Edison's patent lawyers or the endless litigation characteristic of the American system. His company quickly eclipsed Gatling and Gardner, bringing him immense wealth—and realizing his goal of creating a weapon for "these Europeans to kill each other more easily." Once the Maxim gun was deployed in combat, trench warfare became the norm, and the number of soldiers killed by bullets rose dramatically. On July 1, 1916, the Battle of the Somme began. Both sides were equipped with improved versions of the Maxim

---

[9] Teulings, A. (1979). Structure and Logic of Industrial Development: Philips and Electronics Industry. *Social Scientist*, 3–22.

[10] Dominique S. Ritter, Switzerland's Patent Law History, 14 Fordham Intell. Prop. Media & Ent. L.J. 463 (2004). Retrieved April 8, 2024 from: https://ir.lawnet.fordham.edu/iplj/vol14/iss2/3

gun. In the initial offensive by Allied forces, nearly half of the 120,000 attacking Commonwealth troops were killed or wounded, with 20,000 dead—most within the first hour. Total casualties in the battle exceeded one million.

Though the story of Fulton offering steamships to Napoleon can be historically disproven, tales of transformative inventions being ignored are not rare. The Maxim gun is one such example: though it quickly captured the interest of European powers, Americans were initially indifferent. The U.S. Civil War had ended, and only small-scale conflicts with Native Americans remained—conflicts for which the Gatling gun sufficed. Repeated attempts by the Maxim company to sell to the U.S. military failed. It wasn't until the Spanish-American War in 1898 that the U.S. truly recognized the value of machine guns.

Another case is the tank, which emerged during the Battle of the Somme. Young military minds like Heinz Guderian, Charles de Gaulle, and Mikhail Tukhachevsky became fascinated by this new weapon. Meanwhile, older French marshals like Joseph Joffre and Philippe Pétain showed little interest. After the war, Guderian and Tukhachevsky developed groundbreaking theories on mechanized warfare in Germany and the Soviet Union, respectively. In France, Joffre remained fixated on trench warfare and static fortifications, championing the construction of the Maginot Line, with Pétain's full support. Though de Gaulle also advocated for armored forces and opposed the Maginot Line, his ideas were largely ignored.

In 1940, Germany launched a blitzkrieg against France. Swift mechanized units bypassed the Maginot Line and rapidly overran French defenses. France surrendered in just 46 days. At 84, Pétain became Prime Minister and capitulated to Germany. After the war, he was convicted of treason and died in prison.

The histories of the machine gun and tank exemplify the ruthless inevitability of technological change—a central theme of this chapter on disruption.

The two World Wars decimated Europe's young male population, profoundly altering the continent's demographic structure. From 1946 to 1964, the Western world experienced an unprecedented baby boom. Birth rates soared, ushering in a massive new generation. At the same time, the horrors of war gradually extinguished the martial spirit that had defined Europe since the Middle Ages. Anti-war sentiment and a live-for-the-moment lifestyle became dominant among the youth.

It was in this context that Sony rose to prominence, skillfully catering to the needs of the baby boomers. As this generation came of age, Sony released a succession of innovative products—the TR series transistor radios, Trinitron televisions, and the iconic Walkman portable music player—all of which became indispensable entertainment tools in young people's lives.

During the same period, Philips also made outstanding contributions to cultural and entertainment life. The company invented the compact cassette and owned all foundational patents related to both the tape itself and the devices that played it. By the late 1970s, Philips had launched a range of portable tape players, and its subsidiary PolyGram had become the world's largest record label. In the consumer electronics sector and in the development of various media standards, Philips maintained a complex relationship with Japanese firms like Sony—marked by both fierce competition and strategic cooperation.

In 1977, Brazilian-German engineer Andreas Pavel created the first prototype of the personal stereo by combining a cassette player, audio amplifier, and batteries on a belt, which he connected to headphones. He applied for a patent for this "belt-worn sound system" in Italy (patent number IT2162577) and later secured patents in the U.S., Germany, the U.K., and Japan. During this time, he proactively reached out to Philips to pitch his invention. This was nearly 2 years before Sony launched its revolutionary Walkman.

Unlike Napoleon's navy, Philips possessed all the necessary technology, hardware capabilities, and music catalog resources to spearhead the personal stereo era. Yet, the company dismissed Pavel as a lunatic with wild ideas. He was also rebuffed by audio giants like Grundig, Yamaha, and ITT—none of them saw the potential in his invention.

In a 2005 interview with The New York Times,[11] Pavel bitterly recalled:

> *They all said they didn't think people would be so crazy as to run around with headphones, that this is just a gadget, a useless gadget of a crazy nut.*

Pavel never stopped asserting his claim as the inventor of the personal stereo. After Sony launched the Walkman and achieved unprecedented success, he approached the company to demand licensing fees for his patents, but was promptly rejected. After prolonged negotiations, Sony reluctantly agreed in 1986 to pay a small royalty for a few Walkman models sold in Germany—roughly 1% of their profit in that market. Pavel found this unsatisfactory. In 1989, he filed a patent infringement lawsuit against Sony in the UK, formally bringing the matter to court. The patent in question was the UK counterpart to his original application, numbered GB1601447A (see Fig. 13.4 for the diagram). Its claim 1 read:

> *An audio listening apparatus comprising, in combination, (i) a belt for personal wear having attached thereto an audio system including an interconnected stereo audio signal source and stereo amplifier means and a battery power source therefor, and (ii) stereo headphones or earphones electrically connected to the stereo amplifier means.*

Sony brought forth several prior art references in an attempt to invalidate Pavel's patent. These included the Sony TC-153SD, a portable cassette recorder with a shoulder strap and headphone jack, which had been sold in the UK as early as 1977; and the Nakamichi DT-550 by Nakamichi Corporation, which also used a similar strap design. Though larger than the first-generation Walkman and not pocket-sized, these products disclosed every technical feature claimed in Pavel's patent. Even though Pavel's diagrams depicted a belt-worn device, claim 1 itself never explicitly mentioned a "belt."

Pavel's other independent claim, claim 10, built upon claim 1 and explicitly stated the use of a belt carrying the playback device, amplifier, and battery power supply. The key piece of prior art Sony cited against this was the Nagra SNN spy

---

[11] Larry Rohter. (2005). An Unlikely Trendsetter Made Earphones a Way of Life. Retrieved April 8, 2024 from: https://www.nytimes.com/2005/12/17/world/americas/an-unlikely-trendsetter-made-earphones-a-way-of-life.html also available at: https://www.theguardian.com/technology/blog/2005/dec/18/andreaspaveli

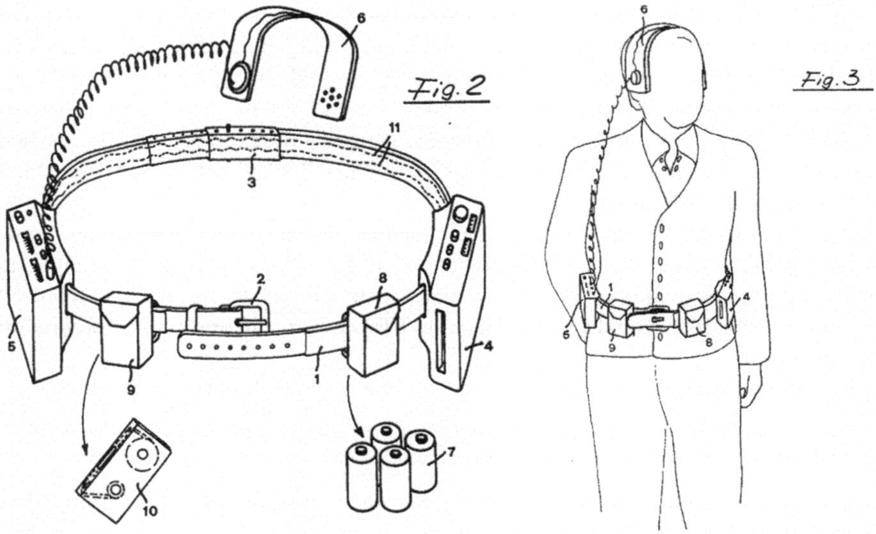

Fig. 13.4 Patent drawing of Pavel's British patent GB1601447A

recorder from Switzerland. Though expensive, the Nagra SNN was compact, similar in size to the Walkman, came with headphones, and featured a loop for belt attachment.

The Nagra SNN, however, was monaural and lacked stereo functionality. But during the trial, a witness for Nagra testified that customers had suggested adding stereo capability to the SNN back in the 1970s. The judge ruled that upgrading the SNN from mono to stereo was not a creative step. Likewise, modifying the Sony TC-153SD's shoulder strap into a belt was not considered an inventive leap, leading to the invalidation of claim 10.

In 1996, Pavel lost his appeal in the UK Court of Appeal and was left with legal fees totaling around $3.68 million, nearly bankrupting him. Yet Pavel refused to give up, threatening further lawsuits in two other countries. Sony, still unable to produce prior art that disclosed all elements of claim 10, remained at risk. In 2001, they finally settled. Though the terms were confidential, European media speculated that the payout might have been several million euros.[12]

Meanwhile, in the United States, an engineer and skiing enthusiast named William J. Hass independently came up with the idea of personal audio before Sony did. In 1977, he applied for a patent on a "personal audio system," which was granted in 1978 as US4070553A (Fig. 13.5). The patent described a bone-conduction headphone system hidden inside a scarf, designed to connect to a radio or cassette player.

---

[12] Estelle Dumout. (2004). Sony Pays Millions to Inventor in Walkman Dispute. Retrieved April 8, 2024 from https://www.cnet.com/news/sony-pays-millions-to-inventor-in-walkman-dispute/

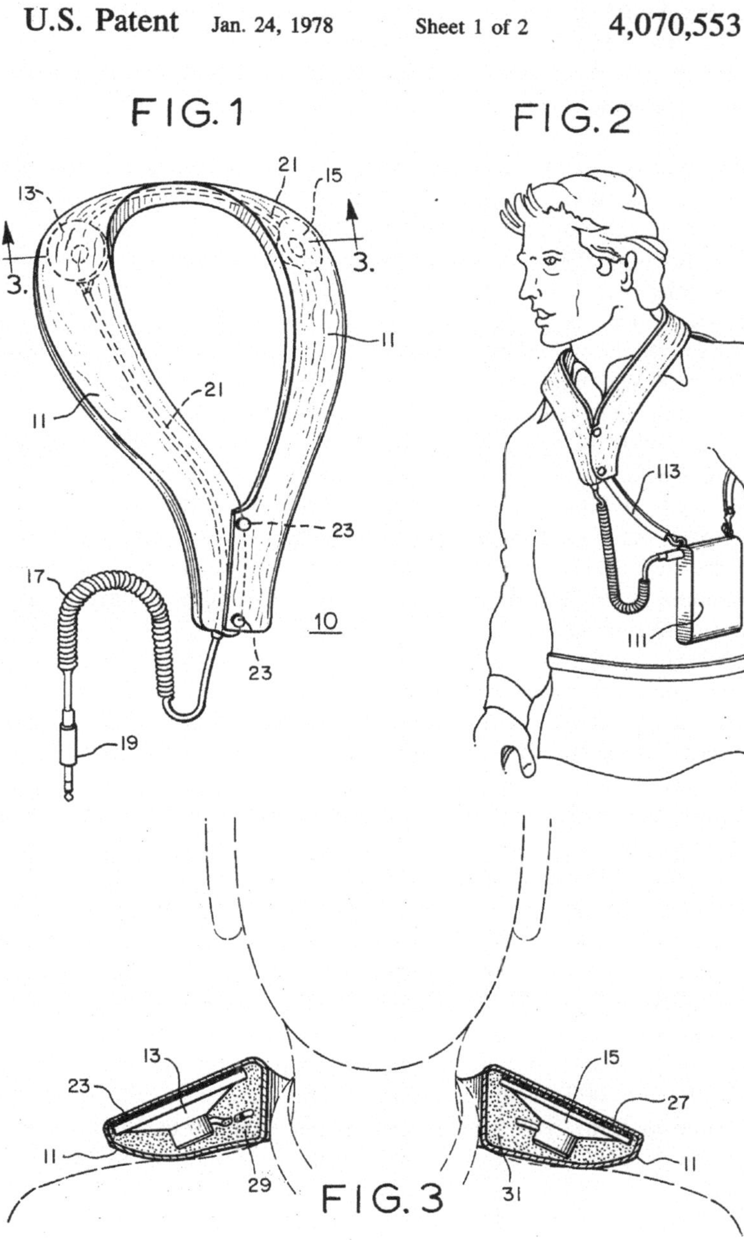

Fig. 13.5 Patent drawing of U.S. Patent No. 4070553A

Unlike Pavel's fruitless attempts, Hass successfully licensed his invention to American company JS&A. In 1978, JS&A commercialized the product under the name "Bone Fone," priced at $69.95. By 1981, over 100,000 units had been sold. While Hass's patent did mention cassette players, JS&A only marketed Bone Fone with portable radios and never launched a tape-playing version. Despite moderate commercial success, the Bone Fone disappeared shortly after the Walkman hit the market.

The evidence from Pavel's lawsuit and the Bone Fone suggests that the Walkman was not a technological leap forward. Its components—headphones, stereo sound, and portable playback—were all existing technologies. Pavel's patent was less a technical invention and more a conceptual blueprint for personal portable listening. Whether it was Sony or Philips, either company was only one simple idea away from creating the Walkman. It's hard not to wonder: if Philips had acquired Pavel's patent and embraced his concept, how might history have changed?

By the late 1980s, Philips was struggling. Though its standard-essential patents continued to generate steady income, its consumer electronics products consistently underperformed. Ambitious projects like the Video2000 VCR, the CD-i player, and the DCC tape system failed disastrously, prompting a series of divestitures over the next two decades. Today, Philips is a much leaner company, focused mainly on medical equipment and personal health—a far cry from its heyday as a globe-spanning industrial titan.

As we've seen, real history often proves more dramatic than fiction. And by the twenty-first century, the same story would play out again—this time, with Sony in the starring role.

At the dawn of the twenty-first century, Sony had amassed formidable technological strength across a wide range of digital domains: digital cameras, digital music, portable audio, personal digital assistants (PDAs), and even personal computers. Much like how Philips once owned the powerhouse music label PolyGram, Sony boasted Sony Music, one of the largest record companies in the world. While the MD series never achieved the iconic success of the cassette Walkman, it remained the high-end market's top choice for portable audio. Facing the challenges of the mobile communications era, Sony partnered with Ericsson to form Sony Ericsson, solidifying its position in the mobile phone market. Meanwhile, its Alpha series digital cameras were making aggressive inroads into the high-end photography segment. Sony also consistently ranked among the top ten recipients of U.S. patents annually, with over 1000 patents granted each year—except in 2003 (ranked 14th) and 2005 (ranked 11th). Several key patent pools managed by Philips—covering standards like CD, CD-R, and DVD3C—also included Sony as a major participant.

What followed is a story well known to anyone interested in technology. In 1997, Steve Jobs returned to Apple, and by 2001, the release of the iPod had catapulted Apple to dominance in the portable music player market. Just as Sony's long-nurtured MD product line was beginning to flourish, it suffered a devastating blow. With the launch of the iPhone in 2007, Sony's flagship consumer electronics—including PDAs, consumer digital cameras, portable audio players, and

Table 13.1 U.S. Patent grants comparison between Apple and Sony (2000–2007)

Year	Apple Patents	U.S. Rank	Sony Patents	U.S. Rank
2001	97	172	1363	7
2002	77	225	1434	8
2003	80	218	1311	14
2004	94	180	1500	9
2005	85	184	1234	11
2006	106	176	1906	8
2007	118	156	1454	10

smartphones—came under crushing pressure. It's worth noting that from 2000 to 2007, despite Apple's explosion of breakthrough products, its annual U.S. patent grants hovered around just 100—a mere fraction of Sony's output (see Table 13.1).

So why did Philips and Sony—companies rich in technical know-how and protected by robust patent walls—both succumb to the revolutionary products of later entrants?

Leo Tolstoy once wrote in *Anna Karenina*, "*All happy families are alike; each unhappy family is unhappy in its own way*." We believe that in product development and market success, all thriving companies share one vital trait: an acute sensitivity to customer needs. Failures, on the other hand, arise from myriad causes. The reasons why Philips missed the boat on personal audio, and why Sony fell short in the smartphone race, differ and deserve careful analysis.

To learn from history, it's helpful to start with the earliest patents for magnetic tape recorders and trace the journey of Philips, Sony, and Apple—examining how some of their most pivotal products and patents succeeded or faltered amid waves of rapid technological change. One lesson emerges clearly: in an age of constant replacement and innovation, clinging to patents or isolated technical advantages isn't enough to guarantee survival. The customer is the key to the door of success.

Ultimately, we find that in today's world, Chinese engineers no longer need to gaze in despair at Western technologies as Hu Linyi once did when he witnessed a steam-powered ship. A pragmatic, customer-centered approach to products can help us leapfrog ahead, break through patent barriers, and challenge Western dominance.

Chapter 14
Tracing the Origins of Magnetic Tape Technology

In 1928, German scientist Fritz Pfleumer invented a magnetic tape capable of recording audio (German Patent No. DE500900C). By 1931, he had built the first prototype, and the following year, he licensed the related patent technology to the German company AEG.

Building upon Pfleumer's invention, AEG designed and produced a magnetic tape recorder named the "Magnetophon." In 1935, they debuted their first model—the Magnetophon K-1—at the Berlin Radio Exhibition. Over the next few years, AEG continuously improved the Magnetophon's sound quality and reduced background noise, eventually achieving a level of fidelity that came remarkably close to that of live radio broadcasts. Patent DE712759C (Fig. 14.1) and DE664759C (Fig. 14.2) are among the technological advancements AEG made on top of Pfleumer's invention.

Despite its technical superiority, the Magnetophon failed to gain commercial traction. As geopolitical tensions escalated in Europe, the device faded from public view and disappeared from the market. Its only remaining major client was the Nazi Party's Ministry of Propaganda.

During World War II, Germany deployed numerous Magnetophon units at various radio stations (Fig. 14.3), using them to broadcast Adolf Hitler's speeches across the country in high fidelity. At the time, most people had no idea what magnetic tape was, nor could they imagine that recorded audio could rival live broadcasts in quality. Hearing what they believed were live addresses from Hitler himself, German civilians and soldiers were greatly uplifted. Due to the strict secrecy surrounding equipment and technology in Nazi Germany, the Magnetophon became a vital propaganda tool for the Third Reich.

The Allies were aware that Nazi Germany could preserve and play back high-fidelity recordings up to 15 minutes long, but had no idea what equipment or methods were used. After the liberation of Paris, U.S. Army engineer Colonel Jack Mullin was assigned to investigate the Nazis' secret recording technology. Mullin scoured hundreds of traditional low-fidelity recording devices across various radio stations from Paris to Frankfurt without success. Just before returning to the U.S.,

Q. Yue, F. Yang, *Governing Innovation: A Global History of Patent-Driven Technological Progress*, https://doi.org/10.1007/978-981-95-5570-3_14

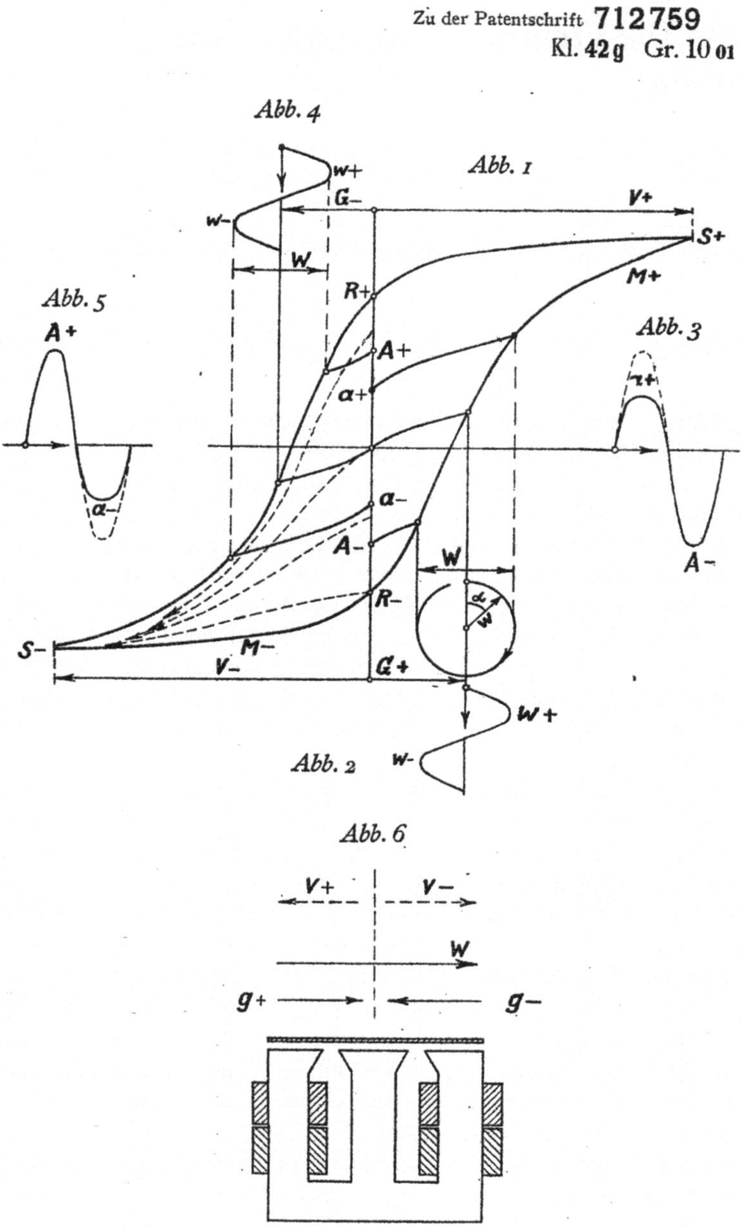

Fig. 14.1 AEG's 1935 patent DE712759C, aiming to reduce non-linear distortion caused by magnetic sound methods. The inventor is Pfleumer himself

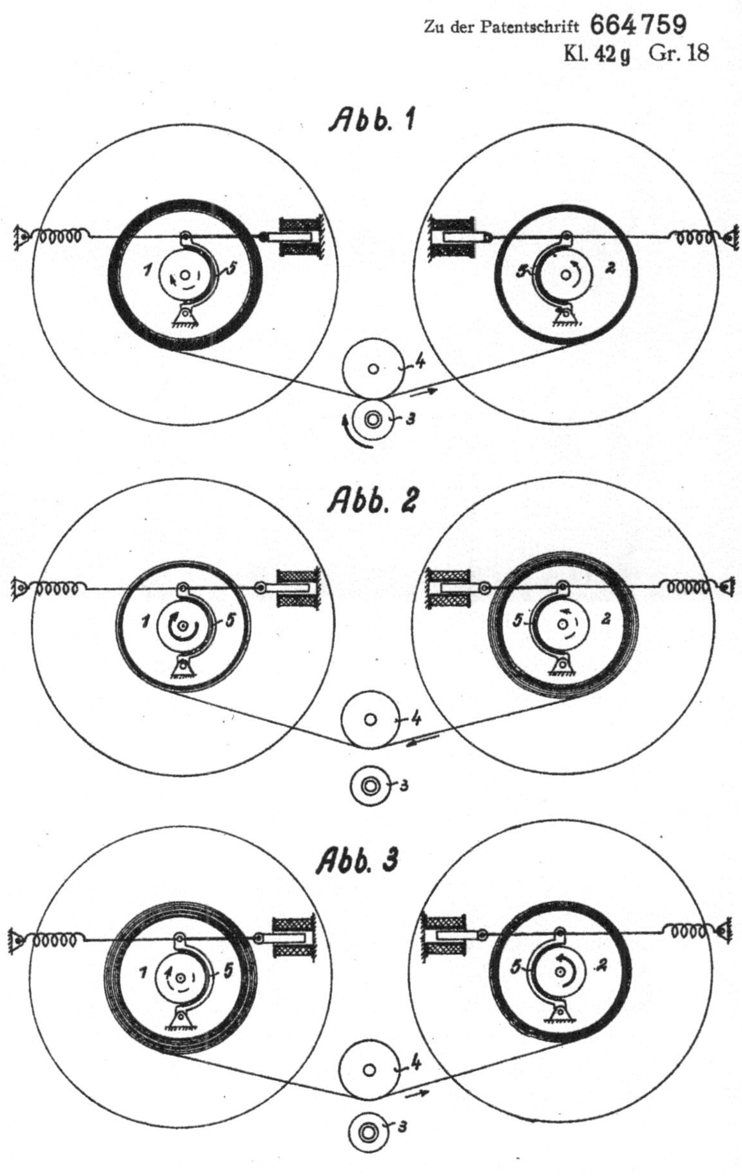

Fig. 14.2 AEG's 1935 patent DE664759C, using three motors to ensure consistent tape speed

he passed through the German town of Bad Nauheim and stumbled upon two suitcase-sized AEG Magnetophon K-4 units.

Mullin brought the two machines and 50 reels of magnetic tape back to the U.S. Over the next 2 years, he painstakingly studied, tested, and modified them. As

Fig. 14.3 Magnetophon used during WWII. (Source: George Shuklin. (n.d.). *Tape unit Ton S.b* [Image]. Wikimedia Commons. https://commons.wikimedia.org/wiki/File:Ton_S.b,_tape_unit.jpg. This image is shared under the Creative Commons CC BY-SA 3.0 license.)

a citizen of a victorious nation, Mullin had no reason to worry about AEG's patent rights. AEG, being a key supplier to the Nazi Party, was under intense scrutiny and sanctions after Germany's surrender. They were overwhelmed and had neither the capacity nor the courage to pursue minor patent issues. Meanwhile, the German Patent Office had been devastated by bombings in Berlin—losing over a third of its buildings, many files destroyed during evacuations, and its workforce scattered. Postwar, German patent affairs were an irretrievable mess.

In a 2005 article, *Business Insider* listed six technologies the U.S. government had "stolen" from Germany during WWII: airborne operations, synchropter helicopters, jet-powered aircraft, cruise missiles, methamphetamine (crystal meth), and rockets.[1] A mere tape recorder seemed insignificant in comparison.

In 1947, Mullin began applying for patents for his improved models (see Fig. 14.4, U.S. Patent No. US2529097A) and ventured to Hollywood to promote them. He conducted demos where a live band would perform behind a curtain, followed by playback of a recording—audiences couldn't tell the difference and were astonished by the technology. Mullin's invention soon caught the attention of Bing Crosby, the celebrated American actor and singer.

[1] David Nye. (2015). 6 things the US stole from the Nazis during WWII. Retrieved April 8, 2024 from https://www.businessinsider.com/6-things-us-stole-from-germans-during-wwii-2015-5

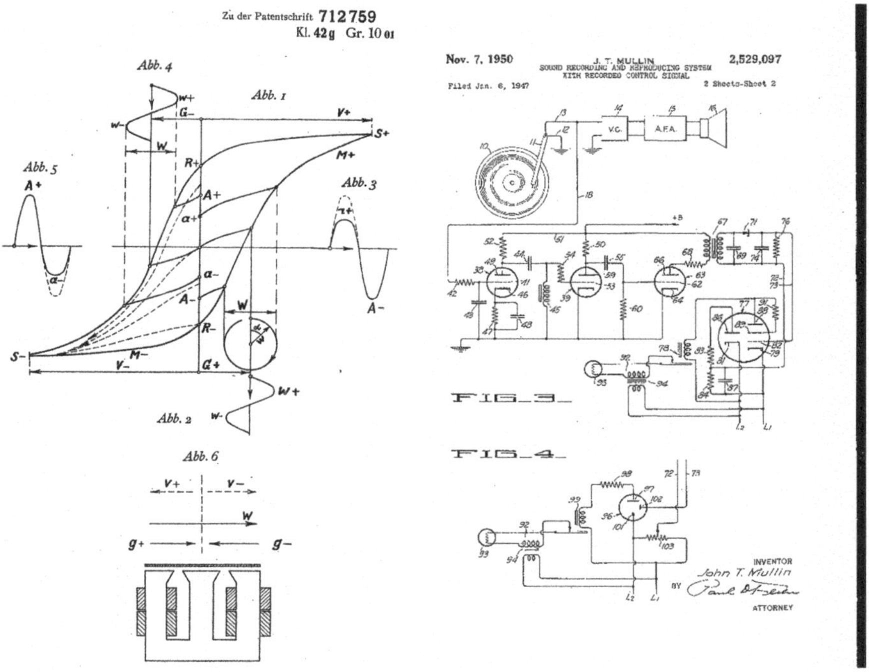

Fig. 14.4 Mullin's 1947 patent for a sound recording and playback system—his first patent

Crosby, fresh off his 1944 Academy Award win for Best Actor, was at the height of his fame. As a singer, he disliked the chaos of live concerts and preferred the quiet of radio studios, making him one of the top radio stars of the time. Crosby, hoping to avoid ever performing live again, invested \$50,000 in a local company, Ampex, urging them to rapidly develop a new recording device. In collaboration with Mullin, Ampex built the Model 200 based on the improved Magnetophon. The device was a hit, and magnetic tape recorders began entering the music industry on a large scale.

Chapter 15
The Birth of the Cassette Tape

The earliest audio tapes were reel-to-reel tapes, which required users to manually wind the magnetic tape onto plastic reels to store sound information. Installing a reel-to-reel tape into a tape recorder was a complicated process, making it difficult for ordinary users to operate.

Around the time when reel-to-reel tapes began gaining popularity, Masaru Ibuka and Akio Morita co-founded Tokyo Tsushin Kogyo in 1946—the company that would later become Sony. During this period, transistors began replacing the bulky and fragile vacuum tubes, leading to smaller and more compact radios and televisions. In 1954, Texas Instruments developed the world's first transistor radio for consumer use, the Regency TR-1 (U.S. Patent No. US2892931A). Just 3 years later, Sony released the even smaller and thinner Sony TR-63, which was only half as long and three-quarters as wide as the Regency TR-1. Its compact form made it highly popular; approximately 100,000 units were sold in the U.S. market that year, helping Sony gain a significant reputation. Sony's early history with transistors will be further explored in the later section on "Core and Frontiers."

As miniaturization became a prevailing trend, the large size of reel-to-reel tapes (see Fig. 15.1) became a major bottleneck in reducing the size of recording and playback devices.

In 1958, RCA in the United States proposed a solution: encase the magnetic tape in a small housing, winding it around a pair of reels within a shell, and design a recording and playback system that could operate the tape within the cartridge directly. This meant users no longer had to touch the fragile tape or manually rewind it—everything would be handled mechanically inside the shell. In appearance, RCA's cartridge tape was already quite similar to what we now know as the cassette tape (see Fig. 15.2).

While RCA's design concept was nearly flawless, the actual product often jammed and failed to gain traction in the market. As shown in Fig. 15.3, its size was also considerably larger than that of the familiar compact cassette.

Q. Yue, F. Yang, *Governing Innovation: A Global History of Patent-Driven Technological Progress*, https://doi.org/10.1007/978-981-95-5570-3_15

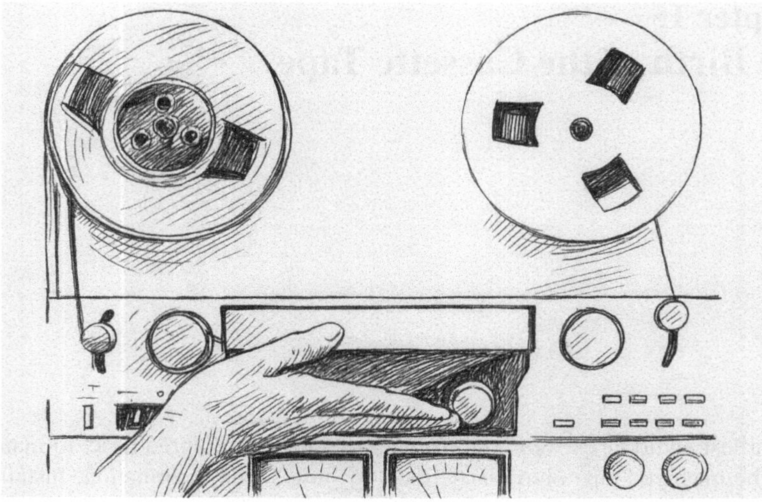

Fig. 15.1 Illustration of reel-to-reel tape operation

In 1960, three American companies—CBS, Zenith, and 3M—jointly introduced another cartridge tape system. Their design featured an ultra-narrow magnetic tape just 3.81 millimeters wide—the same width as the tape in today's compact cassettes. Unlike RCA's dual-reel design, their cartridge housed only a single reel, and users had to manually rewind the tape after playback.[1] The consortium offered licenses at a steep price: a one-time entry fee of $1 million.

In Europe, Philips and AEG (Allgemeine Elektricitäts-Gesellschaft) were the main producers of reel-to-reel tapes. Unwilling to pay exorbitant licensing fees, they decided to go their own way and jointly develop a miniature tape system tailored for European consumers.

As previously mentioned, by this time, Western Europe had fully emerged from the shadow of two world wars. With economies booming and personal wealth increasing, teenagers now had ample pocket money at their disposal. At the same time, the meteoric rise of bands like The Beatles and The Rolling Stones led to a musical generation gap—youths were drawn to strong, energetic rhythms and no longer wanted to listen to their parents' music at home. Portability in music devices became an urgent need of the times.

However, neither the CBS/3M camp in the U.S. nor the Philips/AEG alliance in Europe recognized this emerging demand. Both sides were developing new tape systems primarily for home use, prioritizing sound quality over portability.

[1] Dormon, B. (2013, September 2). *Compact Cassette supremo Lou Ottens talks to El Reg. The Register*. Retrieved March 31, 2024 from https://www.theregister.com/2013/09/02/compact_cassette_supremo_lou_ottens_talks_to_el_reg/

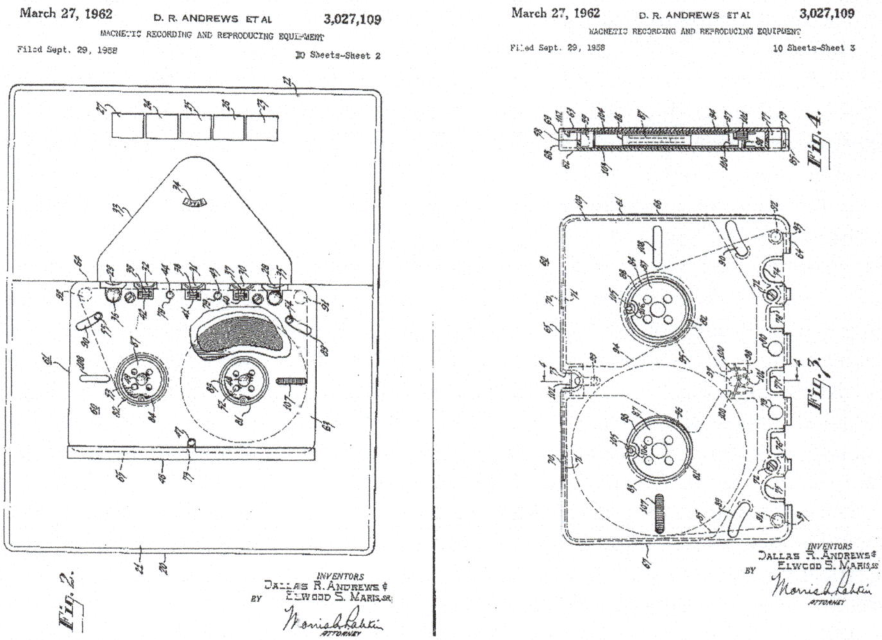

Fig. 15.2 Patent drawing of RCA's magnetic recording and playback device, filed in 1958 (U.S. Patent No. 3027109A)

Fig. 15.3 Comparison between Philips' compact cassette (left) and RCA's cartridge tape (right). (Source: Joe Haupt. (n.d.). *RCA Quarter Inch Tape Cartridge 2A* [Image]. Wikimedia Commons. https://commons.wikimedia.org/wiki/File:RCA_Quarter_Inch_Tape_Cartridge_2A.png. This image is shared under the Creative Commons CC BY-SA 2.0 license)

The first to recognize this need was Lodewijk Ottens (also known as Lou Ottens), an engineer at Philips' Belgian division. Ottens would become the principal inventor of the compact cassette and later a key figure in the development of CD technology. He submitted his idea to Philips headquarters, where it received executive approval. As a result, Philips initiated a new project aimed specifically at creating a portable music playback system.

In this new initiative, the playback speed of the tape was optimized for musical performance, with each side offering a minimum of 30 min of playtime. To control the cassette's size, Ottens built a wooden model of a recorder—small enough to fit into a shirt pocket. He required that the final cassette size must not exceed the dimensions of this wooden mock-up. To meet the demands of portability and compactness, his team incorporated the strengths of both RCA's and CBS/3M's designs: like RCA's tape, the new cassette used a closed-loop dual-reel system allowing bidirectional playback, and like CBS/3M's tape, it used the same 3.81 mm-wide tape running at 4.76 cm per second. Figure 15.4 shows one of Philips' patents for the compact cassette (U.S. Patent No. US3394899A).

During this time, Philips and AEG were still proceeding steadily with their joint cartridge tape project, entirely unaware of Ottens' separate initiative. From Philips' perspective, the compact cassette and the cartridge tape served different purposes and thus were treated as unrelated projects.

Willem Andriessen, a former Philips employee, later recalled that he had been working on the joint cartridge tape project with AEG and was quite satisfied with their progress. But the moment he laid eyes on Ottens' compact and elegant cassette, his first reaction was: "My God, this is it!" He was immediately taken by the cassette's refined appearance and complete technical concept.[2]

Andriessen emphasized that the key to the cassette's success lay in solving a major problem of earlier tape designs: "putting the 'unwieldy tape' into an 'easily handled shell.'" He summarized:

> *All this raises the question: what was and still is the reason for this unique success? Well, the answer is given by the behaviour and reactions of the highest sovereign in the decision about success or non-success in consumer electronics; the consumer!*[3]

In 1963, Philips simultaneously released the compact cassette and the portable tape recorder designed to use it—the EL 3300. Ottens later recalled that to Grundig, this release was tantamount to a declaration of war. Until that moment, Grundig had been completely unaware that Philips had secretly developed a new tape system behind its back.

---

[2]Andriessen, W. (1999). 'THE WINNER'; compact cassette. A commercial and technical look back at the greatest success story in the history of AUDIO up to now. Journal of Magnetism and Magnetic Materials, 193(1–3), 11–16. https://doi.org/10.1016/s0304-8853(98)00502-2

[3]Andriessen, W. (1999). 'THE WINNER'; compact cassette. A commercial and technical look back at the greatest success story in the history of AUDIO up to now. Journal of Magnetism and Magnetic Materials, 193(1–3), 11–16. https://doi.org/10.1016/s0304-8853(98)00502-2

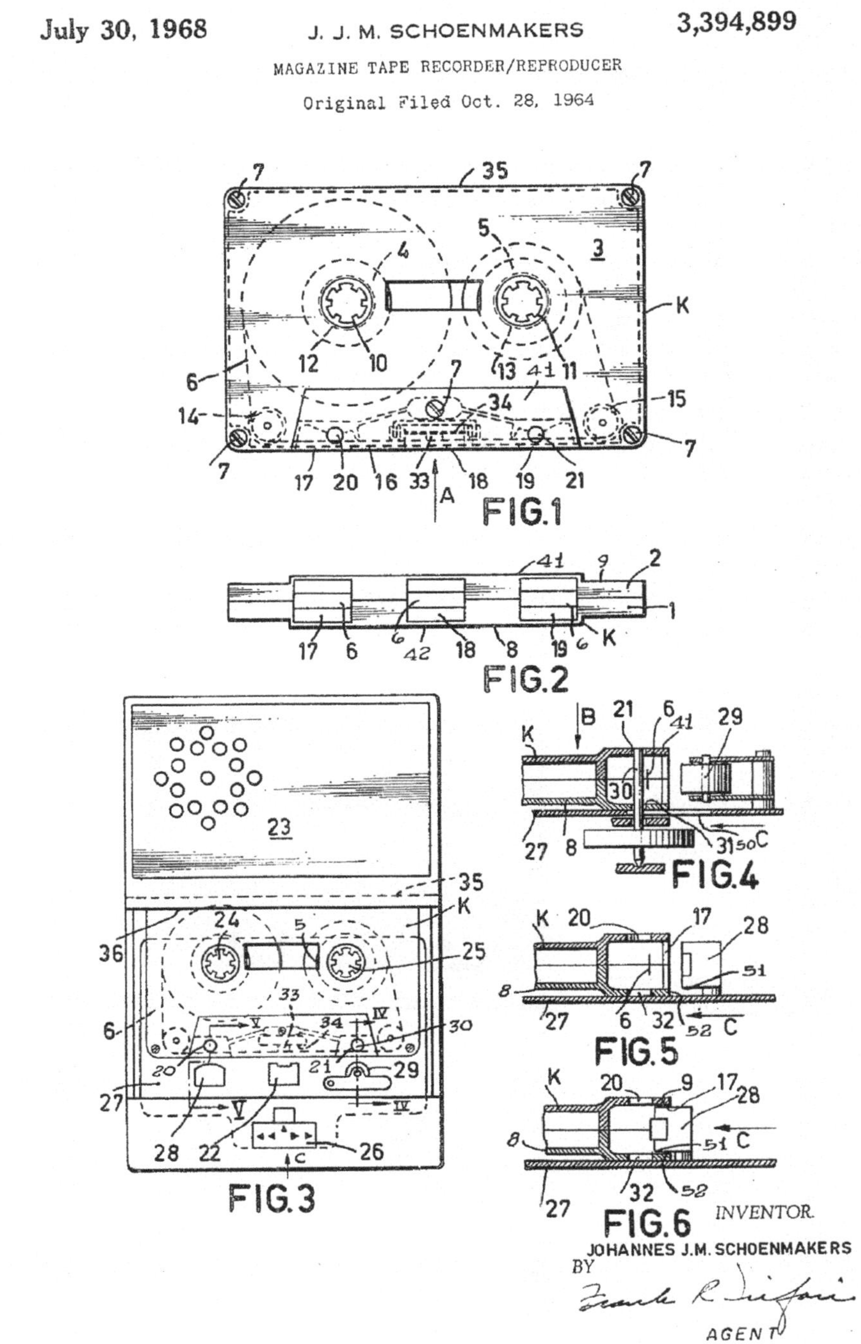

Fig. 15.4 One of Philips' compact cassette patents (U.S. Patent No. US3394899A)

Max Grundig, the head of Grundig, felt deeply betrayed. He immediately terminated the joint project and ordered his team to develop a rival mini tape system modeled after Philips' cassette. He also initiated talks with Japanese manufacturers, including Sony, to seek alliances. In 1965, at a trade show in Berlin, Grundig launched a surprise offensive by unveiling its "DC-International" tape system (Fig. 15.5 shows a comparison between the compact cassette and the DC-International tape). "DC" stood for "Double Cassette," while "International" reflected Grundig's ambition to establish a global standard. Because of potential patent infringement risks, Grundig waived licensing fees altogether.

Philips was caught off guard by Max Grundig's sudden move. Like other European electronics firms, Philips lacked sufficient tape manufacturing capacity and relied heavily on Japanese manufacturers. If Grundig were to partner with a major Japanese player like Sony and quickly flood the market with DC-International tapes, Philips' compact cassette could be stillborn.

Norio Ohga of Sony understood Philips' predicament and leveraged it to apply maximum pressure. According to the "Sony History" section of Sony's official website, Philips had initially intended not to charge licensing fees for the music cassettes themselves but wanted to retain fees on cassette recorders and players. During negotiations, Philips quoted a fee of 25 yen per device. Ohga, calling the fee excessive, demanded free licensing. With the threat from Grundig looming large, Philips

Fig. 15.5 Comparison between the compact cassette (top) and the DC-International tape (bottom). (Source: Hubert79. (n.d.). *Vergleich Compact Cassette DC-International Kassette (ii)* [Image]. Wikimedia Commons. https://commons.wikimedia.org/wiki/File:Vergleich_Compact_Cassette_DC-International_Kassette_(ii).jpg)

reluctantly compromised, lowering the fee to 6 yen per unit. Though a Sony executive had advised accepting the new rate, Ohga refused to budge and even threatened to align Sony with Grundig's camp.[4]

In 1965, Philips finally gave in, waiving all licensing fees and offering the compact cassette technology free to manufacturers worldwide.

This cooperation with Sony effectively sealed the fate of the DC-International tape. Philips' compact cassette went on to dominate the market, becoming synonymous with "tape" in many countries. Yet within Philips' leadership, the decision to waive licensing fees remained a painful memory. The "Sony History" section notes:

> *While regarding Ohga as a crucial partner in the success, Philips viewed him as someone who had caused them to lose a potential fortune in royalties.*

One study observed that Philips always carried a "bitter memory" of the free licensing strategy for compact cassettes. By the time CD standards were established in the 1980s, Philips firmly refused to give anything away for free. In the 1990s, they demanded even higher fees for DVD licenses. By the twenty-first century, licensing fees for DVD players had surpassed their actual manufacturing costs.[5]

Until 1965, Philips had not settled on an official name for its new tape format. Some people referred to it as a "cartridge," while others called it a "cassette." Philips' French division even gave it a nickname—"K7," which is pronounced like *cassette* in French. Starting in 1965, Philips formally adopted the name "compact cassette" as its brand. In China, it came to be known either as "盒式磁带" (*cartridge tape*) or "卡式磁带" (*cassette tape*), and some simply translated it literally as "紧凑式磁带" (*compact tape*). For most Chinese people, however, "cassette tape" became synonymous with "tape."

Before the 1960s, Japan was still in the early stages of postwar economic reconstruction. With low per capita income, the country's consumers could not afford expensive reel-to-reel systems. The emergence of affordable compact cassettes allowed Japanese households to leapfrog the reel-to-reel era entirely—just as China would later skip over VHS tapes and move directly to CDs.

Thanks to the compact design of the cassette, tape-based recording and playback devices could now be significantly miniaturized. In 1966, Philips launched the "Radiorecorder," a portable unit combining a tape player, recorder, and radio—model Philips 22RL962. This machine could not only play music tapes and radio broadcasts but also record radio programs onto blank tapes for later playback. Its design closely matched the one depicted in the patent drawing for DE1937759U (Fig. 15.6). Over time, this kind of all-in-one machine evolved into a recognizable form: a front-facing tape slot in the center, speakers on both sides, and mechanical buttons on the top. In Western countries, it became known as the "boombox,"

[4] *Promoting Compact Cassettes Worldwide* (Chapter 5). (n.d.). In *Sony History*. Retrieved March 31, 2024 from https://www.sony.net/SonyInfo/CorporateInfo/History/SonyHistory/2-05.html

[5] Flamm, K. (2013). A tale of two standards: patent pools and innovation in the optical disk drive industry (No. w18931). National Bureau of Economic Research.

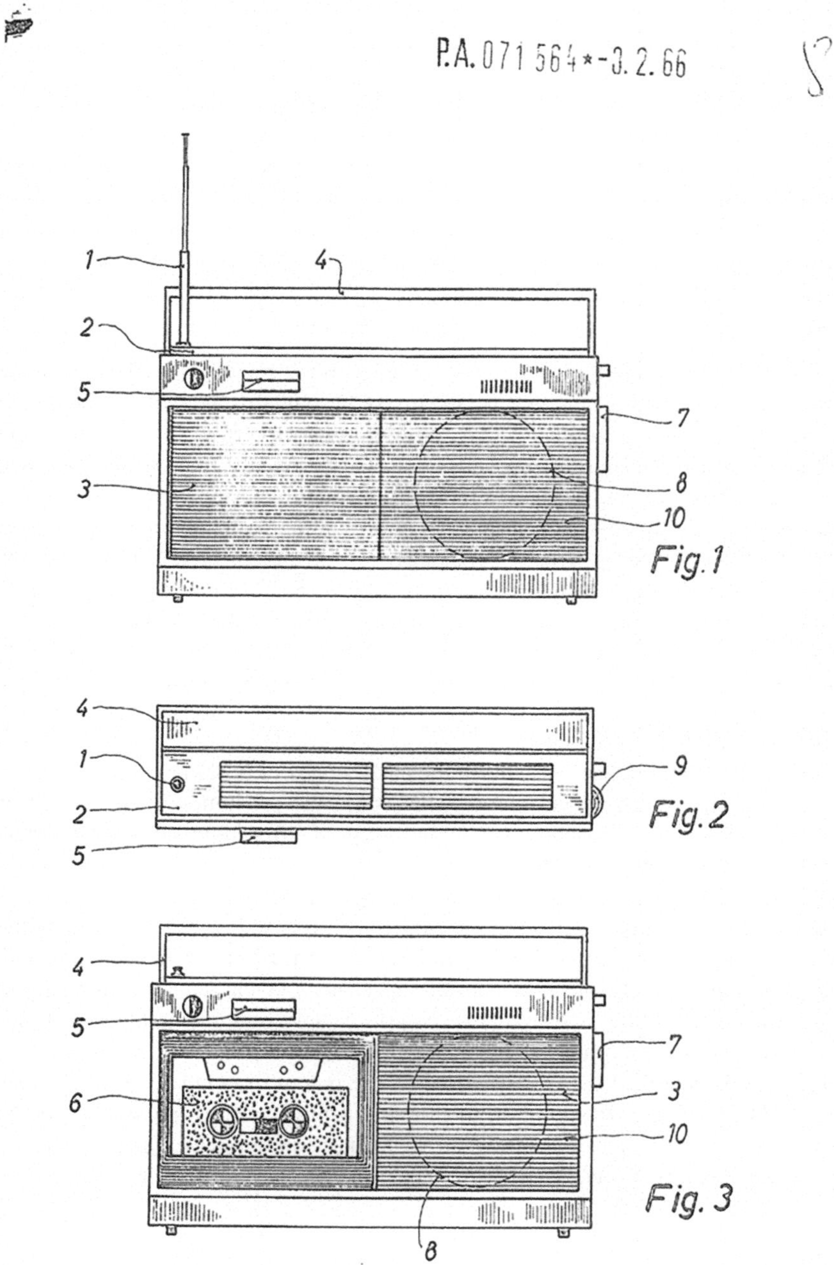

Fig. 15.6 Philips' 1966 patent drawing (DE1937759U)

eventually spawning a distinctive street culture. Even today, one might still see hippies, vagrants, and street dancers gathered around a boombox in cities across the globe.

For China's "post-70s" and "post-80s" generations, the boombox—often called "shouluji" or "luyinji" (portable recorder/player)—was the iconic home audio device of their youth, filled with memories of music and family entertainment.

After inventing the boombox, Philips no longer had a monopoly over the market. Japanese and American giants quickly entered the fray, launching countless iconic products. Though no one questioned Philips' status as the inventor, it did not dominate the boombox market.

With the advent of the boombox, the cassette music playback system had finally fulfilled Ottens' original vision of portability. Meanwhile, miniaturization of recording devices also continued apace. In 1969, Philips introduced the world's first truly portable cassette recorder, the Typ EL 3302. As shown in Fig. 15.7, the unit was already quite compact—roughly the size of a stack of six cassette tapes—just one step away from a true "personal stereo."

Yet even Ottens, with his sharp market instincts, never thought to pair headphones with a portable music player. In 1958, American engineer John C. Koss had already developed the world's first stereo headphones—designed specifically for music listening. Prior to this innovation, headphones were almost exclusively used in military communications and rarely associated with music enjoyment.

By the time Andreas Pavel arrived at Philips with his "belt-worn sound system," Philips had already mastered the standards for cassette tapes and their players. In sound storage technology, it stood alone. In terms of miniaturization, it was also a global leader. Its subsidiary, PolyGram, was on the rise and would later become one of the world's largest record labels.

Fig. 15.7 Philips Typ EL 3302 portable recorder, released in 1969. (Source: mib18 at German Wikipedia. (n.d.). *Philips EL3302* [Image]. Wikimedia Commons. https://commons.wikimedia.org/wiki/File:Philips_EL3302.jpg. This image is shared under the Creative Commons CC BY-SA 3.0 license)

Yet this time, there was no Ottens to respond to the moment. Nor was there a visionary like Ibuka or Steve Jobs to anticipate the future. And so, in a moment of inattention, Philips missed its chance to invent the Walkman. From the release of the Typ EL 3302 in 1969 to the debut of Sony's Walkman, another full decade would pass.

Chapter 16
The Sudden Rise of the Walkman

The question of who should be credited with inventing the Walkman—or the portable cassette player—remains a matter of debate. Andreas Pavel has long claimed to be the inventor of the cassette Walkman. However, according to Sony's official narrative, the inspiration came to Masaru Ibuka during a business trip, as he used a portable recorder to pass the time. The actual leader of the team that developed the first generation of the Walkman was Kozo Ohsone, who later served as Sony's Executive Vice President from 1994 to 1996. Another key figure was Shizuo Takashino, who became Sony's Chief Representative in China starting in 2005 and later served as Chairman of Sony (China) Co., Ltd. For over a decade, he remained Sony's top executive in China. Because of his direct involvement in the development of the first Walkman, some Chinese media have referred to him as the "Father of the Walkman."[1] Sony maintains that the invention of the Walkman had nothing to do with either Andreas Pavel or William Haas. As early as 1969, Sony had already developed a compact and robust portable recorder, the TC-50 (see Fig. 16.1), which gained fame after participating in the Apollo Moon landing mission. According to Sony's official website, the true prototype for the Walkman was the Pressman, a compact recording device developed in 1977. The name "Pressman" referenced the media industry, as it was originally designed as a tool for journalists to record interviews. To expand its functionality, Sony engineers modified the Pressman to create a prototype capable only of playback, not recording. The audio quality of this prototype impressed the engineers so much that they decided to adapt it into a dedicated music playback device.

Meanwhile, another Sony team was working on designing the lightest possible headphones. At the time, the world's lightest headphones weighed about 100 grams, but Sony's engineers aimed to cut that in half—and succeeded in building a prototype. In the spring of 1979, while listening to music on a TC-D5 recorder during a

[1] 索尼高篠静雄:*WALKMAN*的父亲不只我一人 [Sony's Shozo Takashina: I am not the only father of the Walkman]. (2006, September 6). *Sina Tech*. Retrieved March 31, 2024 from http://tech.sina.com.cn/it/2006-09-06/16581123060.shtml

Q. Yue, F. Yang, *Governing Innovation: A Global History of Patent-Driven Technological Progress*, https://doi.org/10.1007/978-981-95-5570-3_16

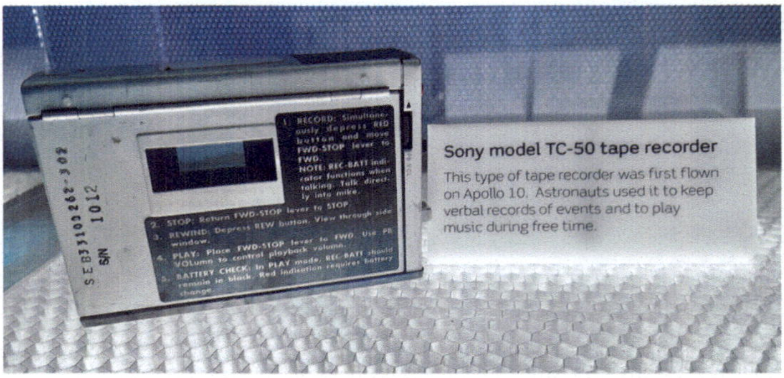

Fig. 16.1 Sony TC-50 portable recorder, released in 1969 (https://commons.wikimedia.org/wiki/File:Sony_TC-50.jpg)

business trip (see Fig. 16.2), Masaru Ibuka had a moment of insight. He decided to combine the Pressman and the headphone project. The result was the world's first true personal stereo—the Sony TPS-L2 (see Fig. 16.3). As noted earlier, Sony apparently believed that the simple combination of a tape player and headphones was not patentable, so the company did not initially file a patent for the Walkman, thereby allowing others to freely imitate it.

In fact, Sony's exploration of compact, high-quality recorders predated the Pressman. As early as 1972, Sony released the TC-55 portable recorder, which became one of the company's flagship products in the early 1970s. This period marked a turning point in Japan's manufacturing industry, as it began shedding its reputation for cheap, low-quality goods and entered the high-end market. In the United Kingdom, the TC-55 was priced at more than three times the cost of Philips' entry-level recorder. The device was solidly built—reportedly twice as heavy as one would expect upon picking it up.

The TC-55 featured a headphone jack on the side of the body. When headphones were plugged in, the external speaker would automatically mute. Although it lacked a stereo amplifier, the TC-55 was already quite close to functioning as a personal stereo. When Andreas Pavel filed a lawsuit against Sony in the UK for Walkman patent infringement, his expert witness presented a modified TC-55—equipped with a stereo amplifier and headphones—to demonstrate that Pavel's patent was technically feasible in the 1970s.

Due to the lack of effective patent protection, Sony's first-generation Walkman quickly faced a wave of competitors. Some notable rival products included the Sanyo M5550, the Akai PM-01, and the Aiwa TPS-30.[2] Each had its own distinct features: the M5550 was smaller and slimmer than Sony's Walkman, while the PM-01 was the first to offer FM radio functionality. To maintain its first-mover advantage, Sony had to keep innovating.

[2] *Walkman History 101*. (2004, August 19). *Vintage Electronics Have Soul—The Pocket Calculator Show Website*. Retrieved March 31, 2024 from https://www.pocketcalculatorshow.com/walkman/

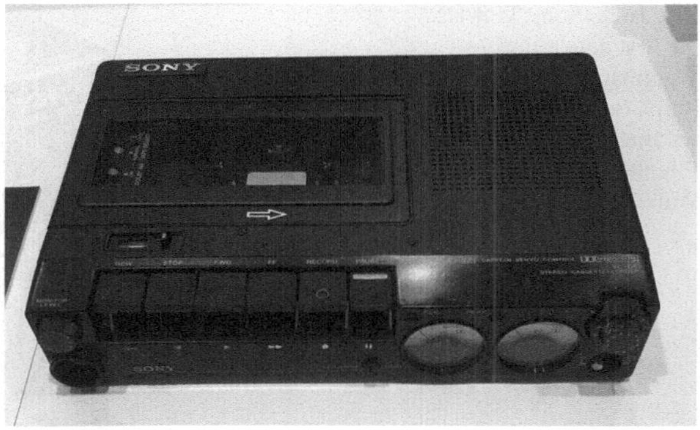

Fig. 16.2 Sony TC-D5 portable recorder (https://commons.wikimedia.org/wiki/File:SONY_TC-D5.jpg)

Sony launched the first-generation Walkman, the TPS-L2, in July 1979, and followed up in 1981 with the WM-3 model. Much like the boombox market, the Walkman market soon exploded with a diversity of products. Major electronics companies from Europe, Japan, and the U.S.—not just Philips but even General Electric—jumped into the fray. However, throughout the 1980s, Sony maintained its dominance through relentless innovation. As shown in Fig. 16.4, between 1979 and 1988, Sony upgraded the Walkman's platform twice and introduced several groundbreaking technologies, including DiscDrive, ultra-flat motors, gumstick batteries, and dynamic bass enhancement circuits.

As shown in Table 16.1, Sony's share of the U.S. and Japanese markets for cassette-based Walkman products in 1989 and 1990 approached 50% in both regions. Researchers Stephen W. Sanderson and Mustafa Uzumeri noted that throughout the 1980s, Sony held about 40% of global Walkman unit sales, and roughly 50% of global sales revenue.[3]

In 1981, Sony began mass-promoting the second generation of the Walkman—the WM-2 series. Sanderson and Uzumeri identified this as the first major technological leap in the Walkman product line. The WM-2 featured a smaller body and lighter weight. The next generation, the WM-20, was only half as thick as the WM-2. To further reduce size, the WM-20 used a 1.5-volt AA battery and consumed only half the power of its predecessor. Sony's R&D team spent 2 years perfecting the product, which finally hit the market in October 1983.

According to a study (see Table 16.2),[4] Sony was the source of numerous revolutionary innovations in cassette Walkman systems during the 1980s. Although almost all of these innovations were eventually copied by competitors, Sony's dominance in the personal stereo market remained unshaken.

Unlike the iPod, the Walkman product line was extraordinarily diverse, offering a wide array of models tailored to meet the varying preferences of consumers in

---

[3] Sanderson, S., & Uzumeri, M. (1995). Managing product families: The case of the Sony Walkman. *Research policy*, *24*(5), 761–782.

[4] Sanderson, S., & Uzumeri, M. (1995). Managing product families: The case of the Sony Walkman. *Research policy*, *24*(5), 761–782.

Fig. 16.3 Sony TPS-L2, the first generation Walkman, released in 1979 (https://commons.wikimedia.org/wiki/File:Original_Sony_Walkman_TPS-L2.JPG)

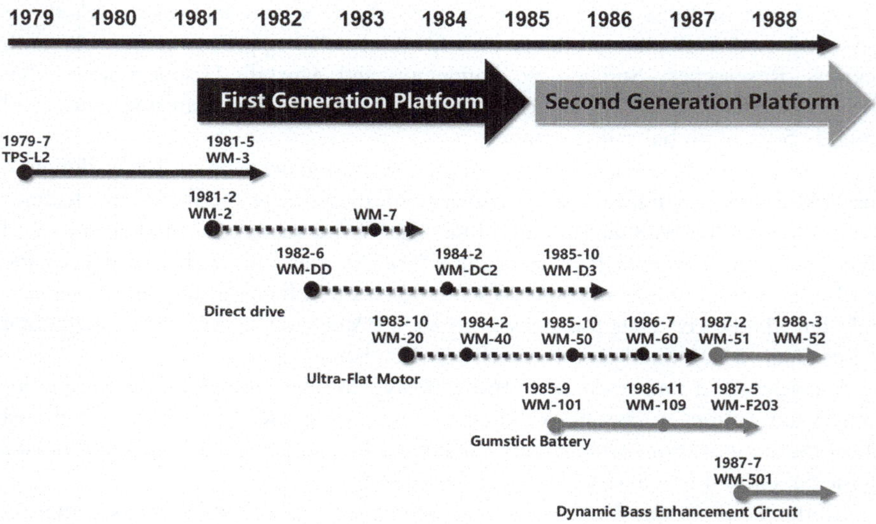

Fig. 16.4 Sony's platform upgrades for the Walkman, 1979–1988 (Fujita, K., Akai, R., & Amaya, H. (2009). Product Family Deployment Strategies under Different Types of Product Variety Design Circumstances. In *DS 58-4: Proceedings of ICED 09, the 17th International Conference on Engineering Design, Vol. 4, Product and Systems Design, Palo Alto, CA, USA, 24.-27.08. 2009* (pp. 215–226).)

Table 16.1 Market Share for Cassette Walkman Products, 1989–1990

Rank	Company	U.S. Market Share	Company	Japan Market Share
1	Sony	45–50%	Sony	46%
2	General Electric	10–12%	Aiwa	20%
3	Panasonic	6–8%	Panasonic	8%
4	Sanyo	4–6%	Toshiba	7%
5	Aiwa	4–6%	Sanyo	6%
	Others	25%	Others	13%

Note 1: U.S. data is based on interviews with industry insiders conducted by Sanderson & Uzumeri (1995) and confirmed by Sony's marketing department. Japanese data was sourced from *Market Share in Japan 1989*, Yano Research Institute Ltd., New York.
Note 2: At the time, Sony held more than 50% of Aiwa's shares and was deeply involved in its strategic operations. Aiwa's Walkman product line can be seen as an extension of Sony's, targeting different market segments, and thus the two companies' market shares can be considered jointly.

Table 16.2 Technological Innovations in 1980s Walkman Systems

Technology	Company	Year	Copied by Competitors
First personal stereo	Sony	1979	Yes
AM/FM stereo	Sony	1980	Yes
Stereo recording	Sony/Aiwa	1980–1981	Yes
FM tuner tape integration	Toshiba	1980–1981	Yes
Auto-reverse	Sony	1981–1982	Yes
Headphone radio	Sony	1981–1982	Yes
Dolby sound enhancement	Sony/Aiwa	1982	Yes
Shortwave tuner	Sony	1983	No
Remote control	Sony	1983	Yes
Built-in speaker	Sony	1983	Yes
Waterproofing	Sony	1983–1984	No
Graphic equalizer	Sony	1985	Yes
Solar power	Sony	1986	No
Preset radio channels	Panasonic	1986	Yes
Dual-cassette technology	Sony	1986	No
TV audio band	Sony	1986–1987	Yes
Digital tuning	Panasonic	1986–1987	Yes
Child mode	Sony	1987	No
Enhanced bass	Sony	1988	Yes

Japan and the United States. Many Japanese consumers lived in large metropolitan areas and commuted daily on public transportation. They tended to prefer models that were compact, high-performing, and easy to recharge. Remote controls embedded in the headphone cord were specifically designed with Japanese office workers in mind—because in a crowded subway car, it was inconvenient for passengers to reach into their pockets to operate the device. In contrast, American users had no such need.

Moreover, Japanese consumers were generally less interested in radio, and the country had far fewer radio stations compared to the United States. As a result, many Japanese-market Walkman models didn't even include FM tuner functionality.

To cater to American consumers' love of individuality, all Walkman manufacturers—including Sony—emphasized diverse and expressive product designs, as opposed to the relatively standardized look of Japanese-market models. Throughout the 1980s, Sony released no fewer than 250 different Walkman models in the U.S. market. Americans preferred to use Walkmans while relaxing outdoors or during physical activity. Listening to music on the beach was a common scenario. Because of this, Sony was the first to launch waterproof models in the U.S., where they were warmly received. In contrast, Japan, despite being an island nation, rarely saw the release of waterproof Walkmans.

The bright yellow waterproof "Sports Walkman" was specifically designed for the American market. Later personal CD players would adopt similar aesthetics. The "Sports Walkman" series remained unrivaled throughout the 1980s, and all competitors who tried to challenge it eventually withdrew from the race.

Although Sony did not patent the idea of combining headphones with a portable recorder, the company still applied for numerous patents throughout the Walkman's product lifecycle. The secret to Sony's dominance in the global personal stereo market lay within these innovations.

Chapter 17
Products as Art

Let us begin with the very first patent associated with the Walkman. The original blue-and-silver design of the first-generation Walkman is iconic—a classic in the history of industrial design. Yet even among seasoned Sony fans, few know the purpose of that conspicuous orange-yellow button on the top of the device.

In a 1979 utility model patent application filed by Sony (patent number JP60017052Y2), inventor Toshio Asai—who later testified as a member of the Walkman development team during the Pavel litigation—outlined this very detail. The patent, filed in Japan on June 21, 1979, was also filed as an invention patent in 13 countries. Its American equivalent is US4395739A, whose diagrams are identical to those of the original 1979 Walkman model (see Fig. 17.1).

In his testimony during the Pavel lawsuit, Toshio Asai explained that Sony did not believe that combining headphones with a portable tape recorder constituted a patentable invention. Nonetheless, Sony registered at least one patent related to the first-generation Walkman and secured protection in key markets.

That patent, however, did not primarily cover the integration of headphones and a recorder. Instead, it protected a now largely forgotten feature of the Walkman: the "hotline." According to Akio Morita, before the product officially launched, he brought a prototype home for testing. While he enjoyed listening to music, his wife expressed displeasure, saying she felt excluded. Morita relayed this feedback to the development team.

As a result, the Walkman was designed with two headphone jacks, enabling two users to listen simultaneously. To allow them to talk while sharing the music, Sony thoughtfully added a feature called the "hotline": by pressing the orange-yellow button on the top of the device (see Fig. 17.2), the Walkman would lower the music volume and activate a built-in microphone, allowing both users to hear each other through their headphones. This feature was dubbed the "hotline button" (referred to as the "talk button" in the patent documentation).

Contrary to Morita's expectations, while the Walkman was an instant hit, the "hotline" feature was largely ignored—few people were inclined to share their music player. The feature was quickly abandoned in later models.

Q. Yue, F. Yang, *Governing Innovation: A Global History of Patent-Driven Technological Progress*, https://doi.org/10.1007/978-981-95-5570-3_17

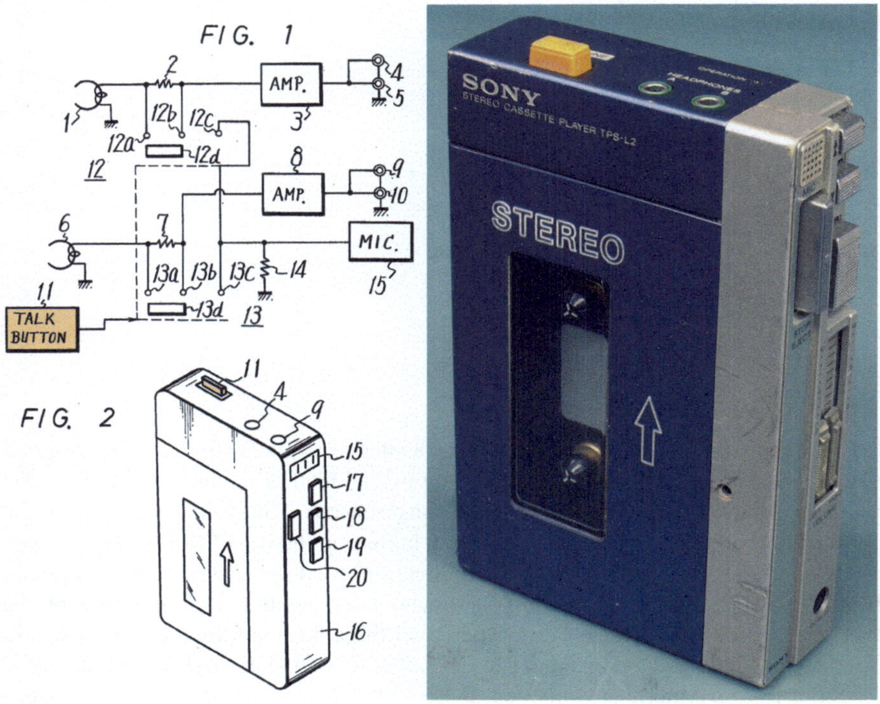

Fig. 17.1 Comparison of patent drawings from US4395739A (left, center) and the first-generation Walkman (right)

In February 1981, Sony released the second-generation Walkman, model WM-2 (also known as the Walkman II), which was smaller and more user-friendly than the original (see comparison in Fig. 17.2). In June 1982, Sony introduced the WM-DD model, incorporating Disc Drive technology adapted from the TC-D5 recorder favored by Ibuka Masaru. This model was designed as a high-end product for audiophiles. In October 1983, Sony released yet another new Walkman, model WM-10 (marketed in the U.S. as the WM-20), which had a completely different appearance from both the WM-2 and WM-DD. The WM-10 and its successors emphasized ultra-thin design and embodied Sony's pride in miniaturization technology.

Throughout the 1980s, most of Sony's cassette-based Walkmans were essentially iterative variations on these three core models. In a scholarly paper, Stephen W. Sanderson and Mustafa Uzumeri referred to these as the "three platforms" of the Walkman (see Fig. 17.3).[1] According to their research, Sony developed hundreds of Walkman models based on these three platforms, more than 250 of which were released in the U.S. alone. Sony engineers essentially mixed and matched modular

[1] Sanderson, S., & Uzumeri, M. (1995). Managing product families: The case of the Sony Walkman. *Research policy*, *24*(5), 761–782.

Fig. 17.2 First generation (left) and second generation (right) Walkman. (**Source:** Birmingham Museums Trust. (n.d.). *Thinktank Birmingham—object 1986S03911.00001* [Image]. Wikimedia Commons. https://commons.wikimedia.org/wiki/File:Thinktank_Birmingham_-_object_1986S03911.00001%281%29.jpg. This image is shared under the Creative Commons **CC BY-SA 4.0** license)

components on these platforms—"like building with LEGO blocks." This platform and modular approach drastically reduced design costs. Even when a particular model failed, the low development cost allowed for generous experimentation. We identified foundational patents corresponding to each of these platforms—the technological foundation of the Walkman's success was embedded in these innovations.

Let us begin with the mainstream WM-2 platform. Released in 1981, the WM-2 introduced a significant design improvement by relocating the tape head from inside the cassette compartment to the lid of the compartment, thereby saving precious internal space. Another major refinement was the repositioning of the control buttons from the side of the device to its front face, which reduced the device's overall volume and prevented accidental operation when carried in a pocket. Sony filed patents to protect these key innovations and established patent coverage in major international markets. While the dual headphone jacks were retained, the "hotline" function was decisively discarded.

On December 29, 1980, Sony submitted at least five patent applications in Japan for the core technological advancements of the WM-2 series. Figure 17.4 (left) shows a common illustration shared among these applications; the image on the right depicts an actual WM-2 unit. One year later, using these Japanese filings as priority documents, Sony filed four corresponding patents in countries including the UK, France, Germany, the US, South Korea, Canada, Austria, Spain, Australia, and

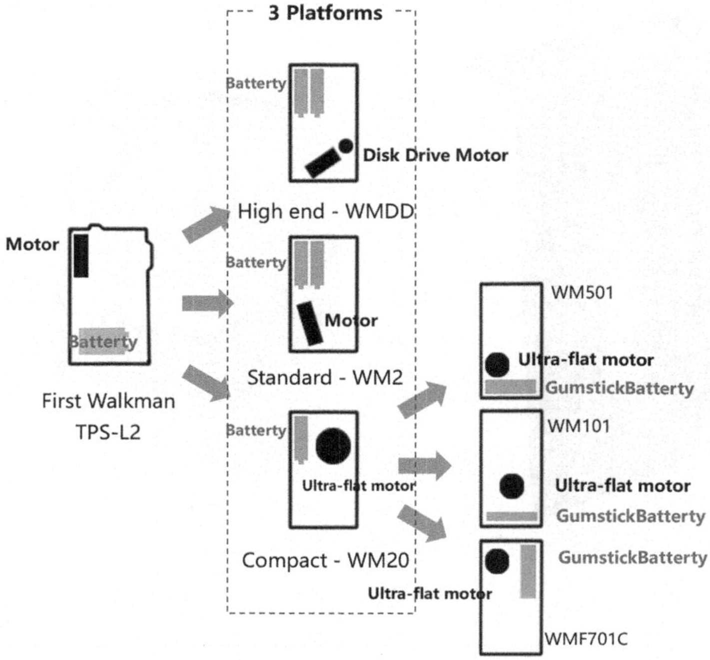

Fig. 17.3 Three Walkman platforms

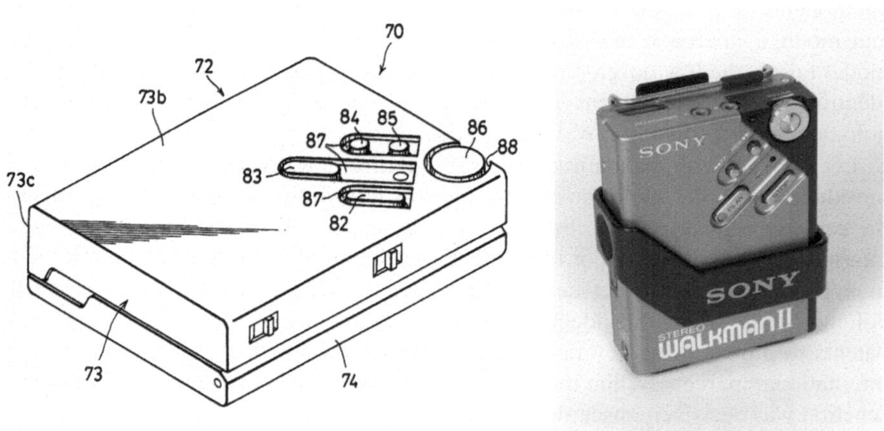

Fig. 17.4 Patent drawing from Sony's December 29, 1980 patent applications (left) and actual WM-2 series Walkman (right)

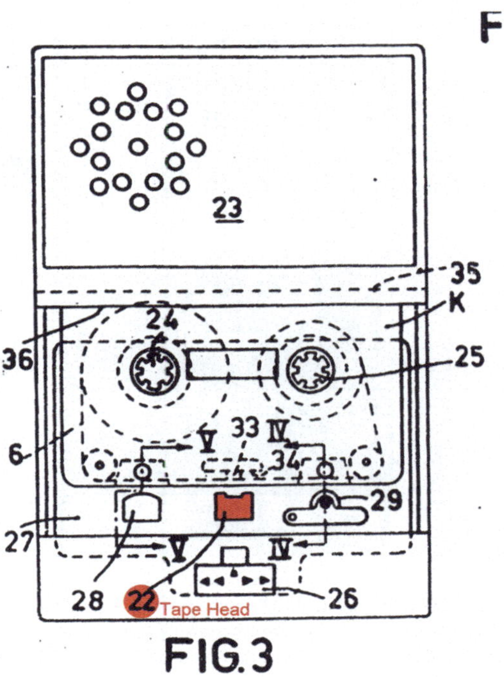

Fig. 17.5 Patent drawing from DE1896300U

Brazil—establishing a global portfolio. Notably, Sony's Walkman-related patents in the 1980s rarely extended beyond these ten countries to others like the Netherlands, Italy, or the Nordic nations.

As previously mentioned, one of the WM-2's key innovations was the relocation of the tape head. Prior to the WM-2, the conventional placement of the tape head mirrored the design found in Philips' 1963 German patent for a "cassette tape recorder/playback device" (DE1896300U). As shown in Figs. 17.5 and 17.6, the tape head was typically mounted inside the cassette compartment, near the control buttons. When the "play" button was pressed, the tape head would lower to make contact with the tape. The original Walkman model followed this same design.

As shown in Fig. 17.7, the tape head placement in the first-generation Walkman closely resembled the original Philips layout. Upon pressing "play," the head would drop down to contact the tape. In the specification of US4511941A, Sony engineers noted that the mechanical linkages between the buttons and tape head, as well as the vertical space needed for the tape head's movement (indicated by the bidirectional arrows in Fig. 17.8), were major bottlenecks in reducing the size of portable recorders. The primary objective of US4511941A was to minimize this spatial requirement.

An earlier American patent filed by AT&T in 1974 (US3909845A) addressed this issue by relocating the tape head to the bottom of the cassette compartment lid (see Fig. 17.9). This design saved internal space, but the tape head remained in constant contact with the tape—even during fast-forward and rewind—posing a risk of tape damage.

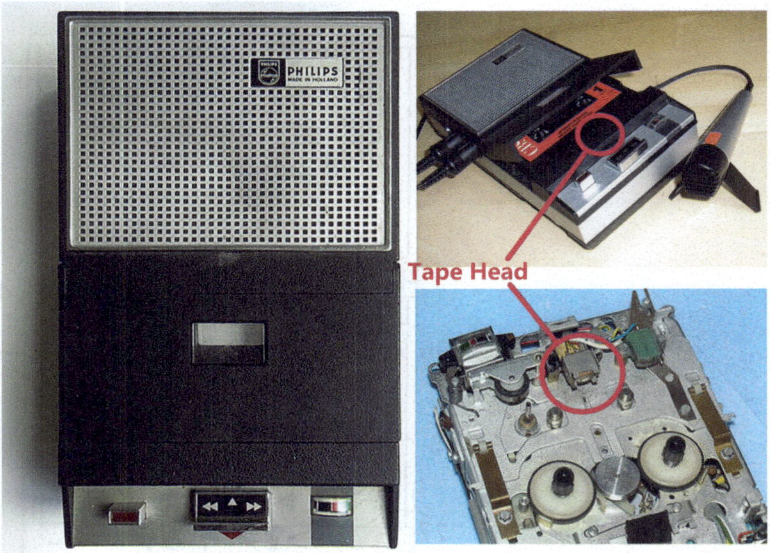

Fig. 17.6 Philips EL-3302 recorder from the 1960s (**Source:** Erkaha; adjustment by user Retired electrician. (n.d.). *Kasetten Recorder PHILIPS (adj)* [Image]. Wikimedia Commons. https://commons.wikimedia.org/wiki/File:Kasetten_Recorder_PHILIPS_(adj).jpg. This image is shared under the Creative Commons **CC BY-SA 4.0** license), with design and tape head placement matching DE1896300U

In US4511941A, Sony avoided simply fixing the tape head to the lid. Instead, they installed an intricate mechanical assembly called the "head mounting unit" (153) on the underside of the lid. Both the tape head (100) and pinch roller (101) were attached to this unit (see Fig. 17.10).

During fast-forward or rewind, the head would not touch the tape. Only when the "play" button was pressed would the head mounting unit pivot slightly in the b' direction (as shown in Fig. 17.10), bringing the tape head into contact with the tape. Pressing "stop" would trigger a torsion spring (163) to return the unit to its original position (see Fig. 17.11).

At the time, some classic cassette Walkmans from Panasonic and Sanyo—such as the RQ-S75 and JJ-P101—featured front-facing control buttons and mounted the tape head on the cassette lid, characteristics similar to the WM-2 series. Whether Panasonic and Sanyo had designed around Sony's patents or obtained licenses remains unclear.

Apart from the aforementioned innovations, the WM-2 series introduced a notable visual change: relocating the control buttons from the device's side to its front. Traditionally, the internal mechanics of a tape recorder rely on a motor that drives several gears to rotate the tape spools. To switch between playback, fast-forward, and rewind modes, the system must adjust which gears or idlers are engaged.

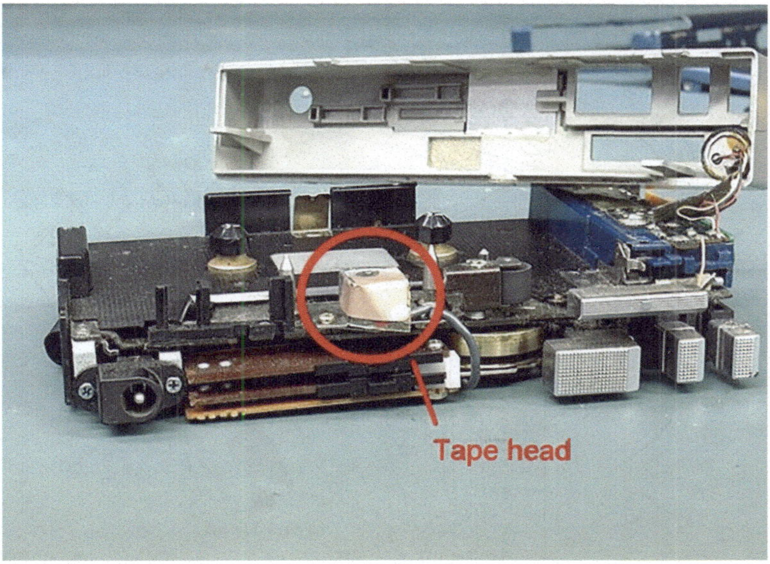

Fig. 17.7 Tape head location in the first-generation Walkman (circled). (**Source:** Dave Jones. (n.d.). *Original 1979 Sony TPS-L2 Walkman Teardown* [Image]. Wikimedia Commons. https://commons.wikimedia.org/wiki/File:Original_1979_Sony_TPS-L2_Walkman_Teardown_(18639098196).jpg. This image is shared under the Creative Commons **CC BY 2.0** license)

Fig. 17.8 Internal layout of the first-generation Walkman TPS-L2. (Source: Dave Jones. (n.d.). Original 1979 Sony TPS-L2 Walkman Teardown [Image]. Wikimedia Commons. https://commons.wikimedia.org/wiki/File:Original_1979_Sony_TPS-L2_Walkman_Teardown_(18477710770).jpg. This image is shared under the Creative Commons CC BY 2.0 license)

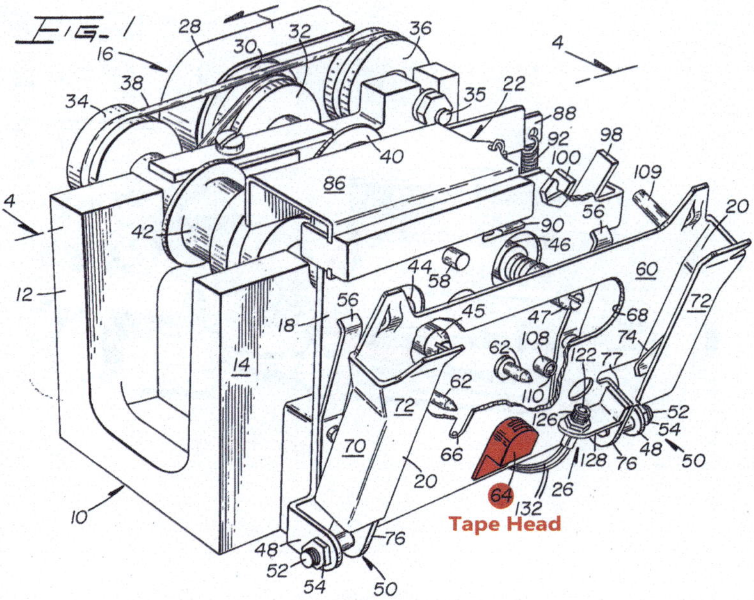

Fig. 17.9 Tape head position in US3909845A patent by AT&T

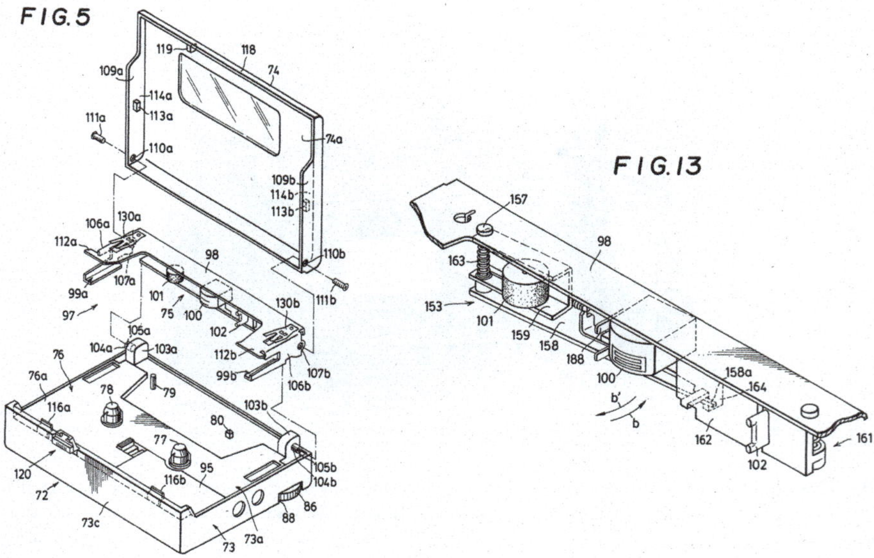

Fig. 17.10 Sony's head mounting unit from US4511941A

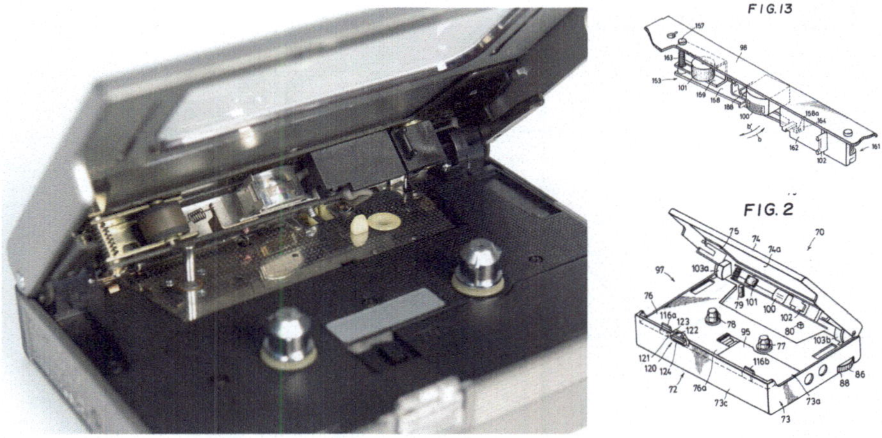

Fig. 17.11 WM-2 interior. (**Source:** Birmingham Museums Trust. (n.d.). *Thinktank Birmingham—object 1986S03911.00001* [Image]. Wikimedia Commons. https://commons.wikimedia.org/wiki/File:Thinktank_Birmingham_-_object_1986S03911.00001%281%29.jpg. This image is shared under the Creative Commons **CC BY-SA 4.0** license.) and comparison with US4491287A patent drawing

Before the WM-2, these gears and idlers were typically arranged on the same plane. The common design involved buttons controlling mechanical linkages that moved gears laterally (parallel to the base of the device) to change modes. For example, in US4010493A—a prior art cited by Sony—the fast-forward and rewind functions were toggled by shifting gear 27 (see Fig. 17.12). When the "rewind" button was pressed, the blue mechanical structure in Fig. 17.12 moved gear 27 into mesh with green gear 20, activating rewind mode. Pressing "fast-forward" moved gear 27 in the opposite direction, engaging a different set of gears to achieve that function.

This configuration required sufficient internal space to accommodate gear movement and associated mechanical linkages, which themselves added to the device's bulk.

In contrast, Sony's design for the WM-2 series enabled switching between playback, fast-forward, and rewind by moving all gears and idlers along an axis perpendicular to the device base—referred to in the patent documentation as "axial movement." As shown in Fig. 17.13, the gears and idlers in the WM-2 were made very thin, and could engage or disengage with other gears by moving up or down just a few millimeters.

The buttons themselves were also reoriented: instead of being pressed sideways, they were now pressed vertically. This allowed the buttons to be located on the front of the device. Since the gears only needed to move slightly, the buttons could have short travel distances, further reducing the device's thickness.

As shown in Fig. 17.14, during normal playback mode, the second drive gear (222) meshes with gear 234, which controls the main tape reel, causing it to rotate

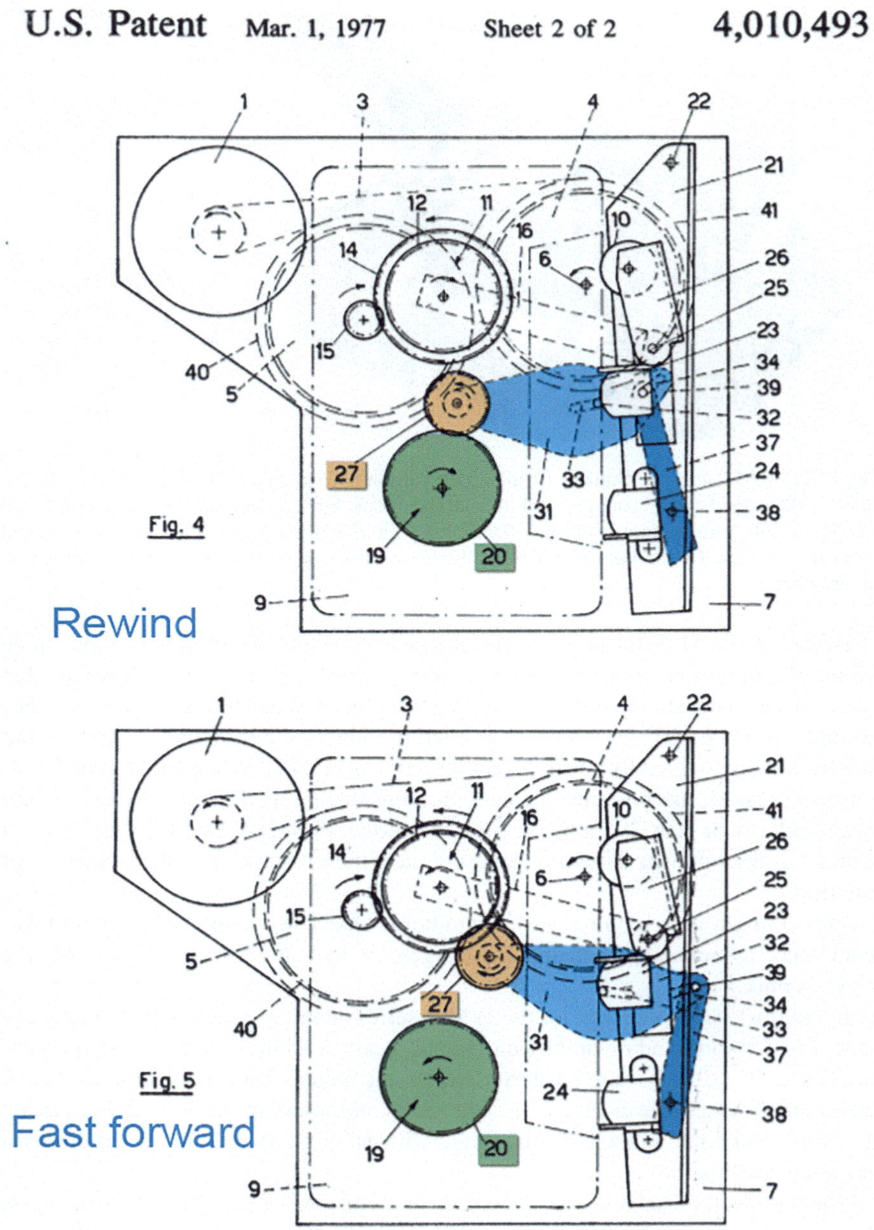

Fig. 17.12 Gear arrangement in US4010493A patent

Fig. 17.13 Axial gear movement in Sony's US4491287A patent

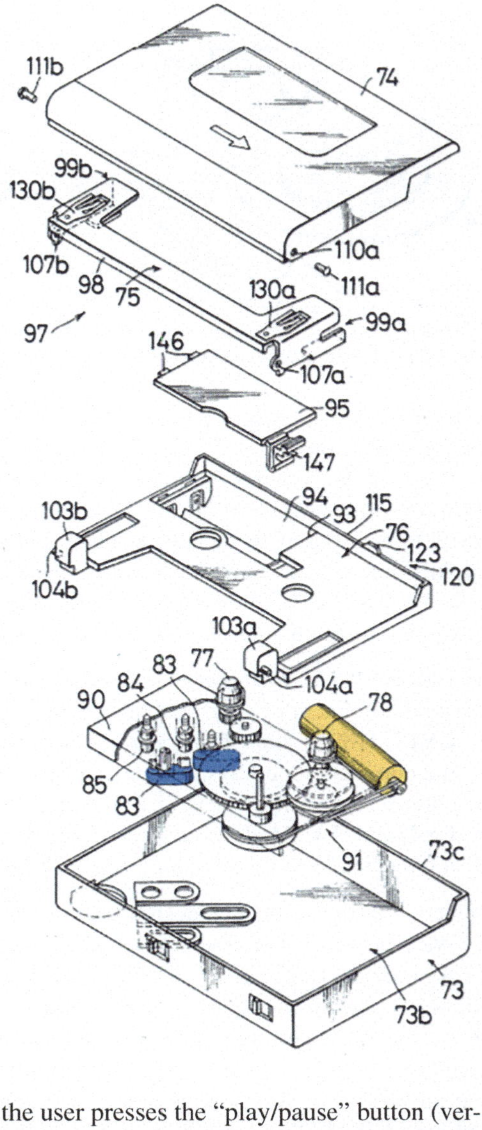

at a steady speed. To pause playback, the user presses the "play/pause" button (vertically aligned with the device base), which causes the underlying component (283) to move downward and latch onto a protrusion (295) on the first drive gear, halting the tape. This action also disengages the tape head.

When the "fast-forward" button is pressed, its connecting lever (253) lifts both operation plates 248 and 245. This elevates the first drive gear (221) to its highest position. The second drive gear (222) disengages from gear 234, while the first drive gear (221) meshes with a smaller gear located beneath gear 234 (indicated in the

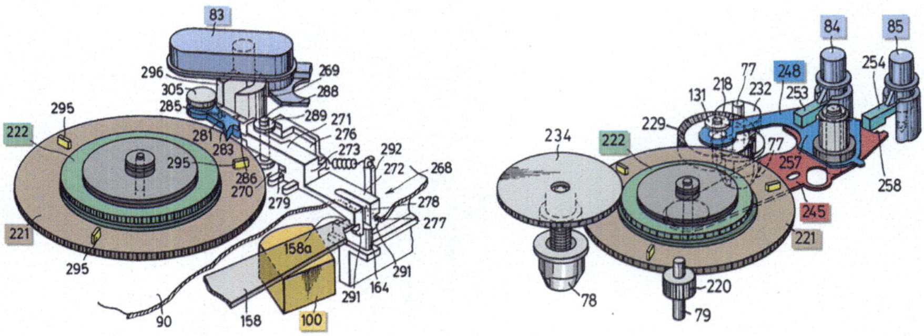

Fig. 17.14 Patent diagrams from US4491287A and US4453189A

dashed lines of the right panel of Fig. 17.14), rapidly rotating the main reel and enabling fast-forward.

Pressing the "rewind" button activates lever 254, which lifts only operation plate 245. As with fast-forward, gear 222 disengages from gear 234. However, due to the restriction imposed by operation plate 248, the first drive gear (221) rises only slightly and instead engages with shifting gear 229. This gear redirects motion to the secondary reel (77), allowing for rewind mode.

Sony registered at least three U.S. patents covering this gear and button system:

- **US4491287A**: *Pushbutton Actuated Cassette Tape Transport*—protecting the axial alignment of gears and button direction perpendicular to the base.
- **US4453189A**: *Tape Transport Mechanism for Cassette Tape Player*—covering the complete tape transport mechanism of the WM-2, including main reel, secondary reel, shifting gear, and drive gear relationships.
- **US4542431A**: *Mode Changing Mechanism for Cassette Tape Player*—protecting mechanisms for switching and locking playback modes.

Next comes the anti-shock design featured in Sony's high-end Walkmans.

We now turn to the anti-shock design featured in Sony's high-end Walkman models.

As previously discussed, the inspiration for the Walkman stemmed from Ibuka Masaru's personal experience of listening to music on the TC-D5 recorder. He preferred the TC-D5 because it employed the pioneering Disc Drive technology, which significantly reduced wow and flutter, ensuring stable and high-quality audio output. In June 1982, Sony applied this same technology to the Walkman, releasing the renowned WM-DD series. Priced higher than other models, the WM-DD was a premium product aimed at audiophiles.

Visually, the WM-DD and WM-2 looked almost identical (see Fig. 17.15), with nearly the same button shapes and positions. In fact, the two models shared most of their casing molds.

In the WM-2 series, the motor used a belt (structure 91 in Fig. 17.16) to rotate the capstan and spool, enabling playback, fast-forward, and rewind functions. This design was simple and cost-effective, but over time, the belt would deform and

Fig. 17.15 Walkman DD (left) and WM-2 (right). (Source: hdboy88. (2017, December 9). *Sony Walkman collection (Sony WM-DDIII, red)* [Image]. Wikimedia Commons. https://commons.wikimedia.org/wiki/File:Sony_WM-DDIII_red.jpg. This image is shared under the Creative Commons CC BY-SA 2.0 license)

produce increased noise and distortion when the device was jostled. In many Walkman repair videos on YouTube, the most common issue in these 30-plus-year-old units is worn or broken belts. Once replaced, these vintage devices can often perform like new.

Back in 1978, Sony had already addressed this issue using Disc Drive technology: instead of having the motor's shaft directly rotate the capstan via a belt, a rubber idler wheel (structures 83, 84, and 85 in Fig. 17.16) was installed on the motor shaft. This idler wheel pressed against a capstan (upper right of Fig. 17.16), and

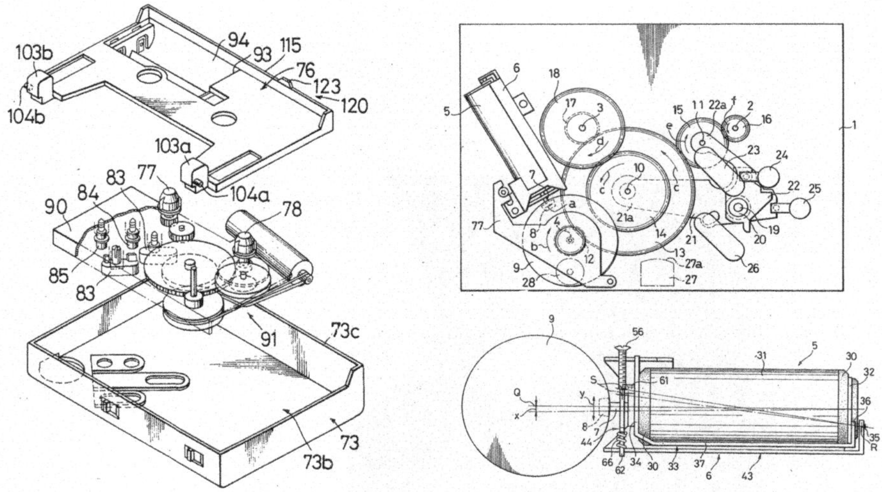

Fig. 17.16 Patent drawing from U.S patent US4701816A

when the idler wheel rotated in direction "a," it would spin the capstan in direction "b" through frictional contact.

This design was far more stable than traditional belt systems. It was this innovation that led Ibuka to favor the TC-D5 (Fig. 17.17), ultimately catalyzing the creation of the Walkman. However, in this setup, both the idler wheel and capstan were made of rubber. Constant contact between them exerted long-term pressure, which could cause deformation.

To address this issue, Sony designed a pivoting motor mount (structure 6 in Fig. 17.16), using light mechanical force to press the idler wheel against the capstan. The contact points between the motor and mount were designated "S" and "R" (see lower right in Fig. 17.16), with a slight offset from the motor's axis.

When the motor was idle, the pressure applied by the idler wheel to the capstan was equivalent to a force of roughly 20–30 grams (compared to the 90–100 grams in earlier models). Once the motor was engaged and its center of gravity shifted, the resulting pressure and friction were sufficient to rotate the capstan. This technology was detailed in Sony's US4701816A patent, filed in March 1984.

In addition to these innovations, Sony also developed the WM-10 series to showcase extreme compactness. The first model, the WM-10 (known as the WM-20 in the U.S.), was released in October 1983. It remained the smallest cassette Walkman on the market until the end of the twentieth century.

The WM-10 featured a drawer-style sliding mechanism, allowing it to be smaller than a cassette case when not loaded with a tape. To save space, it used only a single AA battery. Beginning with the WM-101 in 1985, models in the WM-10 platform started using long, gumstick-style rechargeable batteries. However, due to the fixed size of cassette tapes, even with gumstick batteries, later models could rarely match

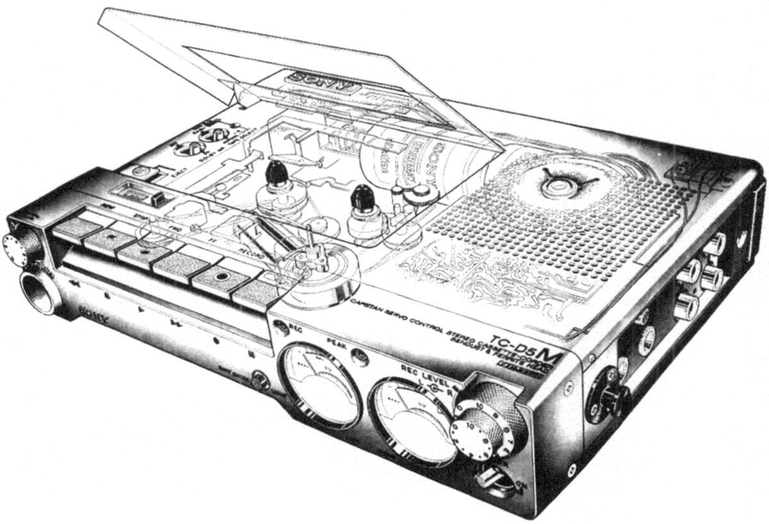

Fig. 17.17 Sony TC-D5. (**Source:** MIKI Yoshihito. (n.d.). *TC-D5* [Image]. Flickr. https://www.flickr.com/photos/mujitra/24107928745. This image is shared under the Creative Commons **CC BY 2.0** license)

the compactness of the original WM-10. In terms of slimness, the WM-10 was Sony's "peak debut."

In an advertisement from around 1983, a Walkman was held delicately with tweezers to emphasize its featherweight design. It's hard not to wonder whether Steve Jobs's famous presentation of the MacBook Air in a manila envelope drew inspiration from this image. Like the MacBook Air in the MacBook lineup, the WM-10 was the lightest, smallest, and most affordable model in the Walkman range.

Inside the WM-2 Walkman, the motor was roughly the size of an AA battery, as shown by the cylindrical component in Fig. 17.18. When powered on, the motor shaft would rotate and drive the capstan via a belt, causing the tape to move and play music.

To achieve a smaller footprint, Sony removed one of the AA batteries in the WM-10, reducing the power source to a single battery (while the WM-2 and WM-DD series used two). Even so, internal space remained tight—there was simply no room for the conventional cylindrical motor.

It isn't difficult to design a flat motor, but Walkman users often listened to music while moving. To ensure the motor remained stable during operation, the motor shaft (labeled 38 in Fig. 17.19) had to be sufficiently long and supported by at least two bearings spaced apart to minimize wow and flutter.

Sony's brilliant solution in the WM-10 was to ingeniously integrate the motor shaft and bearing assembly *inside* one of the tape spindles (labeled 9 in Fig. 17.19). What looked like a simple reel spindle was, in fact, a marvel of engineering.

In U.S. Patent US4630149A, filed by Sony in 1984, claim 1 reads:

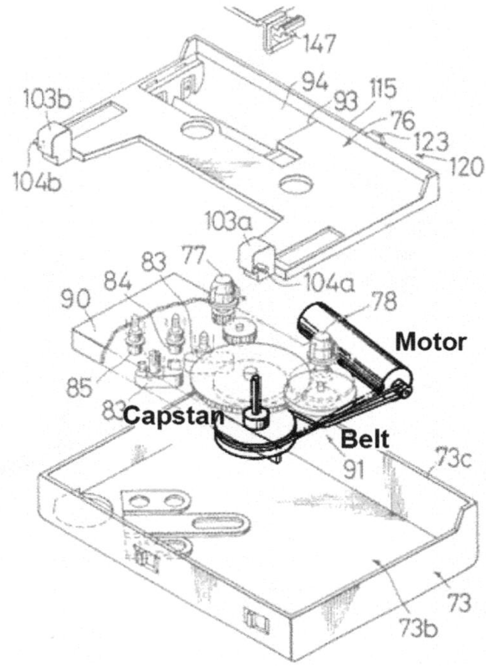

Fig. 17.18 Patent drawing from U.S patent US4491287A

> *...bearing means disposed, at least in part, within one of said spindle means, a portion of said shaft remote from said end portion having the rotor secured thereon extending axially into said one spindle means and being rotatably supported by said bearing means in the latter with said rotor and stator being axially remote from the reel mounted on said one spindle means;*

As seen in Fig. 17.19, component 34 is the rotor, situated above the stator (42). The cylindrical component (38) is the motor shaft. A sleeve bearing (structure 28) is mounted at the upper end of the shaft, inside the assembly labeled 27, and a ball bearing (29) is placed at the lower end near the stator. The spacing between the two bearings ensures rotational stability. This clever arrangement is what made the WM-10's ultra-compact form factor possible.

However, there was a slight catch. The WM-10 only achieved its smallest dimensions when no cassette was inserted. Once a tape was loaded, the cassette compartment had to slide outward like a drawer, and the device became marginally larger than a tape case (see Fig. 17.20). This design was detailed in U.S. Patent US4614991A (Fig. 17.21).

A Chinese counterpart to this patent, CN1009504B, was one of the earliest filed by Sony in China, dated June 19, 1985. Since China only established its patent system that year and had not yet joined the Paris Convention, this document did not cite any priority claims. Nevertheless, based on its content, it is unquestionably the Chinese equivalent of US4614991A.

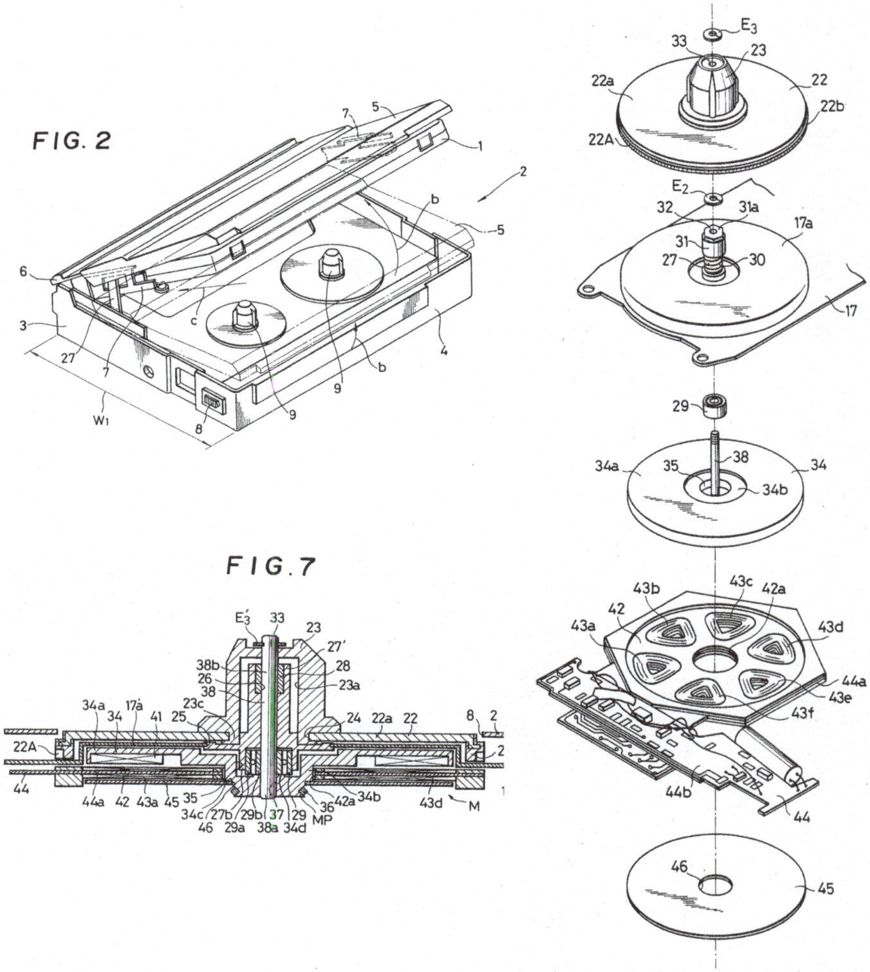

Fig. 17.19 Patent drawing from U.S patent US4630149A

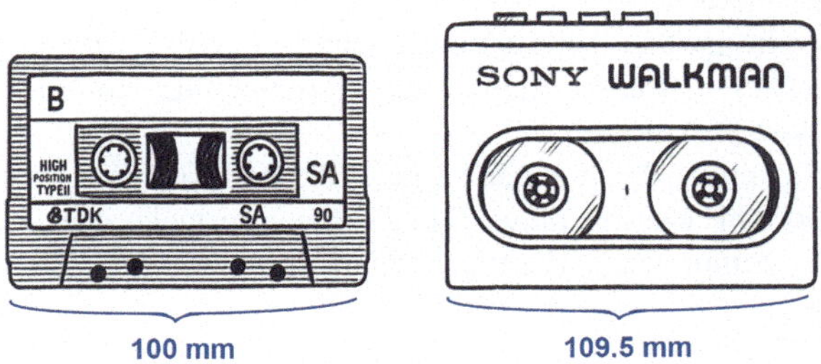

Fig. 17.20 Size of WM-10 (right) and a cassette (left)

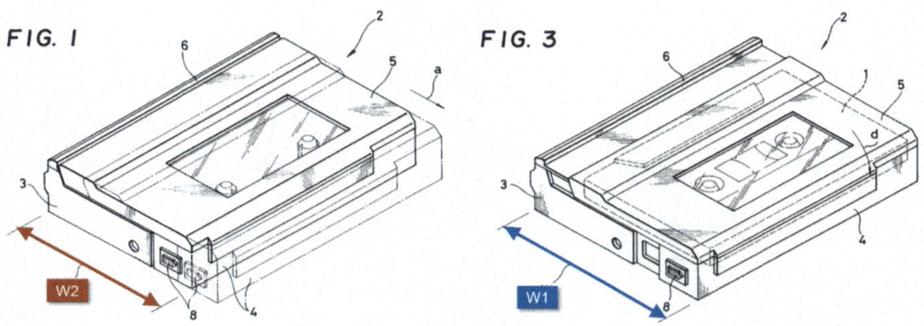

Fig. 17.21 Patent drawing from U.S Patent US4614991A

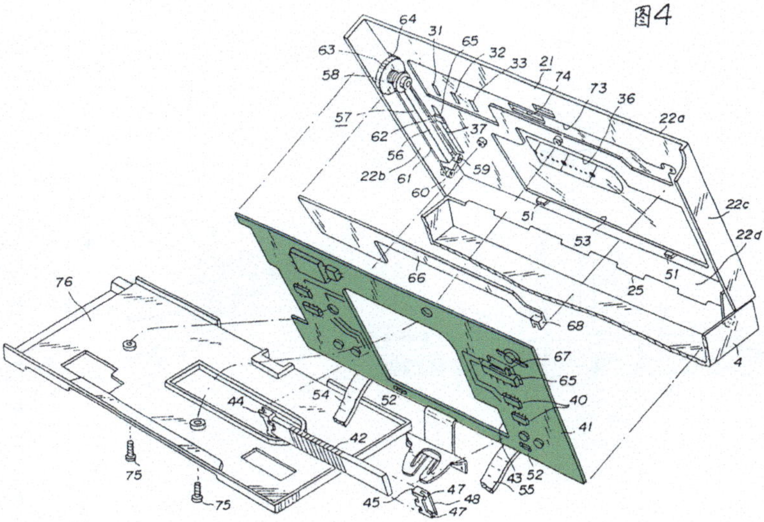

Fig. 17.22 Patent drawing from CN85101051A

As mentioned earlier, the WM-10 only reached its smallest size when no cassette was inserted. But what use is a Walkman without a cassette? Sony's answer: use it as a radio.

In another 1985 Chinese patent, CN85101051A—"Cassette Tape Player with Radio"—Sony added radio functionality to the WM-10, resulting in the WM-F10 model.

Sony's documentation in CN85101051A noted that integrating a radio into a portable cassette player faced a major obstacle: interference from the player's metallic components. The text explained:

> To ensure necessary mechanical strength, components like capstans, motor casings, and chassis must be made of metal... Incoming radio signals are blocked or disrupted before they reach the antenna rod.

Sony's solution was to install all radio circuitry on a printed circuit board (the gray section in Fig. 17.22) mounted inside the cassette lid:

> The radio is housed in the hinged plastic lid of the cassette player... Since the lid requires minimal mechanical strength, it is made of plastic. The rod antenna is also mounted on this lid, maximizing its distance from the internal metal parts.

As noted, the WM-10 series was both the lightest and the most affordable among Sony's Walkman products. This is likely why Sony prioritized patenting the WM-10 platform in China during the 1980s.

Speaking of which, we must mention Sony's early patent strategy in China. In 1985, the inaugural year of China's patent system, Japan filed over 3000 patent applications—more than any other foreign country. The Netherlands ranked fifth, with nearly half of its filings submitted by Philips (see Fig. 17.23).

Sony alone filed 176 patents in China, ranking fifth among all applicants and fourth among foreign companies, trailing only Hitachi, Sharp, and Philips. By contrast, American companies were far less active. Only Westinghouse filed more than 100 patents, while IBM—famously prolific elsewhere—submitted just 12. Among the top ten corporate applicants, more than half were Japanese companies. In the years that followed, these familiar names—Hitachi, Sharp, Sony, Panasonic, and Philips—became household brands for China's post-1980s generation (see Fig. 17.24).

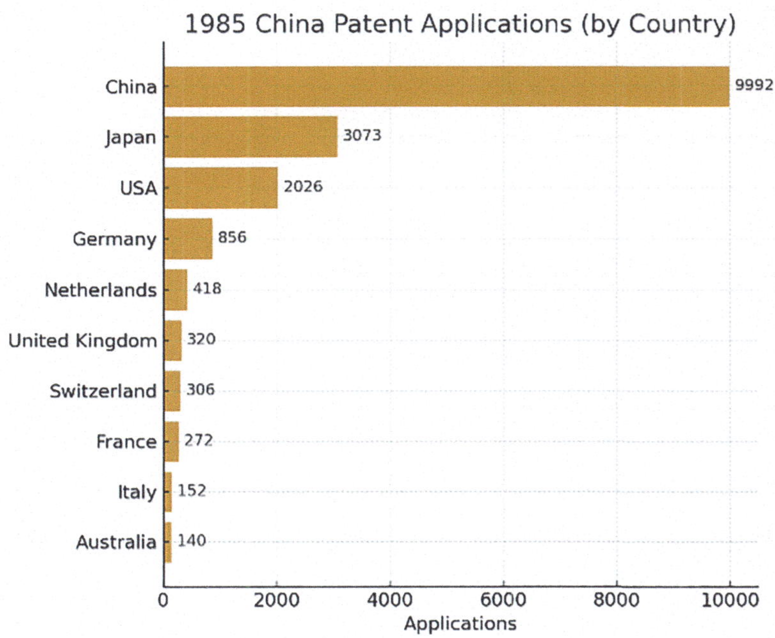

Fig. 17.23 Patent applications filed in China in 1985, ranked by country

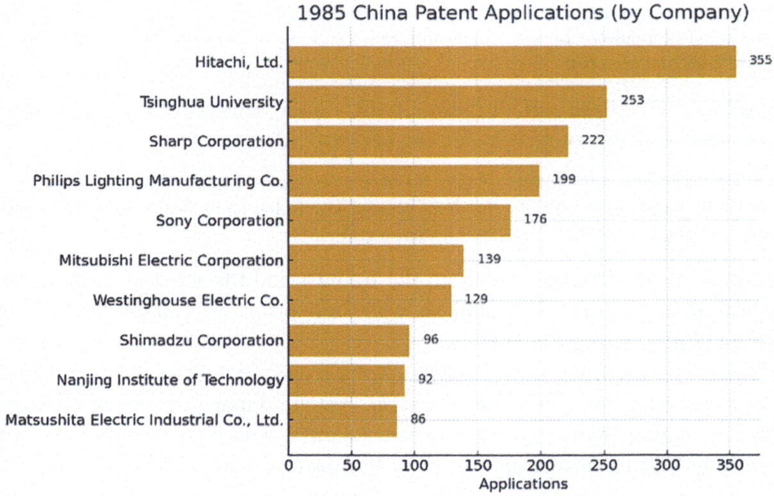

Fig. 17.24 Patent applications filed in China in 1985, ranked by company

Chapter 18
Sony for the Young

British science fiction writer Douglas Adams once offered a tongue-in-cheek observation on people's reactions to emerging technologies in his posthumous book *The Salmon of Doubt*. These quips, popularly dubbed the "Three Laws of Technology," go like this:

1. *Anything that is in the world when you're born is normal and ordinary and is just a natural part of the way the world works.*
2. *Anything that's invented between when you're 15 and 35 is new and exciting and revolutionary and you can probably get a career in it.*
3. *Anything invented after you're 35 is against the natural order of things.*[1]

Though clearly humorous, these "laws" strike a chord because they reflect the perennial generational gap between youth and their elders—a sentiment echoed across both Western and Chinese internet culture.

Magnetic tapes and the cassette-playing Walkman exemplify these "laws" in action. Before the Baby Boomers were born, the piano, phonograph, and radio were the mainstays of middle-class entertainment in the West. For Boomers growing up, records and turntables felt like immutable parts of the world. But cassette technology, which came of age with them, became a "revolutionary invention that changed the world."

Sony has always had a deep relationship with magnetic tape. According to Sony's official "Sony History" archive, when the company was first founded, Masaru Ibuka and his colleagues began developing a tape recorder based on little more than two lines they found in a book called *Audio Engineering*. That book mentioned that in 1936, the German company HEC invented the tape recorder by applying magnetic material onto plastic. These sparse words were all it took to inspire them into action.

---

[1]Adams, D. (2002). *The Salmon of Doubt: Hitchhiking the galaxy one last time*. London: Macmillan.

Q. Yue, F. Yang, *Governing Innovation: A Global History of Patent-Driven Technological Progress*, https://doi.org/10.1007/978-981-95-5570-3_18

For many years, magnetic tape posed no serious threat to the supremacy of vinyl records. In fact, even compact discs took a long time to dethrone vinyl. In 1986, global CD sales reached only 130 million units—a mere twentieth of the 2.5 billion vinyl records sold that year. But in the United States, CD sales had already hit 53 million, about one-tenth of vinyl's total. It wasn't until 1988 that CDs finally outsold vinyl records.

In the 1950s, the primary users of magnetic tape were radio stations. For the average household, reel-to-reel tape machines were too bulky and difficult to operate, and the ability to record wasn't very useful. But for radio broadcasters, the technology was invaluable. This era of tape innovation spurred the rise of numerous small, local radio stations across Europe and America, turning radio into a dominant leisure medium for the Baby Boom generation. It was in this context that Sony's iconic transistor radio was born and found commercial success.

After the Walkman's release, criticism quickly followed, often couched in moral concerns about contemporary culture. Some academics saw the Walkman as an emblem of fake demand created by corporate consumerism. To them, the growing sales of the Walkman reflected how 'false' consumer needs, driven by mass marketing, had replaced humanity's 'true' needs. They believe that corporations manipulated behavior, encouraging people to abandon natural constraints on their desires and succumb to an ever-deepening cycle of hollow gratification.[2]

Another common critique at the time was that headphones and portable music players were isolating technologies. In the 1980s, American conservative philosopher Allan Bloom famously lamented, "As long as they have the Walkman on … they cannot hear what the great tradition has to say."[3]

Bloom believed that technologies like the Walkman had closed off the American spirit (his book *The Closing of the American Mind* became a cultural touchstone). He feared that young students were straying from the likes of Shakespeare and the Bible, thereby eroding a shared cultural identity.

For the post-80 s, post-90s, and even post-2000s generations raised in the digital era, such critiques feel all too familiar. Each transformative product embraced by young people—be it tablets, smartphones, or social media—has faced similar moralizing. In this sense, Douglas Adams' "Third Law" remains eerily prescient:

> Any technology invented after the age of 35 is unnatural and destined to bring doom upon humanity.

As the poet Du Fu once wrote:

> The styles of Wang, Yang, Lu, and Luo were mocked as frivolous in their time. Yet those who scoffed are long forgotten, while their works endure like rivers and seas.

---

[2] Du Gay, P., Hall, S., Janes, L., Madsen, A. K., Mackay, H., & Negus, K. (2013). Doing cultural studies: The story of the Sony Walkman. Sage.

[3] Bloom, A. (1987). *The closing of the American mind: How higher education has failed democracy and impoverished the souls of today's students*. New York, NY: Simon & Schuster.

As generations turned, once-scandalous artists like the Beatles, Elvis Presley, and Bob Dylan came to define their eras. Dylan even received a Nobel Prize in Literature—whether or not the Swedish Academy knew what hit them.

The story of the Walkman and Adams' techno-laws offer one clear lesson: for those working at the forefront of technology, keeping a youthful mindset is crucial. Philosophers may indulge in their moral high ground, critiquing youth culture from a distance, but scientists and entrepreneurs must stay young at heart—embracing change and believing in the power of new technologies.

As the Chinese thinker Liang Qichao once wrote in *On the Young China*:

> The old constantly dwell on the past; the young constantly dream of the future. From nostalgia comes conservatism; from hope comes progress. From conservatism comes stagnation; from progress comes renewal…

When Sony invented the Walkman, music cassettes had yet to fully catch on. Even within Sony, many saw the Walkman as little more than a gimmick—a "tape recorder without recording capabilities." In his memoir, Akio Morita recalled that even inside the company, most people were overly focused on technical details, paying little attention to how the target audience—especially young consumers—actually engaged with music【3】.

So was the Walkman's success merely a fluke? To answer that, let's examine the patent activity of Sony and Philips during the decade before the Walkman's release. A simple comparative analysis of their global patent families from 1969 to 1978, filtered by relevant keywords, reveals telling patterns.

As shown in Table 18.1, Sony was still far smaller than Philips in the 1970s. In the 10 years leading up to the Walkman, Sony had disclosed 4894 patent families globally, while Philips had nearly 13,000. According to research by Fletcher【4】,

Table 18.1 Patent and patent family analysis (1969–1978)—Sony vs. Philips

Keyword	Philips patent count	Philips patent families	% of Total philips families	Sony patent count	Sony patent families	% of Total Sony families
Stor*	2719	1697	13.35%	719	389	7.95%
Portab* OR Miniatur*	265	155	1.22%	429	255	5.21%
Audio	349	219	1.72%	257	257	5.25%
Video	926	492	3.87%	1100	551	11.22%
Stereo	65	34	0.27%	74	74	1.51%
Fidelit*	24	17	0.13%	21	21	0.43%
Playback	391	295	2.32%	373	255	5.21%
Magnetic tape	262	146	1.15%	716	392	7.99%
Magnetic head	261	144	1.13%	502	271	5.54%
Total	**46,127**	**12,714**		**11,029**	**4894**	

*(i.e., it represents any characters/variations).

Sony was only the ninth-largest electronics company by global marketing revenue in 1989, earning $1.69 billion, trailing behind Philips at $2.15 billion.

From the table, we see that Philips far outpaced Sony in storage technologies—holding 1697 relevant patent families (13.35% of its total), compared to Sony's 389 (7.95%). But when it came to audio and video, Sony held its own: 257 and 549 patent families respectively, even outperforming Philips in proportional terms.

This suggests Sony was more tuned in to the consumer electronics sector. For specific audio technologies like stereo and high-fidelity sound, Sony not only paid more attention but also secured more intellectual property. While Philips was the original inventor of the compact cassette, it had clearly relaxed its focus on tape technology by the 1970s. Sony, by contrast, poured significant effort into related innovations.

Sony held 393 patent families related to "magnetic tape" and 271 for "magnetic head"—more than double Philips' count for both—and these categories accounted for over 5% of Sony's total patents. For Philips, the numbers barely exceeded 1%. This demonstrates Sony's commitment to refining a technology that had caught on with young consumers: the cassette tape—affordable, easy to copy, and widely pirated, much like the MP3 in the 1990s.

Sony understood this youthful passion for music and video, and made serious investments in magnetic tape innovations to improve the user experience. That, ultimately, was the key to its success.

Using IncoPat's patent sandbox tools, we can see even more clearly how the two companies diverged. Between 1969 and 1978, Sony consistently invested in "magnetic recording," "cassette housings," and stereo technologies. In color television, Sony's patent portfolio even exceeded that of Philips, with a particular emphasis on chrominance and luminance signal processing. In short, Sony aggressively pursued fields that fascinated young consumers—hardly the strategy of a company banking on luck.

Let's now examine the patent portfolios of both companies under the main classification code **G11B**—which pertains to "information storage based on relative movement between a record carrier and a transducer." In this category, Sony held **1353** patent families, while Philips had only **584**.

Although both companies had a fair number of patents related to video tape, only Sony registered a substantial number of patents tied specifically to "color video" and "color signal" technologies. According to Douglas Adams' "three laws of technology," color televisions were seen by young people at the time as "revolutionary inventions." Targeting "color" as a research direction aligned well with youth-driven demand.

By contrast—just as we described in the "Rise and Fall" chapter regarding IBM—Philips appeared more inclined toward foundational scientific research. We know Philips became globally known for its electric shavers, showing that it was not, like IBM, a company fixated solely on B2B business while shunning mass consumer products. So why, then, did Philips' patent layout skew toward fundamental science?

Chapter 19
Philips in the Ivory Tower

In 1914, the Philips brothers hired a young physicist named Gilles Holst to lead the company's research efforts. Holst had studied under Nobel Prize winner Kamerlingh Onnes and was a key participant in the discovery of superconductivity. In fact, when the experimental material's resistance dropped to zero, it was Holst who personally recorded the reading—making him, in a sense, the first human to witness superconductivity.

At that time, it was rare for scientists to work in industry. The Philips Research Laboratory that Holst established became one of the first commercial institutions to employ a large number of pure researchers. He recruited staff from Dutch universities and research institutes, and within the lab, he fostered a strong academic culture. These researchers drove a wealth of exploration in fundamental science, which played a key role in Philips' development.

As noted earlier, the Philips founding brothers were both a brilliant engineer and a savvy businessman, with deep understanding of both technology and markets. They not only supported Holst's vision of building an academically oriented research center but also had the ability to steer its research agenda to ensure it met practical company needs. In the decades leading up to World War II, the brothers maintained close ties with the research lab, personally overseeing many affairs and actively transforming new findings into marketable products. Some of the early successes—such as the X-ray tube (US1893759A), rotary shaver (US2308920A), and high-pressure mercury vapor lamp (US2094694A)—originated from the lab during this period. Thanks to these diverse technological breakthroughs, Philips expanded into new fields and evolved into a tech giant spanning medical, lighting, and electrical industries.

After World War II, Philips' approach to R&D began to subtly shift. The research center gradually became absorbed in basic science and increasingly disconnected from the company's product divisions. Two key factors drove this transition:

Q. Yue, F. Yang, *Governing Innovation: A Global History of Patent-Driven Technological Progress*, https://doi.org/10.1007/978-981-95-5570-3_19

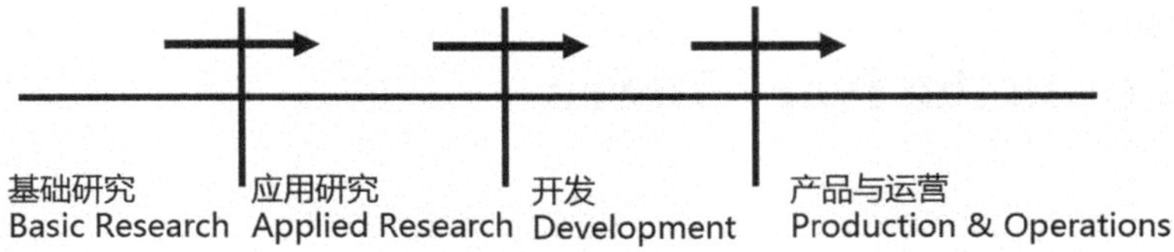

Fig. 19.1 The linear model from the Bush report

Theoretical Foundations

In 1945, Vannevar Bush submitted his report *Science, the Endless Frontier* to President Franklin D. Roosevelt. Widely considered one of the most influential documents in global science policy, the report proposed the now-famous "linear model" of innovation (see Fig. 19.1): basic research leads to applied research, which then leads to development, and ultimately results in products and market applications.

This model was embraced not only by the U.S. government but also by European countries and the broader industrial sector. Many major corporations adopted Bush's framework, elevating basic science to unprecedented importance. For decades following WWII, Bell Labs, IBM, and Philips all pursued academically inclined research paths shaped in part by this linear view.

Leadership Preferences

In 1946, the renowned physicist Hendrik Casimir took charge of the Philips Research Laboratory. Casimir had made significant contributions to quantum physics and maintained close relationships with figures like Niels Bohr, Wolfgang Pauli, and Paul Ehrenfest. He was highly respected in the academic world. As head of the research center, Casimir brought his passion for science to the role, energetically promoting studies in superconductivity, superfluidity, and laser physics—even though these areas had no immediate practical application for Philips at the time.

Casimir believed his primary duty was to encourage individual freedom in research and help scientists escape bureaucratic constraints. He gave generous budgetary support to the lab's projects, ensuring that researchers could work independently. For many years, Philips ranked third in global R&D investment, just behind AT&T and IBM. Casimir firmly believed that "a deeper understanding of natural phenomena" would eventually open the door to new products. He instructed research supervisors to be theory-driven and to treat fundamental scientific inquiry as their core mission. He also encouraged strong ties with academia, and evaluated researchers based on their publications and patent output.

Under Casimir's leadership, the Philips Research Laboratory expanded significantly. Between 1946 and 1965, the number of scientists grew from 157 to 388, while the number of assistants increased from 170 to 991, and administrative staff from 240 to 722.

During this period, Philips' organizational structure underwent major changes compared to the pre-war era. Product divisions (PDs) gained autonomy to determine their own portfolios and directions, and even established their own in-house research teams. These divisions held considerable power within the company. However, they had no influence over the central research lab, which reported directly to Philips' top-level Management Committee and received its budget independently. Research managers largely dismissed the practical concerns of product divisions, leading to increasing friction between the two.

During Casimir's tenure, scientists at Philips Research still contributed several important inventions. From a commercial perspective, two technologies in particular—**the Plumbicon (US3372056A)** and **LOCOS (Local Oxidation of Silicon, US3970486A)**—proved highly successful. Both became industry standards and generated substantial licensing revenue for the company.

The Plumbicon, a lead oxide camera tube, showcased the lab's strength in fundamental science. Before its invention, RCA had developed a television pickup tube based on **photoconductivity**, but the design suffered from significant application issues. RCA ultimately concluded that the problems stemmed from inherent limitations in photoconductivity itself. Scientists at Philips disagreed. Drawing on their expertise in **band structure theory**, they conducted only a few experiments before selecting lead oxide as the optimal material from a long list of candidates. The resulting tube quickly became the industry standard.

However, the process of turning the Plumbicon into a mass-produced product was far from smooth. After designing the prototype, the research center handed it off to the product division, expecting them to commercialize it. But the process proved far more complex than anticipated. Image artifacts, exploding bulbs, and other production issues led to frequent factory incidents. Eventually, the project was kicked back to the research center, which had to dispatch engineers to the manufacturing site to provide ongoing technical support. The product division complained that the researchers failed to understand the transition from prototype to scalable production, further straining their relationship.

From the 1940s to the 1970s, Philips Research spent decades studying the **Stirling engine**, also known as a hot-air engine. At one point, the Stirling team was the largest in the entire research center. Philips accumulated a large number of patents in this area, which remained foundational to industrial Stirling engine research well into the 1990s. A heat map of Philips' patents from 1969 to 1978, generated using the IncoPat database, shows a major cluster on the right labeled "working space/hot gas/working fluid"—evidence of the lab's extensive work in thermodynamic engine research (Fig. 19.2).

Tragically, no product division within Philips showed interest in commercializing these results.

In 1971, the research center, after years of development, excitedly unveiled a new type of camera tube to the product divisions—only to be met with rejection. The product managers dismissed the design as suitable only for small-format images

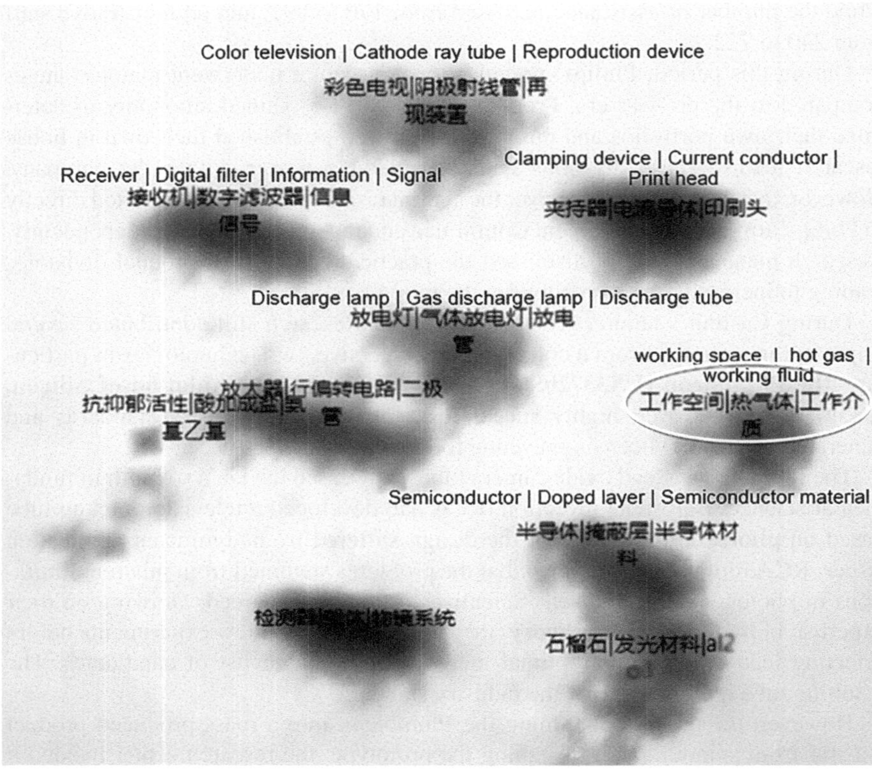

Fig. 19.2 Heatmap of Philips patents, 1969–1978

and declared it had no market potential. The researchers countered that Japanese televisions were predominantly small in size, only to be rebuffed with: "Then we'll teach the Japanese to enjoy large images." The meeting ended in mutual frustration.

In fact, the product division had already noticed Japan's shift toward larger screen sizes and had canceled a related development project accordingly. Yet, there had been no effective communication between the two sides. The research center was indignant over the product team's dismissiveness, while the product team accused the researchers of wasting company resources and delivering nothing of value.

While Japanese firms like Sony were on the rise, Philips' own R&D efforts were increasingly mired in internal dysfunction. This was the dire state of Philips' innovation machinery on the eve of the Walkman's debut.

By the 1970s, the era of continuous economic growth in the Western world was drawing to a close. Large corporations that had once lavishly invested in basic research began to face financial constraints, and skepticism toward Bush's linear model grew. This period also marked Philips' transition from a family-owned business to one led by professional managers. In 1971, Henk van Riemsdijk became the

CEO of Philips—the first to bear a surname other than "Philips." However, van Riemsdijk was still a family insider: he had joined the company's commercial division in 1934 and had married the daughter of one of Philips' founders, Henriëtte Anna Philips.

As the founding family receded into the background, the new leadership began placing greater emphasis on return on investment. Research projects without clear commercial outcomes faced mounting scrutiny.

In 1973, the developed world entered a period of stagflation, and Philips' industry challenges grew more acute. Around the same time, Casimir retired to take up the presidency of the European Physical Society, an organization he had helped establish. He was succeeded by Eduard Pannenborg, who became the third head of Philips Research.

Pannenborg aimed to strike a balance between individual scientific freedom and serving the company's product divisions. He continued to support the research center's independence but argued for a shift away from "technology push" and toward "market pull." He encouraged researchers to think of themselves as part of the industrial world. However, the academic culture that Casimir had firmly instilled proved deeply entrenched, and Pannenborg's attempt at compromise failed to win over the researchers. The schism between the research center and the rest of the company remained unresolved.

During this period, Japanese companies were mounting ever more aggressive challenges. Philips began shutting down factories and cutting low-margin business lines. Even so, between 1971 and 1977, the company maintained relatively stable growth: its workforce held steady at around 380,000 employees, and its revenue climbed from 8.2 billion to 14.1 billion euros.

In 1977, Nico Rodenburg became CEO, marking the first time in Philips' history that someone outside the Philips family led the company. However, Rodenburg did not distinguish himself with strong leadership. While company revenues continued to rise during his tenure, profits remained sluggish. Compounding the situation, the first oil crisis had plunged the world into economic turmoil. In response, Rodenburg initiated Philips' first large-scale round of layoffs to control costs. Between 1977 and 1981—right around the time the Walkman made its debut—Philips' consumer electronics division underwent a massive restructuring that resulted in roughly 36,000 job cuts.

Viewed in retrospect, it's difficult not to see the poetic convergence of events: the Philips family's retreat from leadership, the pains of corporate transition, the stagflation crisis, the oil shock, mass layoffs in the consumer division, and the sidelining of Andreas Pavel and his "belt-worn stereo"—all of these seem to align in a tragic refrain of timing, fate, and fortune.

In 1982, Wisse Dekker took over as CEO. That same year, Philips' V2000 video format failed, and the onslaught from Japanese electronics firms grew even more intense. Beginning in 1984, Dekker broke with Philips' long-standing strategy of technological autarky, launching 21 joint investment projects to bolster R&D. One of the most consequential outcomes of this pivot was the founding of ASML, the now-renowned semiconductor lithography company, formed through a joint venture between Philips and external partners—a name that would later gain prominence in the context of U.S.-China trade tensions.

In 1986, Dekker was succeeded by Van der Klugt, whose tenure was marked by his reassertion of control over the rebellious North American division of Philips. To compensate for currency losses caused by a weakening U.S. dollar, he embarked on a divestiture campaign, selling off 23 business units. ASML was one of the entities spun off during this era.

By late 1989 and early 1990, Philips' profits plummeted. Van der Klugt lost the confidence of the board and was forced to resign in 1990. Under his leadership, Philips had cut 71,000 jobs, roughly one-fifth of its workforce.

In 1989, amid escalating internal and external pressures, Philips initiated another reform of its research center. Under pressure from the company's Management Committee, the center adopted a new "contract research" model, requiring researchers to secure project funding from Philips' production divisions.

In 1991, Carrubba, formerly of Hewlett-Packard, assumed leadership of the research center. He pushed for a mindset shift: researchers should see themselves as "knowledge workers" and approach problems from the customer's perspective. With that, the Philips Research Laboratory—once an exalted academic institution—finally lost its Ivory Tower status.

Chapter 20
Sony at the Crossroads of the Digital Music Era

In her acclaimed book *The Silo Effect*, *Financial Times* senior journalist Gillian Tett coined the term "silo effect" to describe how poor internal communication and siloed operations within companies can stifle innovation. She cited Sony as a textbook example. Tett focused on a 1999 Sony press conference held in Las Vegas that vividly illustrated this issue.

To unveil two flagship products, Sony pulled out all the stops. The event opened with an animated introduction from characters in the then-hit movie *Stuart Little*, followed by a speech from George Lucas praising Sony's contributions to audio and video technology. For the grand finale, Sony CEO Nobuyuki Idei invited guitar virtuoso Steve Vai onstage for a live performance. After the last chord rang out, Vai pulled a sleek new gadget from his pocket—a Memory Stick Walkman, which had just recorded his performance and now played it back. The audience erupted in applause at the sight of this smaller, more portable alternative to the MiniDisc.

But the show wasn't over.

Riding the wave of excitement, Idei revealed yet another music player. This second device, the VAIO MusicClip, had an elegant cylindrical form resembling a fountain pen. Like the Memory Stick Walkman, it could record and play music. But instead of using Sony's proprietary Memory Stick, it featured built-in solid-state storage.

The audience was once again impressed by Sony's design prowess. Yet no one asked the obvious question: Why was Sony launching two nearly identical digital music players on the same day?

As if two new devices weren't confusing enough, Sony soon introduced yet another music player: the Network Walkman. Then in 2004, it unveiled a major upgrade to the MD system—Hi-MD. The new Hi-MD discs offered 1GB of storage and featured backward compatibility with older MDs, thanks to a proprietary technology called Domain Wall Displacement Detection (DWDD). Sony also enhanced PC connectivity and launched a new music management software platform, allowing users to download music from its CONNECT music store.

Q. Yue, F. Yang, *Governing Innovation: A Global History of Patent-Driven Technological Progress*, https://doi.org/10.1007/978-981-95-5570-3_20

Previously, as discussed, Sony had released hundreds of cassette Walkman models during the 1980s, each tailored to different user groups. For example, the Sport Walkman series was a hit among North American fitness enthusiasts. But the Memory Stick Walkman, VAIO MusicClip, Network Walkman, and Hi-MD followed a completely different logic. These were not devices designed around diverse user needs. Instead, they were the products of different Sony divisions, each pursuing its own market goals, resulting in redundant internal competition.

While Sony's departments stumbled at the threshold of the digital era, the tide of MP3 and the internet had become unstoppable.

As one researcher noted, the MP3 format truly entered public consciousness and received widespread media attention in 1997.[1] Even before then, from 1995 to 1997, North America had experienced a surge in online music file sharing, which contributed to the format's growing popularity.

In December 1988, the Moving Picture Experts Group (MPEG)—a subcommittee under the International Organization for Standardization (ISO)—began soliciting proposals for a new audio coding standard. Submissions included "Musicam" from Matsushita, CCETT (France), IRT (Germany), and Philips; "ASPEC" from Germany's Fraunhofer Society, AT&T, and France Télécom; and "ATAC" from Sony, Fujitsu, JVC, and NEC.

In 1991, MPEG adopted a standard based on both Musicam and ASPEC, resulting in the MPEG-1 Audio Standard, which included three layers: Layer 1, Layer 2, and Layer 3.

MPEG Layer 2 quickly gained widespread support—especially from Philips and Matsushita—and became the standard format for Digital Audio Broadcasting (DAB) and the audio component of VCDs. Meanwhile, MPEG Layer 3 received little attention.

As the primary backer of MPEG Layer 3, the Fraunhofer Society refused to give up. They toiled to promote the format through a range of efforts. In 1995, Fraunhofer signed an agreement with Microsoft to integrate MPEG Layer 3 support into Windows. That same year, they released WinPlay3, a music playback software for Windows 3.1.

These commercial moves were crucial to MP3's success, but not decisive. What truly propelled MP3 into mainstream use was a perfect storm of unintended events and unforeseen conditions—a process that no single entity could have orchestrated.

In July 1994, Fraunhofer released two command-line programs—L3Enc and L3Dec—for encoding and decoding MPEG Layer 3 files, selling them for $250. An Australian hacker used a stolen credit card to purchase L3Enc, cracked its source code, built a new user interface, and released it for free under the cheeky name "Thank you Fraunhofer."

Even earlier, in 1993, the Internet had seen the emergence of its first music-sharing communities. The most notable among them was IUMA (Internet

[1] Sterne, J. (2012). *MP3: The meaning of a format*. Duke University Press.

Underground Music Archive), made up mostly of independent bands who shared their music in MPEG Layer 2 format.

After "Thank you Fraunhofer" and WinPlay3 became available, IUMA's users quickly realized that MPEG Layer 3 had a much lower bitrate and smaller file size than Layer 2—perfect for the dial-up internet speeds of the era. As a result, the community rapidly shifted to MPEG Layer 3, and a flood of MP3-encoded music began circulating online.

At the time, record companies paid little attention to digital piracy. The Recording Industry Association of America (RIAA), akin to China's Music Copyright Society, still focused on flea markets and early e-commerce sites when investigating pirated cassette tapes and CDs. It wasn't until 1997 that the RIAA launched its first lawsuits against FTP sites distributing pirated music. Before then, the internet was effectively a lawless frontier.

Around this time, the Fraunhofer Society rebranded MPEG Layer 3 as MP3, changing the file extension from ".bit" to ".mp3." The name's simplicity and catchiness helped fuel its online popularity.

Meanwhile, broadband internet and campus networks were booming across North America. Students used FTP and IRC to share music en masse. CD burners dropped to about $1000, and by 1997, new PCs often included one. Users would burn MP3s onto CDs to store, trade, or even sell. That same year, mp3.com was founded in October, and "MP3" became the second most searched keyword online—just behind "porn." By 1999, MP3 overtook porn as the top search term. That August, Napster, the peer-to-peer MP3 sharing site, was launched.

At its peak, Napster claimed 28 million users. It had no advertising, no revenue model, and didn't even collect user data—yet its user base grew rapidly. Venture capitalists believed that a profit model would inevitably follow such explosive growth, and that Napster might one day rival Microsoft or Google. But dreams met harsh reality: after a string of RIAA lawsuits, Napster filed for bankruptcy in 2002.

The rise of Napster and similar platforms reflected a powerful desire among young people to fuse music, computers, and the internet. Unfortunately, Sony failed to respond to this cultural shift and missed a critical opportunity to align with the digital era.

History did, however, offer Sony a fleeting chance.

At the turn of the twenty-first century, an unexpected event briefly resurrected the struggling MiniDisc (MD), giving Sony a rare glimpse into the public's enthusiasm for MP3. Sadly, Sony's response was overly cautious, and the fleeting opportunity slipped through its fingers.

In 1998, the Australian company Xitel noticed a puzzling trend during market research. They were selling sound cards equipped with digital optical output, originally targeting high-end PC gamers who valued stereo sound quality. But with the rising popularity of MP3, Xitel discovered that most of their customers weren't gamers at all—they were MD audiophiles.

It turned out that some tech-savvy users had stumbled upon a clever hack: by connecting Xitel's optical output to their MD recorders, they could easily transfer

MP3 files from their computers to MiniDiscs. The method quickly spread by word of mouth, becoming the go-to DIY solution for converting MP3s to MD format.

Recognizing the surge in demand, Xitel launched a dedicated accessory in 1999 called the MD-Port, specifically designed for seamless MP3-to-MD transfers. The product was a hit. In the first half of 2000, MD sales in North America tripled compared to the previous year. By February 2001, MD had seemingly come back from the dead, briefly becoming the top-selling digital music player on the market.

Sony reacted quickly. Upon learning of the phenomenon, they struck a deal with Xitel to bundle its connectors with MD recorders. By July 2001, MD commanded 40% of the U.S. digital music player market, and Xitel was selling up to 100,000 MD adapters per month.

But the revival was short-lived. It wasn't a comeback—it was a last gasp.

In late 2001, Apple's iPod debuted and took the world by storm. The digital music player market that Sony had pioneered was now swept away by a competitor who had better grasped the new era's pulse.

Chapter 21
The Dream of Total Domination

In October 1998, Sony released a flash memory card product called the "Memory Stick." Over the next decade, many of Sony's digital products supported Memory Stick but not SD cards—this included digital cameras, camcorders, personal digital assistants (PDAs), Walkmans, and mobile phones. This brought endless trouble to Sony users. Meanwhile, Sony filed numerous patents related to Memory Stick and other flash card technologies. These patents may reveal a unique mindset behind Sony's product design and R&D strategy—which, arguably, contributed to many of the company's failures.

In Japanese patent application JP2001216481A (corresponding to the ungranted U.S. counterpart US20010030883A1), Sony envisioned a "key ring" for conveniently carrying numerous Memory Sticks or other memory cards (see Fig. 21.1). The inventors stated: "In the near future, everyone will own and use several IC memory cards to store various types of information." Based on this prediction, they designed memory cards with holes through which a ring—like a key ring—could pass. According to the patent description, these holes not only made it easier to carry multiple memory cards, but in some examples, also functioned as data protection switches.

In another Japanese patent application, JP2001229352A (ungranted), Sony invented an adapter into which multiple Memory Sticks or other flash memory cards could be inserted simultaneously (see left side of Fig. 21.2). Users could insert several cards into this small device, then connect it to a computer to read the files (see right side of Fig. 21.2).

In JP2000029998A, Sony invented a "record and playback device" based on Memory Sticks or other flash memory. From Fig. 21.3, this device resembles a traditional recorder or DVD combo unit—but instead of using tapes or discs, it allows the user to insert ten flash memory cards, domino-style. Those who grew up in the 1980s or 1990s may find this concept familiar: dual-tape recorders and 3-disc auto-playing CD or VCD changers already existed. If Memory Sticks were to become as ubiquitous as tapes or CDs, a device capable of continuously reading them would seem inevitable.

Q. Yue, F. Yang, *Governing Innovation: A Global History of Patent-Driven Technological Progress*, https://doi.org/10.1007/978-981-95-5570-3_21

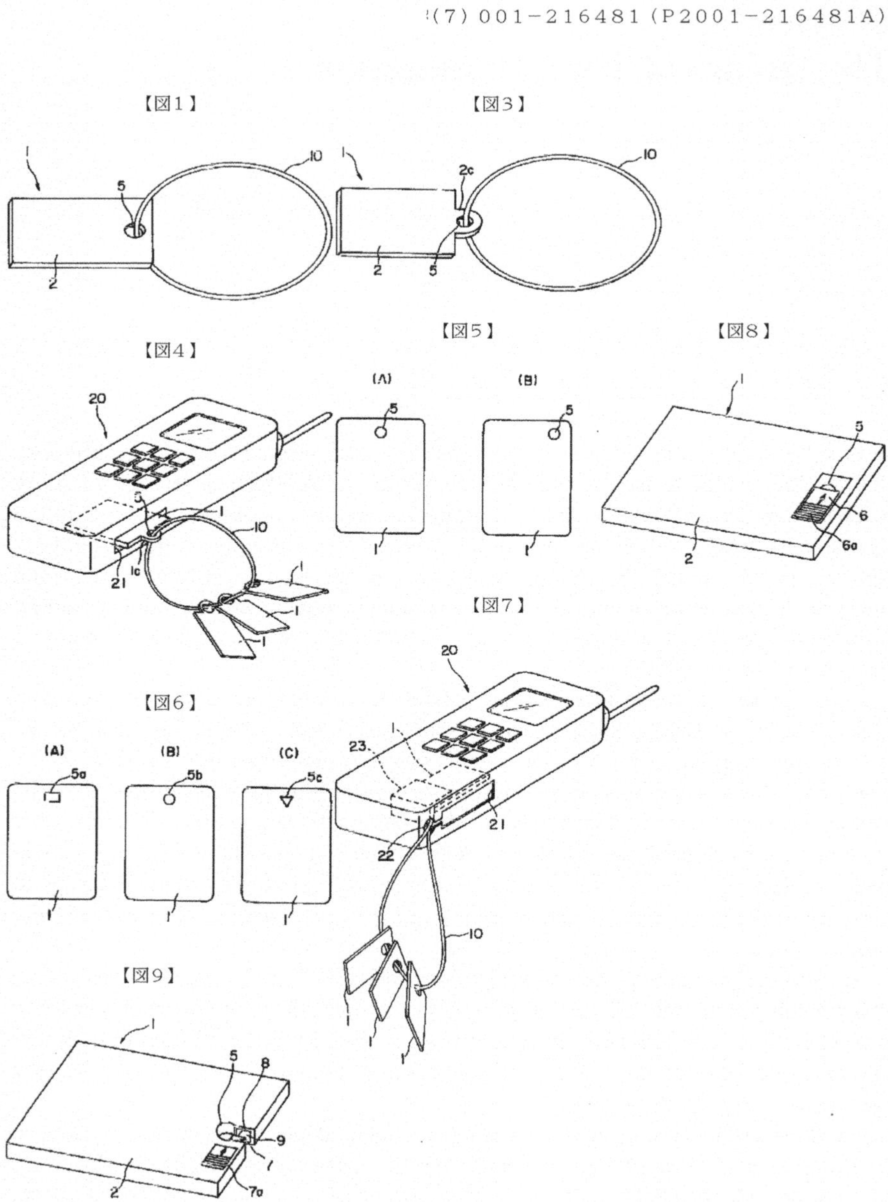

Fig. 21.1 Key ring carrying multiple Memory Sticks or other cards from JP2001216481A

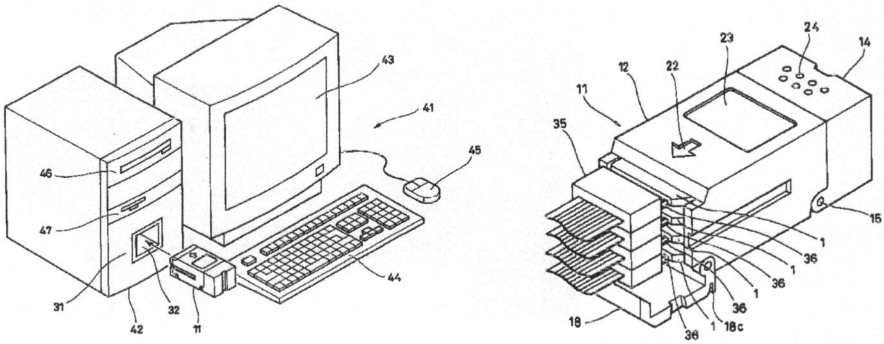

Fig. 21.2 Adapter allowing insertion of multiple Memory Sticks from JP2001229352A

Fig. 21.3 Playback device using multiple memory cards from JP2000029998A

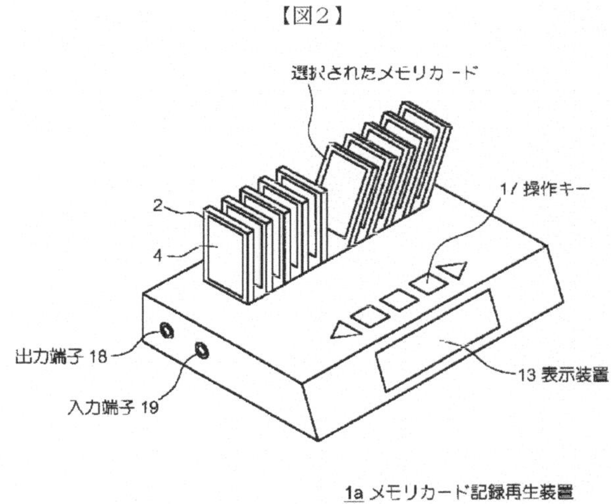

In addition to these patents, Sony filed numerous applications for Memory Stick storage cases, including cases with lanyards (e.g., JP2001222695A) and dual-compatible cases for both Memory Sticks and SD cards (e.g., JP2002166987A). These patents show how much Sony pinned its hopes on the Memory Stick, imagining it as an essential item more common than floppy disks, tapes, or CDs—something found everywhere, used by everyone. This thinking motivated the flood of patents for multi-card usage scenarios. But was there something fundamentally flawed with this line of thinking?

In JP2000032582A, titled "Headphone System," we clearly see the limitation of Sony's mindset. Sony had launched standalone MP3 headphones with built-in flash memory, which are still sold today (see Fig. 21.4). While the product wasn't a huge hit, it did represent a decent technical innovation by eliminating the separate player. The 1998 patent JP2000032582A was an early example of this concept. Unfortunately, this patent was still centered on the Memory Stick (see Fig. 21.5). It describes a "headphone system whose connector section includes multiple memory

Fig. 21.4 Wearable Sony Walkman music player, which is a standalone MP3 headphones with built-in flash memory (https://commons.wikimedia.org/wiki/File:W_Series_Sports_Walkman.jpg)

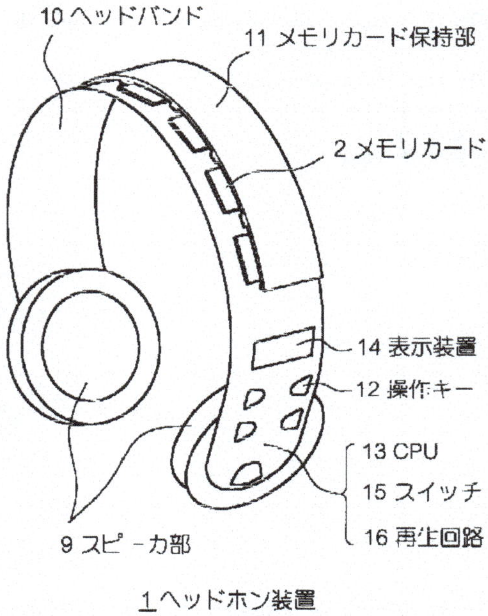

Fig. 21.5 Patent drawing of Japanese patent application JP2000032582A

cards, with a selection function allowing the user to choose a memory card to play from." Despite having the technical capability to simplify the product—to remove the player and keep just the headphones—Sony was unwilling to let go of the Memory Stick. This invention was part of Sony's grand plan to build a Memory Stick empire, not an attempt to free the user's hands. The design centered on Sony's vision, not the user's needs.

Filed in 1998, the patent application JP2000030010A, titled "Remote Control Device," reflects a similar design mindset as the aforementioned headphones (see Fig. 21.6). The remote was designed to control various household appliances, with the relevant information for each device stored on a separate Memory Stick. If a user owned appliances from different manufacturers and one Memory Stick wasn't sufficient, they could simply insert multiple sticks. It's clear that the true aim of these inventions wasn't to make life easier for users, but rather to create more use

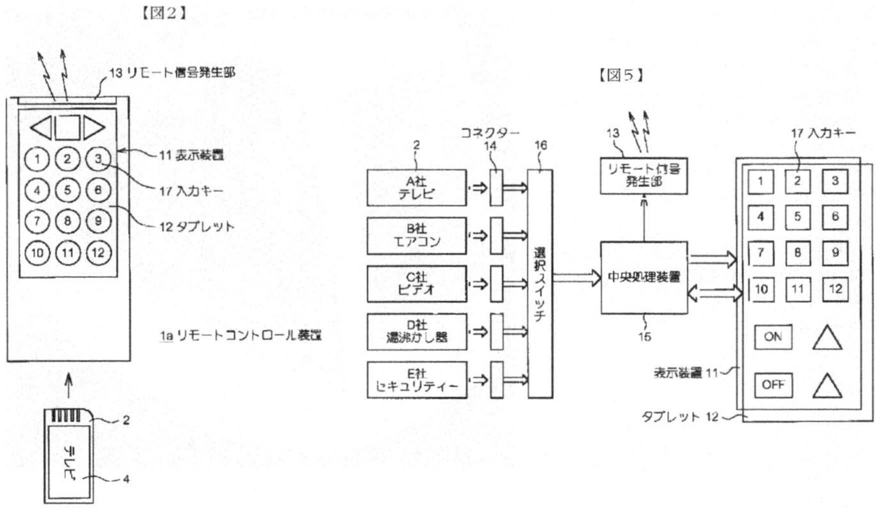

Fig. 21.6 Patent drawing of Japanese patent application JP2000030010A

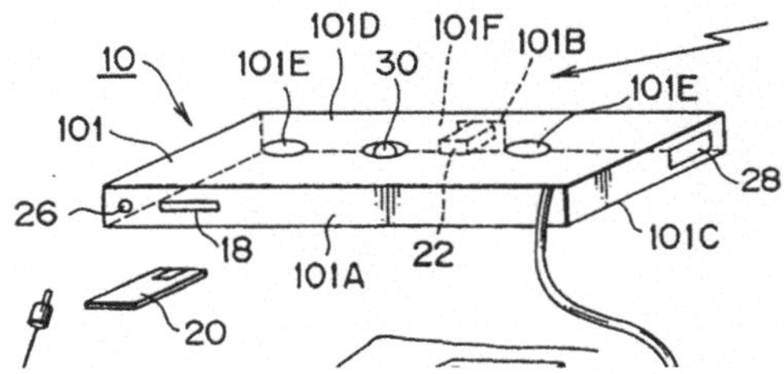

Fig. 21.7 Patent drawing of Japanese patent application JP2001175817A

cases for the Memory Stick—Sony gave no consideration to the inconvenience of users having to repeatedly insert and remove these cards.

Sony's Memory Stick patents also reveal another underlying issue, clearly illustrated in the 2001 patent application JP2001175817A, titled "Cassette Adapter" (Chinese equivalent: CN1300070A, not granted). This invention proposed a flash memory adapter shaped like a cassette tape (see Fig. 21.7). Users could insert a Memory Stick or flash card into this cassette-shaped adapter, which could then be played in a traditional tape deck like any regular cassette. The patent was filed in 2000—at a time when Apple was dreaming of eliminating keyboards, styluses, and floppy drives, Sony was still trying to figure out how to use a tape player to read flash memory.

In 1998, Apple released the iMac G3 all-in-one desktop—the first major product launched after Steve Jobs' return (see Fig. 21.8). This machine marked the beginning of Apple's financial recovery in the late 1990s and was famous for its bold and colorful translucent shell, its round "hockey puck" mouse, and one especially significant innovation: the removal of the floppy drive.

As we know, both Sony and Apple were early producers of 3.5-inch floppy disks. But in 1998, Apple essentially signed the death warrant for the format with the iMac G3, which became a symbol of the floppy drive's obsolescence. While Apple was

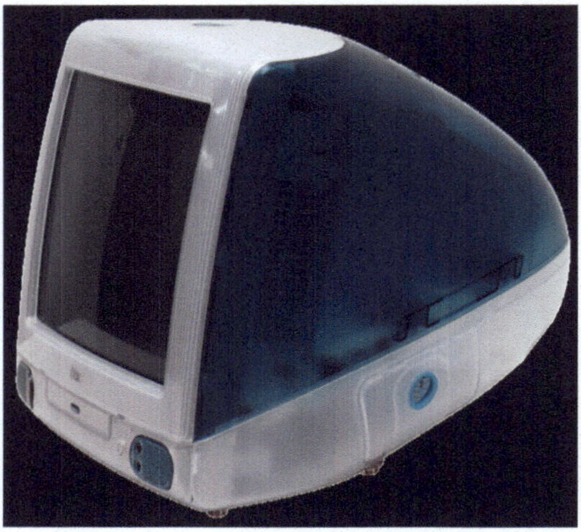

Fig. 21.8 The iMac G3 all-in-one, released in 1998, the first mainstream computer to eliminate the floppy disk drive (Source: Rama & David Fuchs. (n.d.). *iMac G3 Bondi Blue, three-quarters view* [Image]. Wikimedia Commons. https://commons.wikimedia.org/wiki/File:IMac_G3_Bondi_Blue,_three-quarters_view.png)

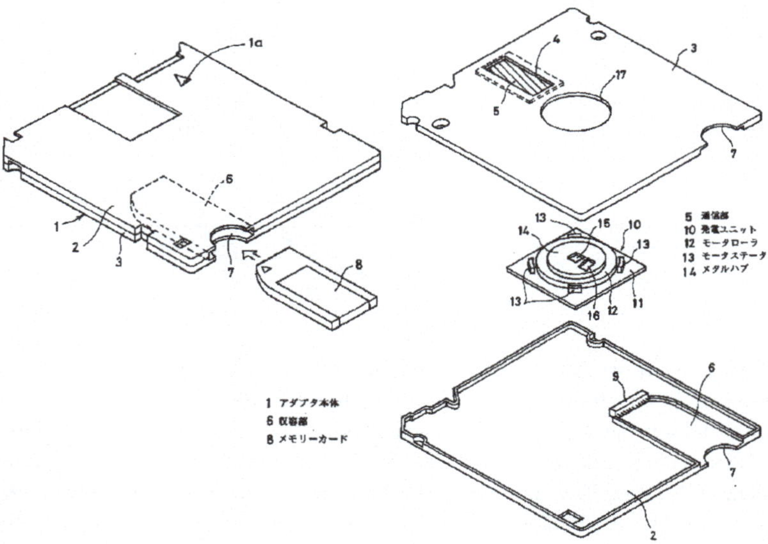

Fig. 21.9 Patent drawing of Japanese patent application JP2000357209A "Memory card adapter"

driving a stake through the heart of the floppy disk, Sony was sentimentally trying to keep it alive.

On June 15, 1999, Sony filed patent application JP2000357209A for a "Memory Card Adapter" that took the shape of a 3.5-inch floppy disk (see Fig. 21.9). Although this application was neither granted nor filed internationally, Sony did manufacture a related product. In October 2000, Sony released the MSAC-FD2M adapter (see Fig. 21.10). Even today, collectors can find these vintage items on Amazon or eBay for around $50 to $100.

From these various and sometimes bizarre patent applications, it's not hard to see the grand vision Sony had for the Memory Stick. While Apple was racking its brain

Fig. 21.10 Sony Memory Stick floppy disk adapter MSAC-FD2M, released around 2000. (Source: Plbcr. (n.d.). Adaptateur Memory Stick [Image]. Wikimedia Commons. https://commons.wikimedia.org/wiki/File:Adaptateur_Memory_Stick.JPG. This image is shared under the Creative Commons CC BY-SA 3.0 license)

to simplify users' lives by subtracting unnecessary features, Sony imagined a world where Memory Sticks were ubiquitous—present in remote controls, computers, mobile phones, home audio systems, car stereos, headphones, digital photo frames, navigation devices... in Sony's vision, all these devices would be equipped with Memory Sticks.

The iPod was a great product because it addressed a real user pain point: copying songs from a computer was a hassle, so Apple designed a high-capacity hard drive to store music and introduced an ecosystem of music management and syncing. Sony's idea of helping users was fundamentally different: they assumed that the Memory Stick would become an essential part of everyday life, and then designed user needs around that assumption—completely overlooking the fact that the assumption itself was the source of user frustration. In Sony's Memory Stick patents, what we see is Sony's vision, not the user's dream.

While Apple aggressively trimmed features—ditching floppy drives, CD drives, digital cameras, MP3 players, headphone jacks, and USB ports—Sony was going in the opposite direction, piling on more and more. Many of its Memory Stick patents were tied to technologies Sony's engineers were already familiar with, such as CDs, floppies, tapes, remote controls, and home audio systems. Through these patents, we also see the stodgy conservatism of an old-school corporate giant.

This patent exemplifies a design philosophy that is entirely Sony-centric rather than user-centric. Filed in 1998, the patent application JP2000030010A, titled "Remote Control Device," reflects a similar design mindset as the aforementioned headphones (see Fig. 21.6). The remote was designed to control various household appliances, with the relevant information for each device stored on a separate Memory Stick. If a user owned appliances from different manufacturers and one Memory Stick wasn't sufficient, they could simply insert multiple sticks. It's clear that the true aim of these inventions wasn't to make life easier for users, but rather to create more use cases for the Memory Stick—Sony gave no consideration to the inconvenience of users having to repeatedly insert and remove these cards.

In 1998, Apple released the iMac G3 all-in-one desktop—the first major product launched after Steve Jobs' return (see Fig. 21.8). This machine marked the beginning of Apple's financial recovery in the late 1990s and was famous for its bold and colorful translucent shell, its round "hockey puck" mouse, and one especially significant innovation: the removal of the floppy drive.

As we know, both Sony and Apple were early producers of 3.5-inch floppy disks. But in 1998, Apple essentially signed the death warrant for the format with the iMac G3, which became a symbol of the floppy drive's obsolescence. While Apple was driving a stake through the heart of the floppy disk, Sony was sentimentally trying to keep it alive.

On June 15, 1999, Sony filed patent application JP2000357209A for a "Memory Card Adapter" that took the shape of a 3.5-inch floppy disk (see Fig. 21.9). Although this application was neither granted nor filed internationally, Sony did manufacture a related product. In October 2000, Sony released the MSAC-FD2M adapter (see Fig. 21.10). Even today, collectors can find these vintage items on Amazon or eBay for around $50 to $100.

From these various and sometimes bizarre patent applications, it's not hard to see the grand vision Sony had for the Memory Stick. While Apple was racking its brain to simplify users' lives by subtracting unnecessary features, Sony imagined a world where Memory Sticks were ubiquitous—present in remote controls, computers, mobile phones, home audio systems, car stereos, headphones, digital photo frames, navigation devices... in Sony's vision, all these devices would be equipped with Memory Sticks.

The iPod was a great product because it addressed a real user pain point: copying songs from a computer was a hassle, so Apple designed a high-capacity hard drive to store music and introduced an ecosystem of music management and syncing. Sony's idea of helping users was fundamentally different: they assumed that the Memory Stick would become an essential part of everyday life, and then designed user needs around that assumption—completely overlooking the fact that the assumption itself was the source of user frustration. In Sony's Memory Stick patents, what we see is Sony's vision, not the user's dream.

While Apple aggressively trimmed features—ditching floppy drives, CD drives, digital cameras, MP3 players, headphone jacks, and USB ports—Sony was going in the opposite direction, piling on more and more. Many of its Memory Stick patents were tied to technologies Sony's engineers were already familiar with, such as CDs, floppies, tapes, remote controls, and home audio systems. Through these patents, we also see the stodgy conservatism of an old-school corporate giant.

Chapter 22
The Petty King of Patents

In our research, we sought to identify the patent technologies most attuned to users—those that genuinely consider the user's perspective. This was no easy task. But if we reversed course and looked instead for patents that stand in opposition to users, that completely disregard user interests, the answer would undoubtedly be Digital Rights Management (DRM) technologies. These are technologies that treat the average user as an adversary—arguably even worse than Sony's labyrinthine Memory Stick patents.

We've previously likened the relationship between users and corporations to Mencius's political philosophy of "the people over the ruler." In a famous passage, Mencius elaborated: "If a ruler treats his ministers like hands and feet, they will regard him as their heart and soul; if he treats them like dogs and horses, they will regard him as a mere stranger; if he treats them like dirt and weeds, they will regard him as an enemy." DRM technologies are, in essence, an embodiment of "treating the people as thieves." And Sony's foray into DRM offers a vivid lesson in the consequences of such a stance.

Copyright protection became a major issue for producers of music, film, and other content as they entered the digital age. In the analog era, copying audiovisual material resulted in noticeable quality loss. In the digital era, however, data is medium-independent and copies retain the same fidelity as the original. DRM emerged in response to this challenge.

During the CD era, record labels commonly embedded computer-readable data into audio CDs as a DRM mechanism. When played on standard CD players or portable CD devices, the computer data would be ignored and the audio would play normally. But if the same CD was inserted into a computer's optical drive, the system would prioritize the computer data and fail to play the audio. A prominent early implementation of this was Cactus Data Shield, developed by Israeli company Midbar Tech (U.S. Patent No. US6425098B1, titled "Preventing unauthorized copying from digital media"). Labels like EMI and BMG employed this system extensively. The anti-user nature of the patent is evident in its description:

Q. Yue, F. Yang, *Governing Innovation: A Global History of Patent-Driven Technological Progress*, https://doi.org/10.1007/978-981-95-5570-3_22

> When unauthorized copying occurs, the auxiliary data cannot effectively correct the altered portions of the encoded data, resulting in essentially unrecoverable errors in the unauthorized copy.

A CD processed with Cactus Data Shield would play fine in a CD player or be viewable in a computer's file explorer. But any attempt to rip the music would result only in distorted noise. Users would likely suspect a problem with their drive or computer, and after repeated failed attempts, they would give up.

As internet use expanded and user data such as habits and preferences gained commercial importance, DRM evolved accordingly. Newer DRM schemes embedded executable programs on music CDs. When inserted into a drive, these programs would auto-run and install monitoring software on the user's system—tracking not only all music CDs inserted, but also uploading user data. A prime example was XCP software developed by First 4 Internet, protected under British patent GB2415826A, which openly described embedding a "monitoring program" that would automatically activate and surveil all "unauthorized access."

Sony heavily invested in DRM, given its ownership of Sony Music and Sony Pictures. In August 2000, Sony Pictures Senior VP Steve Heckler delivered a fiery speech at the Americas Conference on Information Systems, targeting sites like Napster:

> Sony will aggressively block all this. We will develop technology that oversteps user boundaries. We'll block Napster at the source—with firewalls from your cable provider, your telephone company, your ISP. We'll block it from your personal computer.

Sony filed at least 3000 patents related to copy protection, grouped into over 1000 patent families (simple family consolidation). By contrast, Philips filed only 196 such families, and Apple just 53. As noted earlier, Sony's filings peaked in 2003 (see Fig. 22.1), largely to lure movie studios into the Blu-ray camp. After 2006, Sony's DRM patent filings sharply declined, falling to single digits in recent years.

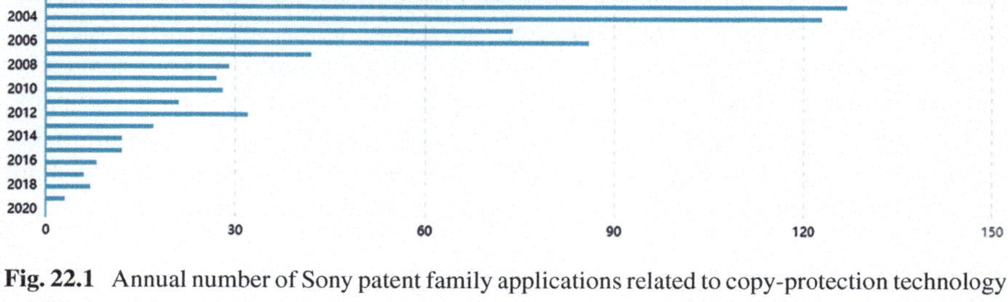

Fig. 22.1 Annual number of Sony patent family applications related to copy-protection technology

From 2005 onward, Sony secretly embedded DRM software on 22 million music CDs. Despite owning numerous DRM patents, the company primarily used technologies licensed from First 4 Internet. Of those CDs, 2 million carried the XCP software (Windows only), and 20 million carried SunnComm's MediaMax CD-3 software (compatible with both Windows and MacOS).

When these CDs were inserted into a computer, the system would display an End User License Agreement (EULA). However, whether the user clicked "Accept" or "Decline," the DRM program would silently copy itself to the machine. On MacOS, users would at least see an installation prompt; on Windows, the installation occurred with no notice. Like many malicious programs, XCP lacked an uninstaller and included a dangerously vulnerable rootkit—soon exploited by trojans circulating online.

It was Mark Russinovich, founder of the software firm Winternals and a brilliant engineer (now CTO of Microsoft Azure), who uncovered and exposed this scandal. After his revelations, Sony released an uninstaller for XCP's rootkit—but added another layer of spyware in the process.

Russinovich wasn't fooled. He published a detailed takedown, showing that Sony's "uninstaller" merely removed the file's hidden attribute without deleting it. Worse still, the tool installed new "dial-home" malware, and users were forced to surrender their email addresses to download it—subsequently receiving Sony's promotional spam. Public outrage exploded. On November 18, 2005, Sony finally released a functional uninstaller and began recalling the affected CDs. Just days later, Texas filed a class-action lawsuit, followed by similar suits in New York and California. Sony ultimately settled out of court, paying substantial compensation and offering DRM-free replacements or cash.

Apple, too, faced its share of DRM woes. In the early 2000s, Apple employed its proprietary FairPlay DRM for copyright protection. Strikingly, it never patented FairPlay—a fact that opened it up to infringement claims. In 2005, patent troll Pat-rights accused Apple of infringing U.S. Patent No. US6665797B1. Company president Peter Chung remarked:

> *This is certainly a patentable technology. If iTunes does not patent it, there must be a very good reason for them not to do so--someone else has patented this.*[1]

Pat-rights aggressively litigated against major firms like Apple, Google, eBay, Samsung, HTC, and Sony. In a recent case, Apple and Sony were co-defendants.[2] The federal appellate court dismissed the appeal by inventor Ho Keung Tse using the doctrine of collateral estoppel—in under six pages.

Beyond patent trolls, Apple faced a prolonged antitrust suit over FairPlay. In January 2005, just as Sony's DRM scandal broke, an iTunes user sued Apple for limiting playback of purchased music to Apple devices. In 2012, the case became a

[1] Jo Best. (2005). Firm seeks piece of Apple pie via iPod claim. Retrieved April 6, 2024 from ZDNet: https://www.zdnet.com/article/firm-seeks-piece-of-apple-pie-via-ipod-claim/

[2] TSE v. APPLE INC., No. 15-1639 (Fed. Cir. 2015). Retrieved April 6, 2024 from https://cases.justia.com/federal/appellate-courts/cafc/15-1639/15-1639-2015-11-05.pdf?ts=1446739315

class-action lawsuit covering most iPod buyers from 2006–2009. Apple narrowly prevailed in 2014.

In response, Steve Jobs published his now-famous 2007 open letter, "Thoughts on Music," urging major record labels to abandon DRM. He claimed Apple had never wanted DRM and was forced into FairPlay by the big four music companies.

In 2007, EMI and Universal Music led the way in abandoning DRM. By year's end, Warner had followed suit. Sony's DRM patent filings that year fell by half. In January 2008, Sony Music also dropped DRM, becoming the last of the "big four" to do so.[3] Notably, Netflix began allowing offline downloads of its streaming content in 2016, marking DRM's true decline. Around this time, Sony's DRM-related filings dropped to single digits per year.

The moral of the story? For Sony and Apple alike, if the strategic path is wrong, more patents only make the failure more complete. A misguided R&D direction—one that protects only corporate interests while ignoring users—is ultimately doomed.

[3] Paul Thurrott. (2008). Sony BMG Plans to Drop DRM. Retrieved April 6, 2024 from https://www.itprotoday.com/sony-bmg-plans-drop-drm

Chapter 23
A New Generation Surpasses the Old

In 1997, when Steve Jobs returned to Apple, Sony received 859 granted patents from the United States Patent and Trademark Office, ranking 9th nationwide—nearly twice the number granted to North American Philips (473). Apple, by contrast, ranked 48th that year with only 206 patents. In 1998, Sony's patent count surged to 1316, propelling it to 5th place, while Apple inched up to 270 patents, maintaining a modest 49th place. Over the following years, Apple's patent grants declined, with its ranking dropping beyond the 170 s, whereas Sony consistently held its position within the top ten (Table 23.1).

Yet during this same period, Jobs completed the foundational work for the iPod—securing the right talent, technology, and software. In 1997, he appointed 30-year-old Jonathan Ive (born 1967) as Senior Vice President of Industrial Design. Ive went on to lead the design of the iPod, iPhone, iMac, and iPad. Today, he is credited as the inventor on more than 2000 Apple patents, including 1593 design patents. Two of these—USD618677S1 and USD593087S1—were central to Apple's design patent lawsuit against Samsung.

Another pivotal figure to join Apple during this period was Anthony Fadell (born 1969), widely known as the "father of the iPod." Fadell had deep ties to Sony, Philips, and Apple. He had worked at General Magic, a company spun out of Apple that focused on touch interfaces in the early 1990s. General Magic counted Sony, Philips, and Apple among its clients and investors; its board included Norio Ohga (Sony), John Sculley (Apple), and George Fisher (Motorola, and later Kodak CEO during its transformation period).

In 1995, Fadell founded the Mobile Computing Group at Philips and served as CTO, developing several handheld smart devices based on Windows CE, including the Nino PDA. In 1999, he left Philips to start Fuse, aspiring to be the "Dell of consumer electronics" with a hard drive–based music player and online music store. Jobs embraced his vision almost entirely. In April 2001, Fadell officially joined Apple and remained a key leader of the iPod division until 2008. Apple currently holds 101 invention patents and 117 applications listing Fadell as an inventor, many related to haptic feedback and data exchange.

Q. Yue, F. Yang, *Governing Innovation: A Global History of Patent-Driven Technological Progress*, https://doi.org/10.1007/978-981-95-5570-3_23

Table 23.1 Number and ranking of patents granted to Apple and Sony, 1997–2001

Year	Apple patents	Rank	Sony patents	Rank
1997	206	48	859	9
1998	270	49	1316	5
1999	169	90	1410	5
2000	86	183	1385	6
2001	97	172	1363	7

Table 23.2 Comparative patent analysis of Apple and Sony, 1997–2001

Keyword	Apple patents	% of total	Sony patents	% of total
Network OR internet OR web	277	52.36%	2721	28.43%
Interface	373	70.51%	3702	38.68%
Interact*	219	41.40%	847	8.85%
Synchro*	108	20.42%	2856	29.84%
Audio	156	29.49%	3652	38.16%
Fidelit*	17	3.21%	112	1.17%
Portab* OR Miniatur*	97	18.34%	1891	19.76%
Stereo	25	4.73%	423	4.42%
Music*	31	5.86%	1266	13.23%
Batter*	46	8.70%	825	8.62%
Total	529	–	9570	–

(i.e., it represents any characters/variations).

iTunes development, meanwhile, was led by Jeffrey Robbin and his team. Before Jobs returned, Robbin had worked at Apple before leaving to develop SoundJam MP, the software precursor to iTunes. In 2000, Jobs acquired SoundJam MP and brought Robbin and his team back to Apple, where Robbin became head of the iTunes project. To date, Apple holds 163 patents and 214 applications naming Robbin as an inventor, primarily in user interfaces, media asset management, and image scaling.

Jobs returned to Apple in 1997 and launched the first iPod in 2001. We conducted a keyword search analysis of patents filed and granted to Apple and Sony during 1997–2001. As Table 23.2 shows, while Apple's patent volume lagged far behind Sony's, its emphasis on human-computer interaction and interface design was far more pronounced.

Undeniably, Sony had a much deeper technical foundation in digital music hardware and maintained overwhelming superiority in audio, data transmission, and battery technologies. However, Apple demonstrated greater attentiveness to users—especially younger ones. Over 70% of Apple's patents involved "interface," and more than half involved "network" technologies. For the new generation, graphical interfaces and the internet were defining features of their tech experience, while plain audio or stereo systems were passé.

Apple captured the hearts of the young. Meanwhile, Sony—like Philips before it—was beginning to show signs of aging.

The journey from bold disruptor to aging giant is often decades long. During this time, Sony and Philips focused their patent strategies on winning format wars in video and optical media, aiming to establish global standards that would guarantee them monopolistic returns. Yet both companies' indecisiveness and political caution were fully exposed in the way those format battles unfolded.

Chapter 24
The Failure of Betamax

After Philips' compact audio cassette achieved widespread success in the consumer market, competitors—including Sony—hoped to replicate that success in the realm of video. Their aim was to develop a cassette- or cartridge-based video tape that could open up an entirely new consumer sector. However, video signals require hundreds of times more bandwidth than audio signals, and magnetic tape cannot simply be made hundreds of times wider or longer. To create a user-friendly video tape system, a host of engineering problems had to be solved. To understand these issues, we must start from the earliest developments in video tape.

As mentioned earlier, Bing Crosby, who disliked performing live, invested in California-based Ampex to support its magnetic tape research. After Ampex achieved breakthroughs in audio recording, it quickly began work on video recording systems. To address the bandwidth needs of video signals, engineers initially tried increasing the number of magnetic heads, lengthening the tape, and speeding up the reel rotation. Jack Mullin, who had brought magnetic tape technology from Germany to the U.S., had already envisioned such a method. Around 1950, he developed and patented this kind of video tape system (patent number US2794066A, see Fig. 24.1), and publicly demonstrated it in 1951.

Mullin's method—involving faster tape movement and longer tape reels—was known as *longitudinal video recording*. The tape needed to travel at a speed of 360 inches per second to properly record video. Even then, a reel with a 45-cm radius could only hold 4 min of footage.[1] The resulting video quality was also poor, lacking in practical utility.

In 1956, Charles Ginsburg's team at Ampex developed a new method called *quadruplex recording*, which used multiple heads. As shown in Fig. 24.2, four magnetic heads were mounted on a spinning, disc-shaped drum that rotated perpendicular to the tape's direction of movement. On the tape itself, the video information tracks

[1] Okuda, H. (2010, November). The dawn of video tape recording and development of the helical scanning system. In *2010 Second Region 8 IEEE Conference on the History of Communications* (pp. 1–6). IEEE.

Q. Yue, F. Yang, *Governing Innovation: A Global History of Patent-Driven Technological Progress*, https://doi.org/10.1007/978-981-95-5570-3_24

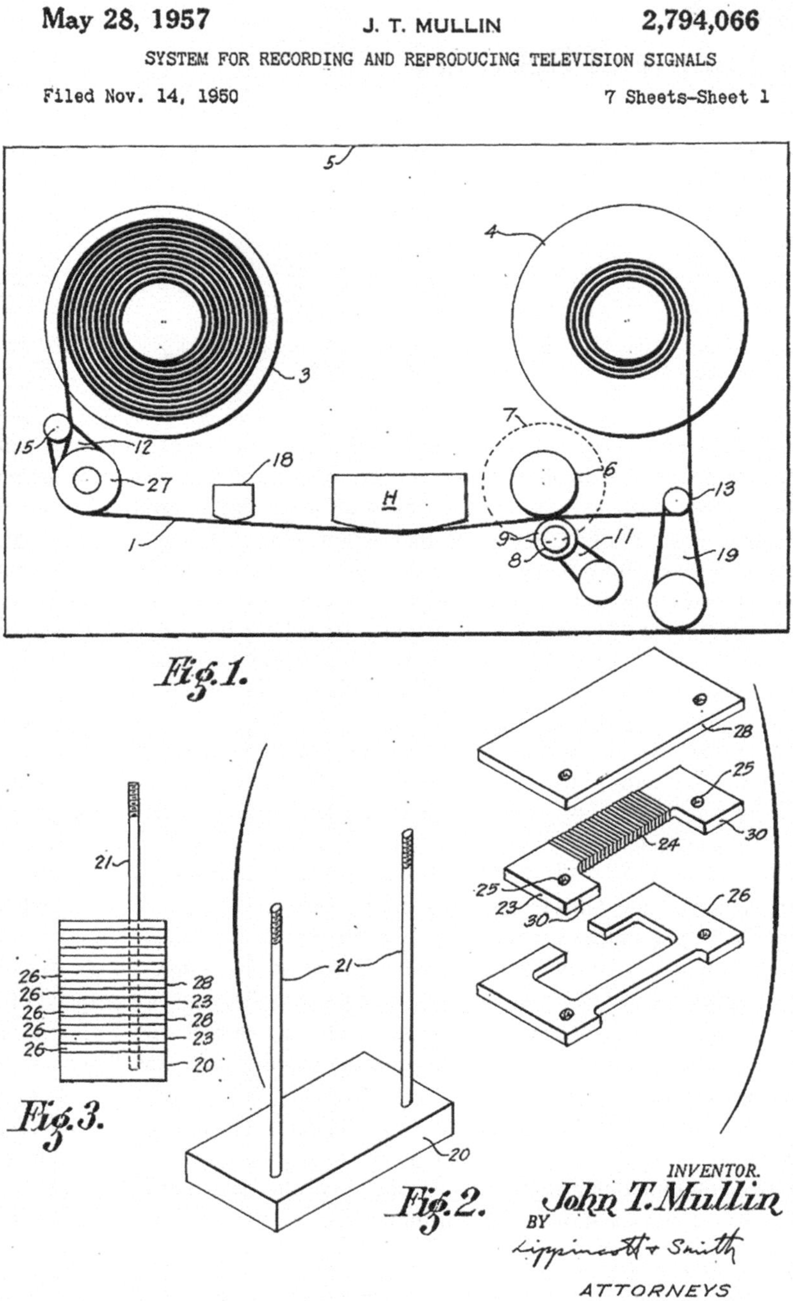

Fig. 24.1 Drawing from US2794066A Patent

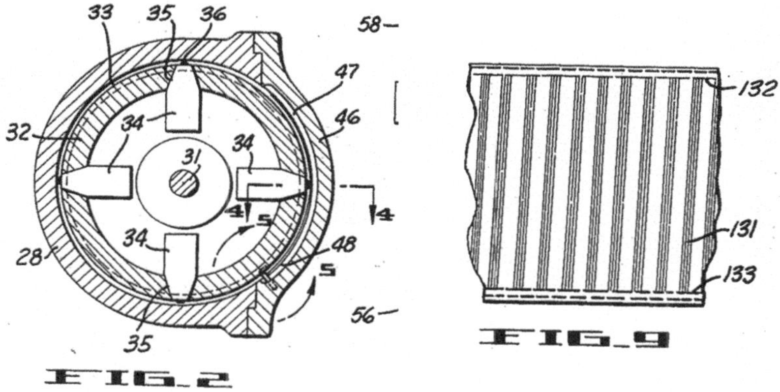

Fig. 24.2 Drawing from US2866012A Patent filed in 1955 by Ginsburg and Others

were also nearly perpendicular to the tape's travel direction (see Fig. 24.3). This design dramatically reduced the required length and size of the tape. For the next several decades, quadruplex technology remained the standard in professional video recording.

The next step in video tape evolution was the *helical scan* method. With helical scan, the magnetic heads no longer rotated perpendicular to the tape direction, but instead at a slight angle—typically less than 5 degrees—so that the tracks on the tape were neither parallel nor perpendicular to its longitudinal axis. Because the tracks formed an angle, each could be much longer than the tape's width. Whereas the quadruplex method required 32 tracks to store one video frame on a 2-inch-wide tape, helical scan needed only two angled tracks, each up to 19 inches long. This innovation significantly reduced the overall size of the tape cassette (see Fig. 24.4).

Helical scan technology itself was developed quite early. RCA filed a patent for it in 1950 and received approval in 1956 (see Fig. 24.5, patent number US2773120A). Ampex quickly licensed the technology from RCA and began developing a new video tape system based on it.[2] Although the core patents for these early video tape technologies were held by Ampex and RCA, it was Japanese companies that ultimately brought them to the mass market. In 1950s Japan, the patent system was still underdeveloped, so Ampex had to find partners to license its patents and establish business relationships. This laid the foundation for Japan's rise in the video tape industry, eventually surpassing the West and bringing video cassettes into everyday homes.[3]

By the 1960s, helical scan technology had reached maturity, but the dominant commercial format remained reel-to-reel tape. As mentioned earlier, Philips'

[2] Lipton, L. (2021). *The cinema in flux: the evolution of motion picture technology from the magic lantern to the digital era*. New York: Springer.

[3] James Lardner. (1987). The Terrible Truth about Japan. Retrieved April 6, 2024 from Washington Post Website: https://www.washingtonpost.com/archive/opinions/1987/06/21/the-terrible-truth-about-japan/ea5f334f-e5f3-4394-9a6f-f23a1d078230/

Fig. 24.3 Schematic of quadruplex recording technology

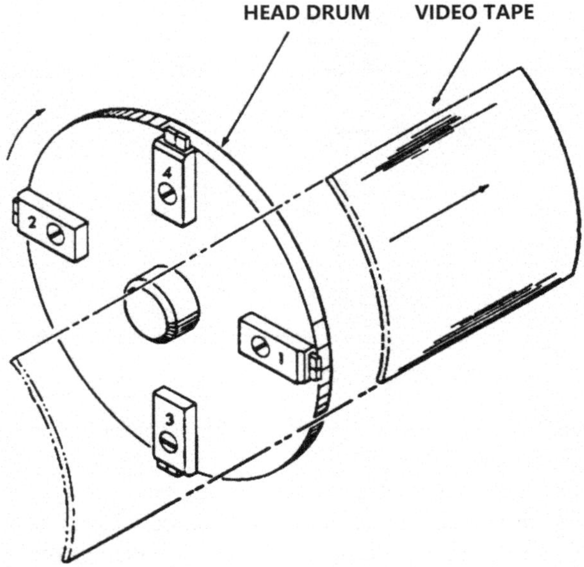

compact cassette had achieved tremendous success in the audio field during this period, supplanting reel-to-reel systems and showing Sony a clear direction forward. Sony was acutely aware that compact size was the key to the cassette's triumph—and miniaturization was precisely what Sony excelled at. This time, Sony wanted to take the lead, developing a superior cartridge-based video tape system before Philips could.

According to Sony's official records, Nobutoshi Kihara, the engineer responsible for developing the video tape system, once believed that creating a video tape system in the form of a cassette, like Philips' audio cassette, was an almost impossible task. He reportedly told Masaru Ibuka: "The construction of a VCR is very complex. It will be extremely difficult to build a machine that will use a cassette tape, let alone in color. You don't understand what you are asking the engineers to do!" But Ibuka was adamant that the new product had to move in that direction: "Look how easy audio tape recorders can be used thanks to the cassette tape! Why can't we incorporate this function into video players? This is the obvious next step that has to be taken in the development of this product."[4]

Under Ibuka's insistence, Kihara's team developed a compact cartridge-style video tape recording system around 1969, known as **U-matic**, and filed patents in multiple countries.[5]

---

[4] Sony Group Corporation. (n.d.). *History: Sony history 2–01*. Sony. Retrieved March 31, 2024 from https://www.sony.com/en/SonyInfo/CorporateInfo/History/SonyHistory/2-01.html

[5] Martin, Jeff (2007). "Curriculum Module 3/4" U-matic Videotape". New York University's Moving Image Archiving & Preservation Program. https://www.miap.hosting.nyu.edu/program/modules/Martin_UmaticTape_old_y.pdf

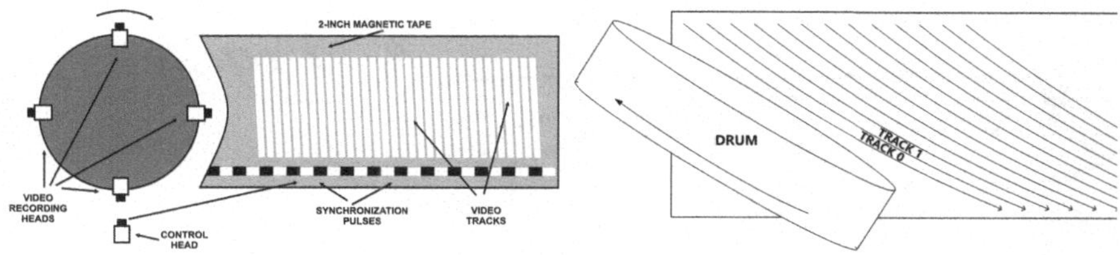

Fig. 24.4 Comparison of quadruplex (left) (Tracks almost perpendicular to the longitudinal axis of the tape. Therefore, track length cannot exceed tape width. [Source: Tecchese. (2009, November 21). *Quadruplex* [Image]. Wikimedia Commons. https://commons.wikimedia.org/wiki/File:Quadruplex.svg. This image is shared under the Creative Commons CC BY 3.0 license]) and Helical Scan Recording (right) (Diagonal tracks can exceed tape width. [Source: Kubanczyk. (n.d.). *Helical tape drive* [Image]. Wikimedia Commons. https://commons.wikimedia.org/wiki/File:Helical_tape_drive.png. This image is shared under the Creative Commons CC BY-SA 3.0 license]) technologies

According to Sony's official documentation, the U-matic system achieved three key technological breakthroughs. The first was solving the color reproduction issue that Kihara had emphasized.[6] Prior to U-matic, video recording typically used the "Y/C signal separation system," where the Y signal represented brightness and the C signal represented color. These were recorded on separate tracks. Engineer Toshihiko Numakura designed a new system called the **Color-Under System**, which recorded both brightness and color information on the same track. This innovation became known as the "Numakura Patent" (Japanese Patent No. JPS5345656B1, U.S. Patent No. US3580990A) and laid the foundation for Sony's later dominance in color video performance.

The second and third challenges that U-matic solved were cassette design and the automatic loading and operation of the tape. As mentioned previously, video recording and playback required multiple magnetic heads, mounted on a large, disc-shaped rotating drum—too large to fit inside the compact cassette shell. Therefore, during playback or recording, the tape had to be pulled from the cassette, guided around the rotating drum and its heads, then returned into the cassette. Kihara achieved this with a complex design involving a series of pins and pulleys, documented in U.S. Patent US3740495A. Because the tape's path roughly formed an inverted "U," the system was named **U-matic**. The cassette design itself was protected under U.S. Patent US3735939A (see Fig. 24.6).

Despite the innovation, Sony lacked confidence in U-matic's mass-market potential. Competing formats were already on the market, and Sony worried that other major companies might establish rival standards that consumers would embrace. To gain a foothold early, Sony signed a 1970 agreement with Panasonic, JVC, and other companies to share U-matic technologies and jointly pursue commercial and

[6] Sony Group Corporation. (n.d.). *History: Sony history 2–01*. Sony. Retrieved March 31, 2024 from https://www.sony.com/en/SonyInfo/CorporateInfo/History/SonyHistory/2-01.html

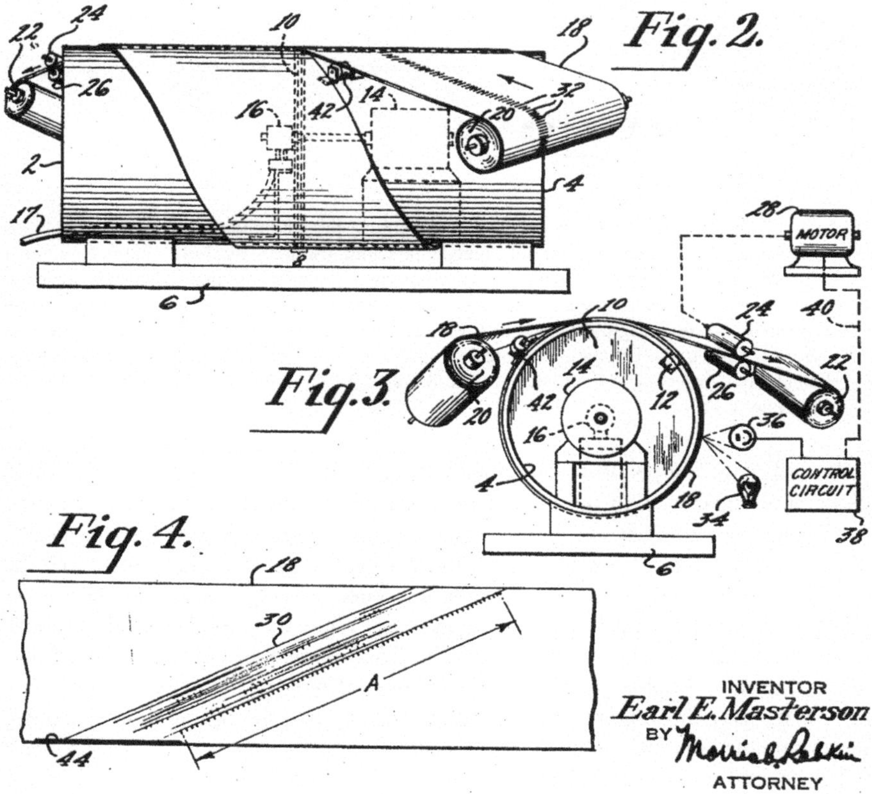

Fig. 24.5 Drawing from US2773120A Patent describing helical scan technology

technical development of the system.[7] Philips, however, did not join Sony's alliance this time.

Although U-matic technology was excellent, it never achieved the same kind of mass-market success as Philips' audio cassette. A U-matic VCR cost around $2000 and weighed 27 kg, while a single blank tape cost $30—far beyond the reach of the average consumer.[8] However, compared to the 16 mm film systems used for television production at the time, U-matic was lightweight and convenient. It unexpectedly gained traction in professional settings such as news media, education, and business, becoming a de facto industry standard. Later, when Philips and Sony

[7] Martin, Jeff (2007). "Curriculum Module 3/4" U-matic Videotape". New York University's Moving Image Archiving & Preservation Program. Retrieved April 6, 2024 from: https://www.miap.hosting.nyu.edu/program/modules/Martin_UmaticTape_old_y.pdf

[8] Tekla S. Perry. (2016). Revisiting the VCR's Origins JVC and Sony transformed the video recording technology pioneered by Ampex into a major industry. Retrieved from: https://spectrum.ieee.org/revisiting-the-vcrs-origins

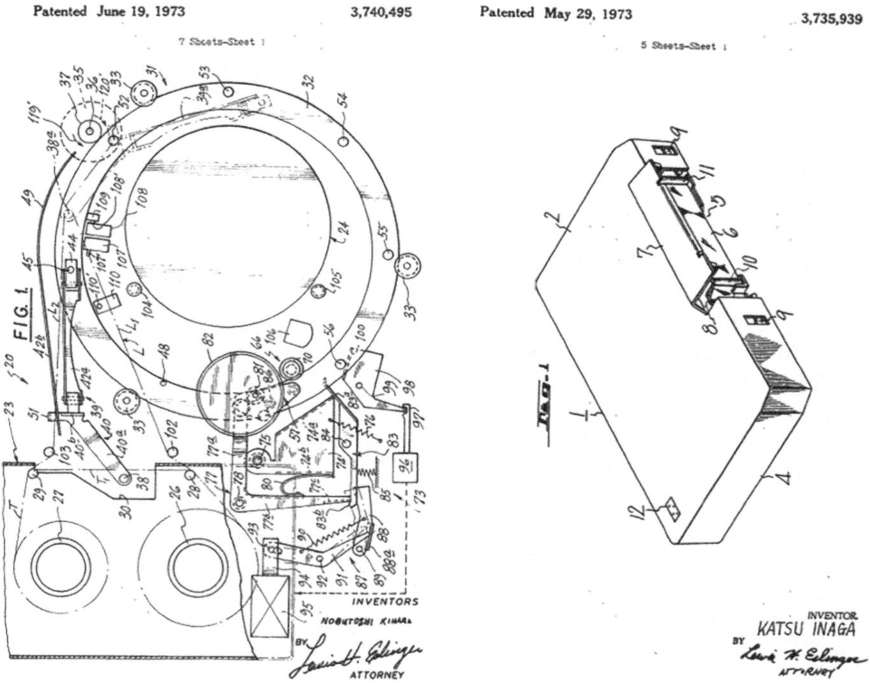

Fig. 24.6 Cassette tape design. (Patent drawing from US3735939A)

co-developed the audio CD, the 44.1 kHz sampling rate used by U-matic had already been widely adopted, and thus became the default sampling rate for CDs.

U-matic's first major customer was the Ford Motor Company (hereafter "Ford"). In August 1972, shortly after U-matic was released, Ford purchased over 4000 machines for internal use—producing videos, training sales personnel, and for external communications. Within a few years, this practice spread widely in the commercial and industrial sectors. That same year, NBC in the United States also began using U-matic recorders. An NBC executive remarked, "S*ony may not have planned exactly this application of their U-matic....but at the moment it is in our view the tape system with the best combination of quality and portability*."[9] The combination of U-matic tapes and recorders, along with Sony's recently released portable video camera—the DV-2400—enabled journalists to cover stories independently with just one set of gear, profoundly impacting the news industry.[10] In the

[9] Martin, Jeff (2007). "Curriculum Module 3/4" U-matic Videotape". New York University's Moving Image Archiving & Preservation Program. Retrieved April 6, 2024 from: https://www.miap.hosting.nyu.edu/program/modules/Martin_UmaticTape_old_y.pdf

[10] Gee, Charles Wesley, "Backpack Journalism in Television Newsgathering: Audience Perceptions of Quality. "PhD diss., University of Tennessee, 2008. Retrieved April 6, 2024 from https://trace.tennessee.edu/utk_graddiss/521

1980s, many Western television broadcasters recorded their original footage using U-matic tapes. Even into the twenty-first century, the task of digitizing massive archives of U-matic footage remained a key initiative for the European Union.[11]

As U-matic's main users became businesses and broadcasters—professional users—Sony adjusted to meet their needs. The system became increasingly complex and expensive, with blank U-matic tapes consistently priced between $25 and $35, which remained unaffordable for ordinary consumers. Furthermore, U-matic was not an exclusive Sony standard, which went against the company's long-term ambitions. Determined to capture the untapped home video market, Sony decided to develop a more lightweight and affordable video recording system. This became the company's strategic flagship product of the 1970s: the Betamax home video system.

Around this time, Japanese companies made a significant breakthrough in helical scan technology, drastically increasing information density on tapes and further reducing tape size. This breakthrough was known as **azimuth recording** (Fig. 24.7 compares traditional helical scan with azimuth recording). Both Panasonic and Sony made key contributions in this area.

It's important to understand that, whether in quadruplex or helical scan systems, the recording tracks on a tape were traditionally spaced apart, with blank areas called "guard bands" in between to prevent **crosstalk**—interference from neighboring tracks. This design had been in use since Jack Mullin's 1950 patent, US2794066A. For over two decades, such guard bands, which made up 30–50% of the tape's area,[12] were standard in the industry, essentially wasting a substantial portion of the tape's storage potential.

Some sources indicate that Panasonic was the first to invent azimuth recording technology in 1968,[13] while Sony also incorporated this technology into its U-matic tape design by 1970. However, when Sony presented this design to its U-matic allies, Panasonic and JVC, both companies objected. As key members of the U-matic alliance, Panasonic and JVC lacked the technical capacity at the time to mass-produce tape systems based on azimuth recording. To accommodate its partners' manufacturing limitations, Sony reluctantly compromised and released a U-matic product based solely on traditional helical scan technology. Though easier to manufacture, this version fell short of Sony's expectations in size and weight.

[11] Martin, Jeff (2007). "Curriculum Module 3/4" U-matic Videotape". New York University's Moving Image Archiving & Preservation Program. https://www.miap.hosting.nyu.edu/program/modules/Martin_UmaticTape_old_y.pdf *"the 2005 PrestoSpace report on preservation issues found that European audiovisual archives considered U-matic reformatting to their top priority (and claimed to have a total of 700,000 tapes in their collections.25) The BBC archive contains 61,000 U-matic tapes that date from 1982–1988—roughly 4% of the total number of items in the archive, but nearly all of the news material for the bulk of the 1980s."*

[12] Ishizuki, N., & Sugiyama, M. (1974). *Electronic still camera* (U.S. Patent No. 3,821,787). U.S. Patent and Trademark Office. Retrieved March 31, 2024 from https://patents.google.com/patent/US3821787A

[13] Luitjens, S. B., & Rijckaert, A. M. A. (1999). The history of consumer magnetic video tape recording, from a rarity to a mass product. *Journal of magnetism and magnetic materials, 193*(1–3), 17–23.

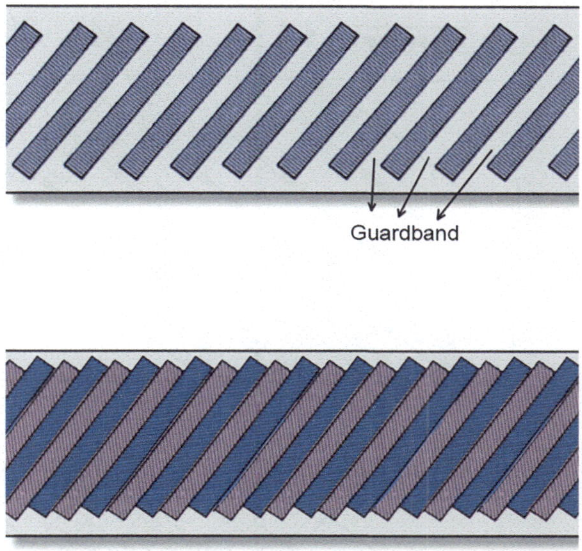

Fig. 24.7 Comparison of helical scan technology (top) and Azimuth recording technology (bottom)

This was one reason U-matic failed to reach the broader consumer market, and it pushed Sony to take a more independent path in the development and promotion of Betamax.[14]

A 1980s article in *The Washington Post* stated: "It was Panasonic that introduced azimuth recording—a clever method of angling the heads to avoid crosstalk between adjacent tracks. But it was Sony that made the next big leap, especially engineer Nobutoshi Kihara, who figured out how to apply the azimuth concept to color video signals and developed the first workable videotape format. If you had to name an 'inventor' of the home VCR, that person would be Kihara."[15]

Based on this information, we identified Kihara's 1971 patent JPS5133690B2 (U.S. counterpart: US3821787A, illustration shown in Fig. 24.8). In the patent description, Kihara wrote: "... a large portion (30% to 50% of the total length) of conventional tape is unused... With the emergence of compact home video recorders, the demand to conserve tape space has grown—first, to lower tape cost, and second, to enable housing the device in as small a space as possible. Therefore, it has become increasingly important to record on more closely spaced, slanted tracks to fully utilize the tape surface. That is one of the primary objectives of this invention." As we've seen throughout this book, Sony's drive for miniaturization is practically genetic. This time, rather than reducing mechanical size, Sony was able to

[14] Michael A Cusumano., Yiorgos Mylnadis, Richard S. Rosenbloom. "Strategic Maneuvering and Mass-Market Dynamics: The Triumph of VHS over Beta." The Business History Review, Vol. 66, No. 1, (Spring 1992) pp. 51–94.

[15] James Lardner. (1987). The Terrible Truth about Japan. Retrieved April 6, 2024 from Washington Post Website: https://www.washingtonpost.com/archive/opinions/1987/06/21/the-terrible-truth-about-japan/ea5f334f-e5f3-4394-9a6f-f23a1d078230/

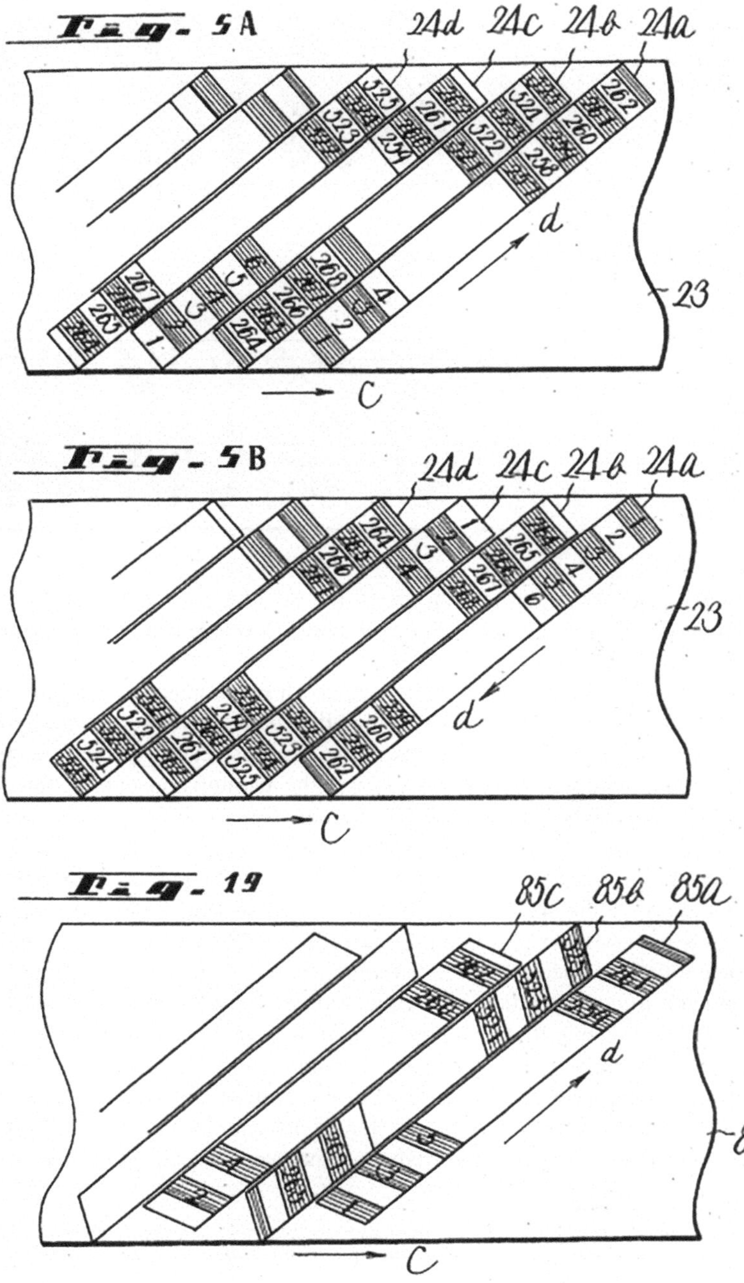

Fig. 24.8 Patent drawing from Kihara's 1971 patent JPS5133690B2

reclaim nearly half of the tape's storage space—an achievement far more impactful than any single ultra-compact device.

Thanks to this space-saving advance, Sony reduced the tape width in the new system from U-matic's ¾ inch to ½ inch. Sony also designed a new cassette and tape transport mechanism for Betamax, which were recorded in patents such as US3925820A.

Meanwhile, JVC—also a member of the U-matic alliance and a Sony licensee—developed its own new home video system called VHS. VHS likewise employed azimuth recording and followed similar principles for color signal processing as U-matic and Betamax. Its tape was slightly larger than Betamax's, and its overall image quality was slightly inferior. The tape loading and transport paths of VHS and Betamax also differed, as illustrated in the respective patents shown in Figs. 24.9 and 24.10.

Figure 24.11 illustrates the tape transport paths of U-matic, Betamax, and VHS, from left to right. The U-matic system uses an inverted "U"-shaped path, hence the name "U-Load." Betamax's path resembles the letter "B," thus "B-Load," which also inspired the Betamax name. VHS, meanwhile, uses a path similar to the letter "M," referred to as "M-Load." Sony's B-Load design is more complex, aiming to reduce pressure on the tape from pulleys and guide pins. In the diagram, the circled areas indicate regions of higher tape pressure. Betamax has only one such high-pressure point, while VHS has two. This design nuance contributes to Betamax's slightly better tape quality. However, the advantage was not decisive. On the contrary, the simpler design of VHS made it easier and cheaper to manufacture—an edge that JVC would explcit fully. Later, when Sony attempted to increase Betamax's recording time, the complexity of the B-Load mechanism became a liability, making internal redesigns far more challenging.

According to Sony's official website, when JVC launched the VHS system, "Sony thoroughly examined the VHS format, and everyone was stunned. The technologies and processes that Sony had voluntarily disclosed to unify the U-matic and Betamax formats had been entirely incorporated into VHS. Despite having offered its foundational patent technologies freely to both Panasonic and JVC, Sony was shocked by what it saw."

VHS was officially released in September 1976, more than a year after Sony's Betamax. During that interval, JVC not only absorbed key technological insights from Betamax but also uncovered one crucial factor that would ultimately spell Betamax's downfall.

Sony had not conducted any market research during the development of Betamax. In interviews with Sony executives conducted by Rosen and others, veteran Sony leader Yasuo Kuroki candidly admitted this oversight.[16] Harvard professor Clayton M. Christensen also discussed the issue:

[16] Rosen, D. E., Schroeder, J. E., & Purinton, E. F. (1998). Marketing high tech products: lessons in customer focus from the marketplace. *Academy of Marketing Science Review*, *6*, 1–17

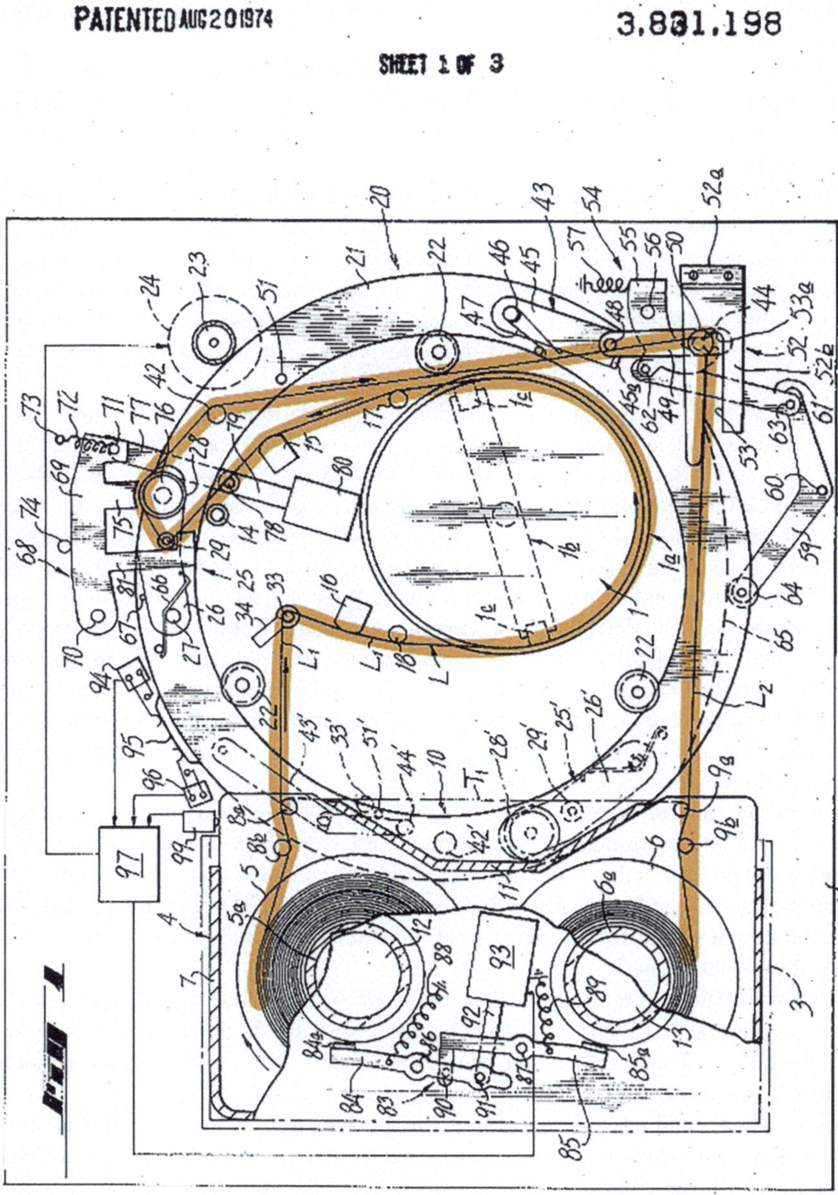

Fig. 24.9 Betamax tape path (Sony's US3831198A patent drawing) (highlighted section shows the tape)

Before 1980, Sony founder Akio Morita and a small group of trusted associates made every new product-launch decision. As a policy, they never did any market research — if a market

U.S. Patent Nov. 29, 1977 Sheet 2 of 3 **4,060,840**

FIG. 2

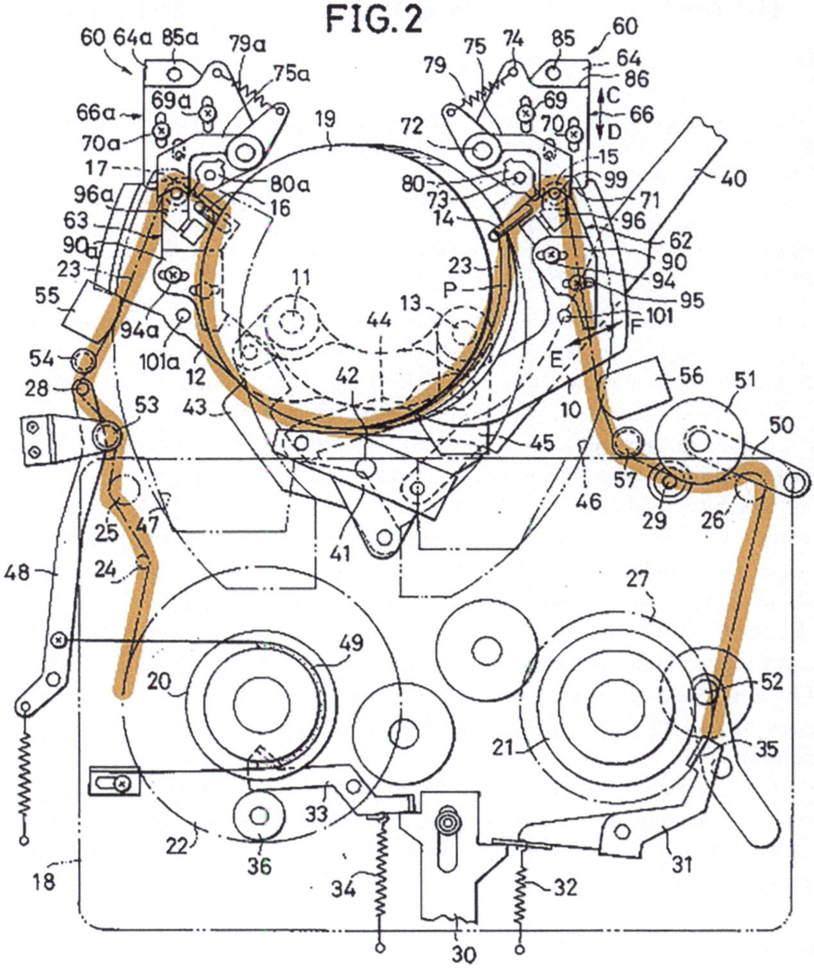

Fig. 24.10 VHS tape path (JVC's US4060840A patent drawing) (highlighted section shows the tape)

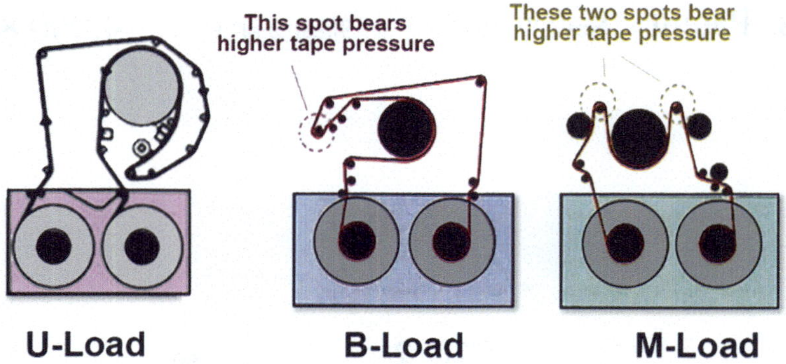

Fig. 24.11 Tape transport paths of U-matic, Betamax, and VHS

> *did not exist, they believed, it could not be analyzed. The group developed an intuitive but practiced process for shaping and launching disruptive businesses. In the early 1980s, Morita withdrew from active involvement in the company in order to focus on political activities. The company began beefing up its marketing functions with M.B.A.s who favored the use of data-driven, analytical processes to assess market needs. Such processes can only identify and shape sustaining innovations — and as a consequence, Sony lost its ability to continue launching disruptive businesses.*[17]

This approach caused Sony to overlook a critical need among American consumers. In September 1974, Sony presented its Betamax design to RCA in hopes of jointly developing the U.S. market. RCA, once a leader in tape technology, was still in the early prototype stage with its own home video system. Impressed by Sony's work, RCA scrapped its own project—but raised concerns about the tape length: Would 1 hour of recording time be enough for American football games or feature films, both of which typically exceeded 60 min?

According to some accounts, the size and recording time of the Betamax tape were determined rather arbitrarily by Masaru Ibuka himself. He once placed a standard American paperback book in the office of Sony's video tape development team and declared that the Betamax cassette must not exceed the dimensions of that book: no longer than 15 cm, no wider than 10 cm, no thicker than 1.5 cm, and capable of at least 1 hour of recording. Betamax was developed in accordance with these specifications, and Sony stubbornly adhered to them even after facing real competition.

In 1974, just before the release of Betamax, Sony executives met secretly with Panasonic leadership on a subway train to discuss the possibility of building a Betamax alliance. Like RCA, Panasonic found the 1-hour recording time unsatisfactory and insisted on a minimum of 2 hours. As for the size, Panasonic was willing

[17] Christensen, C. M., Johnson, M. W., & Rigby, D. K. (2002). Foundations for growth: How to identify and build disruptive new businesses. *MIT Sloan Management Review*, *43*(3), 22. P Retrieved April 8, 2024 from https://sloanreview.mit.edu/article/foundations-for-growth-how-to-identify-and-build-disruptive-new-businesses/

to accept a cassette that was roughly one-third larger than Ibuka's strict dimensions. However, Sony refused to compromise on size and was equally unwilling to sacrifice even the slightest bit of image quality for the sake of longer recording time. After a year of fruitless negotiations, the two sides remained deadlocked.[18] When JVC released the VHS system, Panasonic and RCA both turned down Sony's Betamax proposal and threw their support behind VHS instead.

It was only then that Sony began to grasp the significance of 2-hour recording time. In order to redesign Betamax to support longer recording durations, Sony had to halt licensing negotiations—missing a crucial strategic window. In July 1975, Hitachi expressed interest in Betamax and entered into talks with Sony seeking a licensing agreement. But Sony's executives, chasing technical perfection, rejected the offer. They argued that Betamax wasn't yet refined enough to license to others.

Beyond the issue of recording time, Sony's refusal also had two other motivations. First, JVC was a subsidiary of Panasonic, and Sony still hoped to win over Panasonic to the Betamax camp, thereby neutralizing VHS at its root. Second, Hitachi lacked the manufacturing capacity for video tape systems and had asked Sony to provide OEM (original equipment manufacturer) support. But Sony had an aversion to being an OEM. In early 1976, Akio Morita publicly stated, "Sony is not an OEM," rejecting any suggestion that Sony would help its partners produce Betamax products.

During this period, JVC was aggressively licensing its patents and offering OEM services to help its partners scale production. By this point, Betamax and VHS systems had become so technologically complex that even firms like Ampex, RCA, Hitachi, Sharp, and Mitsubishi lacked the capacity for mass production. Sony's insistence on going it alone stood in stark contrast to JVC's timely support for collaborators, which drew Sharp, Mitsubishi, Hitachi, and others into the VHS camp—leaving Sony increasingly isolated.

Philips did not join the VHS alliance. Instead, it partnered with Germany's Grundig to produce a home video system based on technology dating back to 1972. However, this system was already technologically obsolete compared to what Sony and JVC had developed. In 1975, when JVC brought VHS to the European market, Philips was swiftly overwhelmed. In response, Philips decided to develop a new machine by radically upgrading its outdated technology. In 1980, it released the Video 2000 system, but due to its high cost, late entry, and lack of support from North American Philips, the product floundered—gaining only a fleeting presence in Europe before disappearing from the market.

Between 1977 and 1983, Sony continued to innovate within the Betamax ecosystem, introducing remote controls, half-speed and third-speed recording options, and Hi-Fi audio capabilities. But Panasonic and JVC stayed hot on its heels. Every time Sony achieved a technical breakthrough, the VHS side quickly caught up. Although Betamax retained a slight edge in image quality, the difference was not

---

[18] Tekla S. Perry. (2016). Revisiting the VCR's Origins JVC and Sony transformed the video recording technology pioneered by Ampex into a major industry. Retrieved at: https://spectrum.ieee.org/revisiting-the-vcrs-origins

overwhelming. Furthermore, its smaller cassette, once seen as a strength, turned into a weakness: as Panasonic began developing a VHS tape capable of 4 hours of recording, Sony found it increasingly difficult to extend Betamax's runtime without enlarging the tape—something constrained by the original compact casing.

After nearly a decade of competition, VHS ultimately won the favor of device manufacturers, film studios, and consumers alike, dominating the home video market until the end of the twentieth century. This was the infamous "format war" of the 1970s and 1980s—a cautionary tale of how superior engineering and early market entry may still lose to better strategic planning, broader alliances, and a more intuitive understanding of user needs.

Chapter 25
The Birth of the CD

By the late 1970s, numerous companies had developed compact discs (CDs) capable of storing music. In 1979, Philips unveiled a prototype of its digital audio CD system. Philips and Sony then joined forces, combining Philips' expertise in optical technology with Sony's advanced error correction capabilities to create a brand-new standard for audio storage.

In June 1980, Philips and Sony jointly applied for patents covering the core technologies of the audio CD. These included key patents such as "Method of transmitting error-corrected data" (US4413340A), "Encoding method for binary data" (US4501000A), and "Reflective optical recording medium" (US5068846A). In 1981, the two companies officially released the audio CD standard. That same year, the Japanese Ministry of International Trade and Industry designated the Digital Audio Disc Committee (DADC) of Japan to adopt and promote this standard as the global benchmark.

In 1984, Philips and Sony jointly released the first CD-ROM standard. From 1985 onward, some computer manufacturers began selling CD-ROM drives—commonly referred to as "optical drives"—as accessories for personal computers. In 1989, Philips, Sony, and Microsoft collaborated to propose the unified CD-ROM XA standard, enabling computers to use a wide range of multimedia extensions derived from CDs.

In 1982, Philips and Sony established a patent pool for audio CD–related intellectual property, managed by Philips. Over time, this pool expanded to cover other CD standards developed by both companies.

The original prototype CD had a diameter of 115 mm—a size chosen with reference to the dimensions of a compact cassette. Philips executives believed that the cassette's success stemmed from its small size, and thus the CD should also be kept compact, with its diameter not exceeding the width of a cassette. Sony took a more aggressive stance, initially proposing a disc diameter of just 100 mm.

After prolonged discussions, it was ultimately Norio Ohga's personal intervention that led to the final standard: a diameter of 120 mm. Sony's official "Sony History" page lavishly recounts this decision as a romantic tale of artistic obsession

Q. Yue, F. Yang, *Governing Innovation: A Global History of Patent-Driven Technological Progress*, https://doi.org/10.1007/978-981-95-5570-3_25

and passion. During negotiations, Ohga insisted that Beethoven's Ninth Symphony—his wife's favorite—must fit entirely onto a CD. The longest known recording of the symphony, conducted by Wilhelm Furtwängler in 1951, lasted 74 min. Therefore, Ohga declared, the CD's playback time could not be less than 74 min.

Philips' official site also recounts this story. However, Kees Immink—one of the inventors named on patent US4501000A and a participant in the CD standardization process—expressed skepticism about this narrative. In a 2007 article, Immink offered his own version of events. At the time of the CD standard discussions, Philips' subsidiary PolyGram had already built a CD production facility in Hanover based on the 115 mm prototype. Their production line was already operational. Sony, on the other hand, had not yet established any CD manufacturing facilities. If Sony agreed to the 115 mm standard, it would effectively grant Philips a first-mover advantage in the music market. Immink believed Ohga cleverly used a romantic excuse to pressure Philips into giving up this advantage. "It was not about Mrs. Ohga's great passion for music," he wrote, "but the money and competition in the market of the two partners… The maximum playing time of the CD was settled at 74 minutes and 33 seconds, but in practice, however, the maximum playing time was determined by the playing time of the U-Matic video recorder, which was 72 minutes. Therefore, rather sadly, Mrs. Ohga's favorite Ninth by Furtwängler could not be recorded in full on a single CD till 1988…".[1]

In March 1980, at a CD standards symposium held in Tokyo, Philips accepted Sony's recommendation to set the sampling resolution at 16 bits and the sampling frequency at 44.1 kHz.

According to the Shannon theorem, the sampling frequency must be at least twice the highest frequency of interest—in the case of human hearing, up to 20 kHz—meaning a sampling rate above 40 kHz is required. But why 44.1 kHz specifically? This has long been a topic of online discussion. As previously mentioned, U-matic had already become the mainstream professional video recording system at the time, and the accompanying Sony PCM-1600 audio encoder used a sampling resolution of 16 bits and a sampling frequency of 44.1 kHz. However, Kees Immink, who was directly involved, explained that the reason for choosing "44.1" over Philips's suggestion of "44.056" was quite simple: 44.1 was easier to remember.[2]

In 1982, Philips and Sony jointly established a CD patent pool and launched a licensing program for the initial batch of digital audio CD patents. This pool was later expanded to cover additional CD standards developed by the two companies. In the early 1990s, Philips and Sony partnered with Taiyo Yuden to create a patent

[1] Immink, K. A. S. (2007). Shannon, Beethoven, and the compact disc. *Information Theory Society Newsletter*, *57*(4), 6 Retrieved April 6, 2024 from: http://www.turing-machines.com/pdf/beethoven.htm

[2] Immink, K. A. S. (2007). Shannon, Beethoven, and the compact disc. *Information Theory Society Newsletter*, *57*(4), 6 Retrieved April 6, 2024 from: http://www.turing-machines.com/pdf/beethoven.htm

pool for CD-R (recordable CDs), and later with Ricoh to establish a similar pool for CD-RW (rewritable CDs).

Once the CD standards were finalized, both companies raced to develop CD players. Young and nimble, Sony once again took the lead, completing the world's first CD player—the CDP-101—6 months ahead of Philips. Philips was forced to request a delay in the product's launch. Sony only partially agreed: since Philips had not yet entered the Japanese market, Sony launched the CD player there first. Six months later, once Philips had completed its own player, the two companies released their products globally in tandem.

Haunted by the "bitter memory" of giving away cassette patents for free, Philips was determined not to be the generous "dupe" this time. It sought to transform its CD patents into a steady revenue stream. Having seen Betamax fail in part due to prohibitively high licensing fees that alienated manufacturers, both Philips and Sony took this lesson to heart and set low royalty rates for CD patents.

The early licensing fees for CD drives were not publicly disclosed, but according to some manufacturers, they amounted to about 3% of a product's retail price. In 1994, another manufacturer reported the rate as 2%. By 1995, the fee hovered between 2% and 3%. These relatively low fees proved critical: they encouraged more companies to enter the CD drive manufacturing business and left ample room for competitive pricing. Sony, for its part, vigorously developed portable CD players under the Discman series, which was later folded into the Walkman brand.

For each blank recordable CD sold, the Philips-Sony patent pool charged a fixed fee. In 1982, CD manufacturers were required to pay a base royalty of 3 cents per disc. To account for inflation, the licensing agreement included a clause allowing the fee to be adjusted according to the Consumer Price Index (CPI). By 1995, the royalty had risen to 5 cents per disc. As new technologies dramatically reduced production costs, this 5-cent fee came to represent about one-third of the CD's total manufacturing cost.

Of course, compared to the average retail price of early CDs, this fee was negligible. In 1988, the average price of a CD was as high as $500. By 1990, it had fallen to $300, and by 1995, it dropped below $100—a decline of over 80%. Even so, for early CD content providers, the licensing fee remained a trifling expense.

In 1994, the U.S. Department of Justice launched an antitrust investigation into the Philips-Sony patent pool—one of the earliest antitrust cases involving a patent pool in the United States. Ultimately, Philips and Sony made concessions to reach a settlement, agreeing to allow licensees to apply for individual or partial patent licenses, thereby ending the requirement for mandatory package licensing. The following year, the Department of Justice issued the "Antitrust Guidelines for the Licensing of Intellectual Property," which explicitly prohibited forced bundling or blanket licensing practices in patent agreements. These guidelines were clearly influenced by the Philips-Sony case.

Much like the compact cassette, the CD standard achieved enormous success—far exceeding Philips's initial expectations. Before 1982, Philips and Sony estimated that CD sales would reach 10 million units by 1985, later revising the target upward to 15 million. What neither company anticipated was that by 1985, CD sales had already skyrocketed to 59 million units. Even today, long after cassette tapes have completely disappeared from the market, CD albums and players—though now niche products—continue to survive tenaciously.

In the 1990s, building on their success, Philips and Sony teamed up with Pioneer to form the DVD3C patent pool, which received approval from the U.S. Department of Justice in December 1998. Under the licensing terms of the pool, DVD video players and DVD-ROM drives were subject to a royalty rate of 3.5% of net sales, with a minimum license fee of $5 per unit and a minimum fee of 5 cents per DVD disc. Notably, this provision of a minimum fee had not existed in the earlier CD patent pool.

In 1999, the retail price of a DVD recorder often reached several hundred dollars, so a $15–$20 licensing fee didn't seem excessive. For that reason, the Department of Justice swiftly approved the Sony-Philips patent pool application and even explicitly stated in its report that the fees were very reasonable. However, just a few years later, the royalty fees for DVD recorders became an inescapable burden. In 2007, the global average OEM price of a DVD recorder was $30; in 2008, it dropped to $25, and by 2009, it was only $23. In December 2008, Hisashi Kato of Japan's Mitsubishi Electric reported that manufacturers were paying as much as $17 in royalties per DVD recorder—amounting to 68% of the product's average sale price that year. Of this, $14 went to rights holders in the DVD3C and DVD6C patent pools.

Faced with this heavy licensing burden, Japanese optical drive manufacturers began retreating from the low-end market and pivoting toward high-end products. From CD burners to DVD drives and then to Blu-ray drives, Japanese companies no longer focused on a single storage medium. Instead, they rapidly transitioned to new standards as they emerged, leaving legacy technologies to other countries, including China. Japan's optical drive production peaked in 2001 and then declined rapidly. In this shift, no successors to iconic products like the Walkman ever emerged—whether for CDs, DVDs, or Blu-ray players.

Some researchers have argued that the CD patent pool's pricing strategy helped establish fair market entry conditions and struck a balance between competition and innovation. In contrast, the DVD patent pool imposed higher barriers to entry, increased market concentration, and hindered innovation.[3] This conclusion is not difficult to understand: if companies like Sony and Philips could earn substantial profits from existing CD technologies by simply "sitting back," why would they bother studying consumer needs or investing in the development of DVD players?

[3] Flamm, K. (2013). A tale of two standards: patent pools and innovation in the optical disk drive industry (No. w18931). National Bureau of Economic Research.

Chapter 26
A Hesitant Revolution

In 1983, due to a flood of poorly made, low-quality games from unscrupulous developers, American consumers completely lost faith in video games, leading to a market crash that nearly wiped out the industry. Atari, one of the top companies, suffered massive financial losses. This event became known as the "Atari Shock." Around the same time, Nintendo released its now-legendary Famicom (short for "Family Computer"), which quickly swept the globe and helped revive the American video game market.

Atari's downfall was largely due to the low entry barrier for game development. The company's Atari 2600 console had once dominated the North American market. However, constrained by U.S. antitrust laws, Atari was unable to effectively prevent third-party developers from creating games or producing cartridges for the 2600 platform. This led to an overwhelming influx of low-quality games. The result was a classic case of "bad money driving out good," which devastated the fledgling video game industry.

Learning from Atari's misstep, Nintendo resolved to maintain strict control over game approval. While it couldn't outright ban developers from making games, it could leverage its cartridge patents to prohibit unlicensed game cartridge production for the Famicom within Japan. To ensure quality, Nintendo established an extremely stringent licensing system: initially, each licensee could develop no more than five Famicom games per year, and those games could not be released on other platforms for 2 years after launch. Moreover, developers were prohibited from manufacturing cartridges themselves—Nintendo would produce them, and would determine production volume based on its own assessments.

Under such draconian terms, it was not uncommon for developers to invest heavily in a title only for Nintendo to reject it, preventing production entirely. Their fate lay entirely in Nintendo's hands—developers had no choice but to comply.

Japanese developers tolerated this monopolistic control, but would the U.S. Department of Justice stand idly by?

To avoid violating U.S. antitrust law, Nintendo devised the 10NES lockout chip system and patented it (U.S. Patent No. US4799635, titled "System for Determining

Q. Yue, F. Yang, *Governing Innovation: A Global History of Patent-Driven Technological Progress*, https://doi.org/10.1007/978-981-95-5570-3_26

Authenticity of an External Memory used in an Information Processing Apparatus"). This chip, embedded in game cartridges, verified whether the cartridge was officially authorized by Nintendo. Even if an unauthorized company managed to produce a cartridge, it wouldn't function on the Famicom (called the NES in the U.S.) without the lockout chip.

Interestingly, Nintendo did not install the 10NES chip in Japanese Famicom consoles—only in the NES models sold in the United States. The reasoning is clear: directly banning third-party developers from making cartridges would likely violate U.S. antitrust laws. So instead of banning unlicensed games, Nintendo banned the production of 10NES chips. Atari later reverse-engineered the chip, bypassed the lockout system, and then turned the tables by filing an antitrust lawsuit against Nintendo. The entire saga was a legal and technological rollercoaster, detailed in a comprehensive article on the NetEase website—which this chapter will not further elaborate.[1]

Following the Famicom, Nintendo introduced its next-generation console, the Super Famicom (SFC). Despite the increasing demands for storage capacity, Nintendo insisted on sticking with cartridges. Due to the limited space on cartridges, Nintendo proactively sought a partnership with Sony to develop a CD-compatible version of the SFC.

After several years of joint development, Nintendo abruptly discovered that the contractual terms were highly unfavorable to them. Without prior warning to Sony, Nintendo unilaterally canceled the collaboration. Then, in a dramatic twist, during a major electronics trade show, Nintendo publicly announced a new partnership with Philips to develop a CD-based system for the SFC.

Betrayed and blindsided, Sony decided to develop its own gaming console—this project ultimately evolved into the first-generation PlayStation (PS). Backed by thorough market analysis, Sony opted to use CDs as the game medium and implemented a developer-friendly licensing system with low royalty fees.

Around the same time, Nintendo launched its third-generation console, the Nintendo 64 (N64). Still wedded to the cartridge format, Nintendo introduced a new N64 lockout chip to prevent piracy and retain tight control over game developers.

Cartridges had obvious advantages and disadvantages compared to CDs. They offered faster read speeds and superior anti-piracy measures, but came with significantly higher production costs and much smaller storage capacity—N64 cartridges ranged from 4 to 64 MB. Despite fully understanding these limitations, Nintendo refused to give up its lucrative licensing revenue or its iron grip on developers.

Developers who had long endured Nintendo's authoritarian practices began defecting en masse to Sony. In 1997, even Square, a high-profile developer with a close history with Nintendo, jumped ship to Sony. Aside from the PS's superior performance, Square's new title *Final Fantasy VII* required four 700 MB CDs—vastly exceeding the 64 MB limit of the N64 cartridge. *Final Fantasy VII* became a

[1] NetEase. (2015, June 2). *游戏秘闻录: 震撼业界的雅达利 vs 任天堂之争* [Gaming chronicles: The shocking Atari vs. Nintendo battle]. NetEase Play. Retrieved March 31, 2024 from http://play.163.com/15/0602/18/AR4HF71400314K8I.html

global sensation and played a decisive role in Sony's dominance of the gaming console market.

CDs were only one reason for the PlayStation's success, though a significant one. It's worth noting that the PlayStation wasn't actually the first game console to use CDs. That distinction belonged to Philips' CD-I system.

At the same time Sony entered the gaming industry, Philips was also eager to compete. Leveraging its cooperative relationship with Nintendo, Philips released the CD Interactive system in 1992, known as "CD-I." As the name implies, CD-I stood for "Compact Disc Interactive." The CD-I player could not only play music CDs, but also run videos, games, and educational software stored on CD-I discs.

Philips once expressed ambitious expectations for CD-I, predicting that it would revolutionize television use. The company emphasized that the technology could enable interactive applications for all age groups, such as browsing museums, coloring cartoons, creating animations, learning new skills, or enhancing games and classic stories—possibilities it characterized as virtually limitless.

Clearly, as early as the early 1990s, Philips envisioned a home entertainment system that could unify gaming, video playback, education, and music enjoyment. Even in the twenty-first century, this concept continued to influence devices like Microsoft's Xbox, Apple TV, and LeEco's smart TVs, all attempting to "conquer the living room."

While Philips lacked experience in gaming and software development, from an intellectual property perspective, the company held some strong cards in the early 1990s. First, CD-I was the earliest optical disc game console, and its standards were encoded in the "Green Book" for multimedia CDs, published in 1986—patents owned by Philips. Furthermore, thanks to its partnership with Nintendo, Philips obtained the rights to use certain key Nintendo characters and trademarks, allowing them to develop spin-off titles for flagship Nintendo franchises. This helped win the early support of some die-hard Nintendo fans during the CD-I's launch.

However, CD-I ended in a disastrous failure. By the time production ceased in 1997, this hybrid machine had sold only 570,000 units. The educational software failed to gain traction, and while the games were somewhat better received, the CD-I game controllers were repeatedly ranked among the worst in history (an early CD-I controller is shown in Fig. 26.1). While Sony meticulously crafted the first-generation PlayStation controller, Philips seemed oblivious to the importance of a dedicated gaming interface, releasing a series of oddly shaped remotes instead. Moreover, as a living-room-oriented entertainment device, CD-I did not include a second controller—users had to purchase one separately—and the port for the second controller was inconveniently located on the back of the unit, adding to user frustration.

In its early advertisements, Philips claimed that the CD-I player would retail for around $300. However, the actual price of this ambiguous, neither-here-nor-there gaming machine ended up at a steep $700. This near-deceptive pricing misstep also revealed various issues within Philips' internal coordination.

In the "Rise and Fall" section of this book, we examined Kodak's painful journey of transitioning to digital technology in the 1990s. Philips and Sony, too, faced

Fig. 26.1 An early CD-I game controller. (Source: Evan-Amos. (n.d.). *Philips CD-i Paddle Controller* [Image]. Wikimedia Commons. https://commons.wikimedia.org/wiki/File:Philips-CD-i-Paddle-Controller.jpg. This image is in the public domain)

similar challenges during that decade as they moved from analog to digital systems—a process fraught with difficulty. Ten years after the invention of the Walkman, both Philips and Sony launched new digital audio storage formats, aiming to ignite a new revolution in the portable music market through digital innovation.

In 1991, *The New York Times* published a small news item reporting that Philips and Sony had initiated preliminary contact. By this point, the products each company had developed were nearing release, but both parties had expressed interest in the other's offering. They planned to sign a cross-licensing agreement, allowing either company to benefit even if its own product failed. This reveals that, from a purely technological standpoint, the two competing formats were not markedly different. According to the article, the music market in 1990 was worth $21 billion, with various player devices accounting for around $190 million of that total. For these two embattled electronics giants, finding a new growth point in this sector promised potentially significant returns.

In August 1990, Philips unveiled its Digital Compact Cassette (DCC), which utilized MPEG-1 Layer 1 compression technology and offered audio quality nearly indistinguishable from that of CDs.[2] At the 1992 Consumer Electronics Show (CES) in Chicago, Philips showcased its first DCC recorder. As a partner of Philips, Panasonic's Technics brand and Grundig also introduced DCC recorders soon after. By this time, Grundig was no longer a competitor to Philips: since Grundig's public listing in 1972, Philips had been steadily increasing its shareholding and had become its parent company by 1984 with a one-third controlling stake.

Both MD and DCC reflected a common trait among companies transitioning to digital technology during that era—a lingering attachment to analog systems, much like Kodak's approach with its Photo CD, described earlier in this book. MD and

---

[2] Ratazzi, E. P. (2004). MiniDisc: Successful Innovation or Just Cool Technology?. *An Analysis of Sony's Decade-Long Quest for Success. Rome, NY.*

DCC were hybrid products that supported analog signal recording, illustrating this ambivalence.

This tendency is clearly visible in Philips' patent EP0381266B1 (see Fig. 26.2). The specification acknowledges that analog tape lacks the audio quality of digital tapes but points out that "there is a vast installed base of cassette tapes and compatible recorders in the market… The number of tape recorders is around one billion, and the number of cassettes is several times that." To accommodate this existing user base, Philips did not pursue a pure digital tape system that would constitute a radical break from the past. Instead, it opted for "a gradual market penetration through minimal system adjustments." The new system required "little to no visual difference from existing cassette recorders, apart from the tape head and some electronic components," and it could read and write digital data while still supporting legacy analog cassette tapes "at least for analog-form recordings."

Figure 26.2 in the patent illustrates this design: device 24 is the erase head, and device 25 is the new type of tape head that can both read and write digital data as well as read analog signals.

We have not seen a DCC cassette in person, and it's rare to find references to the product on Chinese-language websites. But based on available photographs (see Fig. 26.3), the new-format cassette appears to be sleek and compact.

MiniDisc (MD), on the other hand, was a "magneto-optical disk" technology. Under the technological conditions of the time, it offered greater potential for rewritability than CDs and showed promise as a tape replacement. The original

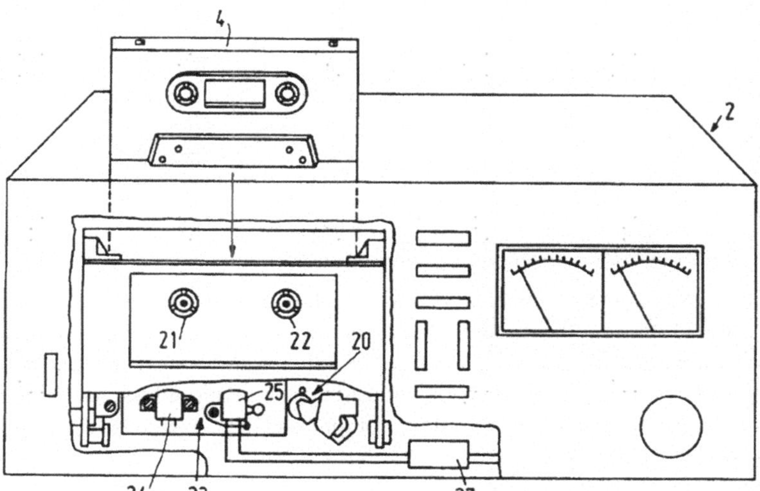

FIG.3

Fig. 26.2 Figure 3 from patent EP0381266B1. Component 24 is the erase head, while component 25 is the new type of magnetic head, capable of both recording/playing back digital information and reading analog signals

Fig. 26.3 DCC Cassette. (Source: DigiAndi. (n.d.). *Philips Digital Compact Cassette open* [Image]. Wikimedia Commons. https://commons.wikimedia.org/wiki/File:Philips_Digital_Compact_Cassette_open.JPGThis image is shared under the Creative Commons CC BY-SA 3.0 license.)

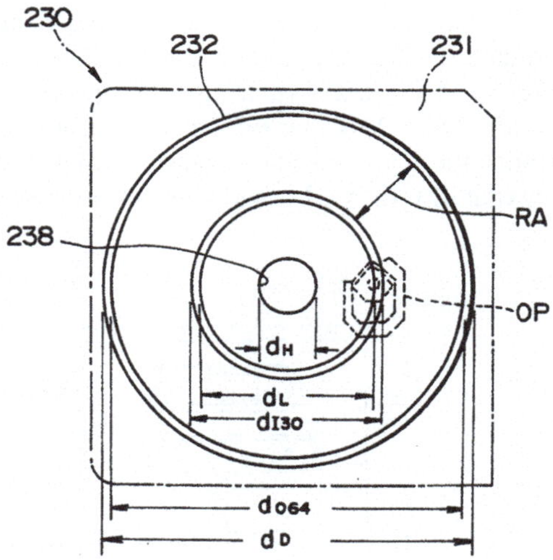

Fig. 26.4 Patent drawing of U.S patent US5400316A

concept of the magneto-optical disc was developed in the early 1980s by Kees Immink, the Philips engineer mentioned earlier. Sony took this idea and carried out extensive miniaturization. In Sony's 1991 US patent (US5400316A, Fig. 26.4), the magneto-optical disc had already been reduced to a diameter of less than 80 mm and was capable of storing at least 130 MB of compressed digital data—one of the foundations of MD technology.

At the time, Sony lacked experience in the digital audio space, and its newly introduced ATRAC (Adaptive TRansform Acoustic Coding) compression technology was still immature. While Philips proudly claimed that its DCC matched CD-level audio quality, MD could only offer audio that was "close" to CD quality.

In reality, early MD audio quality fell noticeably short of CDs; its main advantage lay in its compact disc size.

Sony placed high hopes on MD. Norio Ohga assembled a dedicated task force and elevated the MD project to "corporate status," requiring all departments to collaborate and channel their technological and innovative capabilities into the MD effort. Only Betamax, 8 mm camcorders, and CD players had previously received such organizational priority.

MD's optical pickup system for file directories came from the CD player division, while its anti-shock feature used buffer memory technology also under development for CD players. For core performance and miniaturization, the MD system employed monolithic microwave integrated circuits (MMICs) and heterojunction field-effect transistors (HFETs)—technologies repurposed from an in-car navigation project. Sadly, this kind of unified, cross-departmental effort is rarely seen in today's Sony.

Throughout the development process, Sony's patent team worked closely with the MD development team to establish comprehensive and rigorous intellectual property protections for the format. Thanks to their efforts, the marketing and licensing departments were able to open up MD technology for licensing even before the product launch. By the eve of the MD's release, Sony had already signed 64 licensing agreements: 32 for hardware, 18 for software (i.e., music content), and 14 for blank MD discs. At the same time, Sony's promotional team conducted detailed market research into consumer perceptions of MD versus CD, and into the potential acceptance of MD compared to DCC.

By the end of 1992, both DCC and MD products had entered the market. DCC players were considerably expensive—home stereo systems from Philips and Panasonic retailed around $850. DCC received support from nearly every major record label; by the end of the year, 500 DCC albums had been released. The only major label that refused to support DCC was Columbia Pictures, a subsidiary of Sony.

In its marketing campaigns, Philips promoted DCC as the successor to cassette tape recorders, boasting CD-quality audio and perfect backward compatibility with traditional cassettes.

Sony's first-generation MD player was slightly cheaper at around $650 but still pricey. By early 1993, there were about 300 MD albums available in the U.S., fewer than DCC releases. By September 1993, the number of MD albums had reached 1200, with approximately 750 of them coming from Sony's own music labels.

Faced with two expensive new formats, consumers—burned by prior format wars—responded cautiously. After some time, neither product achieved strong traction in the market, and many consumers chose to wait and see which format would ultimately prevail. A year later, MD player sales in the U.S. had only reached about 50,000 units, while DCC players were estimated to have sold around 30,000 units. Philips quickly lost confidence and admitted in 1993 that both MD and DCC had only attracted audiophile hobbyists. Sony, however, remained bullish, announcing a target of 10 million units shipped by 1995.

After more than a year of technical refinement, Sony's vaunted miniaturization technologies began to show their advantage. In September 1993, Sony released its second-generation MD player, which was 40% smaller and 45% lighter than the original. Meanwhile, the ATRAC compression algorithm had undergone significant improvements, resulting in noticeably better audio quality for the updated MD format. In April 1994, Sony launched a second wave of promotional campaigns in the U.S., cutting the price of MD players and releasing home stereo systems that supported the MD format.

At this point, Philips was clearly faltering. Media scrutiny of DCC was growing, and consumers began to question how much longer the format would survive. In its attempt to maintain compatibility with legacy cassette tapes, Philips faced technical challenges in designing and producing dual-mode heads—capable of reading both traditional analog and new digital tapes—leading to persistently high manufacturing costs and limited economies of scale. Ironically, because DCC players could play traditional cassette tapes perfectly, users tended to continue buying the cheaper analog tapes instead of the more expensive DCC ones. The DCC project never turned a profit. By 1996, Philips had no choice but to officially abandon the format.

Although MD ultimately won the format war, its victory was far from resounding. It never became a global sensation like the Walkman or, later, the iPod. Seven years after its debut, MD player sales remained lukewarm, with only around 1 million units sold in the U.S.—far below Sony's original expectations. Meanwhile, Sony's attitude toward MD began to shift. Unlike its initial full-throttle support, Sony now slashed prices on CD Walkmans while aggressively promoting digital music players that used its proprietary Memory Stick format. Consumers began to sense that Sony might be abandoning MD.

As previously noted, new competitors had already emerged in the market—namely, MP3 players based on solid-state memory. Whether magnetic tape or optical disc, all external storage media were becoming relics of the past.

Chapter 27
Unhelpful Allies

Sony attributed the failure of Betamax largely to its solitary struggle—fighting alone without allies. It lacked the support of film studios and had no exclusive access to valuable content. Senior executives at Sony believed that if they had possessed sufficient film resources to back Betamax, VHS might not have won. Motivated by this reasoning, Sony invested heavily to acquire Columbia Pictures.[1]

This acquisition occurred during the period following the Plaza Accord, when the Japanese yen had appreciated significantly. Talk of the Japanese "buying up America" was rampant. Sony, cautious of American anti-Japanese sentiment, treated its newly acquired business with near indulgence.

According to Sony's official "Sony History" website, films had long been regarded as a symbol of American pride, and Japanese involvement in Hollywood provoked considerable public anxiety. Akio Morita sought to reassure critics by stressing that the Columbia Pictures acquisition was not a cultural invasion and that Sony had no intention of sending Japanese managers into the studio. Norio Ohga reinforced the company's principle of "global localization," noting that post-acquisition success depended on avoiding overt displays of Japanese corporate identity while maintaining the American character of the company. To consolidate the deal, Sony announced the following day a $200 million purchase of Guber-Peters Entertainment and appointed its two founders, Peter Guber and Jon Peters, as co-chairmen of Columbia Pictures.

Though Morita was confident enough to write *The Japan That Can Say No* and was known internally for his autocratic and strong-handed style, he adopted a surprisingly docile posture when dealing with the American acquisition, handing over power to American executives. Nobuyuki Idei commented:

[1] Finkelstein, S. (2006). Why smart executives fail: Four case histories of how people learn the wrong lessons from history. *Business History*, 48(2), 153–170. https://doi.org/10.1080/00076790600576727

Q. Yue, F. Yang, *Governing Innovation: A Global History of Patent-Driven Technological Progress*, https://doi.org/10.1007/978-981-95-5570-3_27

> …We had never really dealt with foreigners. When they made a suggestion, we accepted it eagerly. Our older generation of Japanese harbored an inferiority complex toward foreigners, and Morita was a vivid example of that.[2]

In the end, subservience to Americans did Sony no good. Even though Japan made multiple concessions—signing the U.S.–Japan Semiconductor Agreement and other unequal treaties—American resentment persisted. In 1989, future U.S. President Donald Trump publicly accused Japan of having "systematically sucked the blood out of America—sucked the blood out!"

As for the American executives in whom Sony had placed such high hopes, they had no intention of aligning with the Japanese.

In his article *Sony's Acquisition of Columbia: From Quagmire to Classic*, Wang Yukun summarized the episode as follows:

> Acquisitions are inherently turbulent, involving price negotiations, strategic alignment, organizational design, restructuring, and layoffs. They require a strong, comprehensive team—but at the time, Sony was ill-equipped for such tasks. They simply handed the company over to the Americans and let them manage it in their own way. The team, found through personal connections, was comprised entirely of Hollywood players. The three most influential figures were Mickey Schulhof, president of Sony America; Peter Peterson, Morita's friend and financial advisor for the acquisition; and Peter Guber, who became president of Columbia Pictures (as mentioned earlier). These three worked in unison and played Sony like a fiddle. There was no independent third-party evaluation, no meticulous scrutiny from the Japanese team—Schulhof's judgment became Ohga's judgment.[3]

Throughout the 1990s, Columbia Pictures suffered persistent losses and became a financial burden for Sony, only stabilizing around the turn of the century.

Almost at the same time, Sony suffered another major setback at the hands of its American collaborators. In 1999, Sony and IBM jointly filed a patent titled "Electronic Control System" (U.S. Patent No. US6292718B2). In 2001, the two companies, together with Toshiba, jointly applied for three patents related to computer processor architecture (U.S. Patent Nos. US6809734B2, US7093104B2, and US7233998B2). The description in patent US7093104B2 clearly states that the invention resulted from "activities under a joint research agreement carried out by Sony Computer Entertainment Inc., Toshiba Corporation, and International Business Machines Corporation from March 9, 2001."

This joint R&D project among Sony, IBM, and Toshiba spanned several years. Between 2001 and 2010, they jointly filed 175 patent applications across 64 patent families (i.e., simple family groupings), the vast majority of which were filed between 2005 and 2006 (see Fig. 27.1). IBM itself filed even more patents through this project, possibly over 500 in total. The final product of this effort was a processor named Cell.

The Cell processor was a highly ambitious product that Ken Kutaragi, the "Father of the PlayStation," staked his entire career on. As previously mentioned, Nobuyuki Idei, who lacked personal gravitas, was pushed into the CEO role in the 1990s by

---

[2] 王育琨. (2005). 索尼并购哥伦比亚: 从泥沼到经典. *IT*时代周刊, (3), 108–109.

[3] 王育琨. (2005). 索尼并购哥伦比亚: 从泥沼到经典. *IT*时代周刊, (3), 108–109.

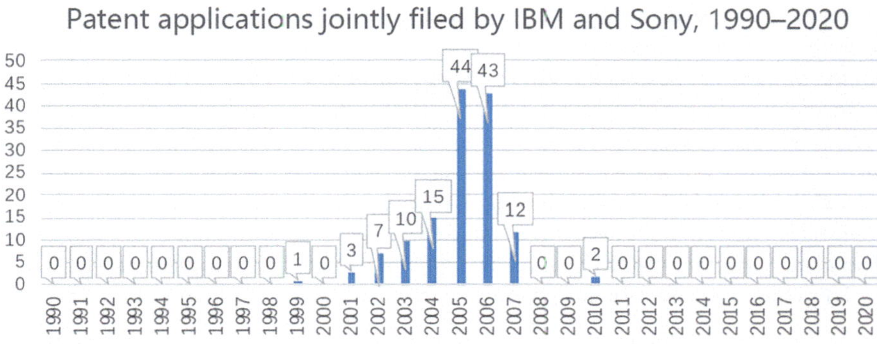

Fig. 27.1 Patent applications jointly filed by IBM and Sony, 1990–2020

Norio Ohga and carried out sweeping corporate reforms, finally gaining control of the board around the year 2000. During this process, Norio Ohga was sidelined and made honorary chairman, losing his grip on Sony's leadership. As Sony's business struggles worsened, Ohga became increasingly dissatisfied with Idei and began stating publicly that "choosing Nobuyuki Idei was a mistake." In this internal schism, Ken Kutaragi sided firmly with Ohga.

In the early 1990s, Ohga had withstood intense internal and external opposition to support Kutaragi's development of the first-generation PlayStation, earning Kutaragi's unwavering loyalty. Kutaragi had a forceful personality, frequently clashing with his superiors. He was known for making rapid decisions and for his authoritarian style. While both Idei and his successor Howard Stringer appreciated his talents, they were also troubled by his rebellious nature. By the early 2000s, with the massive success of the first two generations of PlayStation, Kutaragi had become an Executive Vice President at Sony, overseeing gaming, consumer electronics, and semiconductor operations. With Ohga backing him, he was seen as a potential heir to Sony's leadership. As the chief of Sony's digital business, Kutaragi envisioned a new microprocessor capable of supporting distributed computing. His vision was that this processor would not only power the third-generation PlayStation but also be used across various household devices, thereby creating Sony's future smart home ecosystem.

The development costs of the Cell processor were shared equally by Sony, Toshiba, and IBM, with each contributing one-third. However, in terms of manpower, the project was dominated by IBM: 75% of the development team consisted of IBM employees, while only a few engineers from Sony and Toshiba were involved throughout the process. In terms of intellectual property, Sony allowed IBM to use Cell processor technology in its future products. For reasons unknown, Sony did not prohibit IBM from sharing the technology with Sony's competitors—and IBM took full advantage of this oversight, resulting in Sony suffering a major loss once again.

According to IBM's official website, the Cell processor was described as having a nine-core architecture consisting of a PowerPC-based Processing Element (PPE) and eight Synergistic Processing Elements (SPEs). The PPE, unlike earlier IBM

PowerPC cores, was portrayed as a general-purpose unit tasked with running the operating system, while the eight SPEs were identified as the main computing engine of the chip, designed for high-speed processing.[4]

In the specification of patent US7093104B2, this design is clearly described: "*The basic processing module is a processor element (PE). A PE preferably comprises a processing unit (PU), a direct memory access controller (DMAC) and a plurality of attached processing units (APUs). In a preferred embodiment, a PE comprises eight APUs. The PU and the APUs interact with a shared dynamic random access memory (DRAM) preferably having a cross-bar architecture.*"

This eight-co-processor design is also explicitly claimed in claim 21 of the patent:

> *The processing apparatus of claim 20, wherein the number of said plurality of first processing units for at least one of said processor modules is eight.*

Why eight co-processors? Because having eight meant greater computing power—and because, as Sony Executive Vice President Ken Kutaragi arbitrarily declared: "Eight is beautiful."

This decision, like the butterfly effect, triggered a series of consequences that none of the parties involved could have foreseen.

In the early stages of development, IBM's team had designed the Cell processor with only six synergistic processing elements (SPEs). When the design was already more than halfway complete, Ken Kutaragi abruptly requested that IBM add two more SPEs. His reasoning was stated in a single, enigmatic sentence:

> Eight is beautiful.

David Shippy, the chief architect of the Cell processor, recounted this incident in his book *The Race for a New Game Machine: Creating the Chips Inside the Xbox 360 and the PlayStation 3*.[5] According to his account, IBM's team was baffled and frustrated by this seemingly irrational demand, which came out of nowhere and was never explained. Shippy observed that if Ken Kutaragi had justified the inclusion of eight SPEs by pointing to market expansion or a substantial performance leap beyond the PlayStation 2, the rationale would have been more persuasive. Instead, Kutaragi's remark that "eight is beautiful" struck him as unconvincing. Shippy concluded that the choice was a costly one, eroding IBM's margins and markedly increasing the manufacturing expenses of the PlayStation 3.

At the end of 2002, Microsoft approached IBM to discuss developing a processor for its next-generation gaming console. As we know, Microsoft had launched the Xbox in 2001, entering direct competition with Sony's PS2; to this day, the two companies remain dominant forces in the home console market. While Sony was working on its third-generation console, the PS3, Microsoft was developing the Xbox 360, its second-generation offering.

---

[4] IBM. (n.d.). *Cell processor: A new engine for the digital age*. IBM 100 Icons of Progress. Retrieved March 31, 2024 from https://www.ibm.com/ibm/history/ibm100/us/en/icons/cellengine/

[5] Shippy, D., & Phipps, M. (2010). *The race for a new game machine*. Citadel Press.

When Microsoft came knocking, IBM's newly formed Engineering & Technology Services (E&TS) division saw the opportunity as a golden chance to prove its value. They presented Microsoft with several proposals, all of which were rejected. Eventually, an engineer who had previously worked on the Cell team leaked Sony's design to Microsoft—instantly sparking Microsoft's keen interest.

To be fair, Microsoft had no intention of sabotaging the STI alliance; it simply didn't know that the proposal it had received came from the heart of its biggest rival's next console. E&TS, thrilled by Microsoft's offer—especially given that "services" had been IBM's buzzword since the days of CEO Lou Gerstner—seized the opportunity. Encouraged by E&TS, IBM made a decision devoid of business ethics and concealed it from both Sony and Microsoft.

In his book, David Shippy remarked with evident frustration that IBM's executives allowed the prospect of large profits to cloud their judgment and appeared willing to take whatever measures were necessary to inflate the company's financial results. He described his camaraderie with the Japanese engineers—how they attended creative themed parties, went drinking, boating, skiing, and talked sports together (one of them was even a die-hard Yao Ming fan). Yet, in reality, Shippy used his acting skills to keep them completely in the dark.

In September 2003, IBM and Microsoft signed an agreement to jointly develop the Xbox 360's processor. That year, IBM's business performance was lackluster, and upper management was reluctant to hire new staff. Meanwhile, Qualcomm was aggressively poaching talent from IBM. A mass exodus from the PowerPC processor division left IBM strapped for both funding and personnel, making it impossible to assign a dedicated team to Microsoft's project.

So IBM chose the most convenient solution: having Sony and Microsoft share the same design and development team. Since Sony and Toshiba engineers were already on the team, IBM revealed part of the truth to Sony—that it planned to develop a core for another client and suggested adopting a design suitable for both to avoid spreading the team too thin. While technically accurate, IBM concealed the new client's identity, citing trade secrets.

According to Shippy, Sony agreed without much resistance. After all, IBM had honored Kutaragi's "Eight is beautiful" request, so Sony felt obliged to reciprocate by showing flexibility—"a sign of mutual respect," as Shippy put it. This ironic outcome was something Kutaragi could never have imagined.

Because Sony and Microsoft were higher-paying and more demanding clients, IBM chose to drop a less important potential customer—Apple. The Cell team's supervisor, who had long maintained IBM's relationship with Apple, had introduced the Cell processor early on, and Apple expressed strong interest. At the time, it seemed that with some tweaks to support out-of-order execution, the Cell processor could meet Apple's needs. But once Microsoft joined the picture, the team simply couldn't serve all three companies, and they decided to eliminate out-of-order processing. This completely upended Apple's plans and triggered its fury—becoming one of the catalysts for Apple's decision to abandon PowerPC in favor of Intel.

Shippy recalled that Apple, angered and frustrated, eventually turned to IBM's Server Division to design a different core and promptly devoted considerable resources to pursuing this alternative. When Microsoft held its roadshow in Austin, where the Cell team was based, both Sony and Microsoft remained unaware that their consoles' core processors were being developed by the same group. IBM's secrecy was woefully inadequate. As the collaboration between IBM and Microsoft gradually became public knowledge, Sony and Toshiba engineers began to piece things together through open and semi-open channels. But IBM, as the industry behemoth, could afford to "bully its customers." Having already invested heavily in the Cell project, Sony and Toshiba had no choice but to grit their teeth and continue. IBM, meanwhile, continued using Sony and Toshiba engineers for development.

Shippy recalled that IBM frequently held duplicate debugging meetings—first a confidential session restricted to IBM engineers to address bugs affecting both the Xbox 360 and the PlayStation 3, and then a second session with Sony and Toshiba staff that focused solely on PlayStation 3 issues. He noted that the most difficult part was feigning ignorance during the latter meeting, repeating questions he had already asked earlier in order to conceal his prior knowledge of the answers. Thanks to the convincing performances of IBM staff, Sony's engineers remained oblivious—they had no idea they were also working on their competitor's product.

Sony had started developing its next-generation processor 2 years ahead of Microsoft, aiming to release the PS3 in 2005. But due to IBM's duplicity, Microsoft beat them to market, launching the Xbox 360 in 2005, while Sony's PS3 wouldn't appear until late 2006. Fortunately, Microsoft's early launch didn't guarantee a decisive victory. Rushed to market, the Xbox 360 suffered one of the worst quality control crises in consumer electronics history—nearly one-third of all units failed shortly after release due to overheating.

While Microsoft scrambled to manage its quality debacle, Sony finally launched the much-anticipated PS3 in November 2006—only to shock gamers with its price: $499 for the 20GB model and $599 for the 60GB version. That was a staggering 40% higher than the Xbox 360's launch prices of $299 and $399.

Sony, however, believed the pricing was justified. The PS3 was not only powered by the high-performance, high-cost Cell processor, but also came equipped with a Blu-ray drive. Sony had chosen to use the PS3 as a loss-leader to push Blu-ray technology—an intentional strategic sacrifice aimed at winning the next-gen HD format war. In fact, the PS3 was cheaper than most standalone Blu-ray players at the time; it was a bargain even if used solely as a media player.

Market research firm iSuppli estimated that the cost to manufacture the PS3 was at least between $805 and $840.[6]

For hardcore gamers, tech enthusiasts, and loyal Sony fans, the PS3's pricing was not unreasonable. But for the average American consumer, spending two or three hundred dollars on a gadget might be justifiable, whereas shelling out five or six hundred dollars for a game console far exceeded their expectations. Despite the

[6] Tony Smith. (2006). Sony PS3 'costs at least $805–840' to make. Retrieved March 31, 2024 from The Register: https://www.theregister.co.uk/2006/11/16/isuppli_prices_up_ps3/

Cell processor's impressive performance, its unconventional architecture posed serious challenges for developers accustomed to the ×86 platform.

With both Sony and Microsoft making blunders one after another, neither company emerged as a decisive winner in this generation's console war. Microsoft managed to gain a firm foothold in the home console market, while Sony's gaming division suffered massive losses. In 2007, the long-profitable PlayStation division posted significant deficits, and Ken Kutaragi became the scapegoat. He was pressured into retirement by Howard Stringer, stripped of real power and left with only the ceremonial title of honorary chairman—effectively ousted from Sony's top ranks in an internal power struggle.

The Cell processor, developed with great effort by Sony, ultimately fell far short of Kutaragi's lofty expectations. Internally, he tried to promote the chip for use in Sony's other electronics divisions, but none were willing to adopt it. For most of Sony's product lines, Cell's performance was excessive, its price too high, and its heat output too intense. Externally, the processor's niche appeal left it with few suitors. According to *The New York Times*, in 2004, Nobuyuki Idei and Kutaragi visited California together, attempting to entice Apple into using the Cell processor in its Mac computers by offering access to Sony's technology. However, Steve Jobs dismissed the chip as even less efficient than the PowerPC and flatly rejected the proposal.[7]

The Cell processor's failure epitomized a broader pattern within Sony's product strategy. For years, the company had focused on producing devices with superior sound, higher image quality, and greater performance—backed by grand strategies across its product lines. Yet Sony failed to grasp two critical points: first, consumers don't necessarily seek the best audio, video, or processing specs; and second, consumers are never obligated to pay extra just to help a company fulfill its strategic ambitions.

Ironically, it was the PS3's performance-focused design—which drove up its costs—that won over a small but dedicated group of high-performance computing enthusiasts. From a market perspective, this group was so niche as to be nearly negligible. But in 2010, the U.S. Air Force Research Laboratory made headlines when it purchased 1760 PS3 units to build a computing cluster named "Condor." The Condor cluster achieved 500 teraflops of processing power, making it the 33rd most powerful supercomputer in the world at the time. It was used to analyze high-resolution satellite imagery.

For Sony, this was a case of poetic justice: *Qiú rén dé rén, yòu hé yuàn hū?*—"You sought righteousness and got it; what is there to complain about?"

[7] John Markoff. (2005). What's Really Behind the Apple-Intel Alliance. Retrieved March 31, 2024 from New York Times Website: https://www.nytimes.com/2005/06/11/technology/whats-really-behind-the-appleintel-alliance.html

Chapter 28
Time-Shifting for Others' Gain

Though Betamax ultimately failed, it made one crucial contribution that cannot be overlooked. In the early days, many Americans bought VCRs primarily to record television programs and watch them later at their convenience. In contrast, Japanese consumers bought VCRs mainly to watch movies at home. Panasonic was quick to sense these different consumer needs. In the U.S., it was the first to standardize its videotape playback duration at 2 hours—long enough to record an entire football game. It also actively secured licensing deals with major movie studios to ensure a steady supply of films in every key market. Sony's Betamax tapes, though superior in picture quality, couldn't even record a full-length football or basketball game on a single cassette.

When promoting Betamax, Sony didn't focus on technical specifications. Instead, it marketed the device around an emotionally resonant concept: "time shifting." Before videotape technology, viewers who missed a TV program's broadcast time had no choice but to wait for a rerun. VCRs enabled them to record programs and watch them later, as many times as they wanted. Akio Morita called this practice "time shifting." He recalled:

> When we launched Betamax in 1975, our marketing strategy was to promote this new idea of 'time shifting.' In my speeches, I told people that Betamax was truly something new. 'Now you can hold a television program in your hand,' I said. 'With a VCR, TV becomes like a magazine—you can adjust your schedule to fit it.' That was the concept I wanted to sell.

Time shifting transformed an entire generation's viewing habits, but it also struck a nerve with the film and television industry. Soon, Sony found itself embroiled in one of the most pivotal copyright lawsuits in twentieth-century U.S. legal history.

In November 1976, Universal Pictures filed a lawsuit against Sony in a California district court. Universal's lawyers argued that VCR users were infringing copyrights by recording films and TV shows, and that Sony, as the manufacturer, should be held liable for contributory infringement. That same year, the U.S. had just passed the Copyright Act of 1976—the most significant overhaul since 1909—which

Q. Yue, F. Yang, *Governing Innovation: A Global History of Patent-Driven Technological Progress*, https://doi.org/10.1007/978-981-95-5570-3_28

clarified the doctrine of "fair use," although it wouldn't take effect until January 1, 1978. Nevertheless, Sony invoked fair use as a defense and won support from the district court.

Universal, joined by Disney and other studios, argued that "time shifting" was causing them substantial financial harm. They even garnered support from advertising industry representatives, who testified that viewers could fast-forward through commercials when watching recorded content—thus damaging ad revenues.

Sony countered with testimony from other content providers who claimed that time shifting did not harm them. Supporting Sony were major sports leagues, the NCAA (National Collegiate Athletic Association), and the NRB (National Religious Broadcasters).

Sony's attorney, Dean Dunlavey, argued: "There has never been a case in history where the maker of a legitimate machine was punished because users misused it." The district court agreed and ruled in Sony's favor.

Universal promptly appealed. In 1981, after multiple legal skirmishes, the Ninth Circuit Court of Appeals reversed the ruling. The court wrote:

> Virtually all television content is copyrighted... The fact that some rights holders choose not to enforce their rights does not prevent others from choosing to do so.

The Ninth Circuit's decision sent shockwaves through the fledgling home video industry and was met with celebration by film studios and the Motion Picture Association of America (MPAA). In 1982, after intense lobbying from interest groups, the U.S. Congress began investigating the legal implications of time-shifting technology. Some lawmakers proposed a bill that would explicitly exempt home recording from copyright infringement. Meanwhile, the MPAA launched a parallel lobbying effort to impose royalties on all VCRs and blank videotapes.

At congressional hearings, MPAA president Jack Valenti launched a full-scale attack on home video systems. In a now-infamous quote, he declared, "The VCR is to the American film producer and the American public as the Boston Strangler is to the woman home alone." Much like some U.S. politicians today, Valenti also played the nationalist card, reframing the issue as a U.S.–Japan conflict: "The entertainment industry wasn't invented by Japanese machines—it was created by the American movie business."

Valenti even enlisted the help of legal heavyweights. Harvard law professor Laurence Tribe testified that time shifting infringed on American constitutional freedoms. Tribe argued that time shifting violated the Fifth Amendment's protection of private property, equating home recording with theft of studio assets. He also claimed it violated the First Amendment's protection of free speech: "If creators earn significantly less, they will express less."

In response, the newly formed Home Recording Rights Coalition fired back at the hearings. They contended that any government restriction or surveillance of how citizens used VCRs in their own homes would violate the Fourth Amendment's protection of private spaces.

As both sides clashed in Congress, Sony appealed to the Supreme Court. On January 17, 1984, the Court issued its final ruling in the long-running lawsuit. In a narrow 5–4 decision, Sony won. The majority opinion stated that time shifting was a "non-commercial, non-profit activity" and that it had "no demonstrable effect on the potential market for or value of copyrighted works."

Ironically, once VCRs and videotapes gained widespread acceptance and became household staples in the U.S., the very studios that had once resisted them ended up profiting immensely—while Sony, the pioneer, was left with mere scraps.

As mentioned earlier, by the time Sony won the lawsuit in 1985, Betamax had already lost the format war. In 1986, studios were delighted to discover that home video revenue had surpassed box office receipts. Two years later, Sony began producing and selling VHS tapes, marking its full retreat in the videotape format battle.

The Betamax case had far-reaching consequences. Although Betamax failed commercially, the legal precedent paved the way for DVDs, DVRs, and even computer screen recording. The fierce backlash from the MPAA and studios like Universal showed just how fearful traditional industries could be in the face of hardware-driven technological revolutions.

To support its magnetic tape and CD formats, Sony invested heavily in Sony Music. In response to the Betamax experience, it acquired Columbia Pictures and built its own film studio—hoping to bolster future video formats like DVD and high-definition DVDs. In doing so, Sony gradually became a representative of the very traditional content industry it had once disrupted.

This shift largely explains Sony's eventual inability to reinvent itself. In contrast, Apple, unburdened by legacy formats or entanglements with content industries, entered the internet age with a light touch and blazed ahead.

In 1997, VHS tapes still dominated the U.S. movie rental market. One day, a struggling programmer named Reed Hastings rented the Tom Hanks film *Apollo 13* from a video store. Swamped with work, he forgot to return it on time and racked up $40 in late fees. Dreading his wife's reaction, Hastings was in a foul mood. While venting at the gym, he came up with a subscription-based model—and founded Netflix.

Years later, Sony's carefully cultivated Blu-ray technology triumphed over Toshiba's HD DVD, finally bringing the high-definition disc format war to a close. But just as consumers were still debating whether to replace their DVD collections with Blu-ray discs, Netflix and Amazon launched high-definition streaming movie subscription services. When the 4 K resolution standard emerged—with its 4096 × 2160 pixel format becoming the new industry benchmark—Sony's Blu-ray discs, which only supported 1920 × 1080 resolution, were suddenly rendered obsolete.

As mentioned earlier, the failure of Betamax had dealt a heavy blow to Sony. After much soul-searching, Sony's leadership concluded that one key reason for the defeat was a lack of exclusive content. Seizing the opportunity created by the appreciation of the yen after the Plaza Accord, Sony acquired Columbia Pictures. But under American management, Columbia stumbled repeatedly. The burdens of acquisition-related debt, Japan's post-bubble economic malaise, and managerial missteps left Sony's executives overwhelmed.

To make matters worse, in 1993, Masaru Ibuka suffered a heart attack and fell seriously ill. Just 3 weeks later, Akio Morita collapsed from a stroke while playing tennis, losing both speech and mobility. According to John Nathan's *Sony: The Private Life*, the two founders were housed in adjacent rooms at a rehabilitation facility. During their arduous recovery, they would often visit each other in wheelchairs, sitting side by side in silence, hand in hand. Ibuka and Morita passed away in 1997 and 1999, respectively.

Their sudden illnesses created a power vacuum at the top of Sony. The process of choosing a successor unexpectedly spiraled into a full-blown crisis.

Ibuka had been a brilliant engineer; Morita, a masterful manager. Norio Ohga, the second-generation leader, had joined Sony later but was close to both founders and often seen as their peer. His background as a musician, experience with the Berlin Philharmonic, and personal ties to maestros like Herbert von Karajan gave him a prestige comparable to that of the founding duo.

In 1995, Nobuyuki Idei succeeded Ohga as Sony's president with his support. Compared to the three venerable elders before him, Idei's résumé seemed lackluster. He was the first Sony president to rise from within the employee ranks. Before his appointment, he was also the youngest and most junior member of the executive team.

Upon taking office, Idei faced what could be described as a crisis of legitimacy—"a young ruler without the trust of his ministers." To break the deadlock, he initiated a major restructuring called the "Company within a Company" system. This reform replaced Sony's traditional division-based structure with a subsidiary-based one. Lacking the personal authority of his predecessors, Idei's reforms were not about centralizing power, but rather about large-scale decentralization. This sowed the seeds for Sony's later "silo effect."

Under the long-standing division-based structure, business units were evaluated mainly on sales and profits. Other balance sheet metrics held little weight. If a division's annual sales exceeded ¥100 billion, headquarters would typically assign specific performance targets such as a 10% increase in sales or profit margin, or an additional ¥10 billion in cash flow.

With the shift to the new subsidiary-based structure, Sony's headquarters functioned as a holding company overseeing eight independent operating companies: Consumer A&V (Audio & Video), Components, Recording Media & Energy, Broadcast, Business & Industrial Systems, InfoCom, Mobile Electronics, and Semiconductors. Its game, music, film, and insurance businesses were managed by separate corporate entities.

The new evaluation system adopted investor-style metrics. In addition to traditional sales and profit targets, balance sheet indicators became key performance measures. When subsidiaries sought funding from HQ, they had to present investment cases with detailed ROI (Return on Investment) analyses—just like pitching to external investors.

This heightened financial scrutiny made subsidiary managers highly sensitive to payback periods and the cost of capital. Starting in 1998, Sony began emphasizing shareholder value and incorporated many analytical tools related to it into performance assessments. EVA (Economic Value Added) was introduced as a core metric. Defined as the difference between net operating profit after tax and the cost of capital, EVA tied executive bonuses—at and above the department head level—directly to shareholder value creation.

This strict investment discipline led to the cancellation of many forward-looking, high-profile projects. The most famous casualty was Aibo, the robotic dog project, which was officially terminated in 2006. Its visionary leader, Dr. Toshitada Doi, resigned in protest. Over 100 Sony employees held a symbolic funeral for Aibo. Reflecting on it later, Doi lamented, "Aibo's death signifies the death of Sony's risk-taking spirit."

After 1994, Sony underwent several more rounds of restructuring. In 1998, the eight companies were reorganized into ten. Then, in April 1999, they were consolidated into three "Network Companies": Home Network Company, Personal IT Network Company, and Core Technology & Network Company. These changes reflected Idei's "network" philosophy. He hoped to achieve synergy through a model of "integrated/decentralized management." The performance of these Network Companies continued to be assessed based on balance sheet metrics, with 50% of executive bonuses tied to company results.

The new structure enabled Sony to significantly reduce costs, cut debt, and boost profits. In fiscal year 1993, Sony reported revenues of ¥3.7 trillion, gross profit of ¥1 trillion, but net income of only ¥15.3 billion. By fiscal 1997, revenue had grown to ¥6.7 trillion, operating profit reached ¥500 billion, and net income soared to ¥202 billion. In 1999, *BusinessWeek* named Nobuyuki Idei one of the world's 25 best global managers.

But the shift from ten companies to three Network Companies also brought drastic changes to how Sony made decisions. R&D was no longer centrally managed but was now under the purview of the individual Network Company leadership. A new department—Digital Network Solutions (DNS)—was created at headquarters to oversee emerging network businesses. In 2000, the broadcasting business under the Home Network Company was moved into DNS, forming the Communication System Network Solutions Company, which became the fourth Network Company directly under HQ.

Each Network Company oversaw several independent subunits. For example, the IT company responsible for VAIO personal computers and the Personal Mobile Communications company responsible for phones and digital cameras were now both under the Personal IT Network Company. These subunits effectively replaced the old business divisions—and there were 25 of them in total.

Following the Network Company reforms, the relationship between performance evaluations and compensation became increasingly ambiguous. Different business units under different companies began launching similar products, cannibalizing each other's target markets.

In retrospect, few people today regard Nobuyuki Idei's reforms as a success. His decentralization efforts may have seemed progressive on the surface, but during the great transformations of the digital wave and the internet revolution, such reforms proved woefully ill-timed. The constant reorganizations, moreover, did nothing to strengthen Sony's organizational coherence.

In the analog era, Sony's business divisions operated with relatively little technical interdependence—its TV, audio, and computing departments could function independently with minimal interaction. But in the digital age, the boundaries between these technologies began to blur. Computers could now display television broadcasts, play DVDs, and stream MP3 music. Had Sony's various subsidiaries collaborated closely, the company could have produced the best personal computers, digital audio players, and smartphones on the market. Unfortunately, Sony instead became known for its internal turf wars.

In 2001, Idei sought to develop a new television viewing model. He established the Network Terminal Solution Company (NTSC) under the Home Network Company. NTSC developed a product called "Cocoon," dubbed "a television with a brain." Launched in 2002, Cocoon could automatically record programs based on users' viewing habits, saving them to a local hard drive.

However, once the hard drive filled up, Cocoon would automatically overwrite older content. This made it inconvenient for users who wanted to archive certain programs long-term. So why didn't Sony use DVDs as the storage medium?

The answer lay in its siloed structure: television-related technologies were held by the TV division, hard drives and software by the VAIO company, and DVD technology by the video company. The video company, busy developing its own DVD products, refused to support Cocoon. As a result, when Cocoon launched without DVD functionality, it received praise but poor sales. Much later, Sony released an updated version with DVD support, rebranded as Sugoroku, which found some market success.

But Sugoroku's success came at the expense of another Sony product—the PSX, developed by the Broadband Network Company. PSX was another attempt to redefine TV viewing, integrating a PS2 gaming console with DVD recording capabilities. Poor PSX sales led Sony to slash its price, which in turn affected Sugoroku's market performance. This kind of needless zero-sum competition created endless internal friction and left consumers confused about which product Sony was actually championing.

This is why Sony ended up launching two similar products on the same day: the Memory Stick Walkman, and the VAIO Music Clip Walkman—developed separately by the Audio Division and the VAIO Division, respectively.

A few years later, facing the growing threat of Apple's iPod, Sony decided to develop a new digital audio player. The task fell to its heavily indebted subsidiary, Aiwa. In 2003, Aiwa's project team designed a new music player that supported the industry-standard MP3 format and used USB interfaces. But Sony's Personal Audio Company rejected the plan, insisting on developing products based on its proprietary ATRAC format and sticking with MiniDisc (MD) as the storage medium.

By then, MP3 had already become the global standard, synonymous with digital music. But to the Personal Audio Company, MP3 was still seen as a hostile rival, an existential threat to ATRAC. Just as Aiwa's product neared completion, Sony Vice President Ken Kutaragi personally intervened to shut it down. The Aiwa team was absorbed into the Personal Audio Company and disbanded. The Aiwa brand quietly disappeared from the market.

The conflict wasn't limited to hardware divisions—it extended deeply between Sony's hardware and software departments. A Goldman Sachs analyst once observed:

> *Sony has two DNAs, its two founders, Ibuka and Morita. Ibuka was an engineer and enjoyed playing with technology. Morita, although an engineer himself, was a born entrepreneur. Morita was from a rich family. By his nature, he was interested in making toys for adults. That explains partly why he entered the entertainment business. Even when Ibuka was struggling to develop a Trinitron CRT and incurring huge losses, Morita was constructing a fancy Ginza building. In a sense, both enjoyed their own hobbies, and these two DNAs never mixed. Perhaps, Morita had a very naive idea of synergy, that it would be naturally created if he put these two businesses together, which never happened.*[1]

In the era of cassette tapes and CDs, Sony's hardware and software divisions (referring to music and film content) had a perfectly complementary relationship. But in the digital age, empowered by the internet, everything changed. Of course, Sony's hardware division wasn't stupid—if there was consumer demand, it was more than willing to provide all kinds of tools to facilitate the downloading and sharing of music. This had nothing but benefits for the sales of Sony computers and various music players. But for the software division (i.e., the music company), music files that were easy to download and share also meant they were easy to pirate.

Before Philips launched DCC, Sony had also developed a digital tape format and introduced the DAT (Digital Audio Tape) Walkman. At the time, other music companies were reluctant to release music in digital formats, fearing it would make piracy easier. Sony hoped that Sony Music would provide music content in the digital tape format. But Sony Music's CEO, Schuhoff, sided with the other music companies and refused to provide content—prompting strong dissatisfaction from the hardware division. Professor Sea-Jin Chang of the National University of Singapore noted that Sony has always tended to respect and follow Sony Music's opinion rather than forcing its will. Even after Sony fired Schuhoff and regained control of the U.S. subsidiary, this situation did not change. He believed Sony spent too much time trying to generate synergy between hardware and content, but at most achieved the level of synergy where Arnold Schwarzenegger listens to a song on an MD player in Sony's loss-making film *The Last Action Hero*. According to him, today within Sony, the word "synergy" has already become taboo.[2]

One Sony executive once said: "In-house content businesses are not necessarily assets—they are also liabilities, because we have to protect their interests. We can't do what Steve Jobs did with the iPod. In our view, he sold content too cheaply."[3] In

---

[1] Chang, S.-J. (2008). *Sony vs. Samsung: The inside story of the electronics giants' battle for global supremacy*. John Wiley & Sons.

[2] Chang, S. J. (2011). Sony vs Samsung: The Inside Story of the Electronics Giants' Battle For Global Supremacy. John Wiley & Sons.

[3] Chang, S. J. (2011). Sony vs Samsung: The Inside Story of the Electronics Giants' Battle For Global Supremacy. John Wiley & Sons.

a 2005 interview, Ken Kutaragi also said: "For years now, including myself, many Sony employees have been frustrated that management couldn't create a product like Apple's iPod. Their negativity stemmed from the fear that the music and film divisions would be overly protective of their copyrighted content."

In 2004, Sony followed Apple's lead with iTunes by launching Connect Inc., developing its own music management software and opening the CONNECT Music Store to sell music files online. Sony's hardware and music divisions each assigned one senior executive to serve as co-presidents of Connect Inc., but the result was that the two could never agree on anything.

From the beginning, the CONNECT Music Store was widely criticized. *The New York Times* sharply commented that Sony's new store was "embarrassingly crude," often disconnected, and should be renamed "Sony Disconnect." Songs were priced the same as those on iTunes—$0.99 each—but any track over 7 min cost $1.99 (while all songs in the iTunes Store were priced at $0.99). This petty pricing strategy drew considerable criticism from online users.

Over time, CONNECT grew into one of the largest online music stores, boasting up to 25 million tracks available for download and updating weekly with tens of thousands of new songs. However, all songs were in ATRAC3 format, encrypted with Sony's DRM technology, and could only be played on Sony's own Walkman devices (including MD players) or with Sony's SonicStage software. Ironically, Sony's own in-car and home music systems didn't support ATRAC format CDs.[4]

On March 31, 2008, due to poor market performance, Sony shut down the CONNECT Music Store. To its credit, Sony provided a detailed FAQ and conversion tools to help users convert their downloaded ATRAC3 files into MP3 or WMA formats.

In a 2009 interview, then-CEO of Sony, Howard Stringer, offered his personal reflections on the failure of CONNECT. He said:

> At the time, we thought we would make more money that way than with open technology, because we could manage the customers and their downloads... This approach, however, created a problem: customers couldn't download music from any Websites except those that contracted with Sony. If we had gone with open technology from the start, I think we probably would have beaten Apple Inc of the US.[5]

This interview drew significant mockery—not because Stringer said Sony "would have" beaten Apple, but because he claimed Sony should now seize the opportunity to beat Apple using open technology, when in fact Apple had already abandoned DRM 5 months earlier. Steve Jobs's open letter against DRM had already become widely known, yet Stringer appeared completely unaware.

---

[4]Rob Pegoraro. (2004). Sony's Connect Music Service Offers Fair Pricing, Little Else. Retrieved April 1, 2024 from Washington Post Website: https://www.washingtonpost.com/archive/business/2004/05/09/sonys-connect-music-service-offers-fair-pricing-little-else/aac8a72a-ff1b-4b4f-9dc1-5040b6d6c604/

[5]Jared Newman. (2009). Sony CEO: We Could Have Beaten Apple. Retrieved April 1, 2024 from Technologizer: https://www.technologizer.com/2009/05/11/sony-ceo-we-could-have-beaten-apple/

When Apple launched its online music store, it did not face the same fierce backlash from content creators as Sony did when it promoted Betamax. After 2000, with the rise of online piracy and music sharing, record label sales declined sharply. The RIAA aggressively sued services like Napster into bankruptcy but couldn't stem the broader decline of the traditional music market. In 2002, BMG laid off 1400 employees, Sony Music 1000, and EMI 1800. Industry-wide profit margins fell from 15% to just 5%. While record companies remained skeptical of online music sales, Jobs convinced them. In negotiations, he let record labels keep the bulk of the revenue: from each \$0.99 song sold, Apple kept \$0.33, and the record companies received \$0.66. iTunes Store downloads used the AAC (Advanced Audio Coding) format—whose patent holders included Fraunhofer, Sony, and Philips [6]. But Apple had no qualms about using formats developed by its competitors.

Chapter 29
Philips and the Battle of Formats

After Sony introduced the Betamax format, Matsushita and Philips launched their own competing formats—VHS and Video 2000, respectively. By the mid-1980s, Betamax still retained a 5% market share, but Video 2000 had completely disappeared from the scene.

From the 1970s to the 1980s, Philips implemented what was known as a "matrix management" model. Under this structure, each employee had at least two reporting lines: one to a direct or functional manager, and another to a project manager. These vertical and horizontal reporting lines intersected, forming a matrix. This system allowed companies to organize work around specific projects and gained popularity in the 1970s, remaining a common model to this day.

As one of the earliest multinational corporations to adopt this model, Philips developed an especially complex internal structure. During that period, its organizational matrix consisted of four key components:

- **Board of Management**: Located at corporate headquarters, this nine-member board—with one chairman—set overall corporate policy. Each member oversaw business units aligned with their technical expertise.
- **Product Divisions (PDs)**: Philips' production of goods and components was divided into major divisions, each managed by a senior executive and responsible for global operations without regard to national borders.
- **Corporate Staff Departments**: These included finance, legal, product design and development, marketing support, and HR, providing global services also unconstrained by geography.
- **National Organizations (NOs)**: Subsidiaries or branches outside the Netherlands, each NO managed Philips' operations in its respective country, including production and marketing.

Take, for instance, a Philips radio manufacturing plant in Singapore. It reported to the Singapore NO as its direct administrative superior. Meanwhile, strategic guidance on radio products came from the Consumer Electronics Division at headquarters in Eindhoven, which the plant was also required to follow. For matters related

Q. Yue, F. Yang, *Governing Innovation: A Global History of Patent-Driven Technological Progress*, https://doi.org/10.1007/978-981-95-5570-3_29

to design and outsourcing of components, the plant dealt with the appropriate corporate departments in Eindhoven. This tangled web of relationships forced plant managers to juggle an extensive array of vertical and horizontal interactions, greatly increasing communication costs.

In the format wars of the 1980s, the excessive autonomy of Philips' national subsidiaries proved disastrous. A 1990s research report summarized internal opinions at Philips regarding the failure of Video 2000: some managers believed product developers were too far removed from the market. Others blamed the barriers between R&D, manufacturing, and marketing for delays and budget overruns. Still others noted that subsidiaries around the world made no meaningful contributions to the project and failed to push for success in key markets.[1]

While Video 2000's failure had multiple causes, the refusal of Philips' North American division to cooperate was particularly damaging. Although Philips headquarters launched the Video 2000 system 2 years after Sony's Betamax, it was more technically refined and backed by substantial resources. The company had high hopes for global adoption. Yet the North American subsidiary flatly rejected it, arguing that Sony's Betamax and Matsushita's VHS had already gained a first-mover advantage, with cost, features, and ecosystem support that Video 2000 could not match.

It's important to note that among major multinational tech firms, Philips was one of the earliest to fully embrace globalization. Unlike Japanese, American, or Chinese giants, Philips' domestic Dutch market was far too small to sustain the company's scale. Its overseas subsidiaries were thus granted significant autonomy.

During World War II, then-president Anton Philips fled to the U.S. while his son, Frits Philips, remained in the Netherlands to keep operations running under German occupation. To avoid interference from the occupied headquarters, Philips' North American division gained early independence and legal autonomy.

Armed with this autonomy, the North American arm disregarded the Video 2000 strategy altogether. Not only did it refuse to sell the system in the U.S., it outsourced its entire videocassette product line to Japanese rivals and rebranded them as "Magnavox" for sale in North America.[2] It also sourced TVs from Matsushita under the Magnavox, Philco, and Sylvania brands, and bought consumer CD players from Pioneer. Some researchers described the division at the time as little more than a sales conduit for Japanese-made products.[3]

The complexity extended within subsidiaries themselves. For years, Philips' country-level branches were governed by tri-partite committees of technical, business, and financial heads, rather than a single strong CEO. This tripartite structure had roots in the company's founding: the Philips brothers had divided

---

[1] Johne, A. (1994). Listening to the Voice of the Market. International Marketing Review, 11(1), 47–59. https://doi.org/10.1108/02651339410057518

[2] Barlett, C. A., & Ghoshal, S. (1990). Matrix management: Not a structure, a frame of mind. *Harvard business review*, *68*(4), 138–145.

[3] Czinkota, M., & Ronkainen, I. (Eds.). (2011). *The future of global business: a reader*. Routledge. P518

responsibilities between engineering and sales. Even after later reorganizations in the 1990s installed powerful national managers, the culture of shared authority proved hard to dismantle.

Subsidiaries typically had functional integration mechanisms at three levels:

- **Project Teams** (called "article teams" within Philips): Composed of junior managers from technical and business roles, these teams handled product policy, sales plans, and budgeting. Each had subteams for routine work and special initiatives.
- **Group Management Teams**: Cross-functional groups representing business and technical interests. They met monthly to assess progress and mediate interdepartmental conflicts.
- **Senior Management Committees (SMCs)**: Comprised of senior executives in business, technology, and finance. These bodies coordinated interdepartmental relations within the country and ensured alignment with national strategic goals.

As early as 1979, McKinsey had already identified weaknesses in Philips' matrix model.[4] According to this study, one product manager in Groningen lamented:

> *We feel the disadvantages of the matrix acutely here in Groningen. For one thing, we definitely feel it is a disadvantage not to be responsible for sales—we have people who don't feel the need to sell the product tomorrow, and yet have part of the responsibility for profits. The matrix is too slow—we are in a very turbulent market with great potential, and we have far too many low-cost competitors. We need very short communication lines, quick decisions, alertness—we've got to be able to adapt fast.*

And British scholar Cheryll Barron's commented:

> *At the moment, the Philips organization appears to suffer from all the obvious inefficiencies of the Japanese system with which the matrix has so much in common—endless time spent in meetings and in building personal relationships, and in communication—without enough of the Japanese strengths.*

Due to these organizational flaws, Philips exited the format war early. Though it retained a seat at the table in various audio-video patent pools, its dominance from the cassette and CD eras was never regained.

[4]Tom Peters. (1979). Beyond the matrix organization. *McKinsey Quarterly* September 1, 1979. Retrieved April 1, 2024 from McKinsey Website https://www.mckinsey.com/business-functions/organization/our-insights/beyond-the-matrix-organization

Chapter 30
The Endgame of the Format War

As one of the key inventors behind CD, DVD, and Blu-ray technologies, Sony had long been actively engaged in the research and patenting of optical disc technologies. As shown in Fig. 30.1, Sony's patent filings in this field peaked notably in 2 years: once in 1998 with 209 patent families, and again in 2003 with 194 families. After 2006, Sony's patent activity in optical disc technologies dropped sharply. By 2011, annual filings fell below 50, and starting in 2015, the number dropped to single digits. What exactly happened at those time points?

In December 1994, Sony and its longtime partner Philips jointly released the MMCD (MultiMedia CD) format. One month later, Panasonic and Toshiba announced their own format—SD (Super Density Disc). Sensing yet another format war on the horizon, the IT industry, already weary of repeated format disputes, resisted choosing between two incompatible systems. Under this pressure, the MMCD and SD camps were forced into compromise and collaboration, ultimately merging their efforts into what became known as the DVD format. During this process, both camps formed the DVD6C patent pool. Since the final DVD format incorporated more of SD's technology than MMCD's, this outcome was generally seen as a win for Toshiba and a loss for Sony.

From 1997 to 2003, mainstream consumers in the U.S. and Japan rapidly transitioned from videotapes to DVDs. As seen in Fig. 30.2, U.S. DVD rental and sales revenue soared in the early 2000s, nearly wiping out the VHS market by 2007. During this period, Sony and Philips attempted to secure a greater share of licensing revenue, but failed to win the approval of other patent holders. As a result, the two companies independently formed the rival DVD3C patent pool, a move that further soured relations between Sony and Toshiba.

In 1997, Sony and Philips continued their alliance and began developing a new high-definition optical disc format: DVR-Blue. In that year alone, Sony filed 168 optical disc patents, and in 1998, the number reached an all-time high of 209. In October 2000, Sony and Philips introduced the first DVR-Blue prototype. Six months later, Panasonic unveiled its own HD optical disc prototype. Determined not to repeat past mistakes, Sony and Philips quickly extended an olive branch to

Q. Yue, F. Yang, *Governing Innovation: A Global History of Patent-Driven Technological Progress*, https://doi.org/10.1007/978-981-95-5570-3_30

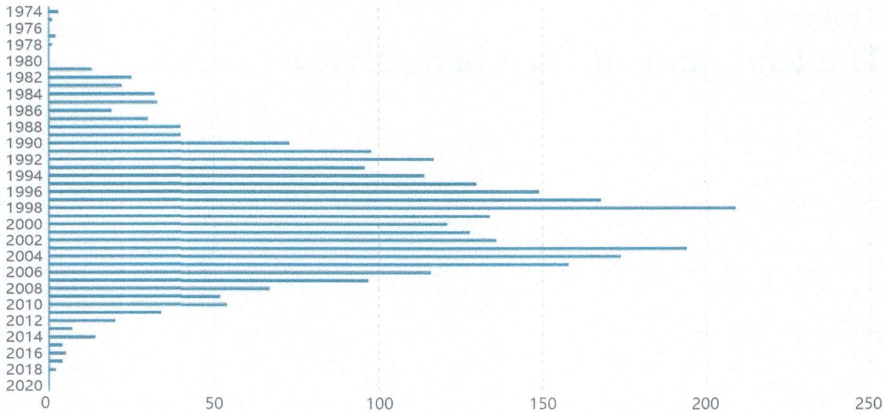

Fig. 30.1 Annual number of optical disc-related patent families filed by Sony

Panasonic, enthusiastically inviting it to join their camp. After Panasonic accepted, a senior executive from the Sony alliance confidently declared: "*Throughout history, when Panasonic supported a technology, that technology won the format war.*"[1]

Following Panasonic's public alignment, major industry players who had been on the fence quickly followed suit, rallying behind DVR-Blue. The only holdout was Toshiba, which now stood alone. The new-generation format war was taking shape. At that moment, the traditional titans of the optical disc industry failed to notice a new disruptor quietly emerging in the U.S. Patent and Trademark Office's database.

In April 2000, Netflix was granted its first patent—U.S. Patent No. US6584450B1, titled "Method and apparatus for renting items." This patent described the subscription-based DVD rental model that launched Netflix: customers paid a monthly fee and could rent DVDs online without late fees, though the number of DVDs they could hold at one time was capped. This cap naturally discouraged customers from hoarding. Interestingly, the patent background even referenced founder Reed Hastings' personal grievance over being fined by Blockbuster, with the opening sentence stating: "*Conventional inventory rental models are typically based upon renting items for fixed rental periods and charging late fees for keeping rented items beyond a specified return date.*"

In June 2001, Netflix filed a second patent, "Approach for estimating user ratings of items" (US7403910B1), beginning a focus on recommendation technologies. However, for the next year or two, many of Netflix's patents were envelope-related designs under IPC code B65D27/06—such as US6966484B2 and US7401727B2, both describing mailing envelopes. At this stage, Netflix was merely a fortunate

[1] Uijl & Henk J. de Vries (2013) Pushing technological progress by strategic manoeuvring: the triumph of Blu-ray over HD-DVD, *Business History*, 55:8, 1361–1384, DOI: 10.1080/00076791.2013.771332

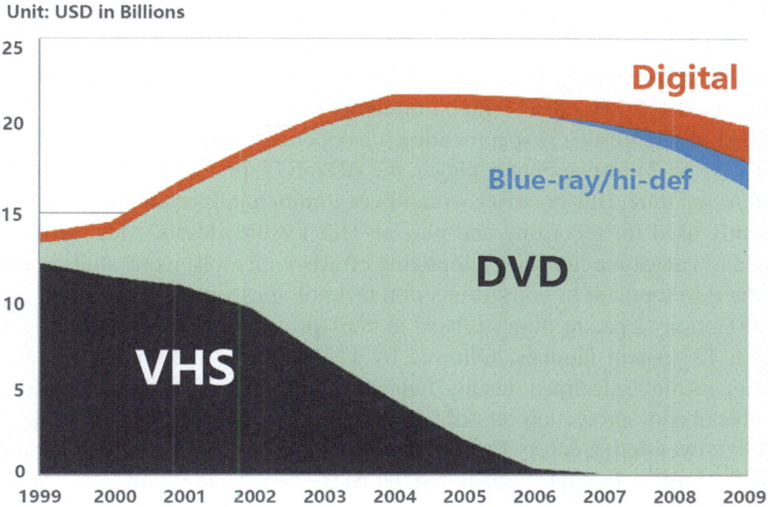

Fig. 30.2 Composition of U.S. home entertainment industry rental and sales revenues, 1999–2009

newcomer profiting from the DVD boom, posing no real threat to giants like Sony or Toshiba.

The rental model described in US6584450B1 was quickly copied by rivals, including Blockbuster—ironically the company that once fined Hastings—and even retail giant Walmart. Yet Netflix took no legal action. As Wired reported, a company spokesperson commented: "*We think that a great business is based on a great service, not a great patent*."[2]

On February 19, 2002, nine companies from the DVR-Blue camp formed a technical standards organization called BDF. DVR-Blue was formally renamed Blu-ray, and Version 1.0 of the Blu-ray standard was released. That August, Toshiba and NEC launched their competing AOD (Advanced Optical Disc) format, later renamed HD-DVD.

HD-DVD retained the same 0.6 mm surface layer as standard DVDs, offering limited technological advancement but allowing reuse of existing DVD manufacturing equipment—saving costs for producers. In contrast, Blu-ray discs featured a thinner 0.1 mm surface, tighter track spacing, and higher production costs. However, they boasted 25 GB of capacity per single-sided layer, surpassing HD-DVD's 15 GB. Both camps tried to ensure backward compatibility with DVDs. HD-DVD drives required only a red laser, while Blu-ray drives needed both a red and a blue-violet laser, further increasing costs.

Movie studios, while taking sides, also tried to resolve the format conflict themselves. In late 2005, Warner Bros. filed a patent for a multi-format disc readable by

[2] Lore Sjoberg. (2004). Netflix Imitators Are Everywhere. Retrieved from Wired Website: https://www.wired.com/2004/02/netflix-imitators-are-everywhere/

both Blu-ray and HD-DVD drives (US20060179448A1, with Chinese counterpart CN101088125B). The disc was described as containing multiple layers, each conforming to a different standard—CD, DVD, HD, or BD. This concept evolved into the short-lived Total HD disc, which ultimately never made it to market.

In 2003, the Blu-ray 1.0 specification was opened for licensing. Sony launched the world's first Blu-ray recorder/player, the BDZ-S77, priced at 450,000 yen (about $3800). At the time, Blu-ray discs weren't yet commercially available, so the device was mainly used for recording and playing HD TV broadcasts. Meanwhile, Sony, Philips, and Panasonic intensified lobbying efforts with Hollywood studios, emphasizing the development of copy-protection technologies—especially BD+.

In 2003, Sony's patent filings related to copy-protection technologies hit a record high with 127 patent families, followed by 123 in 2004—the second-highest year. Thereafter, its filings declined steadily, falling to single digits by 2015. (Fig. 30.3 shows the number of copy-protection-related patent families filed by Sony from 1984 to 2020.)

In 2004, twentieth Century Fox and Disney publicly announced their support for the Blu-ray format. Fox later stated that the BD+ system and its superior anti-piracy capabilities were the key factors behind their decision.[3] That same year, Netflix and Blockbuster began facing off directly in a price war. By this point, two-thirds of American households already owned DVD players—DVD had become a dominant force.

At the beginning of 2005, Blockbuster abolished its late-return fees for DVDs—ironically, it was precisely these fees that had given rise to Netflix in the first place. By the end of that year, Blockbuster had barely reached one million online rental subscribers and suffered a net loss of $590 million. Walmart surrendered and exited the online DVD rental business. Meanwhile, Netflix's subscriber base had reached 3.6 million, and it posted a $42 million profit. That same year, Japan's Ministry of

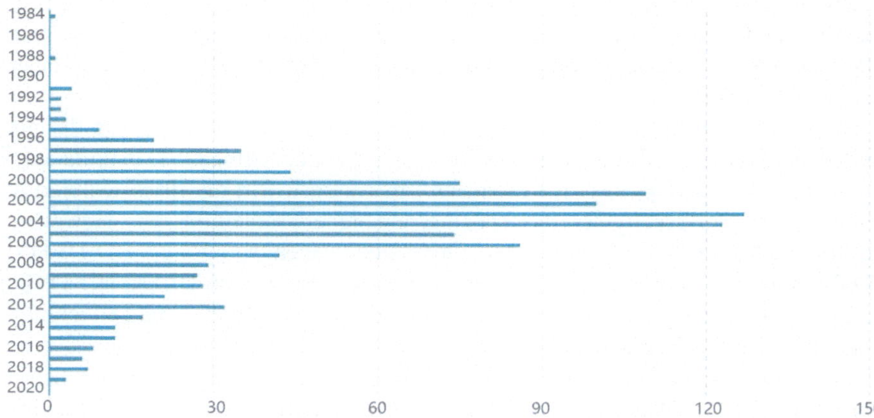

Fig. 30.3 Annual number of copy-protection related patent families filed by Sony

[3] BD+ LLC. (n.d.). *News*. Retrieved March 31, 2024 from http://www.bdplusllc.com/news

Economy, Trade and Industry attempted to broker negotiations between the Blu-ray and HD-DVD camps to unify the HD format. But neither side was willing to compromise, and the talks failed.

On March 31, 2006, Toshiba launched the world's first HD-DVD player, priced at ¥110,000, and began selling it in the U.S. the following month at $499 and $799. On the content side, Toshiba had the backing of Warner Bros. and Universal Pictures. Three months later, Samsung released the first Blu-ray player on the market—the BD-P1000, based on the Blu-ray 2.0 standard, with a suggested retail price of $999.

As mentioned earlier, in late 2006, Sony played its trump card: the PS3 was equipped with a Blu-ray drive. This console could not only read Blu-ray game discs but also play Blu-ray movies, all starting at just $499—cheaper than most standalone Blu-ray players. The format war between Blu-ray and HD-DVD reached its climax. In response, Microsoft released an external HD-DVD drive for its Xbox 360 console, priced at $199. However, by early 2008, only 3.5% of Xbox 360 users had purchased the add-on drive.[4]

That same year, both Apple and Amazon launched their own paid movie download services, while Netflix began to show its potential. Perhaps still nursing a grudge from Blockbuster's past penalty fees, Netflix aggressively pursued its rival, now in steep decline. In April 2006, Netflix was granted a continuation patent titled "Method for renting items to customers" (US7024381B1), derived from its original patent. Immediately after receiving approval, Netflix filed a patent infringement lawsuit against Blockbuster using this patent and its parent.

Netflix also made waves by launching a high-profile challenge. On October 2, 2006, it offered a $1 million prize to any research team in the world that could develop a movie recommendation algorithm that outperformed its Cinematch system by at least 10%.[5] This global competition attracted tens of thousands of teams from over 100 countries. Three years later, a team named "Pragmatic Chaos" claimed the top prize.

In 2007, the high-definition format war continued unabated. Both camps scrambled to win over movie studios. Paramount and DreamWorks joined the HD-DVD camp, declaring exclusive support for HD-DVD releases. Threatened by Sony's aggressive strategy of selling consoles at a loss to promote Blu-ray drives, Toshiba began doing the same with its HD-DVD players, cutting prices to around $500. Analysts estimated that Toshiba's cost per unit was as high as $674. Unlike Sony, however, Toshiba lacked experience in the game industry—where console makers typically sell hardware at a loss and recoup profits through game software—so its financial losses were even more severe.

[4] Brandon Boyer. (2008). NPD: 2007 U.S. Game Industry Growth Up 43% To $17.9 Billion. Retrieved April 1, 2024 from Game Developper: https://www.gamedeveloper.com/game-platforms/npd-2007-u-s-game-industry-growth-up-43-to-17-9-billion

[5] Steve Lohr. (2009). A $1 Million Research Bargain for Netflix, and Maybe a Model for Others. Retrieved April 1, 2024 from New York Times Website: https://www.nytimes.com/2009/09/22/technology/internet/22netflix.html

That year, Warner Bros. finally unveiled a prototype of its multi-format disc, dubbed "Total HD," but by the end of the year, it indefinitely suspended the project.

In early 2007, Apple began selling its Apple TV set-top box. Around the same time, Netflix launched its streaming service, "Watch Instantly." Existing subscribers could watch videos for free, with each dollar spent on DVD rentals earning them an hour of streaming time. In 2007 alone, Netflix invested $40 million in this service—again a loss-leading move.

That same year, Netflix and Blockbuster settled their patent dispute, though the settlement details were not disclosed. Blockbuster claimed the settlement had little impact on its financials. Nevertheless, just 3 years later, the company filed for bankruptcy.

By the end of 2007, the landscape of the format war had begun to take shape. HD-DVD players were cheaper and sold more units than standalone Blu-ray players, but sales of Sony's PS3—equipped with a built-in Blu-ray drive—exceeded the combined sales of all Blu-ray and HD-DVD hardware (see Fig. 30.4). In terms of media, HD-DVD disc sales were clearly lagging behind Blu-ray. The HD-DVD camp responded with aggressive price cuts and promotions: Microsoft dropped the price of its HD-DVD drive from $199 to $179, bundled with five free movies. Toshiba launched an HD-DVD player priced as low as $100.

On January 4, 2008, Warner Bros. abandoned its "dual-format" approach and announced that from May 2008 onward, it would exclusively release movies in Blu-ray format. Its subsidiaries, New Line Cinema and HBO, followed suit. According to Bloomberg, rumors swirled that Sony had paid Warner $400 million to secure this support.[6]

Following Warner's departure, retail giants Walmart and Best Buy removed all HD-DVD discs and players from their shelves. Netflix, now a significant force capable of influencing the outcome, also announced it would cease purchasing new HD-DVD products.

On February 19, 2008, Toshiba officially announced the termination of all HD-DVD-related R&D, sales, and manufacturing, citing "*recent major changes in the market*" as the reason.[7] Though victorious, Sony had paid dearly for its win—its PS3 division bore significant financial losses throughout the campaign.

Yet Sony's triumph was short-lived. In January 2008, Apple released a free update for Apple TV, allowing users to rent and purchase movies directly from iTunes—no longer requiring synchronization with a computer. Steve Jobs proudly proclaimed:

[6] Cliff Edwards. & Ronald Grover. (2008). Blu-ray's High-Definition Knockout. Retrieved April 1, 2024 from Bloomberg Website: https://www.bloomberg.com/news/articles/2008-01-04/blu-rays-high-definition-knockoutbusinessweek-business-news-stock-market-and-financial-advice

[7] CNN. (2008). Toshiba quits HD DVD 'format war'. Retrieved April 1, 2024 from http://edition.cnn.com/2008/BUSINESS/02/19/toshiba.hdd/

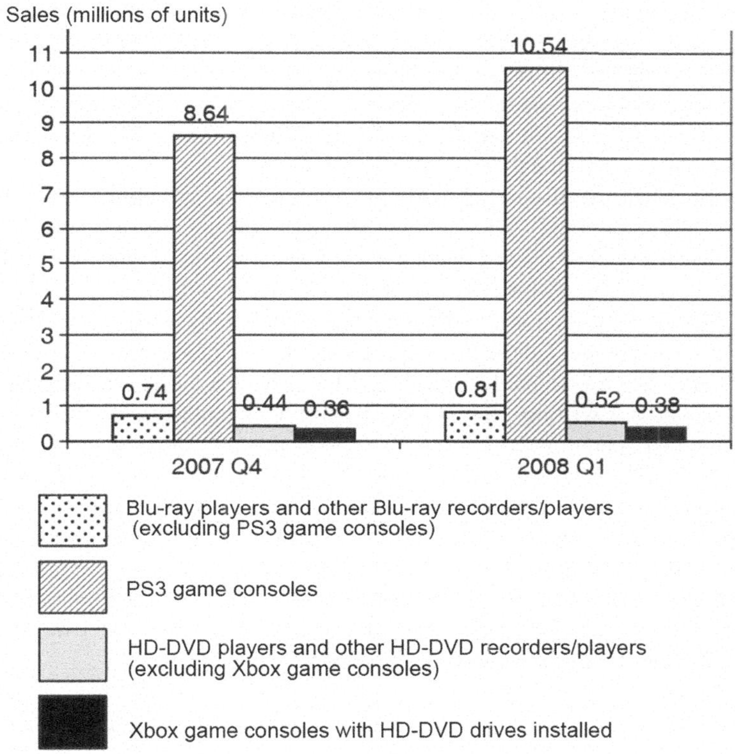

Fig. 30.4 Sales of Blu-ray and HD-DVD hardware in Q4 2007 and Q1 2008. PS3 sales surpassed the total of both Blu-ray and HD-DVD players

With the new Apple TV and iTunes Movie Rentals, movie lovers can rent DVD-quality or stunning HD movies from their couch with just a click of a button... No more driving to the video store or waiting for DVDs to arrive in the mail.[8]

In October 2008, Jobs made the following comment about Blu-ray technology:

Blu-ray is a bag of hurt. I don't mean from the consumer point of view. It's great to watch movies, but the licensing is so complex. We're waiting until things settle down, and waiting until Blu-ray takes off before we burden our customers with the cost of licensing.[9]

Indeed, Blu-ray had not yet gone mainstream. Although it had vanquished HD-DVD, it was still far from replacing DVD as the dominant format. In the year Blu-ray claimed victory, its disc sales were but a fraction of those of DVDs.

[8] Apple Inc. (2008, January 15). *Apple introduces new Apple TV software & lowers price to $229* [Press release]. *Apple Newsroom*. Retrieved March 31, 2024 from https://www.apple.com/newsroom/2008/01/15Apple-Introduces-New-Apple-TV-Software-Lowers-Price-to-229/

[9] Brian Chen. (2008). Jobs: Blu-ray Is a 'Bag of Hurt'. Retrieved April 1, 2024 from Wired: https://www.wired.com/2008/10/jobs-blu-ray-is/

Under the expected trajectory, consumers would gradually shift from DVD to Blu-ray, just as they once transitioned from tape to CD, MP3, and DVD. Sony simply needed to wait for Blu-ray to reign supreme, allowing its patent pool to generate immense royalties, and finally reap the rewards of its vast R&D investment and sacrifices… But this time, it wasn't just the old format facing obsolescence—it was the entire category of optical discs, both old and new.

That same year, Netflix began filing patents related to streaming. The first was US8365235B2, "Trickplay for streaming media." Trickplay—or "trick mode"—refers to fast-forward or rewind playback displays. On physical media like VHS or DVD, this is easy. But with streaming, if the data hasn't yet buffered, users are left waiting. Netflix's solution used still frames to simulate the effect. Amusingly, the patent illustrations included the movie poster for *Apollo 13*—the very DVD that had once cost Hastings a $40 late fee (see Fig. 30.5). This was surely intentional.

As of this book's writing, Netflix had filed 119 U.S. patents related to streaming, 54 of which had been granted. Figure 30.6 shows that this surge began in 2008.

In 2009, Netflix doubled down, filing 13 patents covering streaming acceleration, multiplexing, and video encoding. Coincidentally, Apple filed 13 streaming-related patents that same year. By then, Netflix's streaming library contained over 12,000 film and TV titles—and continued to grow.

In 2010, Apple launched its second-generation Apple TV, abandoning the original Apple TV software for a new iOS-based system. Sales reached 1 million units that year. In an email, Jobs remarked on Blu-ray:

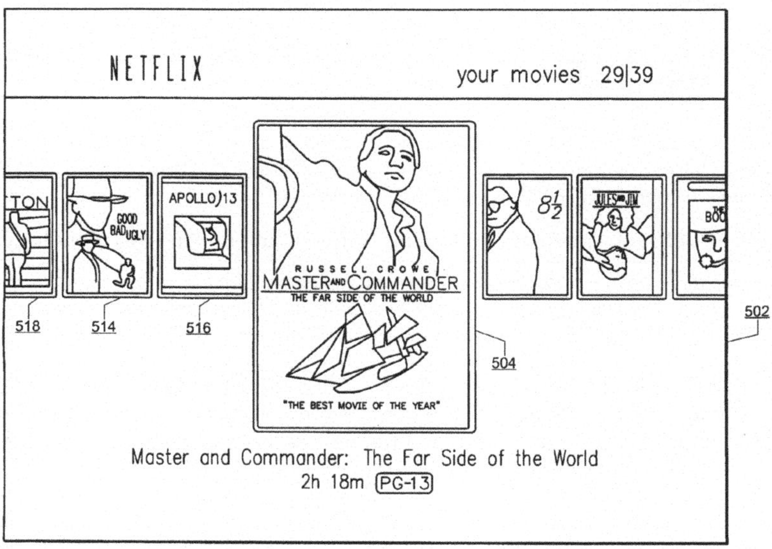

FIG. 6

Fig. 30.5 Poster for *Apollo 13* as shown in US8365235B2 patent

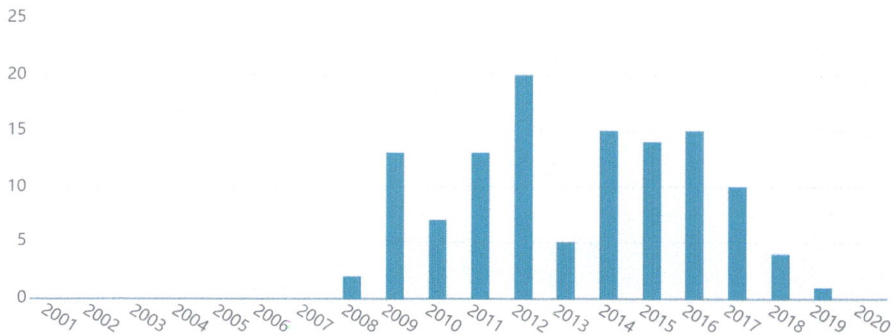

Fig. 30.6 Number of Netflix streaming-related patent filings from 2001–2020

> *Bluray is looking more and more like one of the high end audio formats that appeared as the successor to the CD – like it will be beaten by Internet downloadable formats…No, free, instant gratification and convenience (likely in that order) is what made the downloadable formats take off. And the downloadable movie business is rapidly moving to free (Hulu) or rentals (iTunes) so storing purchased movies or TV shows is not an issue…I think you may be wrong – we may see a fast broad move to streamed free and rental content at sufficient quality (at least 720p) to win almost everyone over.*[10]

As it had done with floppy drives, Apple began removing optical drives from its computers in 2011. That same year, Netflix became the top internet traffic source in the U.S., with 23 million subscribers and $1.5 billion in digital revenue. Amazon also began offering free streaming video to Prime members. Sony's painstakingly developed Blu-ray, like its Memory Stick, faded from the mainstream—relegated to niche use by professionals and tech enthusiasts.

Today, we can clearly see that Jobs' predictions have come true. According to the Digital Entertainment Group's 2019 U.S. Digital Media Entertainment Report, revenue from physical media dropped by 18.29% from 2018 to 2019, down to $3.29 billion. Physical video rentals dwindled to just $250 million—clinging to life. Meanwhile, streaming subscriptions soared by 23.73%, from $12.8 billion in 2018 to nearly $15.9 billion in 2019 (see Table 30.1).

In this final format war, Sony seemed to have done everything right. It invested heavily in R&D, its patents were numerous and solid, and Blu-ray offered better storage and technical breakthroughs than HD-DVD. Sony had learned from Betamax, aligned with Panasonic, secured studio support, and even bundled Blu-ray with its flagship PS3 console. Yet its hard-won victory was steamrolled by the rise of the internet and streaming. As Jobs said, in the age of the web, there's no need to "drive to the store or wait for a DVD to arrive in the mail." Apple, Amazon, and

[10] Josh Ong. (2010). Blu-ray chairman disagrees with Apple chief's assessment of format. Retrieved April 1, 2024 from Apple Insider Website: https://appleinsider.com/articles/10/07/12/blu_ray_chairman_disagrees_with_apple_chiefs_assessment_of_format

Table 30.1 2019 sales breakdown of video media in the U.S. Digital Entertainment Report

Category	Q-42018	Q-42019	YOY	YTD 2018	YTD 2019	YOY
Sell-thru						
Sell-thru packaged goods all	1238.96	1018.30	−17.81%	4029.87	3292.65	−18.29%
Sell-thru including EST	1903.11	1688.60	−11.27%	6487.54	5876.06	−9.43%
Rental						
Brick & Mortar	75.87	61.60	−18.81%	316.99	250.07	−21.14%
Subscription (physical only)	86.02	70.14	−18.46%	369.28	301.21	−18.43%
Kiosk	257.00	206.64	−19.60%	1098.03	884.59	−19.46%
Total rental (excluding VOD)	418.89	338.37	−19.22%	1784.30	1435.87	−19.54%
Total rental (including VOD)	935.80	818.13	−12.57%	3871.29	3393.77	−12.33%
Digital						
Electronic sell-thru (EST)	664.15	670.29	0.92%	2457.67	2583.41	5.12%
VOD	516.91	479.75	−7.19%	2086.99	1957.90	−6.19%
Subscription streaming (SVOD)*	3425.04	4317.15	26.04%	12,848.96	15,897.66	23.75%
Total digital	4606.37	5467.20	18.69%	17,393.62	20,438.97	17.51%
Total U.S. home entertainment spending	6264.22	6823.88	8.93%	23,207.79	25,167.49	8.44%
Box Office in Billions	3.07	2.98	−3.03%	12.09	11.02	−8.86%

Netflix saw the shift and surfed the wave of a new era, while Sony, though technically victorious, ended up losing the war.

Chapter 31
Beginning with *Star Trek*

Whether it's cassette Walkmans or MP3 players, audio and video tapes, CDs, Blu-ray discs, or even digital cameras as discussed in the "Rise and Fall" section—all these devices have either been rendered obsolete or faded into niche use by professionals. The reason is obvious: the ubiquitous smartphone in everyone's pocket. As the final chapter of this section, it is fitting to turn our attention to this little device—the ultimate successor to all the technologies previously discussed.

The invention of the Walkman was the result of a rare flash of genius—an inspired recombination of existing, mature technologies. The smartphone, however, was a different story: the product of countless brilliant minds striving toward a vague goal over many years.

The earliest vision of a handheld digital smart device may well trace back to *Star Trek*, the sci-fi television series adored by America's baby boomers. Unlike many so-called sci-fi action flicks, *Star Trek* was true "hard" science fiction. It featured few action scenes and was rich in plot, exploring themes such as cultural conflict, anthropology, and futurism.

The birth of Apple Inc. was deeply tied to *Star Trek*. Apple co-founder Steve Wozniak once said in an interview:

> *Star Trek had done so much for me. I'm not sure I would've had the inspiration that I had to ever start Apple, and do all these technical things and I don't think I would've had the meaning in my life if I hadn't gone to Star Trek conventions when I was young, and had a car, and could drive to Southern California to attend them.*[1]

Like many Silicon Valley tech companies, Apple had its share of *Star Trek* fans. Numerous biographies of Steve Jobs describe his mesmerizing charisma as a "reality distortion field"—a phrase lifted straight from *Star Trek*. In the 1990s, Apple even named its strategic project to port the Mac operating system to Intel platforms

[1] Steve Wozniak: Star Trek inspired the founding of Apple. (2017, April 20). *TrekMovie.com*. Retrieved March 31, 2024 from https://trekmovie.com/2017/04/20/steve-wozniak-star-trek-inspired-the-founding-of-apple/

Q. Yue, F. Yang, *Governing Innovation: A Global History of Patent-Driven Technological Progress*, https://doi.org/10.1007/978-981-95-5570-3_31

"Project Star Trek." At the launch event for the iPhone 4, Jobs also referenced *Star Trek*, saying while introducing FaceTime:

> *I grew up here in the U.S. ... with The Jetsons and Star Trek and communicators, just dreaming about video calling, and it's real now.*[2]

Star Trek is a science fiction franchise created by Gene Roddenberry. The original TV series premiered in 1966 and has since spawned multiple series, films, novels, comics, and other media adaptations.

The "Communicator" that Jobs referred to first appeared in the original *Star Trek* series from the 1960s. It was a handheld device used by crew members to communicate remotely and transmit information. In retrospect, its appearance resembled nothing more than a clamshell phone.

While it may seem clunky and retro by today's standards, Martin Cooper—the father of the modern mobile phone—publicly acknowledged that the design of Motorola's first commercial phone, the DynaTAC 8000x launched in 1983, was directly inspired by the *Star Trek* Communicator. In 1996, Motorola released the world's first clamshell phone, whose form closely mirrored that of the Communicator and whose name—StarTAC—was a clear homage to *Star Trek* (Fig. 31.1).

Amazon founder Jeff Bezos is another well-known *Star Trek* fanatic—he even made a cameo appearance as an alien in a recent *Star Trek* film. David Limp, Amazon's Senior Vice President, stated that the design inspiration for the Alexa smart speaker came from the voice-activated computer systems aboard the starship in *Star Trek*. Rob Haitani, lead designer of the Palm OS for handheld devices, once said: "*When I designed the UI (user interface) for the Palm OS back in '93, my first sketches were influenced by the UI of the Enterprise bridge panels.*"[3]

As a television series, *Star Trek* obviously made no direct technological contributions to humanity. However, the concepts of touch interfaces, visual UI elements, and even tablet computing owe much to the world it portrayed.

Michael Okuda, *Star Trek*'s graphic designer, once explained in an interview that the early production team had almost no budget. Art director Matt Jefferies received virtually no funds to create futuristic props:

> *[Jefferies] had to invent an inexpensive, but believable solution...The spacecraft of the day, such as the Gemini capsules, were jammed full of toggle switches and gauges. If he had had the money to buy those things, the Enterprise would have looked a lot like that.*[4]

Lacking funds for props like buttons, levers, and switches, the set of the *Enterprise* ended up with a minimalistic design. To save costs, Okuda designed a user interface

---

[2] Krill, P. (2010, June 8). *Poll: Will FaceTime herald an era of video calling? Ars Technica*. Retrieved March 31, 2024 from https://arstechnica.com/gadgets/2010/06/poll-technica-will-facetime-herald-an-era-of-video-calling/

[3] *Palm OS UI sketches influenced by Star Trek Enterprise*. (n.d.). In *PalmAddicts blog*. Retrieved March 31, 2024 from https://palmaddict.typepad.com/palmaddicts/2004/11/sent_in_tricord.html

[4] Foresman, C. (2016, September). *How Star Trek artists imagined the iPad... nearly 30 years ago. Ars Technica*. Retrieved March 31, 2024 from https://arstechnica.com/gadgets/2016/09/how-star-trek-artists-imagined-the-ipad-23-years-ago/

Fig. 31.1 The *Star Trek* Communicator. (Source: Davidbspalding. (2009). *1971 Star Trek TOS communicator replica* [Image]. Wikimedia Commons. https://commons.wikimedia.org/wiki/File:20090704-1971_StarTrekTOSCommunicatorReplica.jpg. This image is shared under the Creative Commons CC BY-SA 3.0 license.), the Motorola DynaTAC 8000x (Source: Redrum0486. (2006). *Motorola DynaTAC 8000X* [Image]. Wikimedia Commons. https://commons.wikimedia.org/wiki/File:DynaTAC8000X.jpg. This image is shared under the Creative Commons CC BY-SA 3.0 license), and the StarTAC phone. (Source: Wermke, R. P. A. (2017). *Motorola StarTac 70 (Made in Germany)* [Image]. Wikimedia Commons. https://commons.wikimedia.org/wiki/File:Motorola_StarTac_70_Made_in_Germany_15.jpg. This image is shared under the Creative Commons CC BY-SA 4.0 license.)

consisting of brightly colored rounded rectangles, printed on transparencies. These were stuck onto prop consoles to simulate screen output.

Most Hollywood sci-fi productions derive spacecraft controls from cars, planes, or even bicycles. But because the *Enterprise* set lacked physical controls, Okuda instructed actors to imagine everything was controlled by software—and simply interact with the “screens.” Actors gracefully gestured at the panels, creating a futuristic look that was logical and convincing—unlike the brawny gear-pulling seen in franchises like *The Avengers*.

The *Enterprise*'s minimalist aesthetic created a strong sense of futurism and left a lasting impression on artists and technologists alike. Ironically, this iconic style was born not from design brilliance alone, but from sheer financial necessity.

Furthermore, *Star Trek* creator Gene Roddenberry explicitly banned the depiction of paper and pens aboard the *Enterprise*, forcing writers to envision new ways to represent futuristic meetings. Initially, the series introduced a clunky touchscreen “electronic notepad.” Later, this evolved into the sleek, handheld PADD (Personal Access Display Device), which could access data and even control ship functions. Because of this, researchers like Bayus et al., in a seminal study on the early PDA market, cited *Star Trek*'s 1960s-era portrayal as the origin of the Personal Digital Assistant (PDA) concept.[5] Fig. 31.2 shows the PADD in use aboard the *Enterprise*.

With this background in mind, it's easier to understand why American technologists—from Martin Cooper to Bill Gates—were so passionately committed to developing handheld and tablet computing. These efforts led to an array of once-groundbreaking yet fleeting products: the Newton, MagicCap, Palm, Windows Tablet PCs, and more. This accumulation of technologies and concepts eventually achieved critical mass under Steve Jobs' direction—and changed the world.

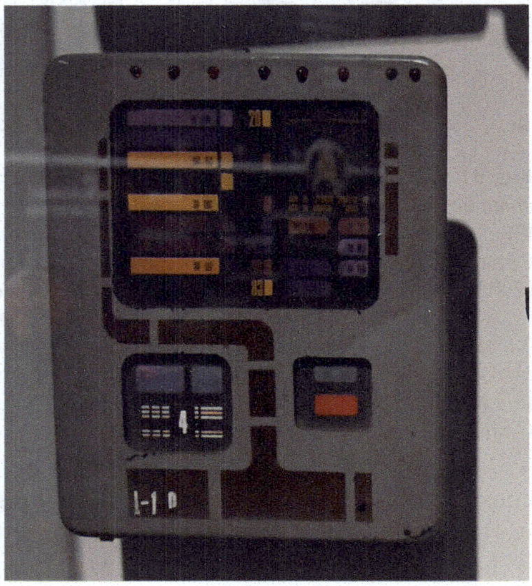

Fig. 31.2 PADD in use aboard the *Enterprise*. (Source: Ross, J. (2016). *Star Trek—The Next Generation props—Star Trek: Exploring New Worlds exhibit at the Henry Ford Museum, Dearborn, Michigan* [Image]. Wikimedia Commons. https://commons.wikimedia.org/wiki/File:Star_Trek_-_The_Next_Generation_Props_-_Star_Trek-_Exploring_New_Worlds_Exhibit_at_the_Henry_Ford_Museum,_Dearborn,_Michigan.jpgThis image is shared under the Creative Commons CC BY-SA 2.0 license)

[5] Bayus, B. L., Jain, S., & Rao, A. G. (1997). Too little, too early: Introduction timing and new product performance in the personal digital assistant industry. *Journal of Marketing Research*, *34*(1), 50–63.

Chapter 32
Newton and General Magic

Anyone familiar with Apple's history knows the story of how Steve Jobs, wielding his famously persuasive "reality distortion field," convinced Pepsi executive John Sculley to join Apple as CEO. In his pitch, Jobs delivered the now-legendary line:

> *Do you want to spend the rest of your life selling sugared water or do you want a chance to change the world?*[1]

As is well known, not long after that, Sculley seized the reins and ousted Jobs from Apple. Under Sculley's leadership, Apple remained profitable for a time; yet perhaps due to being profoundly influenced by Jobs's vision of "changing the world," Sculley became increasingly determined to leave his mark through a truly revolutionary high-tech product. Jean-Louis Gassée, former Senior VP of Product Development at Apple, later remarked that Sculley had an overwhelming desire to attach his name to something big.

By the late 1980s, two of Apple's advanced research projects had caught Sculley's eye: Newton, which focused on personal information management, and Pocket Crystal, which leaned toward communications. Newton was still in the exploratory phase, while the team behind Pocket Crystal wanted to spin off from Apple to develop a groundbreaking device independently.

In 1987, veteran Apple engineer Steve Sakoman sought to start his own company. To keep him at Apple, Jean-Louis Gassée gave him and his team broad latitude to explore wireless networking and handwriting recognition, free from internal interference. Sakoman named the project "Newton," a nod to the apple that allegedly struck Isaac Newton—the inspiration behind Apple's own name. Thanks to the team's efforts, Newton began evolving toward a personal handheld digital assistant. But when Gassée left Apple after a falling out with Sculley, Sakoman resigned as well, casting doubt over the project's future.

[1] Linzmayer, O. W. (2006, March 29). *Steve Jobs' best quotes ever. Wired.* Retrieved March 31, 2024 from https://www.wired.com/2006/03/steve-jobs-best-quotes-ever/

Q. Yue, F. Yang, *Governing Innovation: A Global History of Patent-Driven Technological Progress*, https://doi.org/10.1007/978-981-95-5570-3_32

The other initiative was led by Marc Porat of Apple's Advanced Technology Group. Tasked with finding new directions beyond personal computers, Porat envisioned a handheld smart device, launching what became the Pocket Crystal team. In 1990, Porat wrote a passionate letter to John Sculley, outlining his vision:

> *A tiny computer, a phone, a very personal object ... It must be beautiful. It must offer the kind of personal satisfaction that a fine piece of jewelry brings. It will have a perceived value even when it's not being used... Once you use it you won't be able to live without it.*[2]

Both projects struck a chord with Sculley. Back in his 1987 autobiography, he had introduced the concept of a "Knowledge Navigator"—a portable device capable of browsing libraries, museums, and databases, using voice recognition and wireless communication to gather and organize information. At the time, tech media dismissed the idea as pure fantasy, which stung Sculley. So when Newton delivered a working prototype, he saw it as vindication—proof that his Knowledge Navigator was both visionary and attainable.

Within Apple, Sculley threw his full support behind the Newton project, while also backing Pocket Crystal with significant resources.

In July 1990, Apple invested in a new startup founded by Marc Porat and two key members of the Pocket Crystal team. The company, named General Magic, counted Apple as a major shareholder and quickly attracted funding from Motorola and Sony. Telecom giants AT&T, NTT, and Cable & Wireless soon followed. While flush with capital, General Magic's investor roster—dominated by analog-era titans—would later prove to be a liability. As *Forbes* noted:

> *Its fortunes were tied to five analog technology companies that themselves did not survive the Internet revolution intact.*[3]

General Magic set its sights on developing a communications-enabled personal handheld smart device:

> *We have a dream of improving the lives of many millions of people by means of small, intimate life support systems that people carry with them everywhere. These systems will help people to organize their lives, to communicate with other people, and to access information of all kinds. They will be simple to use, and come in a wide range of models to fit every budget, need, and taste. They will change the way people live and communicate.*[4]

---

[2] Sullivan, M. (2018, July 26). *"General Magic" captures the legendary Apple offshoot that foresaw the mobile revolution. Fast Company*. Retrieved March 31, 2024 from https://www.fastcompany.com/90206157/general-magic-captures-the-legendary-apple-offshoot-that-foresaw-the-mobile-revolution

[3] Charlie Fink. (2018). 'General Magic' Doc Is Full Of Life And Lessons For Inventors. Retrieved April 1, 2024 from Forbes: https://www.forbes.com/sites/charliefink/2018/04/20/general-magic-doc-is-full-of-life-lessons-for-inventors/#320849213e6e

[4] Levy, S. (1994, April 1). *Bill and Andy's Excellent Adventure II. Wired*. Retrieved March 31, 2024 from https://www.wired.com/1994/04/general-magic/

According to a *Los Angeles Times* report,[5] Porat believed these devices would become as ubiquitous as telephones, coming in all shapes and sizes. They would easily send emails and faxes, auto-search phonebook entries, and browse vast digital information. For instance, to plan a trip, users would simply input a date and destination, and the device would fetch flight schedules, hotel availability, and weather forecasts—much like a virtual travel agent.

To realize this vision, General Magic focused primarily on software, aiming to develop an operating system—Magic Cap—for smart handheld communicators. Hardware makers including Sony, Philips, Panasonic, and Toshiba formed the "General Magic Alliance" to manufacture Magic Cap-compatible devices.

General Magic inherited Apple's innovative spirit. Key figures included Dan Winkler (creator of HyperCard), Bruce Leak (head of the QuickTime team), Phil Goldman (from Apple's system software division), and Susan Kare (the iconic designer behind Mac's early fonts and icons). In 1992, a young Apple engineer nicknamed "Android"—Andy Rubin—joined General Magic. Years later, he would be hailed as the father of Android. That same year, Tony Fadell, later dubbed "father of the iPod," graduated college and walked into General Magic's office with a résumé in hand.

Fadell recalled arriving around 8:30 a.m. to find the lobby deserted. Wandering deeper inside, he eventually encountered a few bleary-eyed employees who had clearly worked through the night—and who brusquely told him to leave. Undeterred, he found the co-founders' office and persisted until they offered him a junior role. Reflecting on that moment, Fadell said:

> *I was humbled in the first ten minutes of being there, I was like, Oh my God, this is not like Michigan, these are the smartest people ever, I have got to be working here. Whereas Apple was still making computers, these people were making the next form of a Mac! It was the iPhone, fourteen years too soon. It had a touch screen, it had an LCD, it was not pocketable but it was portable—the size of a book. It had e-mail. It had downloadable apps. It had shopping. It had animations and graphics and games. It had telecommunications—a phone, a built-in modem. I was like, I want to go with them, where they are making this stuff from scratch. I can be part of it.*[6]

The employees of General Magic, like the young Tony Fadell, were fueled by an almost fanatical passion. They worked around the clock, often staying at the office 24 hours a day. As a non-traditional company born out of Apple, General Magic fostered a freewheeling culture full of eccentric ideas. To spur creativity among its engineers, the company even kept a rabbit—mirroring its magician's-hat logo—roaming freely through the office, leaving droppings wherever it pleased. In their

[5] Jonathan Webber. (1993). Pocket-Size Peek into the Future: Technology: General Magic today will unveil its 'personal communicator' software design, which many see as the next big step in consumer electronics. Retrieved April 1, 2024 from Los Angeles Times: https://www.latimes.com/archives/la-xpm-1993-02-08-fi-1272-story.html

[6] *General Magic: An oral history of the influential tech company*. (2018, August 1). *New York Magazine — Intelligencer*. Retrieved March 31, 2024 from https://nymag.com/intelligencer/2018/08/general-magic-oral-history-of-the-influential-tech-company.html

spare time, Fadell and some colleagues built a catapult that accidentally shattered a $6000 windowpane. Fearing the worst, he braced himself for termination at the next all-hands meeting—only to be met with wild cheers and whistles.

Despite his visionary ambition, Marc Porat underestimated the speed of technological advancement. According to *The Los Angeles Times*, Porat believed the personal communicator market would take another 25 years to mature—meaning 2018.[7] In reality, it took less than 14 years from that prediction for the first iPhone to debut. But John Sculley was brimming with confidence. Before Apple and General Magic were truly ready, he helped trigger the birth of a new market.

On January 7, 1992, at the Consumer Electronics Show (CES) in Las Vegas, Sculley built on his Knowledge Navigator vision by publicly introducing the term "Personal Digital Assistant," or PDA. Citing research from Harvard and other institutions, he boldly predicted this new category would balloon into a $3 trillion market by the early twenty-first century. The eye-popping figure caused a media frenzy. Though Sculley later clarified that he was referring to the broader market of "digital convergence," the damage was done. His claim galvanized tech companies across the board. As mentioned earlier in this book, IBM's Simon smartphone was conceived and developed in this climate of PDA fervor.

Apple industrial designer Tim Parsey later reflected, "*In retrospect, Sculley's decision to preannounce Newton was a mistake, because it raised expectations higher than necessary and forced us to commit to a design several months before the product actually shipped.*"[8] Because of this premature unveiling, the Newton team suddenly found itself under immense internal and external pressure—laying the groundwork for its eventual failure.

A simple patent search (see Fig. 32.1) confirms the trend: PDA-related patent filings in the U.S. began to appear in 1993, peaked in the early 2000s, and then sharply declined with the advent of smartphones—plummeting after the iPhone's release in 2007. Today, "PDA" is a term all but forgotten.

At the time, touchscreen technology was still in its infancy. The most viable input method for handheld devices was handwriting recognition using a stylus. Several prominent figures in tech were drawn to this challenge: Pierre Omidyar, founder of eBay, started Ink Development in 1991, specializing in handwriting recognition before joining General Magic to pursue related research. Chinese-American entrepreneur Dr. Kai-Fu Lee was also active in this field during that era.

A keyword search using the expression TIAB = ((handwrit* OR character) (n) recogni*) in the Incopat database reveals a clear spike in handwriting recognition

---

[7] Jonathan Webber. (1993). Pocket-Size Peek into the Future: Technology: General Magic today will unveil its 'personal communicator' software design, which many see as the next big step in consumer electronics. Retrieved April 1, 2024 from Los Angeles Times: https://www.latimes.com/archives/la-xpm-1993-02-08-fi-1272-story.html

[8] Linzmayer, O. W. (2004). Apple confidential 2.0: The definitive history of the world's most colorful company. No starch press. Retrieved April 1, 2024 from: https://theswissbay.ch/pdf/Gentoomen%20Library/Computer%20History/Apple%20Confidential%202.0%20The%20Definitive%20History%20of%20the%20World%27s%20Most%20Colorful%20Company.pdf

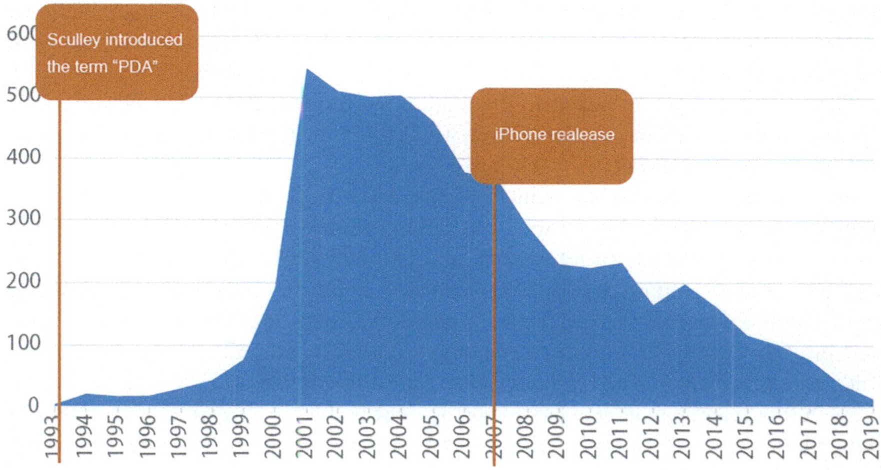

Fig. 32.1 Annual U.S. patent applications related to PDAs, 1993–2019

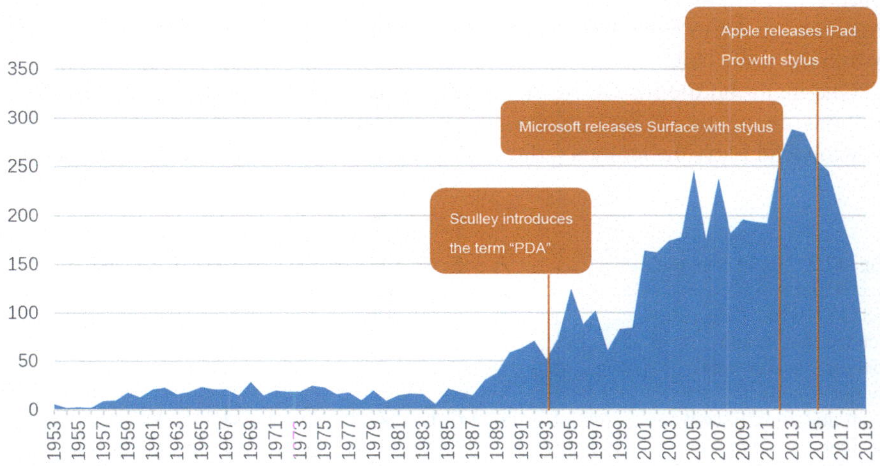

Fig. 32.2 Annual U.S. patent applications related to handwriting recognition, 1953–2019

patent applications beginning around 1993 (see Fig. 32.2)—from 52 filings in 1993 to over 100 by 1995. This uptick was clearly linked to Sculley's PDA announcement, which sparked industry-wide enthusiasm. During this time, no single company filed a dominant number of patents. Apple was granted only 11 U.S. patents from 1993 to 1999. Notably, two of the key inventors on these patents were of Chinese descent—one of them being Kai-Fu Lee.

Magic Cap's two hardware partners weren't idle either. Between 1993 and 1999, Sony filed 49 patent applications in Japan and received 8 Japanese and 4 U.S. patent grants. Motorola secured 19 U.S. patents. Though General Magic itself held no handwriting-related patents, it entered into cross-licensing agreements with both Sony and Motorola.

The undisputed patent leader in the field, however, was IBM—our familiar protagonist from the earlier chapters. From 1993 to 1999, IBM was granted 51 handwriting recognition patents.

Interestingly, the eventual winner of the early PDA race, Palm, held only 4 granted patents during this period. But among them, US6493464B1, which disclosed the Graffiti handwriting system, and US5889888A, which separated alphabetic and numeric input areas, solved the major technical challenge of recognizing handwritten characters on low-power processors. These innovations became standard in early 2000s PDA products. Figures 32.3, 32.4, and 32.5 show the illustrations from those patents and Palm's iconic PalmPilot device.

Even this crude patent analysis makes it clear that tech companies were pouring tremendous resources into the field, and the pressure on engineers at the time must have been intense. Around 1993, Apple itself was facing a host of challenges. Mounting losses forced Sculley to resign as CEO. To save face, Apple retained him as Chairman of the Board and let him continue overseeing the Newton project. During this turbulent period, a Japanese member of the Newton team, Ko Isono, tragically took his own life. Leadership changed hands repeatedly, and morale within the team sank.

One major hurdle for the Newton project was its processor design. Apple spent millions contracting AT&T to develop a custom chip—but the result fell short. Newton lead Larry Tesler then redirected investment to a small British firm working

FIG. 4A

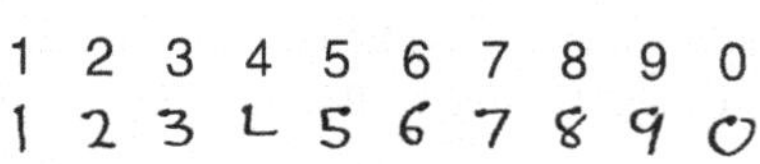

FIG. 4B

Fig. 32.3 Patent drawing from Palm's US6493464B1. This patent defined simple and memorable pen-stroke input methods for each letter and digit. Except for the letter X, every letter and digit could be completed with a single stroke

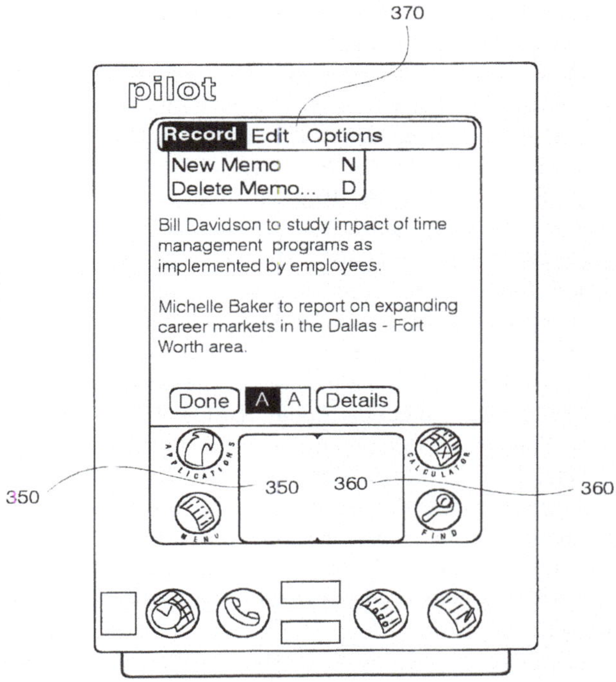

Fig. 32.4 Patent drawing from Palm's US5889888A. This patent separated the letter input area (350) from the number input area (360) to avoid confusion between letters and numbers

Fig. 32.5 Palm's signature product—the Palm Pilot handheld computer. (Source: Ndiggity. (2014, December 21). *Palm Pilot 1000 PDA* [Image]. Wikimedia Commons. https://commons.wikimedia.org/wiki/File:Palm_Pilot_1000.png. This image is shared under the Creative Commons CC BY-SA 4.0 license)

on a low-cost RISC processor. That company would soon become known as ARM (Advanced RISC Machines Ltd.).

On August 2, 1993, after six and a half years of development, the Newton PDA was finally launched (see Fig. 32.6). Around the same time, Apple filed patents related to the device, such as US Patent No. US5563996A (see Fig. 32.7). By then, however, John Sculley had lost interest in staying at Apple. Not long after the launch event, he resigned. In the 10 weeks following Newton's release, 50,000 units were sold, and the device won numerous industrial design awards. Yet the return rate was alarmingly high—about 30%. The most widely criticized flaw was Newton's poorly performing handwriting recognition system.

As Newton entered the market, rival products from other tech giants quickly followed. As discussed in the "Rise and Fall" section, IBM released its Simon device in August 1994, earning it the title of "the world's first smartphone." General Magic, meanwhile, lagged behind Apple, IBM, and Tandy in development.

In September 1994, Sony began selling its Magic Link PDA, which ran on General Magic's Magic Cap operating system. Then in January 1995, Motorola released the Envoy PDA (see Fig. 32.8), also running Magic Cap. Both devices had a similar external design and used Motorola's 68,300 processor. Their graphical user interface mirrored that depicted in General Magic's patent US5689669A (see Fig. 32.9), which used a corridor-and-room metaphor to represent hierarchical navigation.

Although Newton was widely considered a commercial failure, its exterior design still upheld Apple's high aesthetic standards—far superior to the bulky Magic Link. Figure 32.10 displays a lineup of major products from that era: Apple's

Fig. 32.6 The Newton PDA, released in 1993. (Source: Wolfgang Stief. (2012, April 29). *Apple Newton (1993–1998)* [Image]. Wikimedia Commons. https://commons.wikimedia.org/wiki/File:Apple_Newton_(1993-1998)_(6,981,096,718).jpg. This image is in the public domain (CC0 1.0 Universal))

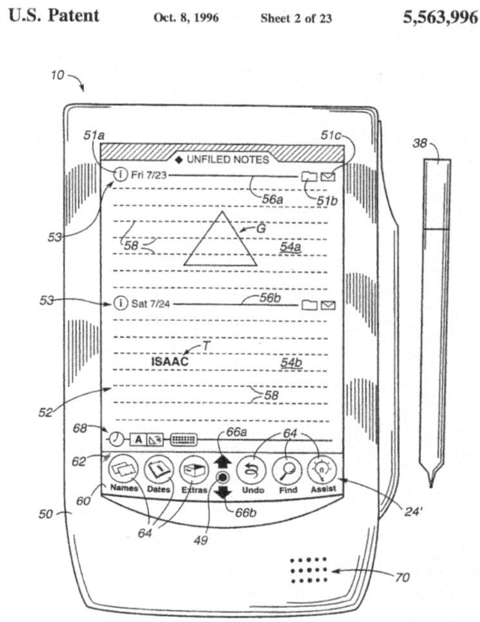

FIG. 2

Fig. 32.7 Patent drawing of U.S. Patent No. US5563996A, filed by Apple in 1993

Fig. 32.8 Human–computer interaction interface of the Sony Magic Link PDA. (Source: Josh Carter. (2010, February 21). *Sony Magic Link* [Image]. Wikimedia Commons. https://commons.wikimedia.org/wiki/File:SonyMagicLink.jpg. This image is shared under the Creative Commons CC BY-SA 3.0 license)

Newton PDA, Sony's Magic Link, Philips' Velo, Motorola's Envoy, and IBM's Simon.

At this point, Philips—another key player in our "Displacement" narrative—joined the fray. In 1995, Philips recruited the then-24-year-old Tony Fadell from General Magic to head a new mobile computing division as CTO, tasked with developing Philips' own line of PDA products. Within just a few years, the PDA market emerged and swiftly entered a phase of intense competition. Between 1993

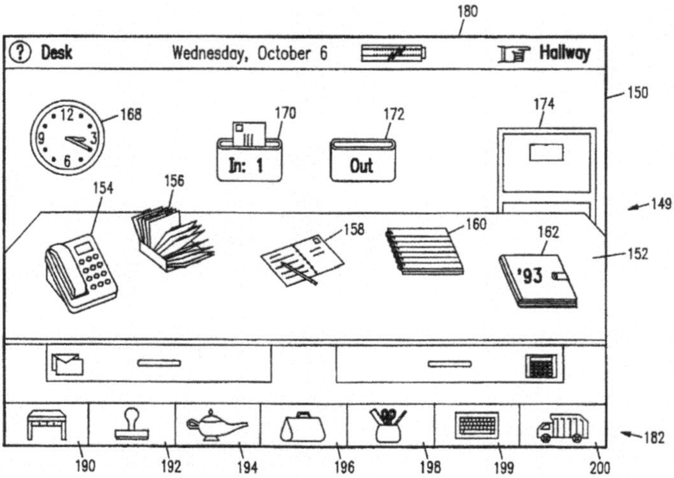

FIG. 3

Fig. 32.9 Patent drawing of U.S. Patent No. US5689669A by General Magic: a graphical user interface using corridors and rooms as an analogy for navigating between hierarchies

and early 1995, eight companies launched their own PDA devices. Table 32.1 compares the features of six major players and their products during this period.

Unfortunately, much of the core technology was not yet ready to support the lofty visions for handheld intelligent communication devices. The era's sluggish communication networks and underpowered CPUs fell short of expectations. Early devices, including the Simon and Newton, failed to impress consumers. Even Philips' later PDAs based on the Windows CE operating system—such as the Velo and Nino—met with lackluster response.

In 1997, General Magic released Magic Cap 3.0, an OS with richer visual elements. But by then, its hardware partners had lost faith. That year, members of the "General Magic Alliance" announced they would cease production of Magic Cap-compatible devices. The dream of a handheld revolution crumbled. Just 10 years later, the iPhone would arrive.

At the turn of the millennium, the PDA market was nowhere near the scale that Sculley had once forecasted. Instead, it was dominated by Palm. As discussed earlier, Palm's two key innovations in user interaction won over early adopters. By 2001, Palm ranked fourth in Google's most-searched brands—behind only Nokia and Sony among consumer electronics (see Table 32.2). If we refer back to Fig. 32.1, we'll see that this same year marked the peak of PDA-related patent filings in the U.S. These two seemingly unrelated facts support each other, illustrating the subtle relationship between patent analysis and the history of technology—encouraging us to delve deeper with renewed confidence.

With the Magic Cap alliance now dissolved, General Magic had no choice but to pivot. This group of "the smartest people in the world" once again chose a path ahead of its time—voice recognition technology. In 1998, the handheld device team

Fig. 32.10 Top row (from left to right): Apple Newton PDA, Sony Magic Link, and Philips Velo. (Source: Blake Patterson. (2016, February 2). *Philips Velo 1* [Image]. Wikimedia Commons. https://commons.wikimedia.org/wiki/File:Philips_Velo_1.jpg. This image is shared under the Creative Commons CC BY 2.0 license). Bottom row (from left to right): Motorola Envoy. (Source: Blake Patterson. (2010, June 22). *Motorola Envoy (running General Magic's "Magic Cap" operating system)* [Image]. Wikimedia Commons. https://commons.wikimedia.org/wiki/File:Motorola_Envoy_%284724647551%29.jpg. This image is shared under the Creative Commons CC BY 2.0 license) and IBM Simon (Source: Mike Mozart. (2014, March 5). *Simon FIRST Smart Phone* [Image]. Flickr. https://www.flickr.com/photos/jeepersmedia/13004426385.This image is shared under the Creative Commons CC BY 2.0 license)

that had built the company was ousted and spun off into a new startup called Icras. That same March, General Magic signed a licensing agreement with Microsoft to provide voice technology. It also attempted to develop voice-assisted functionality for General Motors' OnStar system. Despite occasional backing from major corporations and support from investors, the company never managed to turn a profit. In 2002, General Magic became one of the many casualties of the dot-com bust and ceased operations. By 2004, it was officially dissolved, and its patent portfolio was auctioned off.

From 1990 to 1997 (prior to its pivot to voice recognition), General Magic was granted 17 patents, mostly in the fields of distributed computing, software modems, low-power serial buses, and graphical user interfaces. These patents were cited an average of 145.7 times—indicative of their high quality. In his *Industry Analysis Report* (Vol. 5), Yang Tiejun (2012) categorized smartphone technologies into three main domains: human-computer interaction, low-power design, and applications/

Table 32.1 Comparison of major PDA devices from 1993 to early 1995[a]

Company	Amstrad	Apple	Tandy	IBM	Sony	Motorola
Product	PenPad	Newton	Zoomer	Simon	Magic link	Envoy
Release	Mar 1993	Aug 1993	Oct 1993	Aug 1994	Sep 1994	Jan 1995
OS	Eden Group	Newton OS	Geoworks	DOS	Magic Cap	Magic Cap
Battery	3 AA, 40 hours	4 AAA, 14 hours	3 AA, 90 hours	NiCd, 8 hours	6 AAA, 10 hours	NiCd, 8 hours
Stylus input	Yes	Yes	Yes	Yes	Yes	Yes
Cursive recognition	No	Yes	Yes	Yes	Yes	Yes
Spreadsheet	No	No	No	No	Yes	Yes
Word processing	No	No	No	No	Yes	Yes
Infrared	No	Yes	Yes	No	Yes	Yes
Phone capability	No	No	No	Yes	Yes	Yes
Price	$499	$699	$699	$899	$995	$1000–1500

[a]Bayus, B. L., Jain, S., & Rao, A. G. (1997). Too little, too early: Introduction timing and new product performance in the personal digital assistant industry. *Journal of Marketing Research, 34*(1), 50–63

Table 32.2 Top 10 most-searched brands on google in 2001

Rank	Brand
1	Nokia
2	Sony
3	BMW
4	Palm
5	Adobe
6	Dell
7	Oracle
8	Ferrari
9	Honda
10	Canon

services. General Magic's patents largely fell into these categories: three were related to graphical UIs, five to low-power serial buses, and others to FIR filters, software modems, and code structures. After General Magic's bankruptcy, Microsoft co-founder Paul Allen acquired its patents. These eventually ended up in the hands of the well-known NPE, Intellectual Ventures.

The root cause of General Magic's failure lay in its inability to seize the moment during a pivotal period of technological transformation. Overly accommodating to shareholders and excessively focused on technology while neglecting user needs, the company lost its edge.

Steve Maller, General Magic's 21st employee, summarized the company's downfall on Quora.[9] Like many tech firms dabbling in communications in the 1990s, General Magic had attempted to build its own proprietary communication network—PersonalLink—with AT&T, supporting email, scheduling, and fax without relying on the open internet. It failed to anticipate the inevitable rise of the web. Early General Magic products didn't even support the TCP/IP protocol.

Maller noted that many staffers were internet enthusiasts, and some of the earliest websites were created by General Magic employees. For instance, eBay founder Pierre Omidyar started AuctionWeb during his time at General Magic, which later evolved into the e-commerce giant. However, with AT&T and other telecoms as stakeholders, disruptive networking features were discouraged. AT&T had no interest in enabling a product that might render its traditional services obsolete.

Maller also pointed out that General Magic's leadership were "successful people" whose personal communication habits differed vastly from average consumers. They didn't mind carrying multiple digital devices and failed to envision tools that could replace or unify existing ones.

Ultimately, Maller believed the company was simply ahead of its time. Its ambitions were far too grand. The available CPUs, batteries, and network infrastructure were not remotely capable of supporting its vision.

Investor Alexander Muse also attributed General Magic's failure to conflicts of interest among stakeholders:

> *Marc Porat left product development in the hands of his team, who were at the mercy of their various corporate partners... During the development process, the Internet became a thing and even one of Digital Magic's interns knew that they ought to incorporate the web into their device. The problem? The internet had the potential to disintermediate AT&T, and as a result of their partnership with the carrier, they weren't't able to incorporate the web into their device. The device that Digital Magic built was basically useless and everyone knew it — sadly no one had the balls to admit it to themselves or to Marc. At the end of the day, the fault is Marc's and Marc's alone...*[10]

The tale of General Magic serves as yet another reminder of the ancient Chinese wisdom: "The people are most precious, the state next, the ruler least." Like Apple, General Magic gleaned invaluable lessons. But the company itself was consigned to history, bidding farewell to the field of personal handheld smart devices. Its legacy would live on only through the spark it ignited in a generation of Silicon Valley pioneers.

In a 2018 *Forbes* interview, Marc Porat reflected:

---

[9] Maller, S. (n.d.). *Why did General Magic fail?* [Quora answer]. *Quora*. Retrieved March 31, 2024 from https://www.quora.com/Why-did-General-Magic-fail

[10] Muse, A. (n.d.). *What are the dark secrets of Silicon Valley startups?* [Quora answer]. *Quora*. Retrieved March 31, 2024 from https://www.quora.com/What-are-the-dark-secrets-of-Silicon-Valley-startups

> *One of my proudest, greatest reflections is that Andy Rubin and Tony Fadell, who between them are responsible for every single smartphone in the world, three billion devices, sat next to each other at General Magic.*[11]

In addition to Rubin and Fadell, General Magic also launched the career of eBay founder Pierre Omidyar. The company's three co-founders later thrived in their own ventures, becoming key figures in the mobile internet revolution. Google+'s lead UI designer was General Magic co-founder Andy Hertzfeld. Kevin Lynch, another alum, later joined Apple and became a lead software architect for the Apple Watch. To date, Lynch has filed 90 Apple patents, with 21 granted.

When Steve Jobs returned to Apple in 1997, one of his first moves was to kill the Newton project—partly due to its ongoing financial losses, and partly because it was the "brainchild" of his ousted rival, Sculley. Many key Newton developers moved to Palm, where they continued their dream of building handheld computers. On February 27, 1998, Apple officially ceased development of Newton OS. The project had lasted about 11 years, cost around $500 million in R&D and marketing, and sold only 150,000–300,000 units. It became one of the most painful failures in Apple's history.

Yet as a pioneer in the realm of handheld smart devices, Apple learned deeply from Newton's failure. As Owen W. Linzmayer wrote:

> *Like the Mac before it, the Newton was created largely to satisfy the desire of its designers to own a revolutionary device that only they could conceive.*[12]

Guided by Steve Jobs, Apple would soon rise again—this time driven not by designers' ideals, but by a deep understanding of user needs. It would do so with two revolutionary handheld devices, and in doing so, rewrite history.

[11] Charlie Fink. (2018). 'General Magic' Doc Is Full Of Life And Lessons For Inventors. Retrieved April 1, 2024 from Forbes: https://www.forbes.com/sites/charliefink/2018/04/20/general-magic-doc-is-full-of-life-lessons-for-inventors/#216f63ae3e6e

[12] Linzmayer, O. W. (2004). Apple confidential 2.0: The definitive history of the world's most colorful company. No starch press.

Chapter 33
From iPod to iPhone

Unlike many major corporations that missed the internet boom, Apple under Steve Jobs always had a keen sense of what the youth of the internet age craved. As noted earlier, the hottest term on the internet in 1999 was "MP3." This enthusiasm for online music among early internet users remained strong into the early twenty-first century. In 2003, the top Google search was for pop icon Britney Spears, a title she retained in 2004. In 2005, another pop diva, Janet Jackson, took the crown as the most searched name. During this same period, sales of the iPod were skyrocketing—from just 939,000 units in 2003 to a staggering 22.5 million in 2005 (see Table 33.1).

Interestingly, in 2006, Google's top search rankings underwent a notable shift: none of the top 10 terms were celebrities. Instead, the top spots were held by two now-forgotten social networking sites—Bebo and MySpace (see Table 33.2).

Clearly, with the rapid rise of the internet and broadband access, user tastes were evolving. Socializing, entertainment, and knowledge acquisition had all become essential parts of the new digital lifestyle.

The year 2006 marked the fifth anniversary of the iPod (launched in 2001), the year before the iPhone's release (2007), and the tenth anniversary of Jobs' return to Apple. We can divide the period from 1997 to 2006 into two distinct phases:

- **1997–2001**: Apple's first growth phase, where the iPod was born and became a legend.
- **2002–2006**: Apple's golden years riding the iPod wave, as well as the early years of smartphone development—when Jobs made the bold choice to reinvent the company yet again.

Curiously, Sony had established a joint venture with Ericsson in late 2001 to focus on mobile phones, making the 2002–2006 period a particularly instructive comparison.

For Facebook, 2006 was also pivotal. On September 26 of that year, it opened its platform to anyone aged 13 and older, transforming from a college directory into a full-fledged global social network.

Q. Yue, F. Yang, *Governing Innovation: A Global History of Patent-Driven Technological Progress*, https://doi.org/10.1007/978-981-95-5570-3_33

Table 33.1 Apple's iPod, iPhone sales and total revenue (2001–2008)

Year	iPod Units (in 10 k)	iPod Revenue (Billion USD)	% of Total	iPhone Units (in 10 k)	iPhone Revenue (Billion USD)	% of Total	Apple Total Revenue (Billion USD)
2001	–	–	–	–	–	–	5.36
2002	38.1	0.143	2.49%	–	–	–	5.74
2003	93.9	0.345	5.56%	–	–	–	6.21
2004	441.6	1.306	15.77%	–	–	–	8.28
2005	2249.7	4.54	32.59%	–	–	–	13.93
2006	3940.9	7.676	39.74%	–	–	–	19.32
2007	5163.0	8.305	34.60%	138.9	0.123	0.51%	24.01
2008	5482.8	9.153	28.18%	1162.7	1.844	5.68%	32.48

Table 33.2 Top google searches in 2006

Rank	Search term
1	Bebo
2	MySpace
3	WorldCup
4	MetaCafe
5	Radioblog
6	Wikipedia
7	Video
8	Rebelde
9	Mininova
10	Wiki

Taking Facebook as a benchmark, we see that both Apple and Sony increased their internet-related patent filings between 2002 and 2006, but Apple outpaced Sony. Table 33.3 shows that when searching for keywords related to social networking in the titles or abstracts of patent filings, Facebook led by far, followed by Apple, with Sony trailing.

M:Metrics senior analyst Donovan noted that iPhone users accessed YouTube and Google Maps far more frequently than users of other phones. In January 2008, 30.4% of iPhone users had used YouTube, and 36% had used Google Maps. At the time, only 1% of general mobile users accessed YouTube, and just 2.6% used Google Maps. He explained:

> *While the demographics of iPhone users are very similar to all smartphone owners, the iPhone is outpacing other smartphones in driving mobile content consumption by a significant margin...In addition to the attributes of the device itself, another important factor to consider is the fact that all iPhones on AT&T are attached to an unlimited data plan.*[1]

[1]Comscore. (2008, March 18). *M:Metrics: iPhone hype holds up*. Retrieved March 31, 2024 from https://www.comscore.com/Insights/Press-Releases/2008/03/iPhone-Hype-Holds-Up

Table 33.3 Patent mentions of social network keywords (in TIAB fields)

Company	Keyword	Count	% of total patents
Facebook	Social/community	49	3.54%
Apple		11	0.25%
Sony		33	0.13%
Facebook	Web/internet/online/network	335	24.19%
Apple		576	13.00%
Sony		2024	8.13%
Facebook	Communicat*	253	18.27%
Apple		643	14.51%
Sony		3175	12.76%
Facebook	Messag*	151	10.90%
Apple		186	4.20%
Sony		401	1.61%
Facebook	Personal*/individual*	56	4.04%
Apple		152	3.43%
Sony		725	2.91%
	Total patents	1385	
		4430	
		24,886	

Table 33.4 Mobile content consumption, 2008

Activity	iPhone users	All smartphone users	All mobile users
Read news/info via browser	84.8%	58.2%	13.1%
Web searches	58.6%	37.0%	6.1%
Watched mobile TV/video	30.9%	14.2%	4.6%
Watched subscribed video/TV	20.9%	7.0%	1.4%
Accessed social sites/blogs	49.7%	19.4%	4.2%
Listened to music	74.1%	27.9%	6.7%

According to M:Metrics, after about a year on the market, the typical iPhone user was male, aged 25–34, college-educated, and earned over $100,000 annually. Their mobile habits were vastly different from other users, as seen in Table 33.4.

Why was the iPhone so successful? Product managers and engineers might have different answers—but from a patent perspective, the most obvious reason lies in Apple's dominance in human-computer interaction (HCI) patents. Both in quality and quantity, Apple held a massive lead over competitors like Sony, creating a formidable moat.

Chapter 34
A Brief History of Multi-Touch Patents

From a patent perspective, it's clear that Apple once seriously considered an iPhone design that would repurpose resources from the iPod product line, avoiding the loss of sunk costs. Ultimately, however, this concept was scrapped.

At the inception of the iPhone project, Tony Fadell's iPod team attempted to design an iPhone based on the iPod's iconic click wheel. Their concept maintained the iPod's touch wheel, turning it into a dual-mode device: a music player and a phone. In music mode, the wheel would light up with blue backlighting, displaying controls like "play," "pause," "next," and "previous." In phone mode, it would switch to an orange backlight and display numbers 1 through 9 along with their corresponding letters, mimicking a traditional dial pad. This idea is captured in U.S. Patent No. US7860536B2 (see Fig. 34.1). But Apple ultimately abandoned the concept—even Tony Fadell later admitted that "the worst part was that damn wheel interface. It was never going to work because nobody wants a phone with a rotary dial."

Rejecting this hybrid and clunky click-wheel design, Apple chose instead to go all-in on finger-based interaction—ushering in what we now know as multi-touch technology. At the launch of the first iPhone, Steve Jobs proudly proclaimed, "We've invented a revolutionary new interface. It's called multi-touch."

Bill Buxton, one of the earliest computer scientists to study multi-touch technology, was working at Microsoft Research at the time. After Jobs' bold claim that Apple had invented multi-touch, Buxton published an article on his website summarizing the history of multi-touch.[1] He explained that touchscreen technology dated back to the late 1960s, with early research carried out by IBM and the University of Illinois. The earliest tangible outcomes appeared around 1971, and in 1972 the world's first touchscreen computer—PLATO IV—entered public use,

[1] Buxton, B. (2007, January 12). *Multi-Touch Systems that I Have Known and Loved. Bill Buxton: Microsoft Research*. Retrieved March 31, 2024 from https://www.billbuxton.com/multitouchOverview.html

Q. Yue, F. Yang, *Governing Innovation: A Global History of Patent-Driven Technological Progress*, https://doi.org/10.1007/978-981-95-5570-3_34

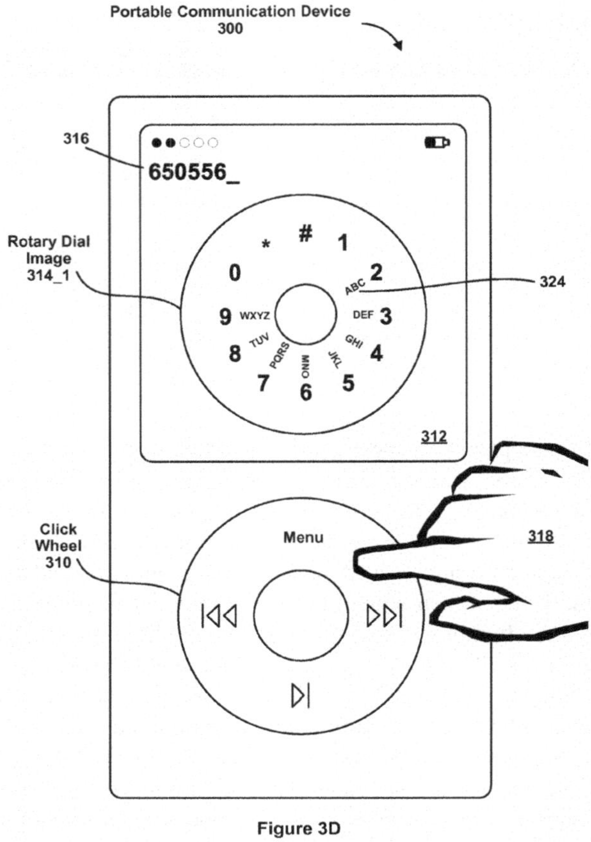

Fig. 34.1 US7860536B2 patent drawing—an iPod-style iPhone

mainly in education. However, these early systems used single-touch input and lacked pressure sensitivity.

In 1984, Buxton and his team completed a study and, in 1985, published a paper titled "A Multi-touch Three Dimensional Touch-Sensitive Tablet," the first academic work to coin the term "multi-touch." This earned Buxton the nickname "Godfather of Multi-Touch."

Though Buxton is credited with pioneering multi-touch and has remained a leading figure in human-computer interaction, patent records show that during the 1990s, his focus shifted away from multi-touch to stylus-based input, which was more in vogue at the time. While at Xerox PARC, Buxton worked extensively on interface technologies. His most influential patent from this era is US6094197A (cited 628 times), which describes a method for combining stylus taps on a virtual keyboard with gestures to simulate modified key inputs—like Shift, Alt, or Ctrl. For example, starting from the on-screen letter "A," dragging the stylus upward would input a capital "A" (see Fig. 34.2).

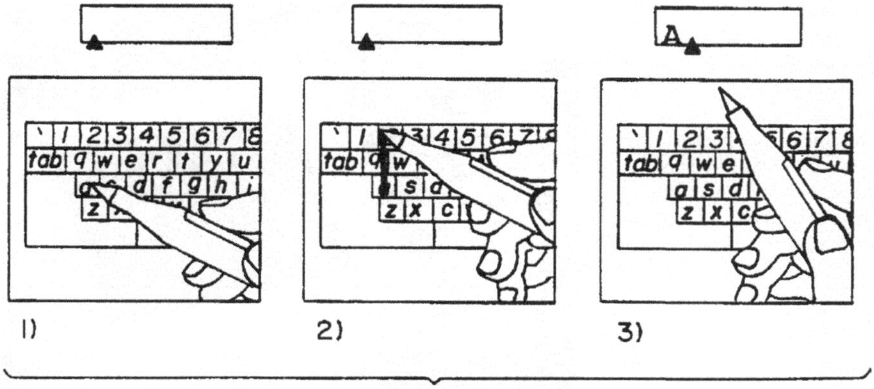

FIG. 8.

Fig. 34.2 US6094197A patent illustration—stylus drag for Shift+A input

It wasn't until after the iPhone and iPad debuted that Buxton returned to the field he helped define. At Microsoft, he filed a new series of patents titled "Touch Screen Control" (e.g., US20160062467A1 [ungranted] and US9223471B2), including several on multi-touch gestures. Clearly, for multi-touch technology to reach the masses, it required not just the spark of innovation from technical pioneers, but also the relentless execution of visionary product designers. In turn, their product successes fueled further technological advancement, creating a virtuous cycle that benefits society as a whole.

Buxton noted that Apple's multi-touch breakthrough owed much to Wayne Westerman's doctoral research. Westerman suffered from severe tennis elbow during graduate school—so debilitating that he couldn't even fill a page with handwriting. To solve his own problem, he invented a wrist-free input device. In 1999, he published his research in his PhD dissertation.[2] Based on this work, Westerman and his advisor co-founded FingerWorks in 1998.

FingerWorks developed a range of multi-touch computer peripherals (see Fig. 34.3) that supported intuitive gestures—like spreading five fingers to maximize a window. Though niche, these devices gained a loyal following among users with repetitive strain injuries. In 2005, Apple acquired FingerWorks, and both Westerman and his advisor joined the company. All product sales ceased. One devoted user, desperate to replace his worn-out FingerWorks pad, offered $1500 for a device that originally retailed at $339, lamenting that "Apple had yanked a critical piece of medical equipment from the market."

In an interview, now an Apple employee, Westerman praised the iPhone:

[2] Westerman, Wayne (1999). Hand Tracking, Finger Identification, and Chordic Manipulation on a Multi-Touch Surface. U of Delaware PhD Dissertation.

Fig. 34.3 FingerWorks' iGesture touchpad. (Source: jpeepz. (2008, May 31). *finger me* [Image]. Flickr. https://www.flickr.com/photos/jpeepz/2559082145/This image is shared under the Creative Commons CC BY-SA 2.0 license)

> The iPhone has a display with multi-touch, whereas FingerWorks had an opaque surface… There are similarities, but Apple clearly took it a step further by applying it to a screen.

Like Apple's later Magic Trackpad, FingerWorks devices had opaque surfaces and internal chips—not suitable for use as transparent screens. To make a transparent multi-touch display, Apple engineer Josh Strickon scoured the literature for solutions. In a twist of irony, his breakthrough came from a 2002 Sony research paper.

"Sony's paper showed how true multi-touch could be achieved using a grid of rows and columns," Strickon recalled. That was the pivotal moment for the project.

The Sony paper[3] described a technology called SmartSkin, which was never intended for the consumer market. Though Sony applied for patents (see Fig. 34.4), the application never reached substantive examination. The technology used a mesh of transmitter and receiver electrodes to measure capacitive changes and detect proximity:

> *The sensor consists of grid-shaped transmitter and receiver electrodes (copper wires). The vertical wires are transmitter electrodes, and the horizontal wires are receiver electrodes. When one of the transmitters is excited by a wave signal (of typically several hundred kilohertz), the receiver receives this wave signal because each crossing point (transmitter/receiver pairs) acts as a (very weak) capacitor. The magnitude of the received signal is proportional to the frequency and voltage of the transmitted signal, as well as to the capacitance between the two electrodes. When a conductive and grounded object approaches a crossing point, it capacitively couples to the electrodes, and drains the wave signal. As a result, the received signal amplitude becomes weak. By measuring this effect, it is possible to detect the proximity of a conductive object, such as a human hand.*

[3] Rekimoto, J. (2002, April). SmartSkin: an infrastructure for freehand manipulation on interactive surfaces. In *Proceedings of the SIGCHI conference on Human factors in computing systems* (pp. 113–120). Retrieved April 2, 2024 from Research Gate: https://www.researchgate.net/profile/Jun-Rekimoto/publication/221514185_SmartSkin_An_Infrastructure_for_Freehand_Manipulation_on_Interactive_Surfaces/links/0fcfd50fde1466ae82000000/SmartSkin-An-Infrastructure-for-Freehand-Manipulation-on-Interactive-Surfaces.pdf. Also available at: https://www2.sonycsl.co.jp/person/rekimoto/papers/chi02.pdf

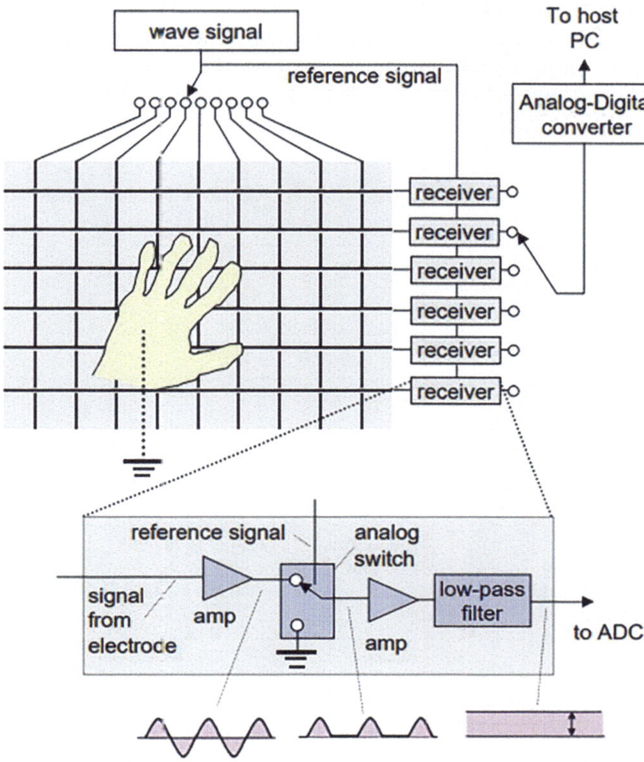

Fig. 34.4 SmartSkin paper illustration

At the end of the paper, the authors even helpfully suggested that "transparent SmartSkin sensors can be achieved using indium tin oxide (ITO) or conductive polymers." Yet, even with this insight, Sony never envisioned developing a multi-touch touchscreen—instead, they built a massive touch-sensitive surface to use as a projection screen.

Although indium tin oxide (ITO) is transparent, arranging it in a grid pattern still compromises overall transparency. To address this issue, Apple's engineers introduced *electrically floating ITO*—sections of ITO that are not electrically connected—into the blank areas between the horizontal and vertical ITO lines. This solution is documented in Claim 10 of Apple's U.S. Patent No. US7663607B2, titled "Multipoint Touchscreen."

During the patent wars ignited by the smartphone revolution, Motorola attempted to invalidate this patent by citing Sony's SmartSkin paper and associated patent applications as prior art. The U.S. International Trade Commission (ITC) initially ruled Claim 10 invalid on the grounds of lacking inventiveness. Apple appealed to the U.S. Court of Appeals for the Federal Circuit, which overturned the ITC's decision, arguing that it had failed to consider the commercial success of the iPhone and other secondary factors that support a finding of non-obviousness.

However, the reprieve was short-lived. In 2014, through an ex parte reexamination procedure, Claims 1, 2, 3, 6, 7, 10, and 11 of the patent were ultimately invalidated.

Chapter 35
Twilight at the Moment of Transition: Sony and Apple from a Patent Perspective

Between 2002 and 2006, Sony filed 291 user interface-related patent applications in the United States, 125 of which were granted, forming a total of 441 patent families globally. During the same period, Apple filed 391 such applications in the U.S., securing 200 utility patents and 19 design patents, with 290 global patent families. Despite Apple's overall patent filings being only about a quarter of Sony's, its user interface patents were nearly equal in number—highlighting Apple's strategic emphasis on user interface design.

Among Sony's user interface-related patents, the most frequently cited is U.S. Patent No. US7456823B2 (see Fig. 35.1), titled "User Interface Device and Portable Information Device." With a priority date of June 14, 2002, it covers innovations in flexible display technology and has been cited 632 times globally.

For its time, this was undeniably a piece of cutting-edge "black tech." The patent explicitly states that the technology is intended for "mobile devices or devices without a mouse and keyboard." Regrettably, Sony has yet to deliver a commercially successful flexible-screen phone to market.

Another highly cited Sony patent is US6998816B2, which pertains to low power consumption, titled "S System and method for reducing external battery capacity requirement for a wireless card." A subsequent key patent is US8150237B2, related to Blu-ray technology. As previously discussed, with increasing bandwidth enabled by advancing communications technology, Blu-ray was overwhelmed by the rise of streaming media and rapidly lost its market relevance.

Further down the list, we find important Sony inventions concerning touchscreens. On July 31, 2006, Sony's headquarters filed a patent application titled "Apparatus and Method for Touch Screen Interaction Based on Haptic Feedback and Pressure Measurement" (Japanese publication number JP2008033739A, ungranted; U.S. counterpart US7952566B2, cited 687 times). This patent introduced haptic feedback when users touched icons on a graphical interface, addressing a key limitation of touchscreens: the lack of tactile feedback when interacting with on-screen buttons. The lead inventor was Ivan Poupyrev from the United States, followed by Japanese engineer Shigeaki Maruyama.

Q. Yue, F. Yang, *Governing Innovation: A Global History of Patent-Driven Technological Progress*, https://doi.org/10.1007/978-981-95-5570-3_35

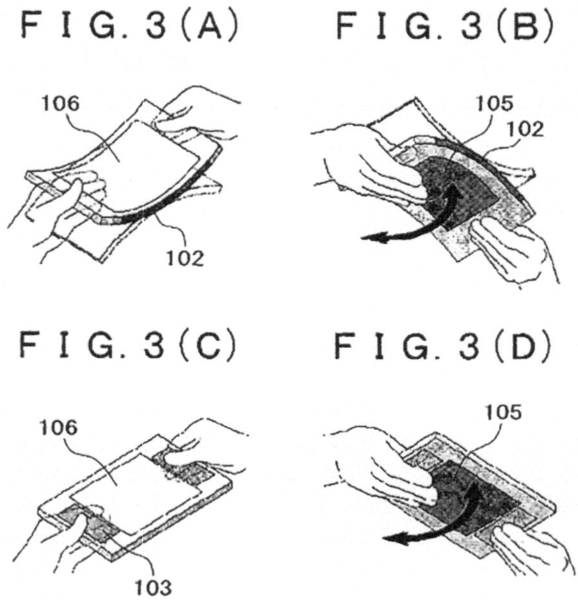

Fig. 35.1 Patent drawing from U.S. Patent No. 7456823B2

In August 2006, Sony Ericsson filed another relevant patent in the U.S., US7791594B2, titled "Orientation Based Multiple Mode Mechanically Vibrated Touch Screen Display." Although its primary focus was on UI redrawing when switching between landscape and portrait modes, it also referenced the challenge of providing tactile feedback for touchscreen operations. This patent lapsed in 2018 due to non-payment of renewal fees.

Given that Sony had already commercialized vibration-enabled game controllers in the 1990s, the idea of using vibration feedback in touchscreens was unsurprising. Unfortunately, while Sony identified the problem early, it did not deeply invest in the field. These two patents never translated into successful products. Ultimately, it was Apple that introduced a vibration motor (Taptic Engine, represented by patents such as US20160211736A1) into its smartphones and marketed it as "3D Touch," making haptic feedback a standard feature in mid-to-high-end touchscreen phones.

Two other Sony patents stand out for their futuristic qualities: US8793620B2, titled "Gaze-Assisted Computer Interface" (cited 314 times), and US8395658B2, "Touch Screen-Like User Interface That Does Not Require Actual Touching" (cited 193 times, including 29 citations by Apple).

In contrast, Apple's most heavily cited user interface patents from this period reflect a different trend: unlike Sony's speculative "black tech," each of Apple's high-impact patents corresponds directly to familiar, tangible features used by everyday Apple customers.

- US8239784B2 (cited 1960 times) is a core multitouch patent covering nearly all Apple touch-enabled devices.

- US7312785B2 (1165 citations) covers the kinetic scrolling effect in Mac and iOS systems.
- US7345671B2 (1142 citations) pertains to the iPod's touch-sensitive click wheel.
- US20060033724A1 (925 citations, not granted) describes the iOS soft keyboard.
- US7509588B2 (831 citations) covers the trembling icon effect seen when deleting or rearranging apps in iOS.

This contrast is thought-provoking: Apple's high-quality user interface patents filed from 2002 to 2006 began rolling out as real products starting in 2007, becoming second nature to every iPhone user. By comparison, Sony's high-quality patents from the same era—like those on haptic touchscreen feedback and gesture-based input—remain futuristic even today. Technologies like gaze-assisted interfaces and flexible screens are still viewed as "black tech," widely discussed yet commercially elusive. These patents remain buried in Sony's technological treasure trove, waiting for the next visionary to rediscover and refine them—perhaps, as with the iPhone, transforming them into world-changing products.

Sony, of course, was not the first consumer electronics firm to explore flexible displays. In 2005, Nokia filed a patent for a "Foldable Electronic Device and Flexible Display Device" (U.S. Patent No. US7714801B2, cited 366 times, with a Chinese counterpart CN101120295A). In 2008, Nokia unveiled the concept phone "Nokia Morph," which featured a flexible OLED screen—marking the first time such technology captured widespread public attention.

Nokia holds 1251 U.S. patents related to user interfaces, of which 481 have been granted—a figure surpassing both Apple and Sony. In comparison, despite being a traditional patent powerhouse, Sony has relatively few user interface patents. This discrepancy is not so much a reflection of Sony's capabilities, but rather indicative of the broader cultural and corporate traditions of Japanese technology companies.

Tony Fadell once recounted a story in an interview with *New York* magazine:[1] In the early 1990s, while working at General Magic, he traveled to Japan to give a presentation at Mitsubishi. During his explanation of the Magic Cap device's modem, he pointed to a slide and said, "This is where you plug in the phone line." In reality, however, the device had no physical modem.

The Japanese asked, "But Mr. Fadell, where is the modem?" Fadell responded that there was no hardware modem—it was implemented via software. The moment he said that, the conference room descended into chaos. Mitsubishi employees erupted in animated Japanese debate, some made urgent phone calls, and soon a large group including the company's CEO joined the room.

Amid the commotion, a typhoon alert went off and Mitsubishi began evacuating employees. Workers filed out of the building, but not a single person in the meeting room left. The Mitsubishi executives insisted on hearing Fadell finish his explanation about the software modem. After he did, one of the Japanese attendees slumped

[1] Adam Fisher. (2018). What Happened to General Magic? Inside the company that invented the iPhone — two decades too early. Retrieved April 6, 2024 from: http://nymag.com/intelligencer/2018/08/general-magic-oral-history-of-the-influential-tech-company.html

into his chair, clutching his head and repeatedly banging it against the table. The moderator explained, "You've just made Mr. Kato's entire department obsolete." A former General Magic employee later remarked with pride: "They were making millions selling modem chips, and we made it so you didn't need a chip at all."

Fadell's story illustrates the software advantage American tech companies held over their Japanese counterparts at the time. Among General Magic's patents is one titled "System and Method for Reducing Processing Requirements of a Modem During Idle Receive Times" (US5995540A), one of the earliest to mention a "software modem" in the title and abstract. It was filed around the same time as similar patents from Motorola, widely recognized as the originator of the software modem.

Many studies have concluded that Japan's software industry lagged significantly behind the U.S. during the onset of the internet age. Former Sony CEO—American-born Howard Stringer—offered his own explanation: Japan's cultural reverence for seniority and its rigid promotion-by-tenure system prevented software engineers from participating in early-stage product development. This, he argued, was a key factor in Sony's decline—and ironically, a consequence of its earlier success.[2]

Sony began as a hardware company, with early innovations focused on miniaturization and sleek, lightweight design. Engineers from the 1980s earned accolades and rose through the ranks. But in the digital era, a new generation of software engineers hit a career ceiling. Stringer observed:

> From the very beginning, we failed to include software engineers in product development. Hardware engineers initiated projects, and software engineers followed later. In a company built on lifetime employment, senior employees held higher positions, while the younger software engineers were at the bottom, climbing up slowly. There was a generational divide.

Stringer noted that once hardware engineers dominated mid- and upper-management, they naturally promoted subordinates with similar backgrounds. Shared expertise and career trajectories made younger hardware engineers more approachable. To these managers, the intangible nature of software made it elusive and hard to grasp. Researchers have argued that this was not merely Sony's internal issue, but a widespread problem among Japanese firms—one of the key reasons why Japan fell behind the U.S. in the IT revolution[3].

To test this hypothesis, we analyzed all of Sony's mobile phone-related patents filed between 2002 and 2006, identifying the top 10 most prolific inventors and their earliest patent filings with Sony (see Table 35.1).

As Table 35.1 shows, six of the ten top patent-holding inventors at Sony filed their first patents in the 1990s—the earliest in 1992. The top-ranked inventor, Yoshiharu Nakajima, and the third-ranked Yoshitoshi Kida, are both experts in LCD technology. Their early patents clearly place them in the hardware engineering category. Fifth-ranked Shigeru Sugaya focused on telecommunications, with early patents concerning telephones and portable terminals. Nobuyuki Matsushita, ranked

---

[2] Cole, R. E., & Nakata, Y. (2014). The Japanese software industry: what went wrong and what can we learn from it?. *California Management Review*, *57*(1), 16–43.

[3] Ibid.

Table 35.1 Top 10 Sony inventors by patent count (2002–2006) and their earliest patents

Rank	Inventor	No. of patents	First Sony patent filing date	First three Sony patents
1	Yoshiharu Nakajima	60	June 18, 1992	JP06003647A (drive method for active matrix LCD); JP06019429A (active matrix LCD device); JP06019430A (color LCD device)
2	Junichi Rekimoto	30	July 10, 1996	CA2180891A1 (3D virtual reality space notification system); US5926116A (information retrieval device); US6055536A (information processing method)
3	Yoshitoshi Kida	27	March 31, 2000	JP4742401B2 (DAC circuit and display device); JP4815659B2 (LCD device); JP2002040464A (LCD panel)
4	Gregory A. Dunko	26	October 30, 2002	US7369868B2 (content sharing via wireless networks); EP1557053B1 (same as above); US7529556B2 (proximity-based contact list)
5	Shigeru Sugaya	26	May 24, 1995	JP08321869A (retrieval method with phone); JP08340377A (portable terminal); JP09114416A (display device and method)
6	Behram Mario Dacosta	20	February 28, 2001	US7293115B2 (internet-aware app updater); US6564286B2 (instant-on non-volatile storage); US7039814B2 (post-processor software decryption)
7	Nobuyuki Matsushita	20	July 7, 1997	JP11024839A (input device); JP11038949A (drawing method and medium); JP11038967A (electronic instrument)
8	Yuji Ayatsuka	20	August 19, 1998	JP2000066800A (image processing apparatus); JP2000122778A (selector and method); JP2000348175A (correspondence point detection)
9	Haruo Oba	18	September 21, 1992	JP06102877A (audio system); USD0356079S (portable color TV); USD0358593S (combo CD/radio/cassette player)
10	William O. Camp	18	August 30, 2002	US7269846B2 (Mobile terminal with AV security); US7269424B2 (distance-sensing receiver); US20040080454A1 (location via DTV signal)

seventh, began with image acquisition technologies. Haruo Oba, ranked ninth, epitomizes the trend: his earliest patents from 1992 and 1993 include designs for audio devices, portable TVs, and CD players.

Only three of the top ten had early inventions leaning toward software—Gregory A. Dunko (4th), Behram Mario Dacosta (6th), and William O. Camp (10th)—and all three filed their first patents with Sony after 2000. Tellingly, their names suggest they may not be Japanese nationals.

This pattern reveals that, within Sony's core R&D operations, veterans of the analog era still held sway.

Table 35.2 Top 10 inventors of Apple's mobile phone patents (2002–2006)

Rank	Inventor name	Number of patents
1	Anthony M. Fadell	25
2	Greg Christie	25
3	Imran Chaudhri	25
4	Jeffrey L. Robbin	24
5	Steven Jobs	20
6	Bas Ording	17
7	Stephen O. Lemay	17
8	Greg Marriott	15
9	Scott Forstall	15
10	Andrew Bert Hodge	14

As a comparison, let's take a look at Apple's top ten inventors of mobile phone-related patents filed between 2002 and 2006 (see Table 35.2).

These names require no deep patent analysis—each is widely recognized in the industry. With the exception of Steve Jobs, nearly all were established computing heavyweights in the IT world.

First is Tony Fadell, whom we've already mentioned multiple times. Tied with him at the top is Greg Christie, who led the iPhone's human interface design team and was one of the main inventors behind the iconic slide-to-unlock patent (US8046721B2). As previously noted, both Fadell and Christie were part of the 1990s wave of early personal smart device development and served on the IBM Simon project.

The third co-top inventor is Imran Chaudhri (b. 1973), who worked at Apple from 1995 to 2017, spending all 19 years on the human interface team. Before joining the iPhone project, he helped design the Mac Dashboard. The decision to have only one button on the iPhone was his idea—whereas Steve Jobs originally envisioned two primary buttons: the familiar Home button and a separate "Back" button.

Fourth on the list is Jeff Robbin, who had worked at Apple since the early 1990s before leaving to develop one of the first MP3 software players for Mac—SoundJam MP. Jobs acquired the software and brought Robbin back into the fold. After integrating music store functionality, SoundJam MP evolved into what we now know as iTunes.

Fifth, of course, needs no introduction—Steve Jobs himself.

Sixth is Bas Ording, a Dutch designer who joined Apple's human interface team in 1998. A computer game enthusiast, Ording drew design inspiration from gaming and loved to incorporate whimsical animations into user interactions. The signature "rubber band" bounce effect seen when scrolling past the end of a list on iPhone was his invention. In 2013, he left Apple and joined Tesla, where he continues to work on UI design.

Ording once stated that he left Apple out of frustration with endless patent litigation—being listed as an inventor on many key Apple patents, he was frequently called to testify in court. "I've spent more time in court than I have designing," he lamented.

Also ranked sixth is Stephen O. Lemay, another UI designer. Eighth on the list, Greg Marriott, was an Apple veteran and a member of the System 7 design team—known as the "Blue Meanies." Ninth-ranked Scott Forstall was one of the primary architects behind Mac OS X and the Aqua UI. Known for his assertive personality and leadership style, he was often dubbed "Mini Steve Jobs." The only one with a background leaning more toward hardware is Andrew Hodge (ranked tenth). With an academic and early career focus in industrial design, he went on to lead Microsoft's HoloLens project and now serves as CEO of Owl Cameras, a smart dashcam company.

From a technical classification perspective, Apple's mobile patents from this era were distinctly more software-oriented than those of Sony or Nokia. Between 2002 and 2006, 31.64% of Apple's 414 mobile-related patent applications (granted and pending) fell under IPC subclass G06F. By contrast, only 17.97% of Sony's 1636 mobile-related patent filings were classified in G06F, and just 15.62% of Nokia's 2266 filings during the same period.

In the domain of human-computer interaction, Apple's patents were also significantly "softer" than Sony's. Between 2002 and 2006, Apple filed 33 U.S. patent applications involving "keyboard" or "keypad"[1]—nearly all of them focused on virtual or soft keyboards. The most-cited among them is US7844914B2 (873 citations), which proposed a single-handed solution for touchscreen text input using an arched keyboard layout triggered from the bottom-right of the screen. Ultimately, Apple abandoned this approach. The second most-cited patent in this group is US9024884B2, covering elements of the GarageBand music software interface.

During the same period, Sony filed 131 U.S. patent applications involving keyboards[2]. Its most influential examples include US20050162395A1 ("Text input in an electronic communication device," 280 citations, ungranted) and US7853863B2 ("Method for expressing emotions in text messages," 224 citations). Though both are software-related, they pertain to conventional keyboard interactions. Only two Sony patents from this era specifically mentioned "virtual keyboard" or "soft keyboard": US7477231B2 and US20030201972A1 (priority filed as JP2003316502A, ungranted). Notably, Apple cited these two patents 18 and 10 times, respectively—more than any other applicant.

In contrast, between 2002 and 2006, Sony filed 117 Japanese patent applications (27 granted) and 50 U.S. applications (21 granted) related to physical, mechanical, or pressure-sensitive buttons, comprising a total of 128 global patent families (simple family grouping)[3]. Using the same criteria, Apple filed only 8 such applications during this period—a striking demonstration of how thoroughly Apple committed to the touchscreen revolution.

Among Sony's physical-button-related patents from this period, one particularly intriguing example stands out: US6560612B1, titled "Information Processing Apparatus, Control Method, and Program Medium." This patent proposed equipping computers and similar devices with a "second display unit" and "second input unit" to facilitate data entry without increasing the device's size or compromising portability. Apple showed exceptional interest in this invention—it was cited 129 times in total, 48 of which were by Apple itself. Notably, Apple's patents related to

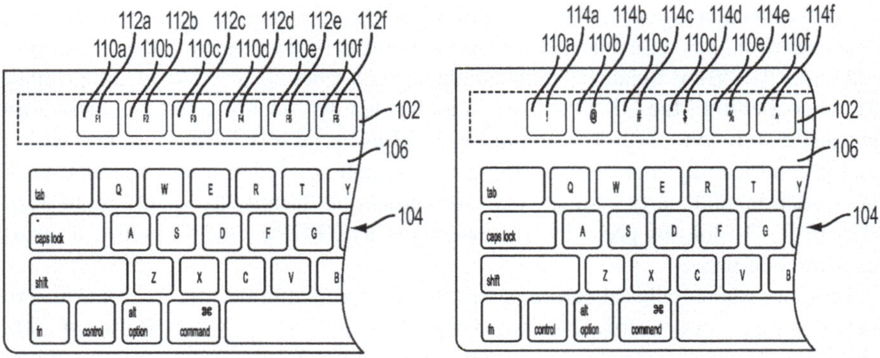

FIG. 2A FIG. 2B

Fig. 35.2 Patent drawing of U.S. Patent No. 9927895B2

Fig. 35.3 MacBook with Touch Bar replacing function keys. (Source: IM3847. (2016, November 18). *Touch bar of 15" MacBook Pro* [Image]. Wikimedia Commons. https://commons.wikimedia.org/wiki/File:Touch_bar_of_15%22_MacBook_pro.jpg. This image is shared under the Creative Commons CC BY-SA 4.0 license)

replacing function keys with a touchscreen on MacBooks—such as US9927895B2, "Input/output device with a dynamically adjustable appearance and function" (see Fig. 35.2)—cite this Sony patent. Figure 35.3 illustrates the MacBook Touch Bar that replaced the traditional F1 to F12 keys.

To what extent were Apple's designs inspired by this Sony patent? While we cannot say for certain, it is clear that companies like Sony possess a treasure trove of brilliant user-oriented ideas—many long buried and overlooked—that deserve deeper exploration and analysis.

Apple's software strengths stand in stark contrast to Sony's hardware prowess. Several factors underlie this divide: Japan's IT workforce tends to be younger, less experienced, and lower paid; meanwhile, the longstanding offshoring of U.S. manufacturing to East Asia also plays a role. As of this book's completion, Apple had filed 273 U.S. patent applications under IPC Section F (mechanical engineering, lighting, heating, weapons, and blasting), of which 116 had been granted.[2] What's particularly fascinating about this group of patents is the high number of Chinese or Asian names among the inventors—far more than in any other batch we've analyzed.

The top two inventors on this list are both ethnic Chinese: Jun Qi (30 patents) and Victor Yin (29 patents). Others include Wei Chen (9 patents), Rong Liu (8 patents), and Amy Qian (6 patents). Around ten other Chinese or Asian inventors each hold 3 or 4 patents. By contrast, famed Apple designer Jony Ive appears on only four patents in this batch. These patents frequently cite prior art from East Asia: 45 cite Samsung patents, 35 cite those from Hon Hai Precision Industry Co. (Foxconn), and 35 cite Sony patents.

Since the early twenty-first century, Apple has shown a special interest in patents originating from Sony's gaming division.

From 2002 to 2006, Sony Ericsson filed 8962 patent applications globally, forming 2990 patent families (simple families). Among these, 59.8% were classified under IPC Section H (Electricity), with subclass H04M (Telephonic Communication) making up 18.09%. The next largest category was G06F (Electric Digital Data Processing) under Section G, comprising 15.45%.

The most influential patent in this collection was US7696718B2, related to "Charging Apparatus and Charging System," with 310 citations—though none by Apple. Next was EP1347361A1, "Text Input in an Electronic Communication Device," with 278 global citations, 113 of them from Apple. In total, Apple cited Sony Ericsson patents 368 times, second only to Samsung (which cited them over 600 times).

During the same period, Sony Computer Entertainment (renamed Sony Interactive Entertainment in 2016) filed 7147 global patent applications, forming 2621 patent families. Of these, 48.53% were classified under Section G, with G06F accounting for 29% and A63F (Video Games) 8.78%.

The most impactful patents included US8547401B2 ("Portable Augmented Reality Apparatus and Method"), US8072470B2 ("System and Method for Providing Real-Time 3D Interactive Environment"), and US9177387B2 ("Method and Apparatus for Real-Time Motion Capture")—the last of which caught Apple's particular attention. Several Apple patents filed after 2011 related to gesture-based interfaces cite this patent.

Unlike many of Sony's highly cited "moonshot" patents, US9177387B2 led to a tangible consumer product: the PlayStation Move, launched by Sony in 2009, a motion-controlled gaming system that competed directly with Microsoft's Kinect. Kinect sold 8 million units in its first 60 days on the market and still holds the Guinness World Record for fastest-selling consumer electronics device. Though PlayStation Move didn't surpass it, it achieved notable success in its own right.

The most frequent citers of these patents include Sony's direct competitors in gaming: Microsoft (over 300 times), Nintendo (nearly 200), IBM (nearly 200), Samsung (over 200), and Apple (over 110). Although Apple cited fewer of these patents than it did Sony Ericsson's, the number is still significant given Apple's lack of a gaming division—highlighting its clear interest in Sony's gaming IP.

It's worth noting that during this period, Sony's gaming division was led by Kazuo Hirai, a user-focused executive beloved by loyal "Sony fans." In 2012, he succeeded American CEO Howard Stringer and returned Sony to profitability in the same year. These facts suggest a recurring truth in the tech world: those who truly prioritize the user often share the same visionary instincts.

Steve Jobs was famously a devoted "Sony fan." Of the 625 patents he filed under Apple, 428 were design patents, while the remaining 197 were utility patents—91 of which were granted. Among these 197 utility patents, 132 cited patents from tech giant IBM; 119 cited Microsoft's patents; 111 cited Sony's; followed by 94 citations of Nokia patents. In comparison, only 57 referenced Samsung patents.

Among his 428 design patents, 278 cited Samsung patents, 251 cited Sony patents, 243 cited IBM patents, 132 cited Nokia patents, and 101 cited Microsoft patents.

Several iconic design patents—including those for the original iPod, iPod Nano, and Mac Mini (such as USD601583S)—cited Sony's US6254477B1 patent, titled "Portable Electronic Device, Entertainment System, and Method of Operating the Same." This Sony patent pertains to gaming consoles and player interaction. Most of its eight cited references also relate to gaming systems, seemingly far removed from Apple's core business.

And yet, Jobs and Apple had an exceptional fondness for this patent: Jobs cited it in 87 of his own patents, and Apple as a company cited it in 170 patents. In contrast, Sony cited its own patent only 41 times. Neither Sony Ericsson nor Sony Mobile ever referenced it. Samsung cited it just once; Microsoft, twice; and IBM, only once.

These figures vividly illustrate Steve Jobs's admiration for Sony—a true "Sony fan" in every sense.

Chapter 36
Summary

As intellectual property professionals, we can assert with some confidence that many of Sony's challenges were deeply tied to the management and utilization of its intellectual property.

As Chang, S.J. (2011) once quoted a former Sony executive: "*An internal content business is not necessarily an asset but it can also be a liability since we need to protect the interests of the content business*."[1] In the same vein, the intellectual property a company holds isn't always an asset. Kodak's obsession with the "razor-and-blades" business model and Sony's fixation on proprietary formats both illustrate a common issue: when companies build products around "the IP we own" and try to monetize that IP at all costs, this mindset can harden into an obsession—one that ultimately becomes a burden. Clinging to such obsessions can turn IP into shackles that prevent companies from adapting to the times, listening to users, initiating internal transformation, and achieving rebirth.

When Sony collaborated with Philips on the compact cassette, the concept of "open innovation" had yet to emerge. And yet, by happy accident, they launched a market-winning product that generated almost no licensing revenue. The Walkman itself lacked patent protection, but it redefined the way people listened to music and created the Sony legend. Apple, for its part, lacked patents for MP3 technology and hard-drive-based MP3 players, yet it too wrote a product success story for the ages.

Perhaps due to the success of the cassette and CD, Sony became deeply attached to the idea of proprietary standards. Yet setting a standard is expensive, and losing in a standards war is even costlier. Maintaining a standard comes with its own price: Sony's Trinitron technology, being wholly incompatible with industry norms, required a costly vertically integrated manufacturing ecosystem. In order to equip the PlayStation 3 with a high-priced Blu-ray drive, Sony delayed its launch and kept its price elevated. As Professor Zhang pointed out, Sony's obsession with its own

[1] Chang, S. J. (2011). Sony vs Samsung: The Inside Story of the Electronics Giants' Battle For Global Supremacy. John Wiley & Sons.

Q. Yue, F. Yang, *Governing Innovation: A Global History of Patent-Driven Technological Progress*, https://doi.org/10.1007/978-981-95-5570-3_36

standards also affected its ability to fairly evaluate technologies or standards from other companies.[2] Often marching to the beat of its own drum, Sony swam against the current of the industry.

Betamax ignited a revolution and provoked unprecedented copyright concerns, drawing fierce opposition from the film industry. Yet those fears proved unfounded: home video systems did not threaten the film business—instead, they became a vital revenue stream for modern movie studios. Sony, in aggressively defending its position against potential partners, effectively resolved the IP risks for its competitors.

The fear expressed by music labels and the RIAA over MP3s mirrored the panic felt by Hollywood during the dawn of the home VCR. Internet users chose MP3, while the music industry tried desperately to suppress it. In this context, Apple aligned itself with consumer preference, launching the iPod—a product deeply integrated with computers and the internet. Sony, on the other hand, sided with the music labels. Its audio division clung tightly to a web of patents related to the ATRAC format. Before the iPod emerged, history had given Sony ample opportunity: it introduced memory sticks, MiniDiscs, MD adapters, and ATRAC3. These were well-engineered, technically sound products—but they neither supported MP3 nor offered any solution for copying or sharing music.

Both Sony's audio division and Sony Music Entertainment viewed MP3 with skepticism. But MP3 was never the real issue.

As Harald Popp, the "father of MP3" and a researcher at the Fraunhofer Institute, once said:

> *The real challenge is the Internet—and it still is. ... It's not MP3—that was always our clear answer: 'MP3 was just a catalyst. If it was an ISO standard it could [just as easily] have been a proprietary scheme.' The technology is out there and you cannot avoid some audio compression ... sometimes illegal ... but that's just part of a technological revolution.*[3]

Looking back, Sony's repeated format wars resemble the rise and fall of dynasties. As the old Chinese saying goes, "A thousand-mile empire ends in dust; each reign has its own ministers." In our age of rapid technological evolution, no standard reigns forever. VHS toppled Betamax, only to be dethroned by DVD. MD overcame DCC, then was crushed by the revolution led by MP3 and iTunes. Blu-ray triumphed over Toshiba's HD-DVD, only to be flattened in turn by high-definition streaming fueled by next-gen networks.

In the face of unstoppable technological change, how can we adapt and survive?

Consumers and customers are the ultimate source of return for all corporate R&D efforts—they are the most important stakeholders a company serves. The author firmly believes: by staying close to its users and letting customer needs guide its direction, a company can ride the waves of innovation and stand tall at the crest of each new technological tide.

---

[2] Ibid.

[3] Sterne, J. (2012). *MP3: The meaning of a format*. Duke University Press. Retrieved March 31, 2024 from https://read.dukeupress.edu/books/book/1663/MP3The-Meaning-of-a-Format

Part III
Core and Frontiers

Chapter 37
Introduction

Many commercial patent databases offer features like "patent sandboxes" or "patent maps" that visualize a company's key technological domains in 3D, much like a sandbox simulation. Exploring the patent histories of century-old tech giants such as IBM, AT&T, Sony, and Kodak through these visualizations became one of the most enjoyable aspects of this writing process. It felt akin to playing historical simulation games like *Civilization*, witnessing the rise and fall of nations on a massive 3D terrain map.

The fate of many companies resembles that of ephemeral empires in history—"rising in flourish, perishing in haste." Some quietly emerge from seemingly obscure technical niches, achieve meteoric success, and carve out a blue ocean market—just like a small tribe rising on the fringes between empires, swiftly conquering territories and expanding its domain. But while some vanish as quickly as they rise, others endure, standing the test of time. Their triumphs and failures alike are valuable historical legacies.

Enterprises that have stood firm for decades—or even a century—inevitably possess a "core business" akin to a political power's strategic stronghold. With a solid core, they can either strike outward or retreat inward. Even if they fail to expand further into new frontiers, a defensible position ensures continued survival and influence. IBM, for instance, dominated commercial computing from punch card machines to the transistor era. Though its dominance was challenged during the PC revolution, by the 2010s and 2020s it still secured a stronghold in cloud computing. Xerox, having once ventured into personal computing and printing with limited success, remained a major provider of business document services in the U.S. Sony, once the undisputed king of consumer electronics, had diversified across audiovisual and cultural domains. Though its flagship products faded during the mobile internet era, Sony retained strength in high-end electronics and gaming—much like Britain or Spain, whose imperial glory faded but whose early accumulation of wealth and culture continues to exert influence.

A company's core business, like a nation's power base, is not immutable. Nokia began as a timber company; Texas Instruments was once a supplier of oil

Q. Yue, F. Yang, *Governing Innovation: A Global History of Patent-Driven Technological Progress*, https://doi.org/10.1007/978-981-95-5570-3_37

exploration tools; Philips started with lighting; HP's first big contract was for audio testing tools used by Disney. In the latter half of the twentieth century, Nokia pivoted to telecommunications and became a mobile phone giant; Texas Instruments shifted toward semiconductors, integrated circuits, and consumer electronics after the invention of the transistor by Bell Labs; Philips expanded into Europe's leading consumer electronics brand; HP rose to dominance through PCs and printers. In recent years, Nokia exited the mobile phone market and returned to commercial telecommunications; Texas Instruments and Philips abandoned consumer electronics, with Philips spinning off its lighting business as Signify and transforming into a professional medical device company.

The evolution of a company's patent map can be likened to the territorial shifts of nations in history. For example, the Ottoman Empire originated on the Central Asian steppes, peaked with a realm spanning the Middle East and North Africa, and once pressured Europe. In 1453, it conquered Constantinople and relocated its core to Anatolia, the chokepoint between Europe and Asia. After World War I, the empire disintegrated, losing its original heartland near the Amu Darya River. Yet Anatolia, now part of Turkey, continues to exert regional influence thanks to its strategic geography.

Similarly, the Kingdom of Prussia traces its roots to the medieval Teutonic Knights, whose earliest base lay in Acre, near Jerusalem. Later, the order acquired territory in Prussia and began engaging in European politics, eventually leading German unification. Prussia became the core of the German Empire, shifting its center to Berlin and Brandenburg. Today, much of historical Prussia lies within Poland, while its cultural and political legacy lives on as part of Germany's status as a technological powerhouse.

As the old adage goes, "The only constant in the world is change." Embracing change is essential not only for nations but also for enterprises. Yet, as businesses shift their core focus, preserving cultural identity and leveraging the technological and branding legacy of their original core remains a critical question.

The governance philosophy of innovation espoused in this book draws from historical evolutions in military, political, and cultural domains. From a metaphysical analysis of such transformations, we derive the core logic behind governing innovation—one that applies to warfare, economics, and civilization alike. The "core" may refer to a military stronghold, an economic hub, or a cultural center. It can evolve or relocate, but it cannot be absent. In history, military forces without a stronghold were like water without a source—perilously fragile. Consider the Knights Templar, rivals of the Teutonic Order: after losing their base in the Middle East, they continued to amass wealth through lending across Europe. But when their largest debtor—the King of France—issued a sweeping decree, most of the order's members were arrested or killed overnight. A century-long power crumbled instantly.

China's history also abounds with peasant uprisings that failed for lack of a base. In the patent world, NPEs that amass patents and extort fees without producing anything are the modern equivalent of roving bandits. They do not invest in industry; they follow the scent of profit, swarming successful operating companies to extract "flesh." Their patents may be ornate yet hollow—mere parade troops with no real firepower—or acquired "mercenaries" with no deep understanding of the technology's origin, prosecution history, or commercial context. Once defeated in litigation, these patents often cannot be revived due to the absence of actual R&D teams to iterate the invention. As discussed in the "Rise and Fall" section, once-great tech companies can devolve into patent trolls if they lose their core business and public support. And like historical bandits, without a stronghold, their "troops" (core patents) dwindle, their "firepower" (patent quality) fades. These trends are often visible in their patent maps.

In contrast, the strategy of building a solid base and expanding cautiously has long been revered in Chinese history. During the Chu-Han Contention, Liu Bang stealthily returned via the Old Road while Chancellor Xiao He secured the Bashu region as the rear base, supporting frontline advances. After defeating Xiang Yu, Liu Bang rewarded Xiao He as the top-ranking civil official, emphasizing the value of the stronghold. Likewise, Zhuge Liang's famous "Longzhong Plan" advised Liu Bei to secure Jingzhou and Yizhou as strategic bases to prepare for northern campaigns—an enduring strategic model. Chinese civilization has stood for 5000 years largely because of its enduring core. With a strong center, it could expand into the Western Regions or the South China Sea, or withdraw inward to preserve peace.

This ancestral wisdom—whether in cultivating a heartland or opening new frontiers—remains deeply instructive. The governance of territory is called nation-building; the governance of a company is called enterprise management; and using patents and intellectual property to guide innovation—that is what we call the governance of innovation.

Traditional Chinese thought on frontiers adheres to a center-oriented worldview that emphasizes a clear distinction between internal and external realms. In ancient times, this was expressed in the "fushi" (服事) system, which conceptualized the world as centered on the royal domain and divided it into "inner服" and "outer服." The "inner服" comprised court officials and noble retainers, while the "outer服" included feudal states and lords. This framework evolved into the "Five服" model in *The Book of Documents* and *Discourses of the States*, and later into the "Nine服" model described in *The Rites of Zhou*. The king's domain—or dianfu (甸服)—was considered the core territory of the state. Surrounding it were semi-autonomous vassal states, and beyond them lay regions inhabited by non-Han peoples, known as huangfu and yifu, often regarded as peripheral or even barbarian. Scholars interpret these models as evidence of a limited-autonomy policy toward frontier minorities during the Spring and Autumn and Warring States periods, known as the policy of *jimi* (羁縻).

The term *jimi* dates back to the Han dynasty. In *Records of the Grand Historian*, Sima Xiangru notes: "It is said that the Son of Heaven governs the Yi and Di peoples by tethering rather than severance." *Han Guan Yi* explains, "A horse is tethered (羁/ji), an ox is muzzled (縻)/mi/—meaning the Four Barbarians are managed like livestock." During the Tang dynasty, jimi prefectures were established in border areas. These regions were not taxed directly—"tribute, taxation, and census rarely reached the Ministry of Revenue"—but were governed by imperial appointees such as Protectors and Governors. Dynasties like the Han and Tang, with their agrarian Han

Chinese core, laid the foundation for China's vast 9.6 million square kilometers by extensively implementing the jimi system and actively expanding into frontier territories.

This chapter explores an innovation governance model aimed at pioneering new domains. The goal is to encourage tech companies to press forward, expand continuously, and avoid decline through inertia or generational turnover. The "core domain" of a business refers to its central line of operations—the foundation of its survival and the heart of its brand and cash flow. Newly developed business areas are referred to as "jimi zones." In the early stages, these zones may not yield profits—just as jimi prefectures once contributed little to state revenue. With sustained support from the company's core technologies, talent, capital, and channels, these jimi zones can eventually attract customers and generate profits, thereby reinvesting in the core. At that point, the jimi zone becomes a "transition zone." If the transition zone eventually surpasses the original core in revenue and user base, the company develops dual cores—or even replaces the old core altogether. Areas not yet explored or developed are considered beyond the company's "business frontier." Companies should determine their R&D strategy by striving to breach this frontier and establish new jimi zones beyond it.

But how can companies identify the direction in which to push their frontiers and create new jimi zones? As the previous chapter emphasized, the foundation lies in a "people-centered" philosophy—putting customer and consumer needs at the heart of innovation. In addition, scientific discoveries from universities, research institutes, or R&D-centric enterprises offer another avenue for expansion. In both cases, patent professionals can play a strategic role.

When Liu Bang was confined to Bashu by Xiang Yu, his advisor Xiao He urged him: "Rule Hanzhong, nurture the people, attract the talented, reclaim the Three Qins—then the empire shall be yours." Likewise, Liu Bei had Zhuge Liang's "Longzhong Plan," and Zhu Yuanzhang had Zhu Sheng's "build high walls, stockpile grain, and bide your time" strategy. The author hopes IP professionals can similarly act as strategic advisors to R&D—not merely as guardians of patents, but as visionaries with a holistic grasp of science, technology, and industry.

More specifically, we propose that companies look to two main sources when aiming to break through their boundaries: technological innovation and user demand.

First, scientific discovery and technological advancement often point the way. Scientists worldwide make breakthroughs every day, but few really understand the commercial implications of their work. Only those with entrepreneurial vision—or scientist-entrepreneurs with keen market instincts—can translate these breakthroughs into consumer-grade products that enhance public well-being. Who knows how many hidden gems lie buried in the patent libraries or academic journals of universities, research institutes, and even competitors, waiting to be uncovered? As discussed in the previous chapter, Apple's touchscreen technology partly stemmed from a niche firm, and a crucial breakthrough came from a Sony research paper. Larry Ellison founded Oracle after reading a paper by IBM. In the first part of this chapter, we revisit the long history from vacuum tubes to integrated circuits, and from photoelectric effects to digital cameras—focusing on the successes and failures of companies like AT&T and Sony with technologies such as the transistor and CCD.

Second, companies can build upon their accumulated expertise in core domains. As a business deepens its core competencies, it can use them to expand beyond the core into new jimi zones. For instance, Bosch started as a small workshop producing ignition systems—its original core. With the advent of the internal combustion era, Bosch leveraged its ignition expertise to develop spark plugs, solving a key problem for the automotive industry. Spark plugs evolved into a transition zone and eventually became Bosch's new core. From this base, Bosch identified additional automotive needs, launched new products, and built multiple core domains. It is now a global automotive solution leader.

Xerox began as a small firm producing photocopying equipment and paper—its original core. Upon securing patents for xerographic technology, it ventured into commercial copiers—a jimi zone that soon became its new core. While exploring this new domain, Xerox was among the first to touch the edges of uncharted terrain: GUIs, mice, and word processors. It secured crucial patents, but failed to commercialize them. These jimi zones never became transition zones, and companies like Apple and Microsoft overtook them. Xerox remains confined to a corner of the commercial copier market.

Third, companies can break boundaries in multiple directions while staying true to their core spirit. Disney began with animation, grounded in themes of childlike wonder and fantasy. To bring these fantasies to life, Disney filed early patents for animation technologies in the 1920s. In the 1960s, it built real-world fantasylands with Disneyland—its second core. In the twenty-first century, CGI from computer animation helped Disney break into live-action films, creating a third core. In recent years, Disney has repeatedly topped U.S. box office rankings. Whether in live-action or 3D animation, Disney's technical prowess is unrivaled. Its commitment to core values, combined with timely innovation and multidirectional expansion, offers a valuable model to emulate.

In 1492, Christopher Columbus discovered the Americas, triggering a wave of colonial expansion from the Old World into the New. To European rulers at the time, the Americas appeared vast and desolate—not scarce, not precious. After France's defeat in the Seven Years' War, King Louis XV ceded over 2 million square kilometers of North American territory to Spain. When Napoleon later reclaimed the land, he sold it to the United States for a mere $1.5 million—just three cents per acre—effectively doubling America's landmass. In 1867, Russia sold Alaska to the U.S. at an even lower rate: nearly 1.7 million square kilometers for just $7.2 million, or two cents per acre. During this era of uncharted New World frontiers, European rulers found boundary expansion relatively effortless—sending convicts, rogues, and exiles to subjugate indigenous peoples and claim swathes of land with ease.

By the late nineteenth century, however, Germany—on the rise—faced a different reality. Though the Earth was vast, the colonial territories had already been carved up by older imperial powers. There were no unclaimed frontiers left to seize. Germany's aggressive push for expansion collided with British and French interests, escalating into World War I and, later, World War II. The result was catastrophic: massive population loss in Europe, widespread decolonization, and the continent's eventual decline. Meanwhile, the United States, watching from across the Atlantic, emerged unscathed and ascended to global dominance.

Today's business landscape mirrors that moment in history. Established markets are tightly held by legacy corporations with decades or even centuries of history. Startups, like pre-war Germany, often find themselves locked in fierce and potentially self-destructive competition. Germany's fate is a stark reminder: launching headlong attacks on entrenched players in mature industries—what we might call "internal overcompetition"—is a perilous strategy.

So where should companies seek to expand?

As discussed previously, Vannevar Bush's 1945 report *Science: The Endless Frontier* had a profound impact on both academia and government. It laid the foundation for America's modern science and technology policy framework. To mark its 75th anniversary, the U.S. government even named its latest research strategy bill *The Endless Frontier Act* in tribute. In the report, Bush argued that governments should fund basic scientific research, confident that discoveries would eventually translate into applied technologies and industries. For enterprises, drawing nourishment from such scientific advances—transforming them into new products and technologies—is a vital path to breaking boundaries.

It must be acknowledged that transforming scientific invention into a consumer-grade product often takes considerable time. Scientists do not always understand how their discoveries might be applied in real-world contexts. This isn't a critique of their pursuit of truth. Were science driven solely by profit, its progress would be stunted beyond recognition. Scientists bear the noble task of uncovering objective laws of the universe, and they should not be shackled by monetary concerns.

In Liu Cixin's short story *The Truth of the Dawn*, humanity makes a groundbreaking discovery in basic science—only to attract the intervention of extraterrestrials. These aliens, though not malevolent, compel humanity to abandon the research, warning that its continuation would lead to inevitable catastrophe. As compensation, they offer to share their knowledge—but at a cost. Those who choose to learn the truth must forfeit their lives. Undeterred, countless scientists, like moths to a flame, board the alien vessel. Upon receiving the answer to the scientific riddle that had consumed them, they willingly accept their fate. It is a moving interpretation of Confucius's line: "If I were to hear the Way in the morning, I could die in the evening."

The Truth of the Dawn is both tragic and beautiful. The scientists who embrace death do so not for glory or wealth, nor even for the betterment of humanity—but for the satisfaction of pure curiosity. In the story, the aliens reveal that Earth became a subject of close observation not when humans discovered fire or split the atom, but when primitive hominids first looked up at the stars with wonder. This pursuit of knowledge, transcending material desires, is among humanity's noblest instincts—and a crucial driver of progress. Scientific effort in this direction deserves our utmost respect.

That said, history also shows that regardless of whether scientists are motivated by material gain, their discoveries often conceal tremendous economic potential—potential they themselves may not realize. These hidden gems lie buried in libraries and patent databases, awaiting visionary entrepreneurs to uncover them.

The story of touchscreen technology is a classic case of industry harnessing academic research to transform human life. In 1985, computer scientist Bill Buxton from the University of Toronto published a paper titled *A Multi-Touch Three Dimensional Touch-Sensitive Tablet*, introducing the concept of multi-touch screens. Widely hailed as the "father of multi-touch," Buxton not only coined the term but also mentored many early pioneers in the field. A decade later, doctoral student Wayne Westerman of the University of Delaware completed his thesis on multi-touch technology and filed several patents. After graduating, he co-founded a startup with his advisor, developing a touchpad capable of recognizing multiple gestures.

In 2002, Sony engineers published a paper describing a capacitive screen with a grid-like antenna capable of multi-touch input on a projected display. By 2005, Apple had acquired Westerman's company and patents. During the development of the iPhone, Apple's engineers drew inspiration from the Sony paper, merged it with Westerman's input technology, and refined the grid antenna design—subsequently filing a new patent (US Patent No. 7663607B2). In 2007, during the first iPhone launch event, Steve Jobs proudly announced: "We've invented a phenomenal new technology. It's called Multi-Touch." At that moment, 22 years had passed since Buxton's initial concept.

Another great example is the path from magnetic audio recording to the invention of the Walkman. In 1888, American inventor Oberlin Smith documented the concept of using magnetic material to record sound. Rather than patenting it, he published his ideas in *Electrical World*. His work languished in obscurity until it was referenced by researchers in 1941. It was not until his 150th birthday in 1990 that Smith was posthumously recognized as the inventor of magnetic recording.[1] In 1898, Danish inventor Valdemar Poulsen independently designed a "telegraphone," the first known device to record sound using magnetic materials (Fig. 37.1). Poulsen patented his invention and received authorization (US Patent No. 661619A; see Fig. 37.2), thereby launching the industry's journey into magnetic sound recording.

In 1928, German scientist Fritz Pfleumer invented a sound carrier containing "finely dispersed paramagnetic materials" that could be used to record audio signals. He filed a patent for it (DE500900), marking the birth of magnetic tape recording. As previously mentioned, German engineers made major contributions to high-fidelity audio recording and playback—a technology that the Nazi propaganda machine heavily relied on during World War II. After Germany's defeat, many high-end recording devices fell into American hands. U.S. companies, led by RCA, began intensive research into magnetic tapes, producing a wide array of designs and patents. In 1968, Philips improved upon American cassette designs and invented the compact audio cassette (e.g., US Patent No. 3394899A), which went on to enter millions of households. In 1979, Sony combined a miniature tape recorder with

[1] Friedrich Karl Engel. (1990). Oberlin Smith and the Invention of Magnetic Sound Recording. Retrieved March 31, 2024 from Museum of Magnetic Sound Recording Website: https://museumofmagneticsoundrecording.org/images/R2R/EngelOberlinSmith2006.pdf

Fig. 37.1 Valdemar Poulsen and his magnetic wire recorder

headphones and released the Walkman—a global sensation that revolutionized how people consumed music and entertainment. By then, it had been over 90 years since Oberlin Smith first published his paper.

In 1887, German physicist Heinrich Hertz discovered the photoelectric effect. In 1905, Albert Einstein published *On a Heuristic Viewpoint Concerning the Production and Transformation of Light*, explaining the photoelectric effect and ushering in the age of quantum physics. Though Einstein is best known for his theory of relativity, his Nobel Prize was awarded not for that, but for his work on the photoelectric effect. This principle has broad applications across many fields—the digital camera being perhaps the most familiar.

Nearly a century passed between the proposal of the photoelectric law and its use in consumer digital cameras. In 1969, Willard Sterling Boyle and George Elwood Smith at Bell Labs invented the Charge-Coupled Device (CCD), which transformed light into digital signals via the photoelectric effect. Their work won them the Nobel Prize in 2009.

In 1972, Texas Instruments designed an "electronic photography system" based on CCD as a light sensor and filed one of the earliest digital camera patents (US4057830A). In 1975, Kodak engineer Steven Sasson built a digital camera and secured a patent for it in 1978 (US4131919A), earning recognition as the inventor of the digital camera. Unfortunately, Kodak failed to appreciate the potential of this innovation and ultimately fell behind in the digital era.

In August 1981, Sony released the MAVICA, an electronic camera that stored analog video signals on magnetic disks and displayed them on standard televisions—the first truly commercialized digital camera. In 1994, Apple launched the QuickTake line of consumer digital cameras, manufactured by Kodak and Fujifilm, priced between $600 and $749. This marked the entry of digital cameras into the mass market. At that point, more than 100 years had passed since the discovery of the photoelectric effect, 35 years since the invention of CCD, and 22 years since

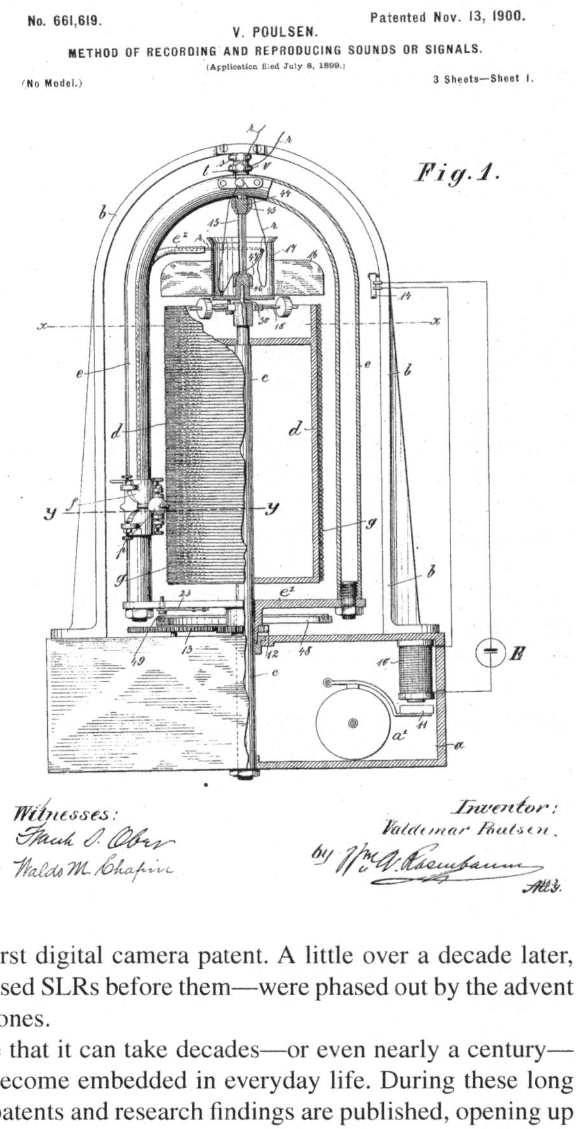

Fig. 37.2 Patent drawing of U.S. Patent No. 661619A (Poulsen's magnetic recorder)

Texas Instruments filed its first digital camera patent. A little over a decade later, digital cameras—like film-based SLRs before them—were phased out by the advent of camera-equipped smartphones.

These stories demonstrate that it can take decades—or even nearly a century—for a scientific principle to become embedded in everyday life. During these long gestation periods, countless patents and research findings are published, opening up endless opportunities for industry. The widespread adoption of a technology can bring transformative change to society.

Take magnetic tape technology, for example. Oberlin Smith's paper laid the theoretical foundation by explaining how audio signals could be converted into magnetic ones. Poulsen's invention marked the first known application of this principle. Pfleumer's patent introduced the first magnetic tape, and Philips established the industry standard for cassette tapes. Compared to these foundational contributions, Sony's Walkman patents seem modest—mostly improvements to magnetic head design or motor placement. Yet it was the Walkman that fundamentally altered lifestyles and entertainment habits. In fact, the Walkman debuted in 1979—just 1 year before the invention of the audio CD in 1980. At that point, cassette tapes had yet to reach mainstream popularity. One wonders: had Sony never invented the portable cassette player and instead gone straight to a CD Walkman, would cassette tapes have faded into obscurity as a failed format?

This raises a profound question: How many promising technologies lie forgotten in scientific literature, never reaching commercialization? How many viable innovations are still buried in archives, waiting for a visionary to bring them to light?

In Liu Cixin's *The Three-Body Problem*, he imagines a terrifying scenario: alien interference from the Trisolarans blocks humanity's progress in fundamental physics, effectively freezing scientific advancement. This setup evokes a deep sense of fear and despair. The recent semiconductor crisis—sparked by U.S.-China trade tensions—has offered China a glimpse of such dread. Yet China's resilience in high-tech fields will not be broken. With Chinese ingenuity and a steadfast pursuit of science, overcoming such technological blockades is only a matter of time.

Among all science fiction works, perhaps the most chilling vision comes from Britain's Games Workshop in the *Warhammer 40,000* universe. In this grimdark setting, human civilization has lost science and retained only technology. All scientific knowledge is monopolized by the Adeptus Mechanicus—a techno-religious cult that interprets every technical action as a sacred ritual. From building starships to maintaining firearms, every procedure is a prayer to the "Machine Spirit." Soldiers clean their guns like modern armies—but in *Warhammer*, they also chant prayers and genuinely believe that wiping a barrel is a form of divine communion.

This fusion of science and superstition is among the darkest and most fascinating aspects of the *Warhammer 40 K* lore. Here, the death of science is not caused by external forces, but by internal regression. Though absurd, *Warhammer's* dystopia resonates with historical precedents: countless civilizations have turned inward and rejected innovation, choosing dogma over discovery. Such outcomes must be avoided at all costs.

In the chapters to come, we will explore stories of scientific innovations making their way into industry through the lens of patents. These accounts will highlight the critical role that science plays in helping companies break through boundaries and forge new frontiers.

Chapter 38
The Long Road from Scientific Innovation to Productization (Part I)

The scientific exploration of semiconductors began in 1833, when Michael Faraday studied the conductivity of silver sulfide. In 1879, Edwin Hall discovered the "Hall effect," noting the presence of positively charged carriers in certain semiconductor materials. Starting in 1907, Konigsberger and others began applying the Hall effect to semiconductors. In 1925, Julius Lilienfeld proposed the concept of the field-effect transistor (FET), securing a patent for his invention (US Patent No. 1745175A, titled "Method and Apparatus for Controlling Electric Currents"), though he never brought it to market. In the 1930s, British physicist Alan Herries Wilson explained the function of band gaps using solid-state quantum theory. Then, in 1947, Bell Labs' William Shockley, John Bardeen, and Walter Brattain invented the transistor and filed multiple patents. This invention sparked rapid development in the semiconductor industry, leading to integrated circuits (ICs), charge-coupled devices (CCDs), and microprocessors—cornerstones of the Third Industrial Revolution.

The transistor is the foundation of modern electronics and one of the most important inventions of the twentieth century. It is the key component of the semiconductor industry and the basis of technologies such as ICs. Bill Gates once said, "If I could travel through time to witness a moment of invention, my first stop would be Bell Labs in December 1947."[1]

Our story begins with the transistor's predecessor. Before the transistor was invented, many of its core functions were performed by vacuum tubes. Though vacuum tubes are not part of semiconductor technology, they were a critical antecedent. The generally recognized inventor of the vacuum tube is American physicist Lee De Forest.

Over his lifetime, De Forest was granted more than 180 patents and became known as the "Father of Radio" and the "Father of the Electronic Age." He ventured into business early on and founded his own company, but unlike Thomas Edison, he

[1] 格特纳. (2016). 贝尔实验室与美国革新大时代. 中信出版集团股份有限公司. p9

Q. Yue, F. Yang, *Governing Innovation: A Global History of Patent-Driven Technological Progress*, https://doi.org/10.1007/978-981-95-5570-3_38

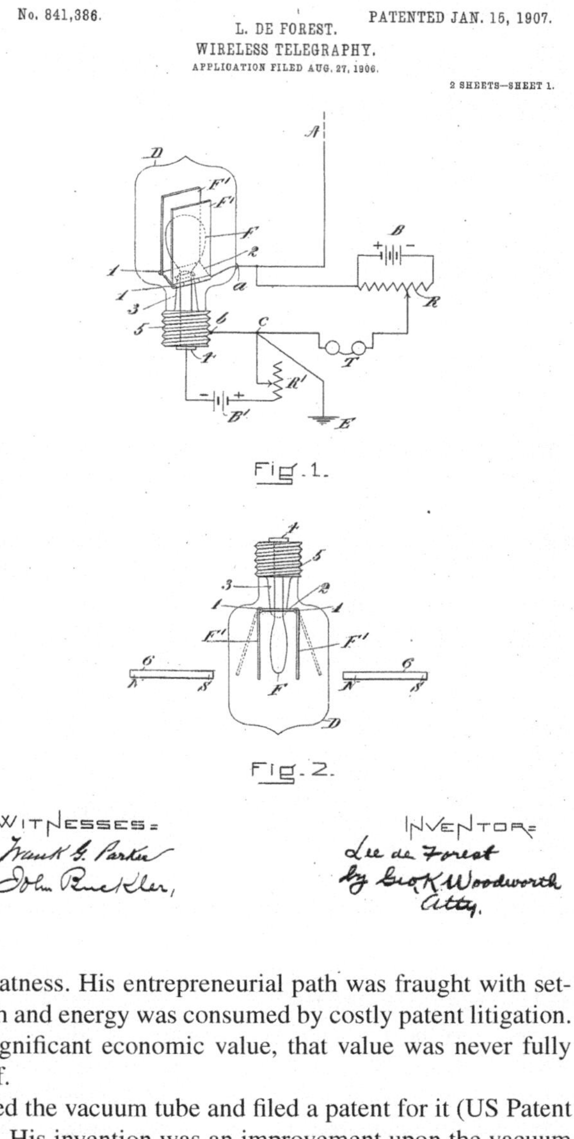

Fig. 38.1 Patent drawing of U.S. Patent No. 841386 ("Wireless Telegraphy," De Forest)

never achieved industrial greatness. His entrepreneurial path was fraught with setbacks, and much of his wealth and energy was consumed by costly patent litigation. Although his patents held significant economic value, that value was never fully realized by De Forest himself.

In 1906, De Forest invented the vacuum tube and filed a patent for it (US Patent No. 841386A; see Fig. 38.1). His invention was an improvement upon the vacuum diode created by British physicist John Ambrose Fleming (US Patent No. 803684A; see Fig. 38.2), by adding a grid to the existing design. Fleming's diode could only convert alternating current (AC) into direct current (DC), while De Forest's triode introduced amplification capability—a foundational feature of modern transistors.

Unlike Edison, who was a self-taught inventor, De Forest was academically trained, holding a Ph.D. in physics from Yale University. Yet even he didn't fully grasp the physical principles behind his invention. He once confessed, "I don't

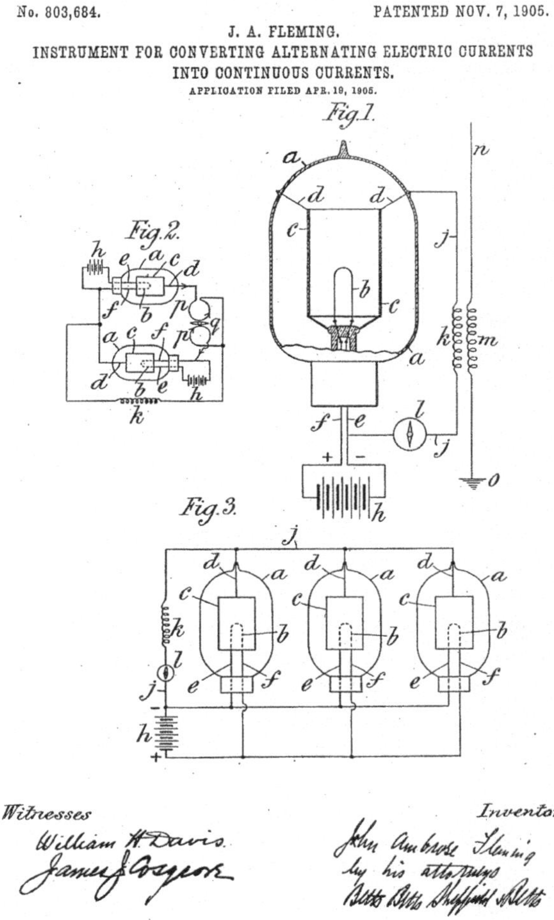

Fig. 38.2 Patent drawing of U.S. Patent No. 803684 (Vacuum diode, Fleming)

know why it works, but it does."[2] Technology historian Tom Lewis wrote, "Though the discovery was accidental, and though he barely understood why his vacuum tube worked, it was his invention nonetheless."[3]

Not only did De Forest fail to understand why the vacuum tube worked, he also didn't foresee what it could be used for. In fact, the term "vacuum tube" wasn't coined by him. Though US Patent No. 841386A is widely recognized as the earliest vacuum tube patent, its title is simply "Wireless Telegraphy." The patent specification opens with: "My invention relates to wireless telegraphy, and more specifically

[2] Paul Litwinovich. (2013). Vacuum in a Tube Enabled Sound from a Box WSHU | By Paul Litwinovich. Retrieved March 31, 2024 from WSHU Public Radio Group Website: https://www.wshu.org/post/vacuum-tube-enabled-sound-box

[3] Adams, M. (2012). *Lee de Forest: king of radio, television, and film*. New York: Copernicus Books. pp98

to an oscillation detector, which may also be used as a tuning device and a static valve, among other applications." In the description, De Forest named his oscillation detector the "audion," derived from the word "audio," emphasizing its amplification of sound frequencies.

In fact, his audion was not a true vacuum tube. The device was filled with low-pressure gas. Claims 13–20 of the patent all specify "a container sealed with a gaseous medium." Because of this gaseous environment, De Forest's audion products were plagued by instability: some worked fine, others didn't function at all. De Forest was baffled. It wasn't until Irving Langmuir, a renowned physical chemist at General Electric, conducted further research that a breakthrough was made—audions worked far better when fully evacuated. In October 1913, GE applied for patents (e.g., US Patent No. 1558437A) to protect this improvement, which specified that "the discharge device is exhausted to the highest vacuum obtainable" and covered "a discharge device... comprising a highly evacuated container." Thanks to Langmuir's work, the vacuum tube truly became a *vacuum* tube.

After De Forest's original vacuum tube patent (US Patent No. 841386A) was granted, he initially used the tube solely to manufacture radio detectors. His company secured some military contracts but saw only modest returns. It was AT&T that first recognized the vacuum tube's true potential. Engineers at AT&T observed that De Forest's invention could amplify telephone signals and enhance long-distance call quality—it could serve as a telephone line repeater. As a result, AT&T approached De Forest with an offer to purchase his seven earliest vacuum tube patents for $100,000.

However, AT&T's acquisition was far from noble. At the time, De Forest was entangled in a stock fraud scandal, deeply in debt, and in desperate need of funds. Learning of his plight, AT&T deliberately stalled the deal, waiting until De Forest was cash-strapped and on the verge of bankruptcy. Then, through a lawyer posing as a representative for an "anonymous client," they offered to buy the patents—at half the original price, just $50,000. Backed into a corner, De Forest had no choice but to accept the cut-rate deal. Later, AT&T would continue to purchase several of his radio-related vacuum tube patents.

AT&T improved upon De Forest's vacuum tube and deployed it widely across its two core businesses: telephony and radio. In 1915, the company enabled the world's first transcontinental telephone call using vacuum tube technology. Later that same year, it achieved transoceanic radio transmission from mainland America to Paris and Hawaii, again with vacuum tube support.

When selling his patents, De Forest granted AT&T the exclusive license for the professional telephony and telegraph markets, retaining rights only for non-professional and experimental equipment. Yet ironically, it was AT&T's recognition of the tube's potential that drove demand for the technology, lifting De Forest's company into a period of growth. Sadly, in his efforts to increase the value of his remaining patents, De Forest soon found himself embroiled in a new patent lawsuit—and much of the money he earned went straight into the pockets of patent attorneys.

In 1913, Edwin Howard Armstrong, building on vacuum tube research, invented the regenerative circuit and was granted a patent for it in 1914 (US Patent No. 1113149A). At that time, De Forest was preparing to sell a portfolio of patents to Western Electric, AT&T's subsidiary. Hoping to raise the selling price, he quickly filed multiple patents related to regenerative circuits and initiated an interference proceeding against Armstrong's patent application. U.S. patent law at the time did not include modern invalidation procedures like IPR; when two inventors claimed the same invention, a formal "interference" proceeding was used to determine priority.

De Forest submitted lab notes showing that he had discovered the basic principles of regenerative circuits as early as 1912 and had used them in radio receivers for the Federal Telegraph Company by early 1913. This may have been true, but De Forest fundamentally misunderstood the circuit's working principles and failed to grasp its significance—until he read Armstrong's patent. In addition to De Forest's notes and witnesses, his legal team uncovered damaging evidence against Armstrong: similar findings had been made earlier by Irving Langmuir of General Electric and Alexander Meissner of Germany.

After years of legal tug-of-war, De Forest ultimately won the case at the U.S. Supreme Court. But the lawsuit had dragged on so long that, by the time the final verdict was handed down, De Forest had sold off most of his company's shares, and the majority of his earnings had gone toward legal fees. As biographer Mike Adams noted, "De Forest won the seven-year lawsuit... Although De Forest Radio Company's annual income had reached $500,000 due to the recent court victory, Dr. Lee De Forest, the true inventor of the radio-frequency triode, was left with a personal income of only $15,000 per year."[4]

Lee De Forest, hailed as the "Father of Radio" and the "Father of the Electronic Age," was one of the most important American inventors after Edison. He also invented sound-on-film motion pictures and received an honorary Academy Award in 1960. Yet unlike Edison or Alexander Graham Bell, he never built a corporate empire. Though his inventions formed the foundation of trillion-dollar industries, he died with just over $1000 in his bank account.

His long-time adversary, Edwin Armstrong, also suffered deeply from patent litigation. To fund his legal battle against De Forest, Armstrong licensed his regenerative circuit patent to 17 companies and later sold it to Westinghouse Electric Corporation in 1920 for $335,000. But like De Forest, most of his earnings were devoured by legal costs. Armstrong also spent years in court battling RCA. In 1954, after a dispute with his wife, he died by suicide. Neither of these two pioneers met a peaceful end, while AT&T and RCA profited handsomely from their patents.

Unexpected new applications for the vacuum tube continued to emerge. Forty years after its invention, John von Neumann's team used 17,468 vacuum tubes to

---

[4]Adams, M. (2012). *Lee de Forest: king of radio, television, and film*. New York: Copernicus Books. p297

build the ENIAC—the world's first general-purpose computer—ushering in the computer age.

De Forest's story illustrates that inventors of foundational technologies don't always understand the full impact of their creations. Even the most significant patent must be paired with industrial application to realize its market value. Whether it was the vacuum tube itself or the regenerative circuits built on it, the inventors were unable to fully anticipate their practical use—and so they missed out on the economic rewards. De Forest's contribution was undeniable, but so too were the contributions of Armstrong, Langmuir, and AT&T. AT&T's relentless development efforts enabled vacuum tubes to flourish across multiple industries. ENIAC's innovative use of vacuum tubes went far beyond anything De Forest could have envisioned, yet its significance rivaled that of the original (and ironically, non-vacuum) vacuum tube patent filed in 1906.

Modern technological history is filled with inventors like De Forest and Armstrong who suffered under the weight of patent disputes. Their stories are often used to criticize the intellectual property system, which is said to benefit corporations, publishers, and capitalists more than inventors and creators. But history also includes many inventors who reaped both fame and fortune: Edison, Alexander Graham Bell, and Chester Carlson (inventor of xerography) were all major beneficiaries of the patent system. These stories show that the economic and social value of patents ultimately lies in product realization. Groundbreaking innovation is crucial—but bringing that innovation to market is what makes all the difference.

Just like the vacuum tube, the transistor patent did not bring substantial financial rewards to its inventors. Instead, it flourished in the hands of outside corporations—what the Chinese proverb might call "a flower blooming inside the wall but fragrant outside."

If the invention of the vacuum tube could be attributed, at least in part, to serendipity, then the transistor was something else entirely. Developed at Bell Labs under AT&T, the transistor was firmly grounded in solid scientific theory. Yet, despite this, Bell Labs had little interest in the commercial prospects or economic value of the invention. As one historian of science noted, "As long as transistors could be manufactured, they could be sold. Even if no one else bought them, the vast Bell empire alone was market enough."[5] This was because, to AT&T, Bell Labs served not just as a research arm for core business operations but also carried another crucial responsibility.

In the first few decades after its founding, AT&T not only used its patent monopoly to dominate the telephone market, but also employed unscrupulous tactics to suppress competitors. These included sabotaging rivals' telephone lines and bribing their suppliers. At one point, AT&T became a public pariah in the U.S., widely regarded as "the enemy of the people." Its cutthroat acquisition of patents from De Forest only worsened its reputation, devoid of any semblance of corporate decorum. Moreover, AT&T long refused to transmit calls from competing carriers across its

---

[5] 格特纳. (2016). 贝尔实验室与美国革新大时代. 中信出版集团股份有限公司. p86

own lines. This recalls the sentiment once popular among Chinese internet users in the early 2000s: "The greatest distance in the world isn't between the ends of the Earth, but between Telecom and Unicom." Back then, cross-network data traffic was agonizingly slow. Two computers using different ISPs—even if housed in the same office building—communicated with glacial latency. Early twentieth-century America suffered a similar plight: many households had to install two or three phones just to reach contacts on different phone networks.[6] And in this fractured market, AT&T—the dominant player—was easily the most despised.

Because of this, the U.S. government's antitrust regulators hounded AT&T for over half a century, with a major legal battle flaring up nearly every decade. To ease the pressure, Bell Labs, a subsidiary of AT&T, frequently offered its cutting-edge technologies to the U.S. military "not for profit, but to cement its indispensability to the federal government." AT&T even founded a lab—Sandia National Laboratories—that had nothing to do with its telecom business. Instead, it focused on nuclear weapons research. Employees at Sandia were often bewildered, asking: "Why the hell are we working on this? It has nothing to do with our mission." The response from upper management? "It will be useful once we get hit with another anti-trust suit."[7]

Thus, although Bell Labs was under AT&T's umbrella, it paid little attention to the transistor's commercial future. As part of an earlier antitrust settlement, AT&T offered transistor licenses to any interested party—for a one-time fee of $25,000.

Initially, transistor applications mirrored those of vacuum tubes, finding early use in the military. Military contracts dominated early demand and spurred advancements in semiconductor technology. "Military requirements largely led to the replacement of germanium by silicon as the preferred material for transistors, due to its superior resistance to thermal drift."[8] Scientific exploration had to be aligned with practical application, with constant adaptation needed to find the right direction.

The transistor's inventors—William Shockley, Walter Brattain, and John Bardeen—were jointly awarded the Nobel Prize in Physics in 1956. Among them, Shockley stands out. In a community of Nobel laureates typically known for their academic devotion, Shockley became infamous as a controversial figure: a vocal eugenicist and racist.

By research credentials, Shockley was universally recognized as a genius. But he was also notoriously irritable, paranoid, and jealous—supremely self-confident and obsessively controlling. At his startup, he evaluated employees and candidates using IQ tests, earning his company a notorious reputation.[9] His suspicion of others ran so deep that he once forced all staff to take lie detector tests over trivial workplace incidents, sparking widespread resentment.

[6] 格特纳. (2016). 贝尔实验室与美国革新大时代. 中信出版集团股份有限公司. p10

[7] 格特纳. (2016). 贝尔实验室与美国革新大时代. 中信出版集团股份有限公司. p10

[8] 约翰·奥顿. (2015). 半导体的故事. 中国科学技术大学出版社. p87 "在20世纪50年代的大部分时间里，年轻的晶体管工业正是依靠军方的支持才存活了下来".

[9] 格特纳. (2016). 贝尔实验室与美国革新大时代. 中信出版集团股份有限公司. p262

Since the fall of Nazism, eugenics and racism have become profoundly taboo in Western society. Yet after his business ventures failed, Shockley descended into a near-obsessive focus on genetics, race, and intelligence. He accused the American academic establishment of neglecting racial intelligence studies, publicly claimed that "the severe deficiency of Black intelligence is primarily genetic," and proposed a voluntary sterilization program for all individuals with IQs below 100. In his seventies, Shockley grandiosely announced he had donated sperm to a Nobel laureates' sperm bank, touting eugenics, and even told *Playboy* that his children had "obviously regressed" in intelligence compared to him—blaming it on the inferiority of their mother's genes.

From this description, one might imagine Shockley as an eccentric recluse or cloistered academic. But in fact, of the transistor's three inventors, Shockley was the only one who recognized its massive commercial potential and took concrete steps to pursue it. He founded the world's first company focused on designing and manufacturing transistors as its core business.[10] By contrast, Brattain and Bardeen failed to grasp the true significance of their invention. After creating the transistor, both men shifted their focus to other fields.

Brattain later recalled visiting Egypt in the early 1960s and seeing a camel herder listening to a transistor radio. It was only then—realizing that the high-tech component he helped create in a lab had journeyed across oceans and reached one of the world's most ancient lands—that he truly appreciated the revolutionary significance of the transistor.[11] Bardeen, at a celebration marking the transistor's 25th anniversary, said: "We knew we were working on something very important and that it would have many uses... but I had no idea what a revolution it would cause—especially the advent of large-scale integrated circuits."[12]

Among the three Nobel Prize-winning scientists, only Shockley foresaw the transistor's enormous economic value. He left Bell Labs and started his own company.

When Shockley registered his company, he originally intended to focus on the production of double-diffused silicon transistors. This was a potentially profitable path, as double-diffused transistors were a Bell Labs technology. But as a scientist, Shockley was driven by a desire for originality. Determined to pursue his own ideas, he redirected the company toward developing and commercializing his own invention—the four-layer diode.

However, his poor temperament and the prolonged, unprofitable R&D process wore down his team. His eight most talented employees eventually resigned en masse and went on to found Fairchild Semiconductor, a company that would become a pillar of the American semiconductor industry and a cornerstone of Silicon Valley.

---

[10] 格特纳. (2016). 贝尔实验室与美国革新大时代. 中信出版集团股份有限公司. p260

[11] Gordon R. Gore. (1998). Playing with bubbles. *The Physics Teacher* 36, 502 (1998);

[12] Riordan, M., & Hoddeson, L. (1997). *Crystal fire: The birth of the information age*. WW Norton & Company. p279

These eight men came to be known as the "Traitorous Eight," and among them were Robert Noyce and Gordon Moore—the future co-founders of Intel.

Scholars have described Shockley's missteps as follows:

> As a theoretical scientist, Shockley viewed manufacturing as an auxiliary to scientific research. But theory was no longer the most critical element of semiconductor technology. Manufacturing required mechanical and engineering prowess, along with the willingness to proceed even in the absence of complete scientific understanding. Shockley's academic background and research experience blinded him to this fact... His company prioritized research, while outsourcing production.[13]

By around 1960, the semiconductor industry had become a booming, high-profit sector. Yet Shockley's company remained fixated on its founder's research, ignoring sales and production, and ultimately failed to turn a profit. Investors lost patience, the company was sold, and it was eventually dissolved.

By that time, transistors had already entered the civilian market. Around 1952, thanks to the invention of the diffusion transfer method, transistor costs had come down, and they found early application in medical devices—specifically hearing aids. Given the unique pricing elasticity of the medical industry, hearing aid manufacturers were willing to absorb the high cost of early transistors.

At the end of 1952, a New York-based hearing aid company, Sonotone, released the Sonotone 1010 model (see Fig. 38.3), priced at $229.50. It was the first commercial transistor hearing aid in history, and one of the earliest commercial applications of the transistor. Because transistors at the time had poor signal-to-noise ratios and produced significant noise, the Sonotone 1010 still relied on two miniature vacuum tubes for signal reception, using a single transistor only at the output stage. In terms of size, the device was comparable to earlier vacuum tube models, weighing approximately 88 g, but it consumed only half the power. As transistor manufacturing improved, more manufacturers began releasing all-transistor hearing aids, quickly replacing vacuum tube hearing aids that had dominated the market for over 30 years.

In 1954, Patrick Haggerty, president of Texas Instruments (TI), took a bolder step. He decided to invest $2 million in the development of a portable transistor radio—a high-risk move, considering TI's total annual revenue was only $25 million and it had no experience in the consumer market.

Texas Instruments had evolved from Geophysical Service Inc. (GSI), a company founded in 1930 to conduct oil exploration using seismic reflection techniques. During the 1940s, GSI leveraged its expertise in signal processing to develop

[13] Holbrook, D., Cohen, W. M., Hounshell, D. A., & Klepper, S. (2000). The nature, sources, and consequences of firm differences in the early history of the semiconductor industry. *Strategic Management Journal*, *21*(10–11), 1017–1041. Retrieved March 31, 2024 from ResearchGate: https://www.researchgate.net/profile/Daniel-Holbrook-2/publication/246850964_The_nature_sources_and_consequences_of_firm_differences_in_the_early_history_of_the_semiconductor_industry/links/63c82668d9fb5967c2e7620c/The-nature-sources-and-consequences-of-firm-differences-in-the-early-history-of-the-semiconductor-industry.pdf

Fig. 38.3 The Sonotone 1010 hearing aid, produced by Sonotone. (Source: Karitxa. (2019, December 29). Sonotone 1010 hearing aid (1952) front view with ruler [Image]. Wikimedia Commons. https://commons.wikimedia.org/wiki/File:Sonotone_1010_hearing_aid_(1952)_front_view.jpg. This image is shared under the Creative Commons CC BY-SA 4.0 license)

submarine detection equipment for the U.S. military, which proved highly profitable. Its "Laboratory and Manufacturing Department," responsible for electronics, ended up outpacing its oil business. In 1951, this department was spun off and reorganized as General Instrument. However, due to a naming conflict with an existing company, it was soon renamed to the now-familiar Texas Instruments.

TI was among the first to recognize the potential of the transistor. In 1952, it paid $25,000 to Western Electric for a license to manufacture and sell transistors. In 1954, after Bell Labs made breakthroughs in silicon transistor technology, TI absorbed the results and became the first company to produce and market silicon transistors. To broaden the transistor's market appeal, TI decided to incorporate it into a popular consumer product—a radio—and use it to showcase the transistor's promise to society at large.

Though TI quickly built a prototype, mass production proved challenging. Lacking manufacturing capacity, TI approached major companies like RCA and Philco to partner on production, but all declined.

Eventually, a company named IDEA took on the project—based on a mistaken assumption. IDEA president Tudor believed that nuclear war between the U.S. and the Soviet Union was inevitable. In such a post-apocalyptic world, he reasoned, portable transistor radios would be essential survival tools. He predicted sales of 20 million units within 3 years. However, nuclear war never came, and miniaturizing the radio proved far more difficult than he had anticipated.

At that time, "miniaturization" was still a foreign concept in the electronics industry. No American suppliers had experience manufacturing such small components. Many vendors, fearing damage to their reputations from possible quality issues, refused TI's orders outright. But after much persuasion, a few pioneering suppliers stepped up: Radio Condenser Company provided the dual variable capacitor; Chicago Telephone Supply built miniature volume control knobs; Jensen Speaker made 2.75-inch speakers; and International Electronics, Inc. set up a dedicated line to produce miniature electrolytic capacitors. None of these firms had ever made products at such small scales before. At TI's 50th anniversary celebration in 1980, IDEA president Tudor noted, "It was the Regency radio that launched the market for miniature discrete components."

Printed circuit boards (PCBs) also found their way into the mainstream thanks to the transistor radio. Austrian inventor Paul Eisler had patented the PCB in 1943, and the technology had been used in World War II for proximity fuzes. After the war, the government declassified it for civilian use. Yet for the next decade, the technology languished—until the push to miniaturize transistor radios finally gave PCBs the industrial foothold they needed.

In November 1954, the first commercial transistor radio—the Regency TR-1—was launched in the United States (see Fig. 38.4), priced at $49.95 (equivalent to approximately $450–500 today). The TR-1 was not heavily patented; its primary intellectual property came from chief engineer Dick Koch's circuit design, which was granted a patent (US2892931A; see Fig. 38.5).

At first glance, the invention seems unremarkable—just an ordinary circuit patent—but its impact on the industry may have surpassed that of many contemporary transistor patents. During negotiations between IDEA and Texas Instruments over patent rights, attorney Frank Mascarich noted that while most transistor-related patents at the time focused on the components themselves, the Regency radio was

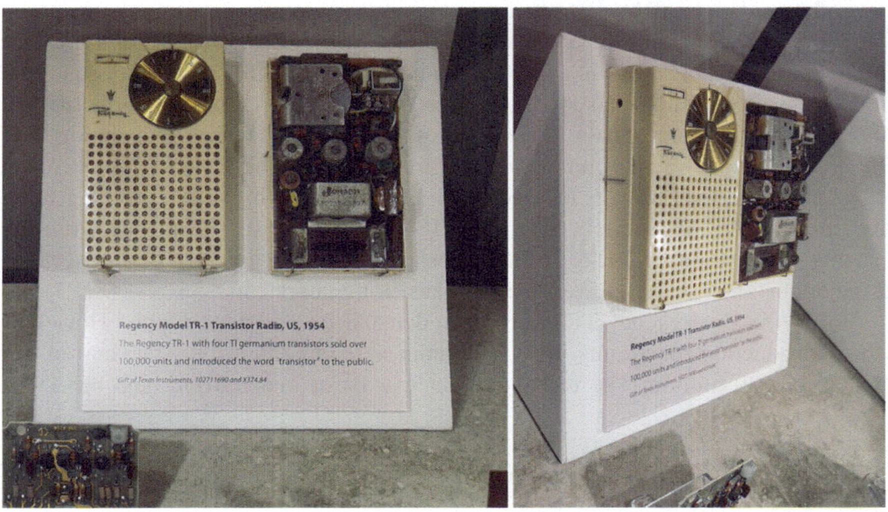

Fig. 38.4 Regency TR-1 radio. (Photograph by the author at the Computer History Museum, USA)

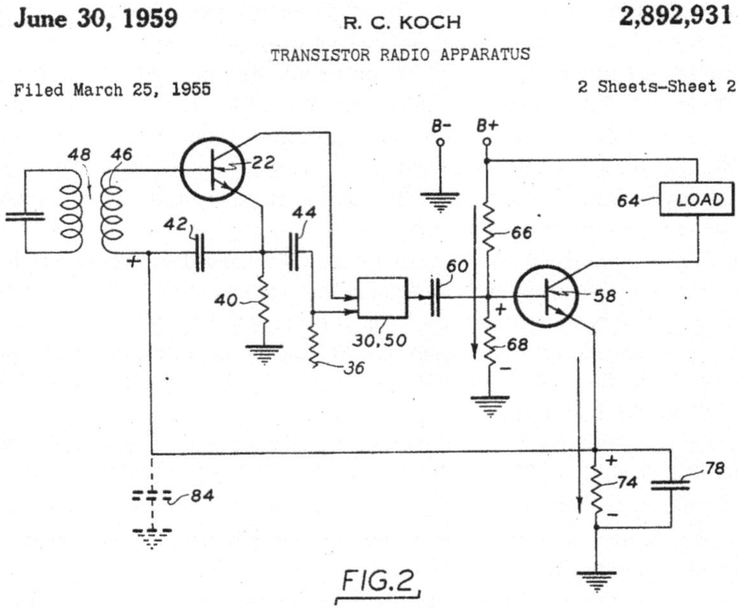

Fig. 38.5 Patent drawing from US2892931A patent

among the few patents concerning actual applications of the transistor. He remarked, "*Regency was a thorn in the side of the industry which in-turn stimulated the practical application of the transistor.*"[14]

One person particularly "stimulated" by this thorn was Thomas J. Watson Jr., the second president of IBM. IBM had recognized the importance of transistors early on. John von Neumann's first patent with IBM (US2815488A, filed in 1954) explicitly mentioned using "crystal triodes" to enhance functionality. After the launch of the TI transistor radio, Watson Jr. purchased several hundred units and distributed them among IBM executives. He also pressured the R&D team, mandating that IBM switch from vacuum tubes to transistors starting June 1, 1958. When some engineers objected, Watson Jr. reportedly silenced them by simply holding up a Regency TR-1.[15] Throughout the 1950s, IBM filed at least 179 transistor-related patents—second only to AT&T's Bell Labs and RCA (see Fig. 38.6)—and became a major customer of Texas Instruments' transistor components.

Following the Regency TR-1's release, demand far exceeded supply. However, due to production constraints, only 1500 units were initially made—far short of the anticipated 20 million. The feared nuclear war never materialized, and the TR-1

[14] Don Pies. (1998). Regency's Development of the TR-1 Transistor Radio. Retrieved March 31, 2024 from http://regencytr1.com/Regency_Early_Years.html

[15] Don Pies. (1998). Regency's Development of the TR-1 Transistor Radio. Retrieved March 31, 2024 from http://regencytr1.com/Regency_Early_Years.html

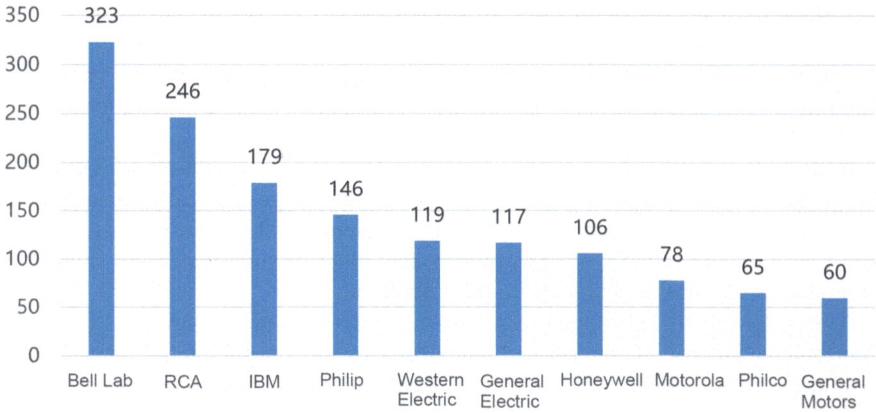

Fig. 38.6 Top 10 U.S. applicants for transistor-related patents in the 1950s

never became an essential survival tool. In the end, only about 100,000 units were produced. Perhaps due to underwhelming profits, TI eventually withdrew from the product line, ceding the transistor radio market. Sony would later seize the opportunity and rise to dominance. Nevertheless, the success of the TR-1 helped Texas Instruments establish a strong foothold in transistor component sales, and the company remains one of the world's leading semiconductor manufacturers to this day.

Because of Sony's dominance in the consumer electronics market, many Americans today mistakenly believe that Sony invented the transistor radio, while Texas Instruments is more commonly associated with calculators.

In 1960, Bell Labs researchers Mohamed M. Atalla and Korean-born scientist Dawon Kahng developed the metal–oxide–semiconductor field-effect transistor (MOSFET), receiving patents US3206670A and US2953486A, respectively. The MOSFET could operate at extremely low currents and voltages, becoming a foundational element in large-scale integrated circuits and microprocessors. However, neither integrated circuits nor microprocessors were Bell Labs inventions.

In the 1960s, Jack Kilby of Texas Instruments filed the earliest patents for the integrated circuit. These patents allowed TI to launch legal actions against companies like Westinghouse Electric, Sprague, and Fairchild Semiconductor, earning it a 1970s-era reputation as "the Dallas law firm".[16] As discussed in the "Rise and Fall" chapter, TI saved itself from bankruptcy in the 1980s largely through patent litigation. Amid sweeping reforms to the U.S. patent system, patent licensing revenue became an increasingly attractive source of cash for tech executives. Still, this is not a business model we endorse.

[16] Ken Shirriff. (2015). The Texas Instruments TMX 1795: the (almost) first, forgotten microprocessor. Retrieved from Ken Shirriff's Blog: https://www.righto.com/2015/05/the-texas-instruments-tmx-1795-first.html

Just as he had done with the transistor radio, TI president Patrick Haggerty wanted to showcase the power of integrated circuits through a new product. Like transistors, early ICs were purchased almost exclusively by the military. Their first major application came in the U.S. military's Minuteman missile development project, but this market was too limited for TI's ambitions.

In the fall of 1965, during a business trip, Haggerty shared a plane with Jack Kilby, the inventor of the integrated circuit. Haggerty suggested Kilby design a new product, offering three ideas: a pocket-sized calculator, like the transistor radio before it; a lipstick-sized voice recorder; or any other product that could showcase integrated circuit usage. Kilby chose the calculator.[17]

After returning to TI headquarters, Kilby assembled a team to refine his vision of a portable calculator. He envisioned a device with a keypad for input, neon lamps for output, battery power, and a size no larger than a book. Early development progressed smoothly except for the output system: neon lamps consumed too much power for battery use. The team ultimately opted for printed output via thermal printing technology and even filed a patent titled "Integrated Heater Element Array and Drive Matrix" (US3501615A) to protect this solution.

In 1966, Kilby's team completed a prototype featuring 18 keys, including the digits 0–9, the four basic arithmetic operations (add, subtract, multiply, divide), as well as buttons for Clear (C), Exchange (E), and Print (P). Although the device had buttons for multiplication and division, in practice it could only perform addition and subtraction—multiplication and division were carried out via repeated addition or subtraction. In 1967, Texas Instruments filed for a patent for the improved version. After a long review process, the continuation application titled "Miniature Electronic Calculator" was granted in 1974 (patent number US3819921A; see Fig. 38.7).

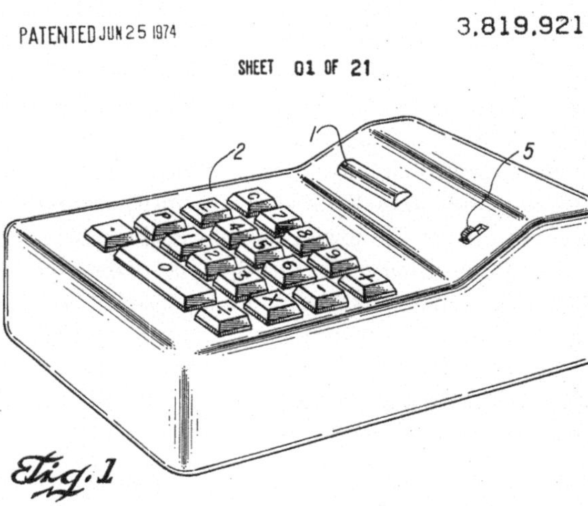

Fig. 38.7 Patent drawing from US3819921A patent

[17] Kathy B. Hamrick (1996). The History of the Hand-Held Electronic Calculator. The American Mathematical Monthly, 103(8), 633–639. doi:10.2307/2974875

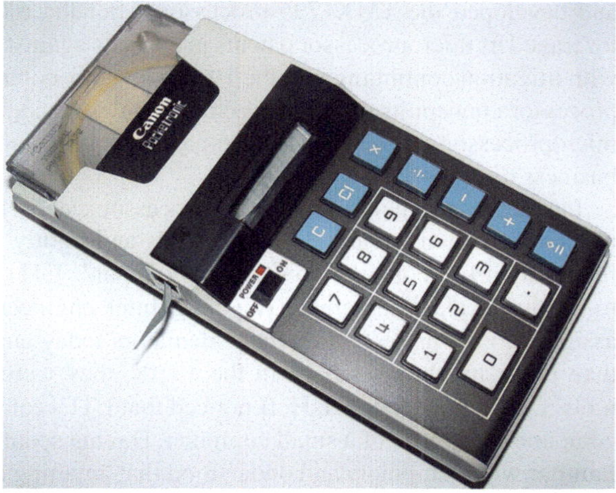

Fig. 38.8 The Pocketronic electronic calculator developed jointly by TI and Canon. (Source: Mister rf. (2016, August 13). *Canon Pocketronic, the first handheld battery-powered printing electronic calculator* [Image]. Wikimedia Commons. https://commons.wikimedia.org/wiki/File:Canon_Pocketronic.jpg. This image is shared under the Creative Commons CC BY-SA 4.0 license)

Since Texas Instruments was unfamiliar with the consumer market, it sought external partners for production and distribution. In 1970, TI collaborated with Canon to release the Pocketronic electronic calculator, which measured 8 inches in length, 4 inches in width, and 1.5 inches in thickness. It retailed for $400 and was an immediate market success (see Fig. 38.8). At the time, traditional electronic calculators were roughly the size of typewriters and averaged over $2000. Against TI's compact and affordable alternative, they stood no chance.

Today, calculators remain a stable source of revenue for Texas Instruments. This is not because TI's products are exceptional, but because it is the designated calculator supplier for the U.S. Department of Education. For nearly two decades, TI has profited from this monopoly without significant innovation. Calculators approved for use in U.S. high schools and standardized tests like the SAT typically retail for around $100, and both their appearance and functionality have remained stagnant for over a decade. According to a 2019 report, calculator sales accounted for approximately 3% of TI's total revenue, with extraordinarily high profit margins. The popular TI-83 Plus retails for $105 but costs less than $20 to manufacture.[18] TI spends about $2 million annually on lobbying, including efforts to maintain its dominance in the education sector. While we fully acknowledge TI's pioneering contribution in inventing the portable calculator, this business model leaves much to be desired.

In terms of microprocessors, Texas Instruments and Intel completed their developments around the same time. TI was first to file patents, including US3757306A ("Computer System CPU"), US4503511A ("Monolithic Integrated Circuit Multifunctional Arithmetic Logic Unit Computing System"), and US4225934A ("Multifunctional Arithmetic Logic Unit in Semiconductor Integrated Circuit"),

---

[18] Maya Kosoff. (2019). Big Calculator: How Texas Instruments Monopolized Math Class. Retrieved March 31, 2024 from Medium.com: https://gen.medium.com/big-calculator-how-texas-instruments-monopolized-math-class-67ee165045dc

and developed the TMX1795 processor. After the rise of personal computers, TI leveraged its microprocessor patents in lawsuits against Dell, Daewoo, and Samsung, with litigation continuing into the 1990s. Intel, by contrast, did not patent the microprocessor concept itself. Its 4004 chip is widely recognized as the first commercial microprocessor and served as a compass pointing countless companies toward new business directions.

Intel's development of the microprocessor clearly demonstrates that only technologies oriented toward customer needs and tightly integrated with industry can realize their full potential. In early 1969, Japan's ETI company commissioned Intel to produce calculator chips. A young Intel engineer named Marcian Hoff was assigned to the project. Although calculators today are cheap—sometimes cheaper than paper notebooks—back in the 1970s, they were considered premium office tools. During negotiations, Hoff noticed that ETI's calculator production costs were comparable to those of a small computer. Having spent years in academia, Hoff was familiar with computers and understood that anything a calculator could do, a computer could do as well. "He went around Intel asking why anyone would pay the price of a computer for something that only offered a small fraction of its functions... This line of questioning exposed both his nerdiness and his naiveté regarding the business side of things."[19]

Eventually, Hoff proposed to ETI a design for a "true general-purpose computing device on a single silicon chip"—essentially the embryo of the microprocessor. Although ETI initially showed little interest in this forward-thinking idea, Intel CEO Robert Noyce gave Hoff his full support. Only after Hoff's team completed the full design did ETI express enthusiasm, signing an exclusive agreement with Intel. Thus was born the 4004 chip. In 1973 and 1974, Intel filed patents based on the 4004: US3753011A ("Voltage-Controlled Bistable Circuit") and US3821715A ("Memory System for a Multi-Chip Digital Computer"). Unlike TI, Intel did not believe that the concept of a "general-purpose computing device on a chip" was itself patentable; these patents merely covered circuit and memory system improvements.

Soon afterward, another company—CTC—commissioned Intel to design a chip, and Hoff once again proposed the idea of large-scale integration. His team developed an 8-bit microprocessor, the 8008, but CTC lost interest. Even Intel's sales team was unenthusiastic—the company's core business at the time was memory chips, which were flying off the shelves, and the microprocessor was still an untested concept.

Thanks to Hoff's persistence, Intel decided to launch the microprocessor as a standalone product and invested heavily in media promotion. Building on the 8008, the company developed the improved 8080 chip and filed a patent for it (US4010449A, "MOS Computer Employing Plural Independent Chips"). At first, microprocessors were niche products purchased only by specialized firms or

[19] 保罗·弗赖伯格 & 迈克尔·斯温. (1985). 硅谷之火—个人计算机的故事. 中国对外翻译出版公司1985年版

hobbyist tinkerers. But it was these hobbyists who gave birth to an entirely new industry—one that would fundamentally transform human life.

One of the most important customers for the 8080 chip was IBM. In 1981, IBM released the epoch-making IBM PC, choosing the 8080 as its microprocessor. Another notable customer was a 16-year-old Bill Gates and Paul Allen, who bought an 8080 processor for $360. They designed a machine that could record traffic data and founded a company called Traf-O-Data to sell their invention—a precursor to what would later become Microsoft. When the 8080 was launched, Bell Labs was still diligently pursuing basic scientific research for humanity, while Intel, by focusing on product development, had inadvertently planted the seed for what would become the dominant Wintel alliance in the personal computing industry.

Japanese tech writer Dr. Yoshio Nishimura, in *The Rise and Fall of the Japanese Electronics Industry*, offered this insight into Intel's success:

> Intel's research and development was guided by the 'principle of minimal information'—that is, to estimate a rough answer to a problem and proceed heuristically. If that didn't work, they'd go back to square one and try another method, deliberately avoiding extensive research just to fully understand the issue. The fewer inputs needed to solve a problem, the better. This gave an edge to R&D carried out on production lines, where engineers could experiment by changing methods, adopting new approaches, or adding new steps.[20]

David Tennenhouse, who once served as deputy head of Intel's research division, also acknowledged in an interview:

> *Historically, IBM did its own research, and Intel historically did not... And much of the research we did do was intended to help us understand the other research out there, whereas IBM was trying to commercialize its own research.*[21]

These insights may explain why Intel lacked the awe-inspiring fundamental science patents of Bell Labs or IBM. But it remains, unquestionably, a great technology company.

The vacuum tube was invented by Dr. Lee de Forest, but it was AT&T and other companies that realized its massive commercial potential. The transistor was born in AT&T's Bell Labs, and it helped transform Texas Instruments from a military electronics supplier into a world-class semiconductor company. MOSFETs also originated at Bell Labs, but it was Texas Instruments and Intel—with their inventions of the integrated circuit and microprocessor—that truly unlocked their potential. Over the following decades, these inventions gave rise to many more outstanding semiconductor and electronics companies. These cases demonstrate how important scientific breakthroughs can unleash extraordinary power in the hands of visionary entrepreneurs—and may even help a company redefine itself and expand into new frontiers.

---

[20] 西村吉雄. (2016). 日本电子产业兴衰录. 人民邮电出版社2016年版

[21] Lim, K., Chesbrough, H., & Ruan, Y. (2010). Open innovation and patterns of R&D competition. *International Journal of Technology Management*, *52*(3/4), 295–321. Retrieved March 31, 2024 from Berkeley University Website: https://corporateinnovation.berkeley.edu/wp-content/uploads/2014/10/Chesbrough_etal_Rivalry.pdf

Indeed, the post–World War II rise of Japanese tech firms was often fueled by adopting American scientific discoveries and translating them into pioneering products or services. Sony was a prime example.

Sony (still known in the early 1950s as "Tokyo Tsushin Kogyo K.K.") initially sold tape recorders for the Japanese domestic market. At the time, due to the high cost of tape and electronic components, Sony's recorders were mainly purchased by schools, and sales were modest. Founders Masaru Ibuka and Akio Morita were constantly looking for new product lines. In 1948, they learned of Shockley's transistor research and became deeply intrigued.

In 1952, Western Electric, a subsidiary of AT&T, planned to license its fundamental transistor patents. Ibuka, then visiting the U.S., got wind of this and requested a meeting with Western Electric's patent manager. But at that time, "Tokyo Tsushin Kogyo" was a little-known firm, and Western Electric turned him down citing a busy schedule. Ibuka had to call on personal connections to finally gain access. The following year, Sony signed a licensing agreement with Western Electric. During discussions, Western Electric made it clear that transistors were unlikely to succeed in consumer products—the only successful civilian application at the time was in hearing aids. The licensing fee was a one-time payment of $25,000, which amounted to about 10% of Sony's total assets back then.[22]

Sony could afford the $25,000 fee, but Japan had strict foreign currency controls at the time. Spending that much in U.S. dollars required approval from the Ministry of International Trade and Industry. The officials didn't understand what a transistor was, nor did they have much faith in Sony's future. Ibuka had to lobby tirelessly, and only after 6 months did he secure permission.

Akio Morita later recalled the hardships of developing high-frequency transistors in his autobiography *Made in Japan*:

> Bell Labs' transistor used two indium dots welded to either side of a germanium plate—germanium as the cathode, indium as the anode. We figured, since electrons move faster from the cathode, reversing the polarity might generate high-frequency signals. In other words, we tried switching from a 'positive–negative–positive' (PNP) to a 'negative–positive–negative' (NPN) structure. But we couldn't find the right materials—indium was off-limits, as it was used for anodes. We tried gallium and antimony, but the results weren't promising.

Sony's young engineers attempted to make new transistors using phosphorus doping, but according to information they had, Bell Labs had already tried and failed with that method. At the time, Sony's semiconductor experts viewed American technology, especially anything from Bell Labs, as virtually gospel.

Morita wrote:

> Still, one member of our team insisted on pursuing the phosphorus doping method. As he increased the amount of phosphorus, results kept improving. During a meeting, he cautiously reported his findings—while everyone else had little progress to show. The head of the high-frequency transistor project told him, 'If you believe in this approach, forget what Bell Labs said. Go ahead and give it your best shot. If it fails, I'll take full responsibility.' The phosphorus method succeeded. Building on that, we continued our research and finally developed the high-frequency transistors we had dreamed of.

---

[22] 约翰·奥顿. (2015). 半导体的故事. 中国科学技术大学出版社. p110

Table 38.1 Sony's five U.S. transistor patents from the 1950s

Patent no.	English title	Primary function
US2959719A	Semiconductor d[373]evice	A transistor with relatively low base extension resistance
US2890159A	Method of etching a surface of semiconductor device	Method for manufacturing NPN transistors using phosphorus doping
US2950385A	Transistor oscillator	A wide-frequency-range transistor oscillator, especially useful for reducing the pull-in phenomenon in shortwave conversion circuits
US3150017A	Doping a pulled semiconductor crystal with impurities having different diffusion coefficients	Method for producing high-frequency NPN or PNP transistors
US3070465A	Method of manufacturing a grown type semiconductor device	Method for manufacturing high-frequency NPN transistors using phosphorus doping

In March 1957, Sony released its own portable transistor radio, the TR-63. All transistors and key components were designed and manufactured in-house. The product became a global hit. Smaller than Texas Instruments' Regency TR-1, the TR-63 still didn't quite fit in a pocket. So Morita devised a clever trick: he had special shirts made for all Sony salesmen—with pockets slightly larger than standard—to ensure the TR-63 would fit snugly. Priced at $39.95 (roughly $360–$400 today), the TR-63 was far cheaper than any American rival. Given Japan's low labor costs at the time, mass production allowed the price to fall rapidly. By 1962, the TR-63 cost only about $15 (approximately $150 today), causing the U.S. radio market to collapse entirely.

In the 1950s, Bell Labs applied for and was granted 356 transistor-related patents, Texas Instruments obtained 73, while Sony had a mere five U.S. patents in this field (see Table 38.1).

As Table 38.1 shows, nearly all of Sony's five U.S. transistor patents from the 1950s were geared toward transistor radios, covering phosphorus doping methods and the manufacturing processes for high-frequency transistors. Although the quantity and technical depth of Sony's semiconductor patents paled in comparison to those from Bell Labs or Texas Instruments, these seemingly modest patents were precisely what brought transistors out of the lab and into everyday homes. They transformed transistors into ubiquitous components of daily life—so common that most people were unaware of their presence.

There's no doubt about Bell Labs' contributions to basic science, but the company that sparked transformational change in industry was Texas Instruments, while Sony emerged as the ultimate victor in the consumer market. For Japan, Sony's success held especially profound significance. Riding the momentum of Sony's breakthrough, a wave of Japanese electronics manufacturers flooded the American market, decimating domestic U.S. radio producers. This early foothold helped Japan gain a strategic advantage in the coming U.S.-Japan semiconductor rivalry. Meanwhile, Sony itself gained global renown through its radio products, stepping into entirely new arenas and transforming from an obscure Japanese tape recorder manufacturer into a world-famous consumer electronics giant.

In 1959, Sony pushed the transistor further by incorporating it into televisions, releasing the world's first transistor television—the TV8-301. This was a portable black-and-white TV powered by two 6-volt lead-acid batteries, featuring an 8-inch screen and priced at $249.95. Though compact and portable, the unit was expensive and prone to malfunctions. Sales were disappointing, and the product was discontinued by 1962.[23] Still, the launch marked Sony's ambition to expand beyond transistor radios and challenge the frontier of cutting-edge consumer electronics. Before long, Sony developed its unique technology and product line—based on patented inventions from a U.S. Nobel laureate—the Trinitron picture tube.

[23] Bryan Gardiner. (2014). As Sony's TV Business Crumbles, a Look Back at its Most Iconic Sets. Retrieved March 31, 2024 from Wired Website: https://www.wired.com/2014/05/the-greatest-sony-televisions/

Chapter 39
The Long Road from Scientific Innovation to Product (Part II)

Tracing its origins, Sony's Trinitron technology can be linked back to the invention of famed physicist Ernest Lawrence. Lawrence, who won the Nobel Prize in Physics in 1939 for inventing the high-energy particle cyclotron, was a central figure in the Manhattan Project and a national treasure of the United States. The 103rd element on the periodic table, Lawrencium, is named in his honor.

In 1951, Lawrence filed patents for his invention, the "gridded color picture tube" (Chromatron, also known as the "Lawrence tube"), including US2692532A, US2614231A, and US2744952A. He also established a company to commercialize picture tube technology. This small firm boasted an impressive roster of scientific talent: Lawrence himself, Nobel Chemistry Laureate Edwin McMillan, and future Physics Laureate Luis Alvarez all served as technical advisors. At the time, the U.S. National Bureau of Standards was in the process of establishing color television standards, and Hollywood giant Paramount Pictures—eager to pivot into television—provided substantial investment to Lawrence's company. The team seemed to have every advantage: timing, resources, and expertise.

Color picture tubes operate by using electron guns to fire electron beams that excite phosphor materials on the inside surface of a display screen, producing colored images. Lawrence was arguably the most knowledgeable person in the world about electronics at the time. When developing the cyclotron, he had boldly declared, "We are going to smash the atom," and then led a team that accomplished exactly that. Alvarez and McMillan were equally formidable, with their major academic achievements also in particle physics. The word "electron" appeared 278 times in patent US2692532A and 61 times in US2614231A, a testament to Lawrence's intellectual rigor.

Unfortunately, this team of the world's most accomplished electronics experts failed to create the best picture tube. RCA's "shadow mask" imaging system ultimately prevailed over the Chromatron, becoming the industry standard.

The shadow mask system used three electron guns to fire beams through a perforated metal plate at the back of the screen, activating phosphors to produce color images. Lawrence's design, by contrast, replaced the perforated plate with a fine metal wire grid. This allowed greater beam transmission, making the resulting

Q. Yue, F. Yang, *Governing Innovation: A Global History of Patent-Driven Technological Progress*, https://doi.org/10.1007/978-981-95-5570-3_39

image more vivid and luminous. To physicists, the Chromatron was a more "elegant" design. McMillan even remarked later that it operated on the same principle as his synchrotron.[1]

Confident in their invention, Lawrence's team presented it to Paramount—but they drastically underestimated manufacturing costs. The delicate wire grid was difficult to mass-produce and prone to instability under high-voltage current. One team member later admitted that they had spent too long in government-funded labs and lacked any sense of business reality. After all, the world only needed one of McMillan's synchrotrons, so any elaborate design could be justified. But a consumer television set needed to be manufactured in the hundreds of thousands—if not millions—to be viable. After RCA's shadow mask was adopted as the industry standard, the three Nobel laureates left the company one by one. Their shares and patents were acquired by Paramount. Though the Chromatron's color quality was unmatched, it never made it to mass-market production.

In 1961—10 years after the Chromatron's first patent application—Masaru Ibuka and Akio Morita saw a demonstration of the technology at a trade show in New York. Captivated by its vivid color rendering, they immediately purchased the patents from Paramount. Back in Japan, Ibuka refined the Lawrence design, creating Sony's proprietary Trinitron system. Though Trinitron TVs were still costly, they were far cheaper than Chromatron-based sets and delivered bright, vibrant images that were especially suitable for outdoor viewing.

Akio Morita summarized Sony's Trinitron improvements in his autobiography *Made in Japan*:

> We combined three electron guns into a single unit capable of emitting three separate beams. Instead of a complex lens assembly, we used a single large lens for focusing. To further reduce size and enhance performance, we avoided both the wire grids of the Chromatron and the shadow masks used by other manufacturers. Instead, we adopted a simple metal aperture grille—a metal plate etched with many fine vertical slits.

The result was impressive. He wrote:

> Our compact Trinitron color TV carved out a niche market with no real competitors. Americans love barbecuing in the yard, enjoying lunch outside, or relaxing in hammocks. To watch television under such conditions, 'portability' and 'high screen brightness' become essential—and our product delivered both.

Figures 39.1 and 39.2 depict the wire mesh described in Lawrence's US2692532A patent and the metal plate described in Sony's US3437482A patent, respectively.

From 1968 to 1994, over 100 million Trinitron televisions were sold, and more than 200 million Trinitron picture tubes were produced and sold, with production continuing until 2008. Alongside the Walkman, the Trinitron picture tube stands as one of Sony's two most important inventions, emblematic of the shift from the stereotype of "cheap and low-quality" to high-quality Japanese manufacturing.

[1] Roebke, J. (2019). Ernest Lawrence's brilliant failure. *Physics Today*, *72*(3), 32–39. Retrieved March 31, 2024 from Physics Today: https://physicstoday.scitation.org/doi/10.1063/PT.3.4162

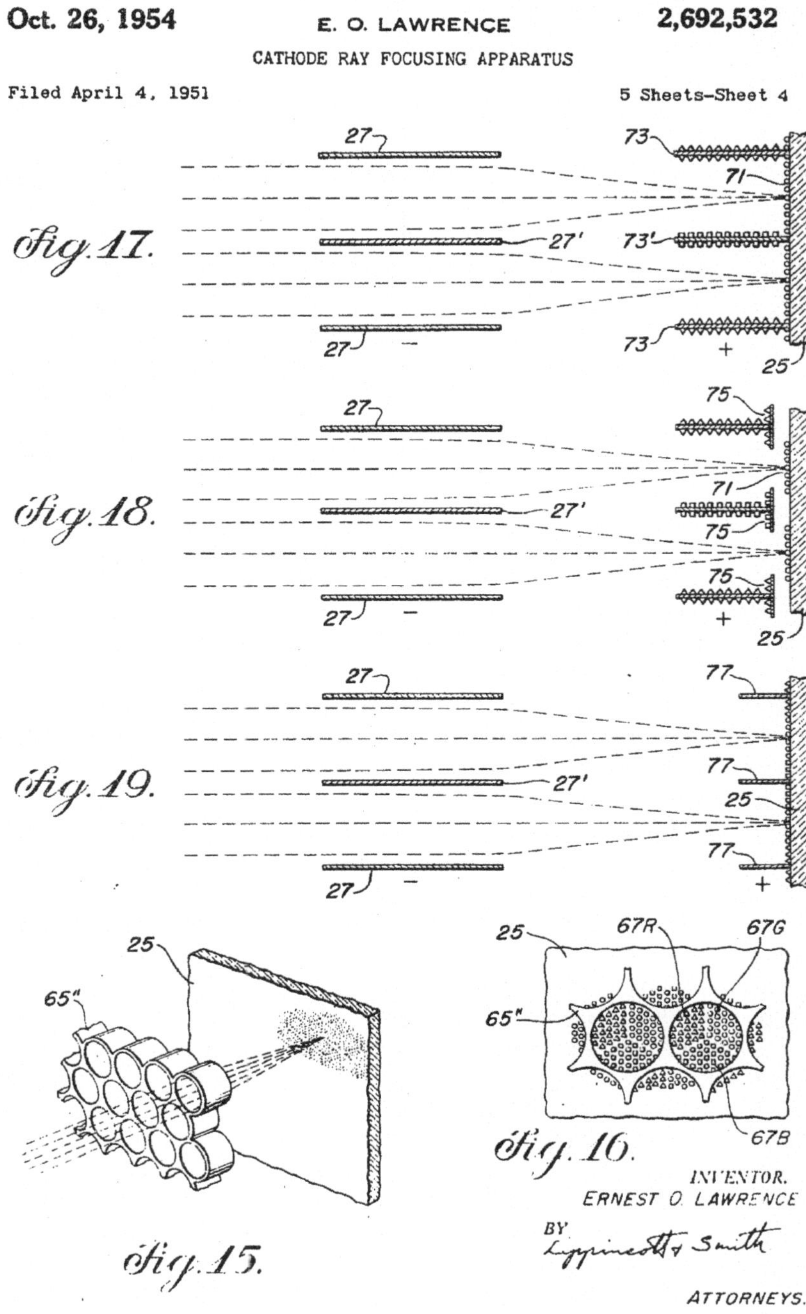

Fig. 39.1 Wire mesh from Lawrence's US2692532A patent

Fig - 9

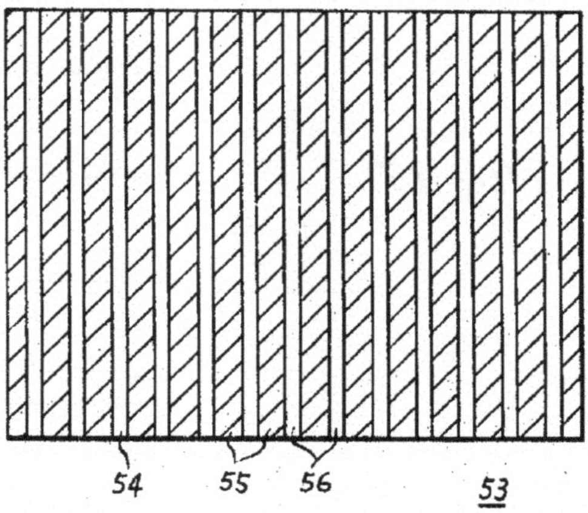

Fig. 39.2 Metal plate from Sony's US3437482A patent

The stories of Sony's transistor radio and Trinitron picture tube show that for a cutting-edge invention to truly change the world, it requires not just technological, material, and process advancements, but also a deep understanding of user needs. The transistor was a revolutionary innovation, and the Chromatron a remarkable scientific achievement, both born of the intense effort of top-tier physicists. Yet these "high and mighty" technologies only became part of everyday life through consumer products developed by a mainstream electronics company.

As noted earlier, Sony began developing its own picture tube technology in 1961, marking its expansion from audio into video. In fact, Sony's push into video was multifaceted. While its radios gained international fame, Sony originally made its mark in Japan with tape recorders, gaining considerable expertise in magnetic tape technology. At the same time it was entering the TV market, Sony also focused on video recording via magnetic materials.

As discussed in the Part "Disruption and Revolution" celebrity Bing Crosby had earlier sponsored important work on tape-based audio recording. In 1951, he asked his chief engineer John Mullin whether it was possible to record television broadcasts on magnetic tape, just like audio. Mullin believed it was theoretically feasible, and Crosby funded a lab to explore the idea. His company, Bing Crosby Enterprises (BCE), made significant breakthroughs in this area, and other companies like the BBC, Ampex, and Toshiba also made progress. However, the resulting products were generally bulky and prohibitively expensive. In 1956, Ampex released the VRX-1000—the first commercially successful video tape recorder (VTR)—which used a quad-head recording system and sold for $50,000.

The quad-head system had a major flaw: it could only play back at standard speed—it didn't support slow motion, fast-forward, or rewind. Japanese firms, including Sony, began applying for patents addressing these limitations. In 1960, Sony engineer Nobutoshi Kihara (also the lead designer of the Walkman and dubbed "Mr. Walkman" by the media) filed Sony's earliest video recording patent in the

U.S., titled "Apparatus for Combining and Separating Signals Using a Synchronous Switching Device," which proposed an all-transistor switching system for magnetic video recording. Throughout the 1960s, Kihara filed numerous video recording patents, many of which directly tackled the inherent limitations of quad-head systems. For example, in September 1965, he filed a patent titled "Apparatus for Recording and Playback of Video Signals at Normal, Slow-Motion, or Still Images" (JP40059263, later filed in the U.S. as US3509274A), clearly aiming to overcome the constant-speed-only limitation.

Ultimately, Toshiba's development of the helical-scan cassette tape recording system addressed the major flaws of quad-head VTRs and was adopted by Sony. Through this process, Sony amassed extensive technical expertise in video recording.

In August 1965, Kihara's team released the CV-2000 VTR, which used reel-to-reel magnetic tape to record television broadcasts and is generally regarded as the first home video recorder. Although the unit cost over \$695—a steep price for most households—it was clearly targeted at the consumer market. The model name "CV" stood for "consumer video," and it featured a compact design and was far cheaper than comparable offerings from other manufacturers. This product marked a pivotal step in Sony's move into the video market and laid the groundwork for the later Betamax format war. Although Sony ultimately lost that battle and VTR technology faded, it was Sony who first opened the home video recording market in 1965—a battlefield it pioneered.

In 1967, building on the CV-2000, Sony released a product that amazed the world: the DV-2400 Video Rover portable camcorder (Fig. 39.3). Just as the Walkman was essentially a portable tape recorder with headphones, the DV-2400 added one already-mature component to the CV-2000—a battery-powered black-and-white camera. This simple combination of camera, recorder, and battery created the world's first consumer-grade portable video recording system, known in

Fig. 39.3 Sony DV-2400 Video Rover. (Source: Mwf95. (2005, May 10). *Sony AV-3400 Porta Pak camera (consumer grade Portapak)* [Image]. Wikimedia Commons. https://commons.wikimedia.org/wiki/File:Sony_AV-3400_Porta_Pak_Camera.jpg. This image is shared under the Creative Commons CC BY-SA 4.0 license)

English as a "portapak."[2] Though miniature cameras and video recorders had existed for years, like the Walkman, it took a savvy entrepreneur to combine them into an easy-to-use system for everyday users.

Here we choose to use the English term *portapak* instead of the Chinese "便携式摄像机" (portable camcorder) because by the 1980s, another kind of portable video camera—the *camcorder*—had emerged. In dictionaries, both *portapak* and *camcorder* are loosely translated as "portable camcorder," but there is an important distinction between them: a *portapak* consists of a portable video camera and a separate dedicated video recorder—two separate units—because at the time, technology had not yet advanced enough to integrate them into a single portable device. The *camcorder*, which emerged in the 1980s, integrated the video tape recorder into the camera body itself, further enhancing portability.

Due to historical circumstances, China did not fully experience these two stages in the development of personal video cameras, and thus the Chinese language lacks a clear distinction between the two. The *camcorder* is also one of Sony's major contributions, which we will discuss in detail later.

To be fair, Sony's first-ever *portapak* was not very user-friendly. Its storage medium was not the cassette tape familiar to us today, but rather the older reel-to-reel magnetic tape. As discussed in the "Replacement Chapter," reel-to-reel tape required users to manually thread the tape through the heads and spools, which could easily get tangled. Additionally, the dynamic image quality of early portable cameras was poor—fast movements resulted in blurry images. The lens also couldn't be pointed directly at the sun, or it would damage the sensitive components.

Despite these flaws, the DV-2400 Video Rover was revolutionary, marking the beginning of video cameras entering everyday life. For the first time in history, non-professionals could shoulder a camera and record whatever scenes and stories they wished—in their yards, kitchens, or on city streets. The DV-2400 Video Rover and its successors hold a special place in cultural history, playing important roles in the 1960s and 1970s U.S. anti-war protests, civil rights movements, avant-garde art, and underground cinema. Works like *Button Happening* by "the father of video art" Nam June Paik, and Bruce Nauman's *Walking in an Exaggerated Manner Around the Perimeter of a Square* were filmed using Sony *portapaks*.

In recent years, the information revolution led by the internet and smartphones has ushered in an era of universal creativity. In just a decade or so, technologies such as BBS forums, novel sites, blogs, Weibo, Q&A platforms, WeChat public accounts, and short videos have continuously excited creators. Today's internet users are fortunate to live in an era of rapid technological renewal, with ever-emerging tools and platforms offering endless opportunities to creators and entrepreneurs alike. But over half a century ago, such changes were rare—a person might only witness one or two major shifts in their lifetime.

[2] Keen, B. (1987). 'Play it again, Sony': The double life of home video technology. *Science as Culture*, *1*(1), 7–42. p25

Following the release of Sony's first *portapak*, many companies began imitating Sony, developing new portable video equipment. In 1980, American inventor Jerome Lemelson received the first patent for a *camcorder*. In 1982, JVC and Sony both officially launched the first *camcorder* products—15 years after the *portapak* was introduced. By comparison, the time span from the launch of first-generation video-sharing platforms like YouTube to the rise of short video apps like TikTok (launched in 2016) is only about 11 years—during which time modern internet users have witnessed the rise and fall of blogs, twitter posts, Instagram, and WeChat Moments.

Sony's 1982 *camcorder* was sold at a discounted price and primarily targeted at journalists. As mentioned earlier, in the early 1980s Sony was aggressively promoting the Betamax format, and the Betamax-based *camcorder* became a strategic weapon in the format war. Sony launched its first consumer-oriented *camcorder* in 1983. Initial sales were decent, but plummeted after Panasonic released its VHS-format *camcorder*. The result of the format war was a classic "winner-takes-all" outcome, with the VHS camp emerging victorious and Sony's Betamax camcorders gradually fading from the scene.

Though Sony's videotape camcorders ultimately failed, the company later achieved great success in the realm of digital still and video cameras—thanks in large part to contributions from American scientists.

In 1969, Bell Labs scientists Willard S. Boyle and George Smith invented the Charge-Coupled Device (CCD) image sensor. Initially called "charge bubble devices," CCDs were designed to transfer electrical charge along the surface of a semiconductor, originally intended as a memory device. Later, another Bell Labs scientist, Michael F. Tompsett, discovered that due to the photoelectric effect, the surface of a CCD could accumulate charge when exposed to light, and that these charges could be used to capture and store electronic images. In 1971, Tompsett successfully used a CCD to capture an image, marking the birth of the CCD as an imaging device.

Subsequently, companies such as Fairchild Semiconductor, RCA, and Texas Instruments began researching the new technology. Fairchild was the first to develop a commercial-grade CCD image sensor in 1974. In 1975, Steve Sasson used Fairchild's CCD to create the world's first digital camera.

Kodak's upper management did not take Sasson's digital camera seriously. Although they filed a patent, they did not pursue further research. During this period, however, Sony's Kazuo Iwama recognized CCD's potential and led Sony to invest heavily—financially and technologically—in this nascent semiconductor component.

Kazuo Iwama, known as "the father of Japanese semiconductors," visited the U.S. four times after Sony acquired the transistor patent. During each visit, he toured semiconductor factories by day and wrote detailed reports at night from memory in his hotel room, mailing them back to Japan. These reports totaled 256 pages and became the foundational document for Sony's entry into the semiconductor industry. Iwama passed away in 1982. To honor his contributions to the development of CCD applications, Sony embedded a CCD chip into his tombstone.[3]

According to Sony's official website, thanks to its success with the transistor radio, Sony was arguably the world's leading semiconductor company in the 1950s. But by the 1970s, its semiconductor business had fallen behind American competitors and stagnated. At that time, Kazuo Iwama returned to Japan from Sony's U.S. subsidiary and declared, "The semiconductor team is dead." In a bid to revive the business, he placed his hopes in CCD technology and elevated CCD development to a formal project.

Iwama gave his CCD team one critical directive:

> Within five years, we must create a CCD-based camera priced under 50,000 yen. We're not competing with other electronics manufacturers in this field. Our true competitor is Eastman Kodak.[4]

In the decade following the invention of the transistor, the United States far outpaced other countries in related patent filings, while Japan had virtually none. However, after the invention of the CCD, the number of CCD-related Japanese patents quickly surpassed those from the U.S. According to IncoPat search results, there were 1941 global patents related to CCD or "Charged Coupled Device" during the 1970s. Among them, Japan accounted for 646 patents, compared to just 398 from the United States. Sony alone filed at least 165 of these, ranking first among all applicants (see Fig. 39.4).

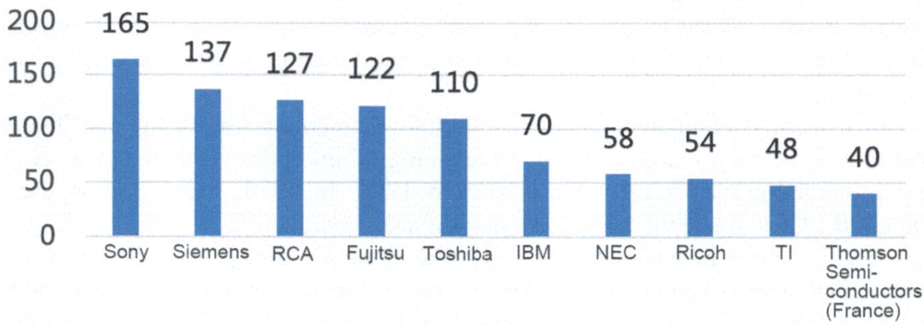

Fig. 39.4 Top 10 CCD patent applicants worldwide in the 1970s

[3] Herms, A. (2012). The CCD sensor: A semiconductor circuit for capturing images. On the Nobel Prize in Physics awarded to Charles Kuen Kao, Willard S. Boyle, and George E. Smith (II). *Contributions to science*, *7*(2), 117–124. Retrieved March 31, 2024 from: https://www.raco.cat/index.php/Contributions/article/download/76610/408092

[4] Sony Corporation. (n.d.). *History*. Sony. Retrieved March 31, 2024, from https://www.sony.com/en/SonyInfo/CorporateInfo/History/SonyHistory/2-11.html

Sony's first CCD-related patent was filed in early 1975 for a "solid-state image element" (patent number JPS59037629). Subsequent filings included "television cameras" in the UK (GB1501017A, GB1501018A), "solid-state cameras with multiple image sensors" in the U.S. (US4016598A), and "solid-state cameras" in Canada, Germany, and France (CA1026455, DE2514155, FR2266404). Of the 165 patents, at least 69 included the word "camera" in the title, and most were "television cameras" based on their specifications. This indicates Sony's early and determined focus on applying CCD technology to consumer electronics—specifically, consumer cameras—and positioning itself against Kodak, the reigning champion of the field.

Ironically, half a century later, Kodak has divested most of its former businesses and become a niche imaging company. Sony, despite ups and downs, still holds a leading position in global imaging, supplying companies such as Huawei—but oddly enough, Sony's own mobile division is excluded from that clientele.

Sony spent 200 million USD over 11 years developing CCD cameras.[5] Kazuo Iwama was repeatedly asked at board meetings when Sony would see a return on its CCD investments. Already advanced in age, his reply was resolute, if somber: "*Maybe not while I'm alive*."[6] Iwama passed away in 1982, before digital cameras and video cameras triumphed over their analog predecessors.

Sony's R&D team increased CCD resolution from 2000 pixels to 8000, then to 70,000, and finally to 120,000 pixels by 1978. In 1979, Sony's R&D investment reached 20 billion yen, but the company still had no consumer-grade product—only a new CCD chip (model ICX008) sold independently.

In 1981, Sony filed Japanese patent JP56212055A for a "solid-state image pickup device," which entered the U.S. in 1982 as US4541016A—one of Sony's earliest digital camera patents. On August 25, 1981, Sony released the MAVICA digital still camera. As noted in the Chap. 12 A Supplementary Note: The Yellow Giant, this camera posed a major threat to Kodak. Powered by three AA batteries, the MAVICA could output images matching the highest quality of televisions and mainstream monitors at the time. Sony designed a complete ecosystem to support the camera, including lenses, Mavipak floppy disks, and a dedicated playback device called the Mavipak Viewer.

This camera represented the culmination of Sony's CCD and video technology. In fact, since the release of the *portapak* in the 1960s, most of Sony's numerous "camera" patents had been for "television cameras"; it had filed very few patents for still-image cameras, indicating that Sony lacked deep technical roots in photography and could not rival Kodak in that field. Technically, the MAVICA was more like a video camera capturing still frames from video signals.[7] This is evident in the

[5] Magoun, A. B. (2007). *Television: the life story of a technology*. Bloomsbury Publishing USA. p127

[6] David Manners. (2010). The Romantic IC. Retrieved March 31, 2024 from Electronics Weekly Website: https://www.electronicsweekly.com/blogs/mannerisms/genius/the-romantic-ic-2010-11/

[7] Mary Bellis. (2020). The History of the Digital Camera. Retrieved March 31, 2024 from ThoughtCo.: https://www.thoughtco.com/history-of-the-digital-camera-4070938

US4541016A patent specification, which addresses problems encountered when using image sensors "typically used in video cameras" to capture still images—such as image contamination from charge leakage and flickering when photographing moving subjects. The patent proposed an optical shutter to control exposure time and eliminate "unwanted charges previously retained in the vertical shift register."

In the digital camcorder field, Sony released the CCD-V8 camcorder in January 1985, featuring a CCD image sensor (model ICX018) with 250,000 pixels. It recorded on 8 mm film, so its storage medium was still analog. Nevertheless, this marked the beginning of CCD becoming the mainstream image sensor in camcorders. In 1991, Logitech launched the first fully digital still camera, the Logitech Dycam. Kodak followed with the Kodak DCS 200 in 1992, and Apple released its first color digital camera, the Apple QuickTake, in 1994.[8] That same year, Sony produced 10 million CCDs, 85% of which were used in digital camcorders, giving Sony a 40% market share in the digital camcorder industry.

Sony's long investment in digital imaging finally paid off. Over the next decade, Sony and Canon shared the spotlight, achieving major success in both digital still and video camera markets. Around 2010, Sony ranked second only to Canon in digital camera sales (see Fig. 39.5), while Kodak, though still in the top ten, teetered on the brink of bankruptcy.

Fig. 39.5 U.S. digital camera market share, 2009–2010. *Sales in millions of units.* (Kazuyo Sawa & Mariko Yasu. (2011). Sony, Nikon Narrow Gap to Canon With New Digital Camera Models. Retrieved March 31, 2024 from https://www.bloomberg.com/news/articles/2011-04-15/sony-nikon-narrow-gap-to-canon-with-new-digital-camera-models. Also available at: https://nikonrumors.com/2011/04/15/2010-worldwide-digital-camera-market-shares-by-vendor.aspx/)

Vendor	2010	2009
Canon	19	19
Sony	17.9	16.9
Nikon	12.6	11.1
Samsung	11.1	10.9
Kodak	7.4	8.8
Panasonic	7.6	7.6
Olympus	6.1	6.2
Fuji	4.9	5.4
Casio	4	4.7
Pentax	1.5	1.7
Vivitar	1.2	0.7
Other	6.7	7

[8] Mary Bellis. (2020). The History of the Digital Camera. Retrieved March 31, 2024 from ThoughtCo.: https://www.thoughtco.com/history-of-the-digital-camera-4070938

Regrettably, neither Canon nor Sony became the ultimate victor in this wave of technological transformation. Though Sony still holds a strong position in digital imaging, the digital camera market is no longer the booming frontier it once was. Starting around 2010, the proliferation of camera-equipped smartphones caused the market to shrink dramatically. According to 2019 data from the Camera & Imaging Products Association (CIPA), global sales of digital cameras from top brands (including Sony, Canon, Kodak, Nikon, Olympus, and others) fell by 87% over the course of a decade (see Fig. 39.6). During this period, Sony failed to launch a groundbreaking, integrated product like it once did by combining camcorders with recorders or headphones with tape players. As discussed in the "Replacement Chapter," we've already analyzed the reasons behind Sony's failure and won't repeat them here.

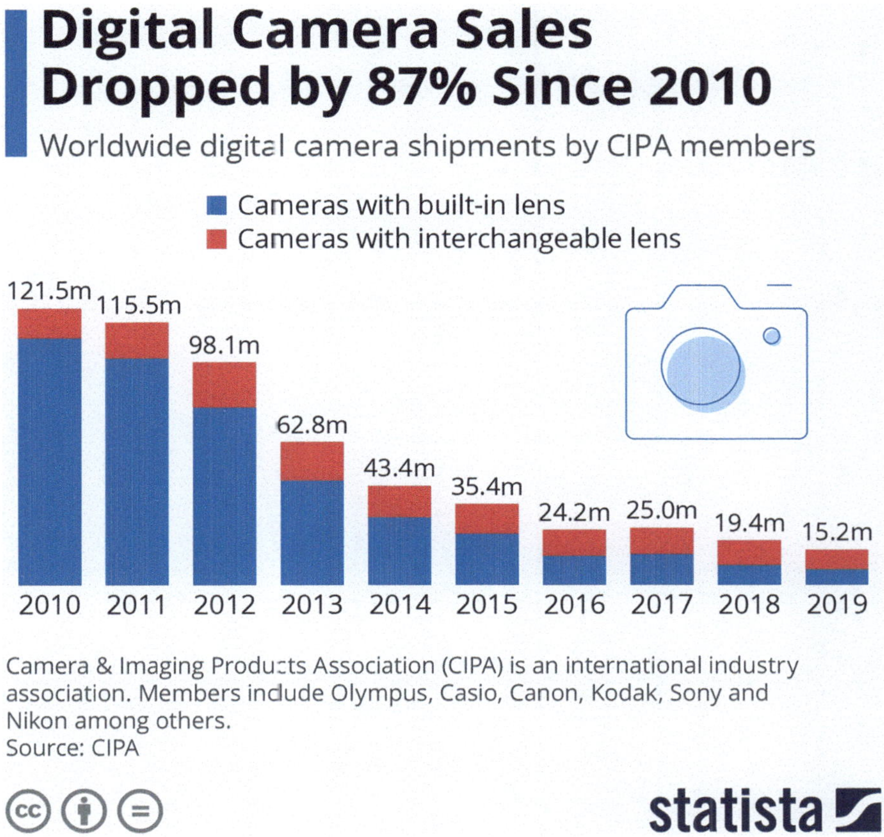

Fig. 39.6 Global digital camera sales plummeted 87%, 2010–2019. (Source: Statista. (2020). *Digital camera sales dropped by 87% since 2010* [Infographic]. Statista. https://www.statista.com/chart/5782/digital-camera-shipments/. This image is shared under the Creative Commons CC BY-ND 4.0 license)

In the "Replacement Chapter," we also discussed how one of Apple's key inspirations for its multi-touch technology came from a paper published in an academic journal by a Sony engineer. In that story, Sony played the role that Bell Labs had once held, while Apple became the new Sony—turning uncommercialized research into world-changing consumer products. A deeply ironic turn of events.

Chapter 40
The Formation of the Bosch Core

In 1876, German scientist Nikolaus Otto invented the four-stroke internal combustion engine and founded the Deutz Company. Two pioneers of the automobile industry—Gottlieb Daimler and Wilhelm Maybach—were once employees of this company.

Following the commercial success of the four-stroke engine, Daimler and Maybach envisioned using Otto's invention to create a new form of transportation powered by internal combustion. However, Otto had no interest in this direction. Consequently, Daimler and Maybach left Deutz and started their own ventures, making monumental contributions to the automotive industry. Their names—Daimler and Maybach—live on as legendary figures.

After their departure, Otto's Deutz Company continued to focus primarily on stationary engines for factories and mills. More than a century later, Deutz remains a prominent German enterprise, producing a full range of engines and reporting €1.4 billion in revenue in 2017. By contrast, Daimler, now the parent company of Mercedes-Benz, had a revenue of €172 billion in 2019—over 100 times that of Deutz.

Nikolaus Otto's son ventured even further: he founded the Bavarian Aircraft Engine Works, abbreviated as BFW, which primarily produced aircraft engines. In 1917, BFW was reorganized into Bayerische Motoren Werke—BMW, the car company we know today.

After leaving Deutz, Daimler and Maybach continued refining Otto's engine, increasing its rotation speed in hopes of developing a gasoline-powered motor vehicle. However, the title of "automobile inventor" went to Karl Benz, who was the first to build a gasoline-powered three-wheeled vehicle (see Fig. 40.1). He applied for a patent in January 1886 and received authorization the same year (Patent No. DRP37435, with "DRP" denoting the German Empire Patent). This vehicle, later known as the Benz Patent-Motorwagen Number 1, saw a production run of 25 units and sold for 600 gold marks—equivalent to over $4000 in 2020. Much like the internet industry a century later, the automotive industry was still in its infancy at the time, with few able to foresee its immense potential.

Q. Yue, F. Yang, *Governing Innovation: A Global History of Patent-Driven Technological Progress*, https://doi.org/10.1007/978-981-95-5570-3_40

Fig. 40.1 The first three-wheeled automobile designed by Karl Benz

Benz's invention faced numerous practical challenges. At the time, European roads were in poor condition; there were no gas stations, no vehicle insurance, and no service centers. To promote her husband's invention, Benz's wife, Bertha, undertook a long-distance journey of over 100 km with their two sons—without informing Karl. Along the way, she bought gasoline at pharmacies, repaired mechanical issues using common household items like hairpins and garter straps, and sought help from a blacksmith when the chain broke. To honor the world's first female driver, the route she traveled was named the "Bertha Benz Memorial Route," a well-known tourist path in Baden-Württemberg for decades.

Bertha's journey demonstrated that the automobile was a viable means of transportation, drawing public attention and yielding great economic rewards for Karl Benz. However, one critical issue remained unresolved: ignition.

Benz's ignition system was powered by batteries. Without an onboard generator to recharge them, the batteries would eventually deplete, preventing the engine from starting and severely limiting travel range. Furthermore, Benz's ignition system closely resembled Daimler's, prompting Daimler to sue for patent infringement in 1896. Benz was forced to pay licensing fees, and despite how closely the names "Daimler" and "Benz" are associated today, the two pioneers never exchanged a single word in their lifetimes.

Perhaps for these reasons, Karl Benz once described ignition as "the problem of problems" in the automotive industry.[1] And it was ultimately Bosch that solved it.

[1] Robert Bosch GmbH. (n.d.). *Ignition Coils Trade Brochure* (p. 2) [Brochure]. Bosch Autoparts. Retrieved March 31, 2024 from https://www.boschautoparts.com/documents/647135/1488055/Ignition%20Coils%20Trade%20Brochure.pdf

Fig. 40.2 The earliest Bosch ignition device

Coincidentally, in the same year that Karl Benz invented the automobile (1886), Robert Bosch opened a "Workshop for Precision Mechanics and Electrical Engineering" in Stuttgart—this was the precursor to the Robert Bosch Company. Soon after opening, a customer asked him to build an ignition device for an engine, modeled after Deutz's internal combustion system. Deutz's system wasn't patented, so Bosch could have copied it outright.

But copying was not Robert Bosch's style. Instead, he improved upon the Deutz design by replacing the bulky bar magnet with a smaller, more stable U-shaped magnet, enhancing the stability and performance of the ignition device (see Fig. 40.2). The improved product quickly became Bosch's flagship offering. In 1888, the "Workshop for Precision Mechanics and Electrical Engineering" produced nine ignition units; by 1891, annual output had surpassed 100, accounting for over 50% of the small company's revenue.[2]

From today's perspective, Bosch had already seized the opportunity presented by the internal combustion engine and stood at the forefront of a new industrial age. Like the early pioneers of the semiconductor and digital revolutions, Bosch rode the wave of the times, gazing out over a vast blue ocean of opportunities.

People often ask whether companies that rise with one wave of high-tech innovation can remain standing through the next. Bosch did.

The improved ignition device developed by Bosch, like Deutz's system, could only be used in low-speed stationary engines. In this system, the armature

[2] Fastnacht, K. (2010). Robert Bosch: His life and work. *Journal of Bosch History. Supplement, 1*. Journal of Bosch History. Retrieved March 31, 2024 from: https://slub.qucosa.de/api/qucosa%3A2908/attachment/ATT-0/?L=1 also available at: https://amicale-citroen.de/wp-content/uploads/2012/09/journal-of-bosch-history-part-1.pdf

reciprocated within the magnetic field to generate current and produce a spark in the engine's combustion chamber. As mentioned, Deutz's engines were mainly used in factories and mills, and they were large, heavy machines with a rotational speed of around 120 RPM. The movement speed of the ignition device's armature matched the engine speed. With technological advances, stationary engine speeds increased to 200–300 RPM, but since the increase was modest, Bosch's ignition device still functioned adequately.

Automobile engines, however, were a different story. They were smaller and ran at much higher speeds. In 1897, British automobile manufacturer Frederick Richard Simms approached Bosch to collaborate on designing an automatic ignition system for his car. A Bosch apprentice volunteered to test Simms's three-wheeled vehicle—only to crash into a neighbor's pile of empty barrels. It was then that Bosch engineers realized that the vehicle's engine rotated at over 1800 RPM, far beyond what their current ignition device could handle. At such high speeds, the heavy armature in Bosch's ignition system would experience extreme wear and tear if forced to operate within a magnetic field.

Numerous pioneers in the automotive industry—including Karl Benz and Daimler—were striving to solve the ignition problem for car engines. Even the legendary inventor Nikola Tesla attempted to develop an ignition system and filed a U.S. patent for it in 1898 (Patent No. US609250A), but it was never commercialized successfully.

Eventually, in 1897, Bosch devised a new solution that avoided the drawbacks of forcing a heavy armature to move at high speeds. Rather than moving the armature within the magnetic field, Bosch's design fixed the armature in place and introduced a metal sleeve around it (see Fig. 40.3). The metal sleeve swung like a pendulum within the magnetic field to generate current. Bosch filed a patent for this solution in June 1897 (Patent No. GB189715411A).

In the autumn of 1897, Bosch installed its electromagnetic ignition device on a car engine for the first time. This approach theoretically solved the ignition problem for high-speed internal combustion engines, but it wasn't flawless. The system still relied on a complex mechanical setup to interrupt the circuit and generate a spark in the combustion chamber. Because engine designs varied widely, each required a customized ignition system. Moreover, the mechanical components were prone to damage and required regular maintenance.[3]

Bosch did not stop there. In 1902, engineer Gottlob Honold developed the first ignition system suitable for mass production and filed for a patent—U.S. Patent No. US802291A. Unlike earlier designs, Honold used ceramic materials for insulation and heat-resistant alloys for the electrodes. He placed two discrete electrodes (marked e and f in Fig. 40.4) inside the combustion chamber to create a spark. Karl Benz's so-called "problem of problems" in the automotive industry had finally been definitively solved.

[3] Robert Bosch GmbH. (n.d.). *High-voltage magneto ignition*. Bosch. Retrieved March 31, 2024, from https://www.bosch.com/stories/high-voltage-magneto-ignition/

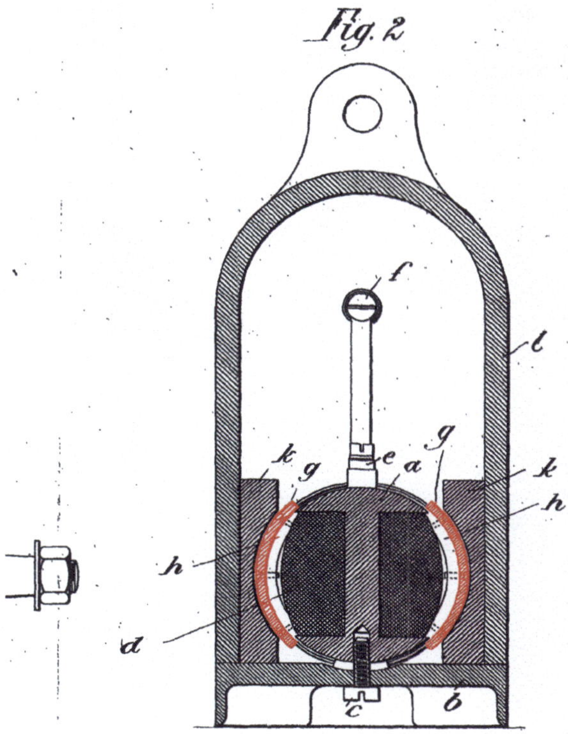

Fig. 40.3 Patent drawing from Patent GB189715411A—The darkened section shows the aforementioned metal sleeve

Today, although Bosch has diversified into many industries, it remains a solution provider to automotive problems—a company that, for over a century, has addressed the many challenges of the automobile sector.

Between 1897 and 1902, Bosch manufactured fewer than 50,000 electromagnetic ignition devices. But after Honold's invention entered the market, Bosch's order volume surged. By 1912, Bosch had produced over 1 million ignition systems. These devices were far from cheap. Around 1930, an ignition system for a medium-sized car sold for about 200 Reichsmarks—approximately one-tenth the price of a small car and twice the monthly wage of a typical Bosch worker.[4] Even today, Bosch produces more than 300 million spark plugs annually.[5]

From low-speed ignition devices for stationary engines in mills and factories to high-speed systems for automotive engines, Bosch leveraged its core technology to expand its business boundaries and enter a new, opportunity-rich industry—joining

[4] Kuhlgatz, D. (2005). Bosch Automotive: A product history. Retrieved March 31, 2024 from: https://amicale-citroen.de/wp-content/uploads/2012/09/journal-of-bosch-history-part-2.pdf

[5] Robert Bosch GmbH. (n.d.). *High-voltage magneto ignition*. Bosch. Retrieved March 31, 2024, from https://www.bosch.com/stories/high-voltage-magneto-ignition/

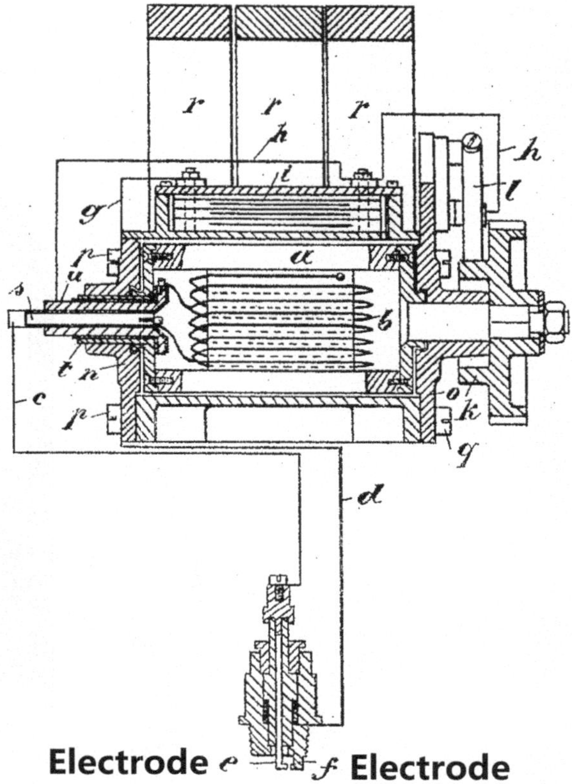

Fig. 40.4 Patent drawing from U.S. Patent No. US802291A

the fast lane of automotive development. Much like IT companies that rode the wave of the internet revolution in the early twenty-first century, Bosch entered a period of rapid, explosive growth.

Next, Bosch would push beyond its existing boundaries in two new directions, to adapt to a changing global landscape.

Chapter 41
Bosch's Expansion Beyond Its Core

On the eve of World War I, Bosch had already grown into a sizable international enterprise, largely on the back of its ignition systems and spark plug sales. Nearly 90% of its revenue came from foreign markets. By 1912, Bosch's factory employed more than 4000 workers, and its global sales reached approximately 33 million German marks.[1] However, the political climate in Europe during the 1910s was growing increasingly tense. Conflicts between Germany and the aging colonial empires of Britain and France were becoming irreconcilable. In 1914, the First World War broke out. The relentless wheels of history rolled across German soil—and Bosch suffered severe blows in the process.

The 3-year war decimated an entire generation of Europe's youth and left the company's major markets in shambles. In 1917, Germany was defeated, and Kaiser Wilhelm II abdicated. From that year on, Bosch patents no longer bore the "DRP" prefix, which had denoted "Imperial Patent." Bosch's factories, offices, and even its brand and trade name in the United States were seized by the American government and ultimately ended up in the hands of a U.S. company (as discussed in the "Disruption and Revolution" chapter). The U.S. appropriation of German intellectual property could, perhaps, be considered a time-honored tradition. For many years afterward, American companies brazenly sold competing products under the Bosch name while the original Bosch was powerless to intervene. It was not until the late 1920s that Bosch finally reclaimed the rights to its own trademark.[2]

After the war, Britain and the U.S., eager to maintain a balance of power in Europe, deliberately supported Germany's recovery—ironically accelerating its economic resurgence. Bosch seized this opportunity to redouble efforts in its core business. The company not only led several innovations within the automotive industry but also gradually expanded its operations into new sectors.

---

[1] Kuhlgatz, D. (2005). Bosch Automotive: A product history. Retrieved March 31, 2024 from: https://amicale-citroen.de/wp-content/uploads/2012/09/journal-of-bosch-history-part-2.pdf

[2] Reif, K. (2014). Fundamentals of automotive and engine technology. *Springer: Bosch professional automotive information*. p16

Q. Yue, F. Yang, *Governing Innovation: A Global History of Patent-Driven Technological Progress*, https://doi.org/10.1007/978-981-95-5570-3_41

We can categorize Bosch's key innovations into two types:

1. **Internal industry boundary expansion**—innovations that leveraged its existing expertise to solve new problems within the automotive sector;
2. **External boundary expansion**—technological applications that extended beyond automotive, allowing Bosch to enter or even create entirely new industries.

Within the automotive industry, Bosch contributed numerous crucial improvements and inventions. These included a redesigned electric horn in 1921, an improved windshield wiper in 1926, and the invention of the diesel injection system and pneumatic braking system in 1927. Bosch even designed a prototype of the Anti-lock Braking System (ABS) as early as 1936, though it would take decades of further research and development before the technology was mature enough to become an industry standard. Through these advances, Bosch transformed from a manufacturer of ignition systems and spark plugs into a full-fledged provider of automotive solutions. Remarkably, in every niche it entered, Bosch maintained a world-leading position—an achievement sustained over nearly a century.

Based on our search using IncoPat, Bosch's early patent applications in Germany were noticeably sparse—only single digits during the 1900s. This could be due to the destruction of many German patent documents during World War II, or perhaps because early German patents used the "DRP" prefix rather than "DE," making them harder for some databases to recognize and record. Between 1900 and 1910, Bosch held 70 granted patents in the United Kingdom and 68 in France—figures far higher than those in Germany or the U.S. For this reason, in the following sections, we'll use the UK patent data as a reference.

As shown in Table 41.1, nearly all patents filed by Bosch during the first decade of the twentieth century pertained to ignition systems.

Among these patents, one notable example is a device for lubricating the breaker cam of a magneto ignition system, filed in February 1909. As mentioned earlier, Bosch's early ignition systems produced sparks through mechanical contact breaking, with the breakers controlled by a cam that required lubrication. Other moving parts, such as the armature, also needed proper lubrication. This invention clearly emerged as a byproduct of Bosch's core ignition system R&D.

Beginning with this camshaft lubrication device, Bosch filed a series of patents relating to lubricating pumps. In 1910, the company filed UK patent GB191022857A (see Fig. 41.1) for an "Improved Method and Apparatus for Distributing Oil to Lubrication Points," and GB191023661A for a "Method and Device for Monitoring Oil Supply from a Lubricating Pump," which used piston pumps and sliding valves to precisely control the drip flow of lubricating oil. Not long afterward, Bosch slightly expanded its product line by introducing an oiler for large vehicle engines and stationary motors—devices capable of accurately regulating oil delivery.[3]

[3] Kuhlgatz, D. (2005). Bosch Automotive: A product history. Retrieved March 31, 2024 from: https://amicale-citroen.de/wp-content/uploads/2012/09/journal-of-bosch-history-part-2.pdf p23

Table 41.1 All patents filed by Bosch from January 1, 1900, to December 31, 1910

Publication No.	Title	Filing Date
GB190000712A	Improvements in Ignition Gear for Explosion Engines	1900/1/11
GB190007108A	Improvements in or connected with Magneto-electric Machines	1900/4/17
GB190116062A	Improvements in Electric Sparking Devices for Internal Combustion Engines	1901/7/12
GB190116063A	Improvements in and connected with Ignition Gear for Internal Combustion Engines	1901/7/12
GB190117249A	Improvements in Magneto-electric Machines	1901/8/28
GB190201359A	Improvements in Ignition Gear for Explosion Engines	1902/1/17
GB190501324A	Improvements in Terminal Rods or Conductors of Electric Machines	1905/1/21
GB190511773A	Improvements in Magneto-electric Ignition Apparatus	1905/6/8
GB190523132A	Improvements in Magneto-electric Ignition Apparatus for Internal Combustion Engines	1905/11/10
GB190601507A	Improved Current Collector or Connection for Magneto-electric Machines	1906/1/19
GB190626982A	Improvements in Electromagnetic Ignition Devices	1906/11/22
GB190701840A	Improvements in Sparking Plugs for Internal Combustion Engines	1907/1/24
GB190707269A	Improvements in Sparking Plugs for Internal Combustion Engines	1907/3/26
GB190715136A	An Improved Contact-breaker for Electrical Ignition Apparatus for Internal Combustion Engines	1907/7/1
GB190724288A	Improvements in Contact Breakers for Electrical Ignition in Internal Combustion Engines	1907/11/2
GB190726435A	Improvements in Electrical Ignition Apparatus for Internal Combustion Engines	1907/11/21
GB190726436A	Improvements in Sparking Plugs for Internal Combustion Engines	1907/11/21
GB190800897A	Testing Installation for Electromagnetically Operated Sparking Plugs	1908/1/14
GB190800898A	Improvements in Bearings for Electro-magnetic Apparatus	1908/1/14
GB190804137A	An Improved Method of and Apparatus for Starting Internal Combustion Engines by Electrical Ignition	1908/2/24
GB190804257A	Improvements in Electrical Ignition Devices for Internal Combustion Engines	1908/2/25
GB190805298A	An Improvement in and relating to Knife Edge Bearings	1908/3/6
GB190806157A	Improvements in Apparatus for the Electric Ignition of Large Internal Combustion Engines	1908/3/24
GB190807125A	Improvements relating to Shuttle Armatures for Dynamo Electric Machines	1908/3/31
GB190808190A	Improvements in Interrupters for Electrical Ignition	1908/4/13
GB190813527A	Improvements in Ignition Apparatus for Internal Combustion Engines	1908/6/25
GB190814698A	Improvements in Carbon Brush Holders for Magneto-electric Igniters	1908/7/10

(continued)

Table 41.1 (continued)

Publication No.	Title	Filing Date
GB190815262A	Ignition Apparatus for Large Internal Combustion Engines	1908/7/18
GB190820288A	An Improved Terminal for Magnetoelectric Ignition Devices	1908/9/28
GB190821568A	An Improved Device for Fixing Carbon Brush Holders	1908/10/12
GB190823362A	An Improvement in Carbon Brush Holders	1908/11/2
GB190825547A	Improvements in and relating to Magneto-ignition Apparatus	1908/12/4
GB190825855A	An Improvement relating to Magneto-electric Ignition Apparatus	1908/12/5
GB190828442A	Improvements in the Insulation of the Layers of Wire of Electrical Coils	1908/12/21
GB190828548A	An Improvement in Magneto-ignition Apparatus	1908/12/24
GB190903511A	Improvements in Armatures for Magneto Induction Apparatus	1909/2/12
GB190903759A	A Device for Lubricating the Interrupter Cams of Magneto Ignition Apparatus	1909/2/15
GB190915985A	Improvements in Sparking Plugs for Internal Combustion Engines	1909/7/8
GB190918457A	Improvements in Ignition Devices for Internal Combustion Engines	1909/8/11
GB190918534A	Multiple Distributor for Electric Ignition Apparatus	1909/8/11
GB190920656A	Improvements in Current Distributors for the Ignition Devices of Internal Combustion Engines	1909/9/9
GB190922479A	Improvements in the Insulation of Windings in Electrical Apparatus	1909/10/2
GB190925236A	Improvements in Sparking Plugs for Internal Combustion Engines	1909/11/1
GB190927103A	Improvements relating to Carbon Holders in Magneto-ignition Apparatus, Current Distributors or the like	1909/11/22
GB190930049A	An Improvement in Sparking Plugs for Internal Combustion Engines	1909/12/23

What may seem like minor technologies and techniques, such as lubrication systems for ignition devices, would, through years of accumulation, later blossom into key innovations.

In 1892, German engineer Rudolf Diesel invented the diesel engine (patent number DRP67207). Though his invention achieved some commercial success, it faced numerous technical challenges and harsh criticism. Diesel later suffered financial ruin and mysteriously disappeared aboard a ferry crossing the English Channel on the eve of World War I. His technology remained largely undeveloped at the time of his death.

After Diesel's disappearance, the diesel engine gradually regained interest in the industry thanks to successive improvements. By the 1920s, some experts began to believe that diesel engines might 1 day replace gasoline engines as the heart of the next-generation internal combustion vehicles.

Diesel engines differ from gasoline engines in that they ignite fuel through heat and high pressure rather than spark plugs. If diesel engines were to replace gasoline engines, this would spell disaster for Bosch, whose core products included ignition

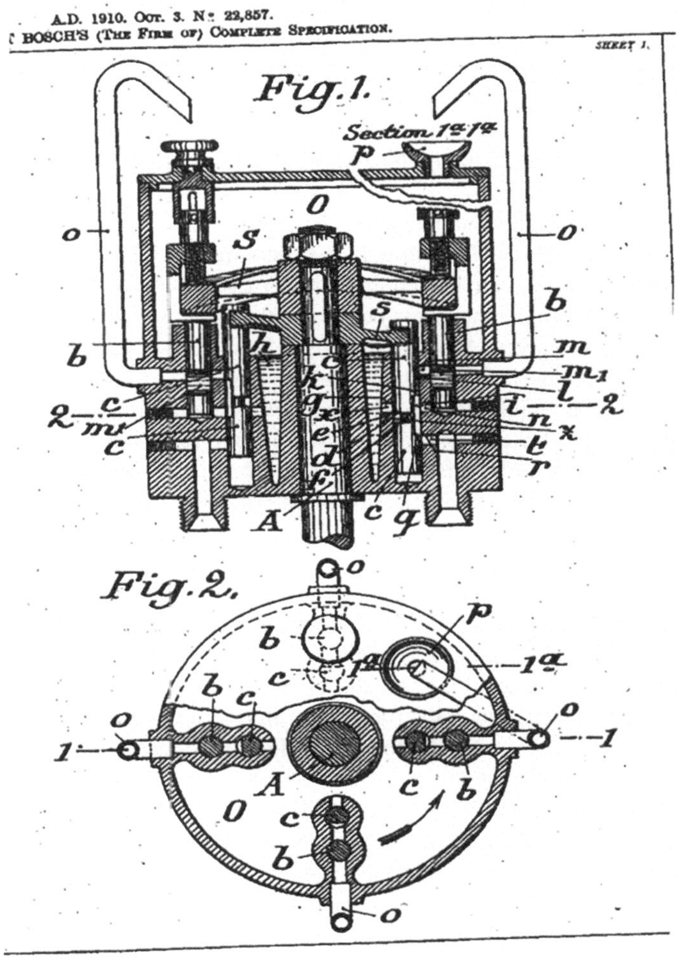

Fig. 41.1 Patent drawing from Patent GB191022857A

systems and spark plugs. In that scenario, Bosch's story might have been more fittingly told in the "Disruption and Revolution" chapter of this book. But diesel did not replace gasoline; instead, Bosch developed new solutions tailored to the diesel engine, solving major technical challenges for the industry and simultaneously opening up new growth avenues for itself.

Bosch responded quickly to the diesel challenge. In 1922, the company began developing diesel fuel injection systems—injecting fuel directly into the combustion chamber. Bosch's experience in this area stemmed primarily from its work on oil drip systems in the 1910s. As noted in *Bosch Automotive—A Product History*, "*Bosch was able to benefit from its previous experience in the development of lubricating pumps. These pumps, also known as Bosch oilers, were capable of delivering precise quantities of fluid under high pressure to specific points in the engine – which is virtually what a fuel-injection pump does.*"[4]

In 1926, Bosch released its first prototype diesel injection systems. By the end of 1927, it had delivered around 1000 units. Soon after, Bosch made significant innovations in diesel engine governors. On May 29, 1930, Bosch filed U.S. patent US1974851A for an "Internal-Combustion Engine Governor," one of its most cited patents from that period.

As seen in Fig. 41.2, the system uses a cam-controlled piston pump, with a piston and sliding valve regulating fluid flow in the pressure pipeline—virtually identical in design and operation to the key components of Bosch's early lubricating oilers.

By the 1930s, diesel engines were becoming the standard for trucks and agricultural vehicles across Europe. Most of them used Bosch diesel injection systems. For example, the Mercedes-Benz 260 D, the world's first mass-produced diesel passenger car, was equipped with a Bosch fuel injection system.

The components of a diesel injection system require extremely high precision and cannot be assembled by end-users. As a result, Bosch sold complete systems, not just parts. These included the injection pump, fuel lines, supply pump, fuel filter, nozzles, nozzle holders, and glow plugs for cold starts. Bosch's long-standing expertise in precision mechanical engineering ensured the tight integration of these parts.

In 1986, Bosch launched the world's first electronically controlled diesel distributor pump and inline pump. In 1997, in collaboration with Mercedes-Benz, Bosch introduced the Common Rail diesel injection system, which went on to become the dominant technology for twenty-first century diesel engines.[5] Over nearly a century, Bosch remained a top global supplier of diesel engine systems.

Yet diesel injection systems were not Bosch's first "system-level" product. According to *Bosch Automotive—A Product History*, the company's transformation from a mere parts manufacturer to a comprehensive solutions provider began with its development of vehicle lighting systems in the 1910s.[6]

From 1900 to the 1910s, Bosch's patents were mainly focused on ignition systems and spark plugs, with IPC classifications typically in H01 and F02. Even including the lubricating oilers, the vast majority of Bosch's early patents were

[4] Kuhlgatz, D. (2005). Bosch Automotive: A product history. Retrieved March 31, 2024 from: https://amicale-citroen.de/wp-content/uploads/2012/09/journal-of-bosch-history-part-2.pdf p32

[5] Electronic Specifier. (2011). 125 years of Bosch—Invented for life The diesel passenger car turns 75 Diesel engines offer high potential. Retrieved March 31, 2024 from Electronic Specifier Website: https://www.electronicspecifier.com/news/analysis/125-years-of-bosch%2D%2D-invented-for-life-the-diesel-passenger-car-turns-75-diesel-engines-offer-high-potential

[6] Kuhlgatz, D. (2005). Bosch Automotive: A product history. Retrieved March 31, 2024 from: https://amicale-citroen.de/wp-content/uploads/2012/09/journal-of-bosch-history-part-2.pdf p15

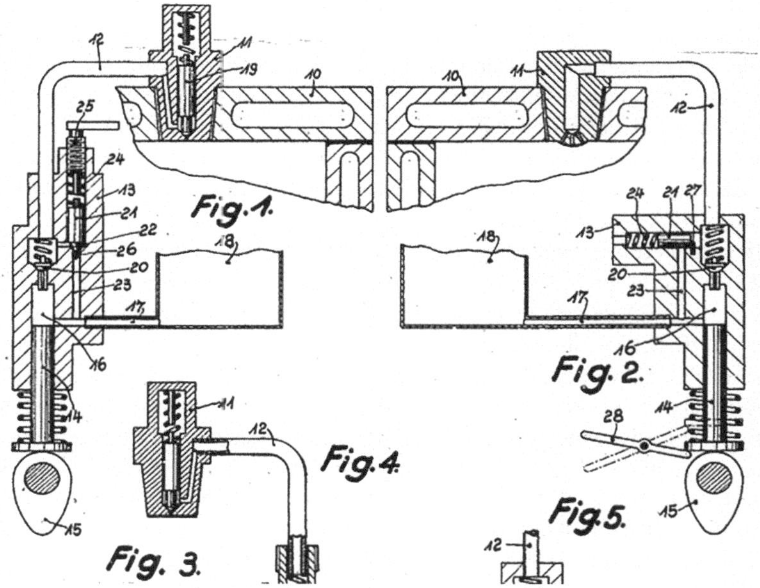

Fig. 41.2 Patent drawing from U.S. Patent US1974851A. When piston 14 moves downward, pump cylinder 16 forms a vacuum, then suction valve 17 opens, and fuel is drawn into pump cylinder 16. Next, as piston 14 moves upward, the fuel inside the pump cylinder is forced into the pressure pipeline 12. When the pressure reaches a certain level, needle valve 19 rises, and fuel is sprayed through nozzle 11 into the combustion chamber. After injection begins, the pressure inside the pipeline continues to rise. Control valve 21 begins to overcome the spring force of spring 24, and once valve 22 opens, the pressure pipeline connects with the return passage 23 leading back to suction valve 17, thereby preventing additional fuel from entering the nozzle. By adjusting screw 25, the preload on spring 24 can be regulated, thereby adjusting the injection pressure and ultimately the quantity of injected fuel

engine-related. Around 1912, Bosch's product line was still extremely narrow—limited to ignition systems, spark plugs, and oilers.[7]

Between 1914 and 1915, Bosch's R&D ventured into a new domain, with the company filing its first UK patents related to automotive lighting (e.g., GB191325497A, GB191505184A, GB191508047A, and GB191508180A). In 1915, Bosch filed its first UK patent related to car horns, GB191504603A. Soon after, Bosch introduced its own headlight and horn products.

Bosch was not the inventor of the automotive headlamp. By the time Bosch began filing lighting-related patents, headlights were already widely adopted. In Germany, laws passed in 1909 mandated that all cars be equipped with two front lamps. Similar regulations were emerging across much of Europe. But unlike

[7] Kuhlgatz, D. (2005). Bosch Automotive: A product history. Retrieved March 31, 2024 from: https://amicale-citroen.de/wp-content/uploads/2012/09/journal-of-bosch-history-part-2.pdf p15

today's electric bulbs, the mainstream headlamps of that era were acetylene lamps (see Fig. 41.3).

Using acetylene lamps as automotive headlamps came with several inconveniences: the ignition and replacement process was cumbersome, the light had poor penetration, and battery life was insufficient for long-term operation. To solve these problems, Bosch introduced a dedicated automotive generator to power its headlights, selling it as part of a bundled system. Patents GB191505184A and GB191508047A reflect this solution.

Both patents relate to headlight circuits and switches. In each, the power source for the headlamps includes not just a battery but also a generator driven by the vehicle's engine. When the car stops and the engine is no longer running, a low-voltage cutoff device disconnects the generator from the circuit. GB191505184A adds a switch for the cutoff device: when the switch is on and the generator, battery, and headlamps are connected, the cutoff device remains in the circuit; in other configurations, it is disconnected. GB191508047A introduces a safety mechanism into the circuit's switch design to ensure that when the headlamps are on—regardless of whether they're powered by one or two sources (battery and generator)—they cannot be turned off without completing a specific operation. This prevents users from accidentally switching off the headlights while driving at night.

In 1913, Bosch began selling complete headlamp systems. In addition to the headlights, each set included a generator, voltage regulator, and battery. This new business model laid the foundation for Bosch's identity as a "solutions provider." Starting in the 1920s, Bosch also developed lighting systems for bicycles and motorcycles. By the 1930s, its product range expanded to include fog lights, high beams, tail lights, and brake lights. Bosch quickly became the world's leading

Fig. 41.3 Left—A miner's acetylene lamp (not for automotive use, but shows the flame); Right—Acetylene lamp mounted on a vehicle, fitted with a glass cover. (Source: User: Herzi Pinki. (2007, June 3). *Carbide lamp on a bicycle* [Image]. Wikimedia Commons. https://commons.wikimedia.org/wiki/File:Carbide_lamp_on_a_bicycle.jpg. This image is shared under the Creative Commons CC BY-SA 3.0 license)

supplier of automotive lighting systems—a position it held until the lighting division was spun off in 1999. The generator originally used in these lighting systems evolved into a standalone product line and remains a key business segment for Bosch today.

Patent filings show clearly that when Bosch launched its first lighting system, it had virtually no technical background in lighting per se. At the time, Bosch did not produce lightbulbs or acetylene lamps and had no products or patents directly related to lighting. Instead, its strengths lay in engines, motors, electrical circuits, and mechanical systems. Yet Bosch managed to leverage its expertise in electrical motors, integrate off-the-shelf components, and adopt an innovative business model—capitalizing on laws that mandated automotive lighting—to carve out a new product line.

In addition to building on existing competencies, Bosch also explored end-user needs, developing products from the consumer's perspective. Its earliest UK patent related to lighting, GB191325497A, exemplifies this approach. Whether early headlights used bulbs or acetylene lamps, they often needed to be replaced, and users could easily scratch or soil the reflective concave mirror behind the bulb, diminishing the light output. Bosch's patent proposed a solution: fix the reflective mirror and the glass cover together as a single, inseparable unit, and attach the rear of the headlamp to a hinge so the whole unit could tilt downward for easy bulb replacement from the back. Patent GB191505866A improved the mounting of auxiliary reflectors on the glass cover to prevent them from blocking the light. From today's perspective, these patents may seem simple, but they show Bosch stepping outside its comfort zone. The innovations focused entirely on the user experience—unrelated to ignition systems, engines, or motors—and were aimed not at Bosch's traditional B2B customers but at end consumers.

A similar approach is evident in Bosch's automotive horn patents. Bosch didn't invent the car horn—that credit goes to Miller Reese Hutchison, an engineer working under Thomas Edison, who developed it in 1908. Shortly afterward, Bosch began filing horn-related patent applications. In UK patent GB191504603A, filed in 1915, Bosch pointed out that car horns emit sound through diaphragm vibrations, and these could be distorted when combined with vehicle body vibrations, making the sound unpleasant. To address this, Bosch borrowed the concept of the pipe organ and designed the horn as a vertical cylinder with a closed top and an open bottom. Patent GB184155A later improved the horn's mounting system. Claim 1 describes: "A horn with diaphragm mounting means, in which the horn is attached at one end to a leaf spring or a bundle of leaf springs, and the other end is fixed to a rigid support as described in the specification." The design aimed to place the diaphragm and horn opening as far apart as possible, minimizing interference from vehicle body vibrations.

In 1921, Bosch launched a distinctive car horn design, as shown in Fig. 41.4. The sound-producing component was mounted at the top and fixed to the car's front using a flexible fixture, while the horn's opening faced downward. The patent drawing appears in Fig. 41.5. This product marked the beginning of Bosch's

Fig. 41.4 1920s Bosch car horn advertisement

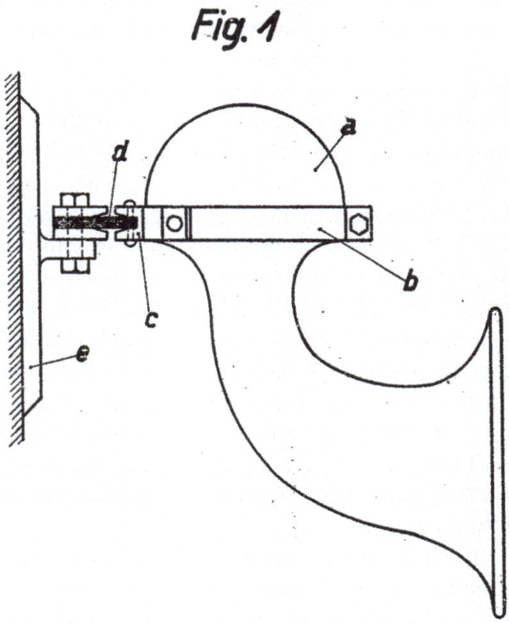

Fig. 41.5 Patent drawing from Patent GB184155A

century-long commitment to car horns, and Bosch has remained a key supplier of horns to most major automakers ever since.

Bosch's horn and lighting system ventures followed a different expansion path than the diesel injection system. The latter was a result of applying existing Bosch technologies to new fields. In contrast, the horn and lighting products were developed from scratch based on deep insights into user needs. These products exemplify the various directions a company can take to push its boundaries. As mentioned, Bosch managed to establish a firm foothold in every field it entered, maintaining industry leadership for nearly a century.

Bosch's long-standing dominance in various automotive subfields was not only due to its early market entry but also to its steadfast commitment to innovation.

In the 1920s, automobile speeds steadily increased, reaching over 80 kilometers per hour. However, brake technology struggled to keep pace with this acceleration. In 1927, Bosch obtained a license for a pneumatic braking system patent from the Dewandre company. Building on this, Bosch launched its own braking system, which reduced stopping distance by one-third. Bosch continued refining its braking technologies, and by 1936, the company designed a "mechanism to prevent locomotive wheels from locking," for which it filed a patent in Germany (Patent No. DE671925C). This patent is widely regarded as the prototype of the modern Anti-lock Braking System (ABS).

In the 1930s, the mechanical control capabilities of the industry were still limited. Although Bosch had secured the patent, the technology of the time wasn't advanced enough to commercialize the invention. It wasn't until the 1960s, with the rise of semiconductor technology and rapid advances in electronics, that it became possible to precisely and effectively control mechanical systems and modulate brake pressure.

In 1969, Bosch resumed development of the ABS system. In December of that year, the company filed a new patent for a "vehicle braking system to prevent wheel lock-up" (German Patent No. DE1961039A1; UK Patent No. GB1328127A).

At the time, Teldix GmbH, a company based in Heidelberg, Germany, had already been researching for 5 years and had developed the world's first anti-lock braking system capable of independently controlling all four wheels. They called it "ABS." The key component of the Teldix system was its low inertia valve, which could produce 60 pulses per second—more than ten times the capability of similar devices produced by companies like Bendix and Dunlop. Teldix was a joint venture between Telefunken and Bendix, initially founded to manufacture avionics for the F-104 fighter jet. The low inertia valve itself was originally designed for aerospace applications. Although Teldix conducted extensive testing and its ABS system could function reliably, its electronic components lacked sufficient durability.

In 1973, Bosch acquired a 50% stake in Teldix and began joint development of the ABS system. Leveraging its strengths, Bosch focused on enhancing the durability of the electronic components. That same year, Bosch filed a patent for an "anti-lock braking system with monitoring circuits" (US Patent No. US3874743A). In 1975, Bosch took full control of the ABS project. In 1978, building on the collaborative work of the two companies, Bosch released ABS2, which became the industry standard.

Chapter 42
New Frontiers: From Hair Clippers to Power Drills

From a long-term perspective, the automotive industry of the early twentieth century paralleled the internet industry of the early twenty-first century as a symbol of high-tech advancement, with an overall upward trajectory despite occasional setbacks. Just as the internet sector faced crises in 2003 and 2008, the automotive industry also endured several major downturns.

The economic crisis of the 1920s dealt a severe blow to the burgeoning automotive industry. In response, Bosch implemented a product diversification strategy, successfully transforming from a small auto parts supplier into a modern multinational electrical engineering group. Among the new products launched at the time were household appliances, personal care devices, and power tools—the last of which proved to be particularly groundbreaking. It created an entirely new market segment, to the point where the Bosch name became virtually synonymous with power tools. Today, Bosch remains a global leader in the power tool industry.

On Bosch's official "History Blog," there is an article titled *"The First Bosch Power Tool – a Hair Trimmer."* It recounts how, amid the 1920s automotive crisis, Bosch was forced to explore new product lines. Engineer Ernst Eisemann designed a clever electric clipper. Robert Bosch quickly ordered it into production, using Bosch-manufactured materials for the casing and a Bosch-made motor. It was produced and sold by the Bosch subsidiary Eisemann.

Although the article does not detail the clipper's technical features, we found a relevant Bosch patent from this period (Patent No. GB306028A). As shown in Fig. 42.1, the patent is fairly simple, with all claims focusing on the use of insulating materials.

The product was named *Forfex*. It achieved some commercial success, though it wasn't revolutionary. However, *Forfex* inspired Bosch to pursue the development of lightweight power tools, eventually opening up a whole new market. At the time, Bosch engineers were fully engaged in developing new diesel injection systems. They realized that the same motor mechanism used in *Forfex* could be applied to improve their grinding tools and screwdrivers. Within 6 weeks, they modified the rotor circumference, armature coil diameter, and winding count of the *Forfex* motor,

Q. Yue, F. Yang, *Governing Innovation: A Global History of Patent-Driven Technological Progress*, https://doi.org/10.1007/978-981-95-5570-3_42

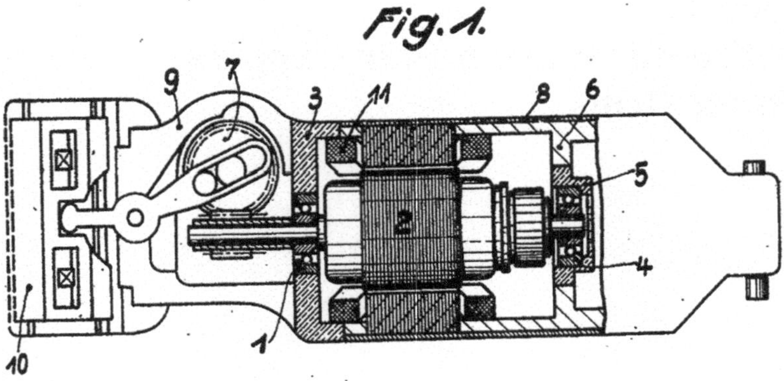

Fig. 42.1 Patent drawing from Patent GB306028A

resulting in a motor with an idle speed of 16,000 rpm and 30 watts of power (compared to *Forfex*'s 8–10 watts), capable of operating for hundreds of hours.

Structurally, there was little difference between the clipper and an electric grinder—both relied on a motor to drive a rotating shaft, which in turn powered the cutting or grinding component. The arrangement of components such as the gear housing, bearing brackets, field coils, and motor casing was also largely the same. The main difference was that the clipper drove blades, while the power tool drove grinding wheels or screwdriver bits.

Unlike the clipper, Bosch's power tools featured a motor shaft that drove the main spindle via two pulleys. The motor shaft and the tool-bearing spindle were not on the same axis, which brought the spindle closer to the handle for better support. Additionally, a fan disc was mounted on the rear of the driven pulley to aid motor cooling, with multiple vents in the casing to discharge hot air. These innovations were detailed in Patent No. US1990035A (see Fig. 42.2), demonstrating that Bosch's power tools were purpose-built with user needs in mind, not just an assembly of existing parts.

As previously noted, Bosch remains a world leader in the power tool sector. During World War II, over half of Bosch's factories were bombed. As a supplier to the German military, Bosch faced post-war sanctions and suffered major losses. The company began rebuilding slowly—on one hand, producing spark plugs for Allied vehicles, and on the other, recycling scrap helmets and metal to forge cast-iron pots for sale.

Perhaps due to the experience and customer base gained from forging cookware, Bosch turned its attention to kitchens and began developing electric kitchen appliances. In 1951, Bosch filed French Patent FR1034398 (see Fig. 42.3) and launched the *Neuzeit I*, a multifunctional electric kitchen appliance, the following year. Like Bosch's multifunction power tools, the appliance used a rotating motor as its core, with interchangeable attachments for slicing, chopping, mixing, and grinding. Bosch continuously improved the product—enhancing cream-whipping

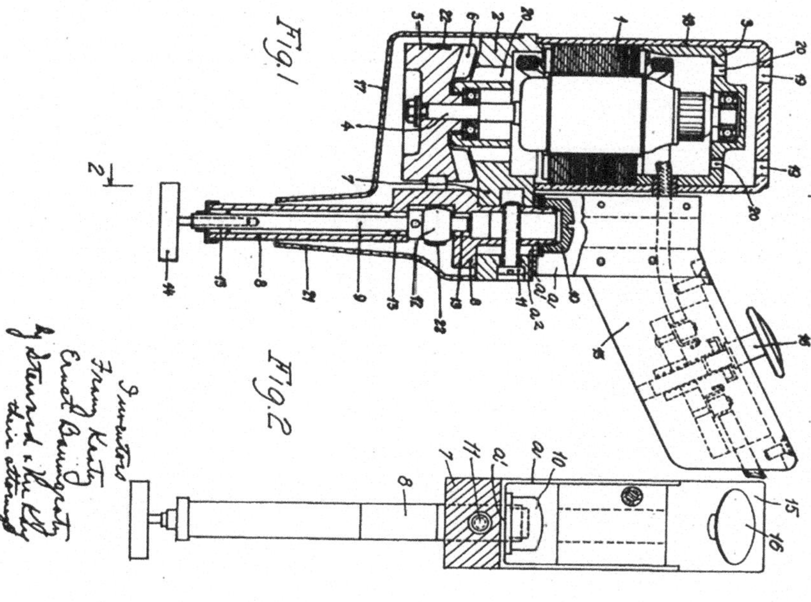

Fig. 42.2 Patent drawing from Patent US1990035A

functionality (Patent No. FR1116305, 1954) and adding centrifugal juicing capabilities (Patent No. FR1208920, 1958). This product line has remained in Bosch's portfolio to this day. In 1953, Bosch developed larger, more powerful rotary motors and began filing patents for washing machines (e.g., Patent No. DE1036802).

Today, Bosch's power tools, kitchen appliances, and household electronics fall under its Consumer Goods division, which in 2023 generated €19.9 billion in revenue—22.2% of the Bosch Group's total income.[1]

Bosch's rise was a product of the Second Industrial Revolution. However, the company didn't remain fixated on that era's core driver—engines—nor did it content itself with being merely an engine parts supplier. Instead, it seized the rapid development of the automotive industry as an opportunity for expansion. As the automotive sector boomed, Bosch also ventured into new markets such as household appliances and small domestic devices.

Building upon its foundational business in engine ignition systems, Bosch consistently pushed the boundaries and explored new business fields. Some of these were spin-offs from its core technological expertise (e.g., oilers), some were born from deep understanding of the automotive industry and end-user needs (e.g., lighting systems and car horns), others emerged as solutions to major industry

[1] Robert Bosch GmbH. (2021). *Annual Report 2020* (p. 50) [Annual report]. Retrieved March 31, 2024 from https://assets.bosch.com/media/global/bosch_group/our_figures/pdf/bosch-annual-report-2020.pdf

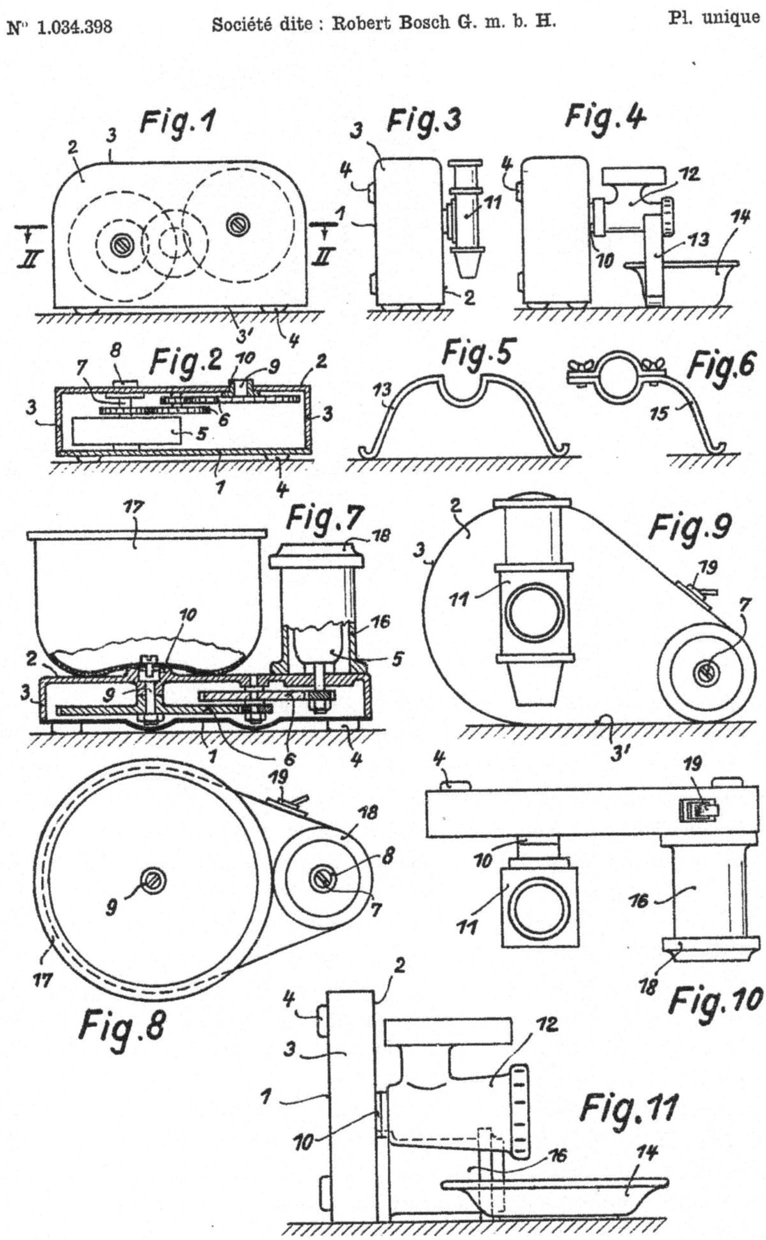

Fig. 42.3 Patent drawing from Patent FR1034398

challenges (e.g., diesel injection systems), and still others represented entirely new markets (e.g., power tools).

The author categorizes Bosch's ventures into *core zones*, *tributary zones*, and *transitional zones*, based on the number of users served and the profitability each brings to the company. The most competitive businesses and top-tier products or services form the core zone. In its early years, Bosch's ignition systems for low-speed engines were the cornerstone of its business—the original core zone. In the twentieth century, ignition systems for automotive engines became the new core. During this period, Bosch's experiments in automotive and power tools represented tributary zones—tentative expansions beyond its established domain.

Once new ventures gained traction and delivered returns, tributary zones evolved into transitional zones. When transitional zones matured into vital, leading-edge operations, they became new core zones. Today, formerly pioneering efforts—like automotive solutions, consumer electronics, and power tools—now constitute Bosch's core, accounting for over 80% of its total revenue.

Bosch's ascent in the 1920s coincided with America's economic boom under "Coolidge Prosperity," fueled by construction, automobiles, and electricity. Bosch's two major expansions—in the automotive and electrical sectors—aligned perfectly with the zeitgeist. Most long-lasting corporate giants in the West have followed a similar path: innovate within their core business, then build further innovations on that foundation, gradually extending their reach into new territory.

It's important to distinguish the concept of "boundary expansion" described here—grounded in patent and product development—from the notion of corporate *diversification*. In the 1960s, the U.S. experienced a wave of corporate diversification, its first major postwar boom in conglomerate formation and multi-sector mergers. By the 1980s, however, this trend reversed: U.S. companies returned to *specialization*.[2]

Mergers in the 1960s were typically friendly, cash-light, and stock-heavy, often involving firms unrelated to the acquirer's main business. In 1959, 22.8% of Fortune 500 companies operated in a single business line; by 1969, that number had dropped to 14.8%. Meanwhile, companies without a dominant business rose from 7.3% to 18.7%.[3]

This trend was driven by several factors: abundant cash flow from a booming economy, inflated stock valuations, and managers' reluctance to issue large dividends. Using overvalued shares to acquire promising startups seemed natural. Additionally, strict antitrust enforcement during the 1960s and '70s, as exemplified by consent decrees against firms like IBM and AT&T, deterred acquisitions within the same industry. As a result, many large firms avoided intra-industry mergers and instead sought growth outside their core sectors.

Investors at the time also favored diversified acquisitions. Studies show that in the 1960s, acquiring unrelated companies tended to increase a firm's market value—by an average of $8 million—while acquiring competitors decreased value by an average of $4 million.

---

[2] Shleifer, A., & Vishny, R. W. (1991). Takeovers in the '60s and the '80s: Evidence and implications. *Strategic management journal*, *12*(S2), 51–59.

[3] Shleifer, A., & Vishny, R. W. (1991). Takeovers in the '60s and the '80s: Evidence and implications. *Strategic management journal*, *12*(S2), 51–59.

This diversification wave was largely a product of its era and policy environment—and history soon discarded it.

The 1980s brought another postwar merger wave, larger and more hostile. By 1989, 143 of the 1980 Fortune 500 companies had been acquired at least once. These takeovers were often resisted fiercely. We'll later recount how Disney nearly succumbed to a hostile bid and emerged reborn through the famous "Save Disney" campaign.

Policy shifts played a key role. With Reagan's rise, U.S. antitrust regulators dramatically reversed their stance on patent enforcement. As discussed in the "Rise and Fall" section, this change helped Texas Instruments rebound through patent litigation and turned patent licensing into a major cash cow for IBM. Meanwhile, antitrust scrutiny of intra-industry mergers relaxed, prompting big firms to buy up rivals. Many over-diversified companies shed unrelated divisions, often under pressure from activist investors. Kodak's ill-fated acquisition of Sterling Drug is a classic case: despite high costs, the market reacted with a $2 billion drop in Kodak's valuation.

While some claimed the 1960s diversification wave optimized resource allocation within firms, others argued it reflected managers prioritizing personal ambition over corporate welfare. The 1980s proved that blind diversification had failed. Once-aggressive conglomerates began shedding excess. Even those unaware of their missteps were "helped" by external buyers to refocus.

Bosch's boundary expansion, by contrast, was rooted not in financial engineering or policy pressure, but in technological evolution and sustained innovation. Each new product line grew directly from Bosch's expertise—either as a novel application of core tech, or as a solution grounded in user needs. It's this continuous, internally driven innovation that allowed Bosch to secure leadership in new sectors and remain resilient over a century.

True diversification, like training inner strength in martial arts, must come from within. If improperly absorbed or wielded, borrowed strength can turn against you.

Chapter 43
The Formation of Xerox's Core Zone: From Copying Techniques to Xerography

Xerography was the invention of a patent attorney. For many, this fact alone may raise eyebrows—after all, when an intellectual property professional invents, drafts, and patents a technology, then sells it to non-practicing entities and reaps huge profits, it inevitably invites suspicion of being a patent troll. Indeed, history's earliest known patent troll was a patent lawyer, and the first recorded case of such a troll suing a practicing company occurred over a century ago.

In 1879, George Selden, a New York patent attorney, foresaw that the automobile could become the dominant mode of transportation. He filed a patent application for a "road engine" (U.S. Patent No. 549160A), crafting broadly defined claims to cover a motorized vehicle comprising "driving wheels and steering mechanisms suitable for road travel; a compression-based hydrocarbon fuel engine with one or more power cylinders; a suitable liquid fuel tank; a drive shaft connected to the propulsion wheels and arranged to rotate faster than them; a clutch or disconnect mechanism; and a suitable carriage body for transporting persons or goods." Selden never built a single car. Aware that engine technology at the time was still rudimentary and the auto industry underdeveloped, he deliberately delayed prosecution, waiting for the right moment.

At the time, patent applicants had 2 years to respond to communications from the U.S. Patent Office, and procedural rules were still immature. Selden exploited every available delay, stretching out the process until 1895—when the industry had finally begun producing practical engine technologies. His granted patent now loomed as a Damoclean sword over emerging car companies.

In 1899, Selden assigned the patent to Electric Vehicle Company (EVC), owned by entrepreneur William C. Whitney. EVC was a pioneer in the nascent electric car sector and one of the largest American automakers at the turn of the century. But EVC soon found itself embroiled in quality scandals and outmatched by internal combustion engines. It pinned its hopes on patent licensing as a new revenue stream and launched a series of lawsuits, demanding a 5% royalty on every car sold.

In 1900, Selden and EVC sued Winton Motor Carriage Company, then the largest automaker. Winton spent heavily trying to invalidate Selden's patent but failed.

Q. Yue, F. Yang, *Governing Innovation: A Global History of Patent-Driven Technological Progress*, https://doi.org/10.1007/978-981-95-5570-3_43

In 1902, major auto companies decided to "join if you can't beat," forming the Association of Licensed Automobile Manufacturers (ALAM) and negotiating with Selden and EVC. They agreed to pay only 1.25% in royalties and were granted the right to enforce Selden's patent themselves. Winton joined too, trading its legal troubles for a seat at the table. ALAM quickly became a cartel, controlling who could enter the automobile market.

In 1903, Henry Ford founded Ford Motor Company and applied for a Selden license. ALAM, seeking to prevent new competitors, cited Ford's prior business failures as grounds to reject the application—and threatened to sue. Ford proceeded with production anyway, advertising: "Only $900—while the Trust sells the same car for $1,500!" ALAM then sued for patent infringement.

The Ford-ALAM case dragged on for 8 years, concluding in 1911. The court upheld the validity of Selden's patent but ruled that Ford had not infringed. The judge interpreted the claims narrowly, limiting them to outdated Brayton-type engines. Ford—and all other modern automakers—used Otto-type engines, and thus fell outside the patent's scope. The ruling noted: "Had he [Selden] possessed greater foresight, he might have claimed a much broader invention… As with many inventors, he had an idea of the result he wanted but chose the wrong path… He was swayed by what were thought to be the superior features of the Brayton engine. Had he realized the advantages of the Otto engine, his patent would have covered contemporary automobiles. But he did not."

The case garnered significant media attention. Public resentment toward monopolies turned Ford into a "folk hero" after his victory. Ford himself later wrote in his autobiography: "No advertisement we ever made could compare to this case." EVC went bankrupt in 1907 and faded into history.

This episode yields two important lessons: First, patents crafted solely from legal maneuvering pale in value compared to those grounded in real products and customer needs. Second, America's flirtation with shady patent tactics dates back over a century. For Chinese patent professionals, mimicking these tricks is like bringing a short blade to a long-sword duel—it will only disrupt the domestic market and lead nowhere internationally. As the saying goes, "a heavy sword has no edge, yet strikes with great power." For true technology companies, patent trolls are but a minor nuisance; product strength is king.

Chester Carlson, the inventor of xerography, was also a patent attorney—and a long-serving corporate IP expert. But unlike George Selden, Carlson dedicated decades of tedious research to his invention. His breakthrough ultimately revolutionized office productivity.

Chester Carlson came from a poor background, and both his education and early career were fraught with difficulties. Because of his family's financial hardship, he had to work three part-time jobs while in school, which severely disrupted his studies and forced him to repeat a year of high school. After many twists and turns, he finally made it into his dream school, the California Institute of Technology. At the time, tuition at Caltech was about $260 per year—a sum Carlson couldn't afford—so he had to borrow money and work part-time, graduating under a heavy burden of debt. To make matters worse, Carlson earned his bachelor's degree in physics during the onset of the Great Depression, one of the bleakest economic periods in American history. After sending out over 80 job applications, he was finally hired by Bell Labs, where he later transferred to the patent department and embarked on a long career in intellectual property. In 1933, as the Depression dragged on, Bell Labs implemented mass layoffs, and Carlson lost his job. Some time later, he managed to secure a stable position as head of patents at a company called P. R. Mallory.

Although Carlson had long since left the world of R&D, he never lost the heart of an inventor. Unlike patent professionals who turned into so-called "patent trolls," Carlson never intended to use his legal expertise to dream up inventions for a quick profit—though he did eventually make a fortune from his invention. His work arose from a real, pressing need: solving the most painful bottleneck in patent work—document copying.

Today's patent professionals enjoy paperless offices equipped with dual monitors, high-speed internet, and powerful patent database tools. For the younger generation, it's hard to imagine the conditions under which China's first wave of patent workers labored in the 1990s and early 2000s: tiny, blinding CRT monitors and poor internet connections. Even harder to picture is what patent work looked like 100 years ago, without computers or the internet, when materials had to be retrieved in person from the patent office—and without even a reliable photocopier! Back then, the dominant copying technology was called "Photostat," a cumbersome process that took several minutes to reproduce a single page. Still, it supported a modest industry—typists were professional workers, and despite the crudeness of the process, using Photostat to copy a document was far easier than hiring a typist and layout artist to recreate it by hand.

Figure 43.1 shows a 1913 advertisement for the Photostat system. The ad boasts that the documents produced by Photostat were "error-free," needing no proofreading, and were identical to the original. Today, no copy machine could get away with such a claim—if a copy isn't identical to the original, it isn't really a copy. But before Photostat, documents could only be duplicated by typists, and reproducing images or formatting required artists and layout specialists. In most cases, simply achieving content consistency between original and copy was a luxury. So despite its limitations, Photostat represented a groundbreaking technological advance.

At the time, the predecessor to Xerox—Haloid Photographic Corporation (hereafter "Haloid")—was a company centered on Photostat technology. The yellow giant Eastman Kodak was also a pioneer in the copying industry, engaging in related business through a patent license from Haloid.

Photostat technology was essentially a photographic method: its core component was a camera that exposed an image directly onto a roll of photosensitive paper, usually 110 m long and requiring manual cutting. After a 10-s exposure, the image gradually appeared on the paper, which was then air-dried or machine-dried. The entire copying process took about 2 m. Moreover, since the process involved direct exposure without using intermediate film, the resulting copies appeared as negatives: white text on black background if the original was black text on white paper. If the user wanted a positive copy, they would have to place the negative copy into the machine and repeat the process—another 5 min of drying included—yielding a

Fig. 43.1 A 1913 advertisement for the Photostat

conventional-looking copy. The whole thing was cumbersome and expensive due to the need for specialized paper.

To develop a more practical copying method, Carlson continued his patent work by day, attended night classes at New York Law School, and spent his remaining hours conducting experiments in his kitchen. Because his work often involved crystalline sulfur as a photosensitive material, the kitchen was pervaded by the rotten-egg smell of burning sulfur. His wife couldn't take it and eventually forced him to rent a separate apartment to continue his experiments.

In 1938, Carlson finally succeeded: he created the world's first xerographic copy. As a seasoned patent attorney, he didn't hesitate to file patents on his breakthrough. On October 6, 1942, he was granted his first patent, U.S. Patent No. 2297691A (see Fig. 43.2), titled "Electrophotography"—a term later translated as "xerography," though it has also been rendered as "electrophotographic copying," "electronic photography," or "electrophotographic imaging." The patent document spanned ten pages, with three full pages of claims totaling 27 items—testament to Carlson's professional thoroughness. Two years later, Carlson was granted a second patent for his copier itself (U.S. Patent No. 2357809A, see Fig. 43.3). He then attempted to commercialize his invention, but companies like RCA, Kodak, GE, and IBM showed no interest. Carlson lamented: "*How difficult it was to convince anyone that my tiny plates and rough image held the key to a tremendous new industry*."[1]

Eventually, the Battelle Memorial Institute, an independent research organization, recognized the potential of Carlson's invention. They funded his research and formed a team under his leadership to continue development.

Meanwhile, the first company in the U.S. to manufacture copiers using Photostat technology was Rectigraph. Haloid had originally been one of its suppliers, providing chemicals and specialty paper. In 1935, Haloid acquired Rectigraph and became a major player in the copier market, selling copying paper and machines, and producing photo paper for companies like Kodak. After the acquisition, Haloid filed multiple patents related to copier mechanisms, especially those focused on automatic feeding (e.g., US2035558A, US2022891A) and cutting (e.g., US1998324A, US2216629A) of specialty paper.

On December 7, 1941, the Japanese launched a surprise attack on Pearl Harbor, triggering the outbreak of the Pacific War. As the saying goes, "When the cannons roar, gold flows by the ton." While World War II brought tremendous hardship to European companies like Philips and Bosch, it presented endless opportunities for the United States, which was just emerging from the Great Depression. Once the country entered a state of war, America's war machine kicked into full gear: airplanes, tanks, and warships rolled off production lines at an unprecedented rate in human history. At the same time, the U.S. military carried out a staggering volume of documentation work.

Japanese Admiral Isoroku Yamamoto had visited the U.S. several times before the war and had witnessed firsthand the crushing economic potential and industrial power of the United States. Though he was the main architect of the Pearl Harbor attack, he had always opposed war with the U.S., fully aware of the terrifying scale of American industrial mobilization once set into motion. After the U.S. joined the war, the extravagance of the American military stunned even its seasoned British and French allies. With an openhanded supply of chocolate and canned meat, American soldiers enjoyed a vacation-like life on the war-torn European continent.

[1] Gross, D. Betting the Company: Joseph Wilson and the Xerox 914 from Forbes Greatest Business Stories of All Time. Retrieved March 31, 2024 from: https://www.stephenhicks.org/wp-content/uploads/2012/01/forbes-xerox.pdf p2

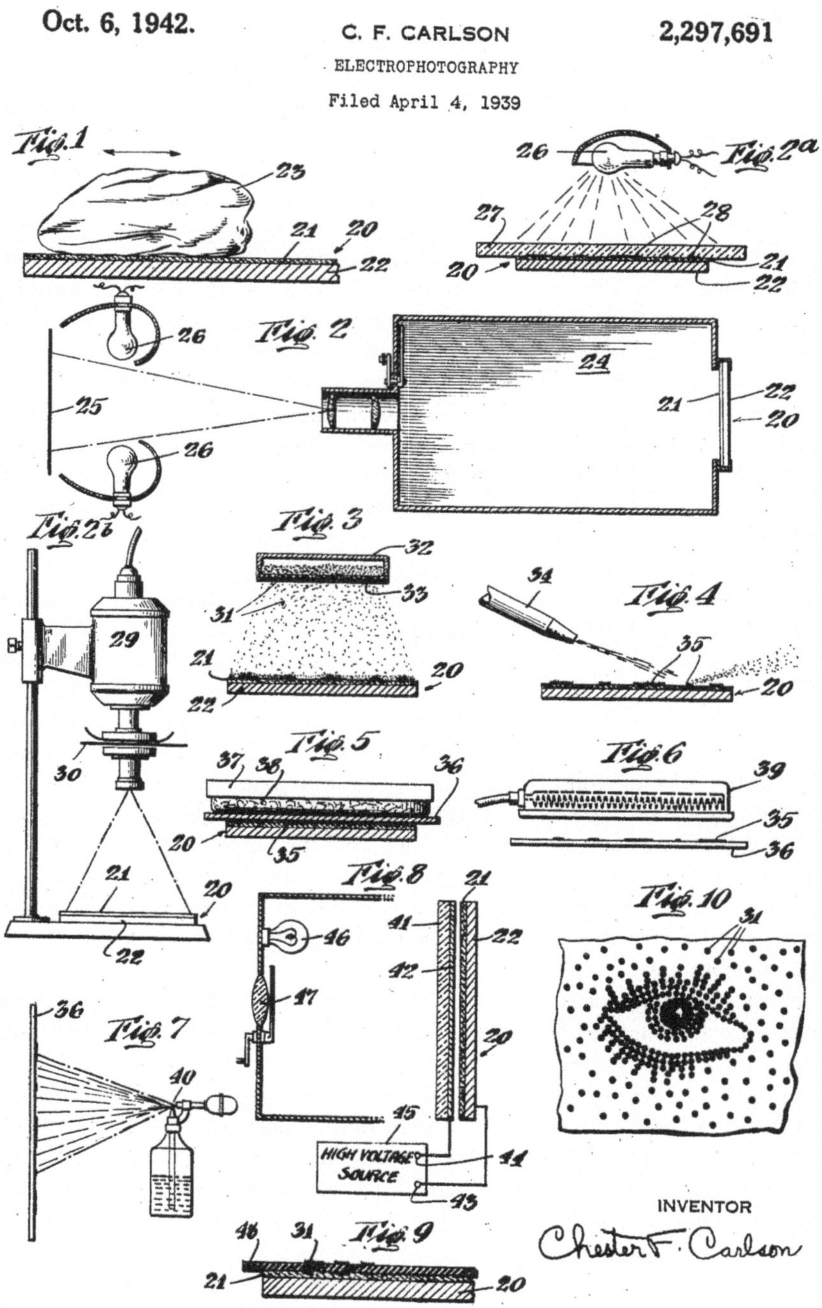

Fig. 43.2 Patent drawing of U.S. Patent US2297691A (electrostatic copying process)

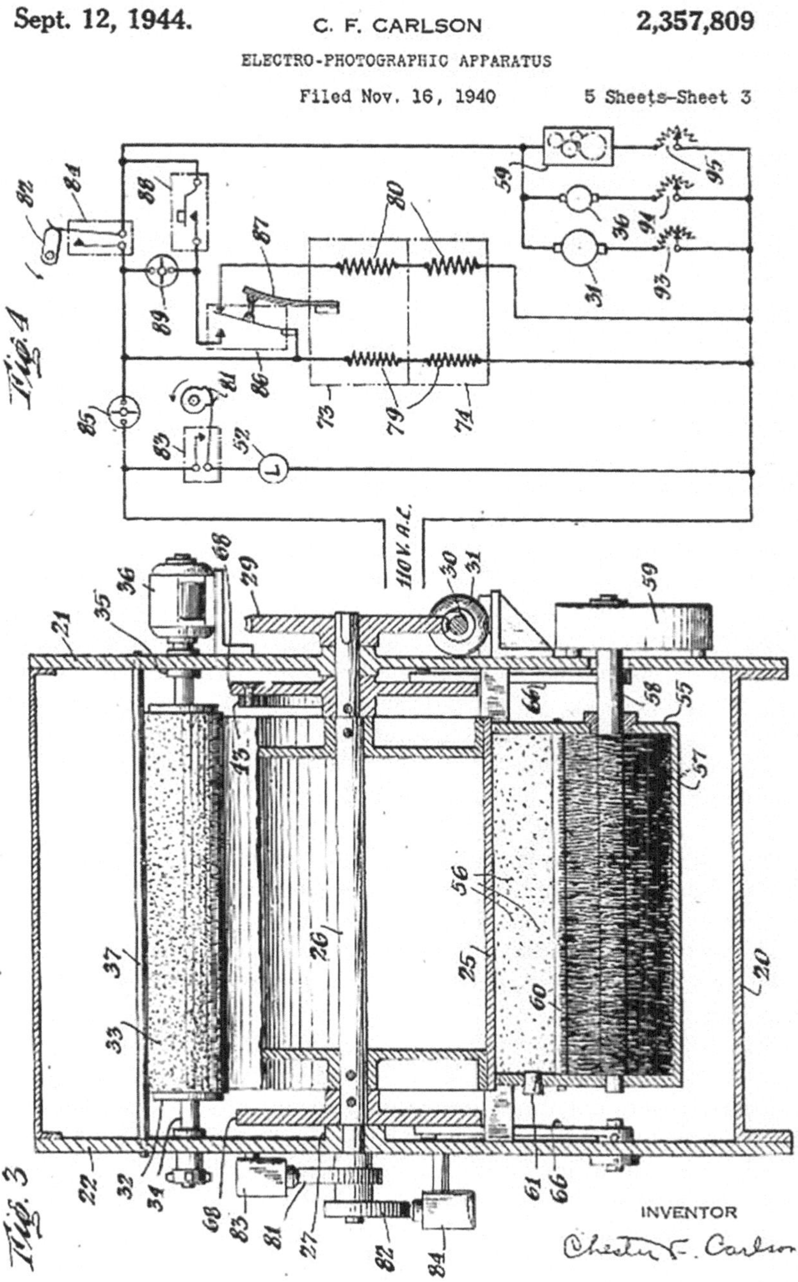

Fig. 43.3 Patent drawing of U.S. Patent US2357809A

Military spending was lavish across the board, and the cost of document-related consumables was a mere trifle by comparison.

For a small company like Haloid, however, document supply was a gold mine. In 1947, Haloid had revenue of $7 million and profits of around $138,000, making it a fairly profitable enterprise. But once the war ended, Haloid's orders plummeted back to normal levels, and the company had to pin its hopes on new technologies.

By chance, John Dessauer, head of Haloid's research and engineering division, came across Chester Carlson's invention in a monthly bulletin published by Kodak. When Haloid's president, Joseph C. Wilson, learned about it, he immediately began negotiations with the Battelle Memorial Institute to acquire Carlson's patents. According to the agreement reached between the two parties, Haloid would pay Battelle $25,000 annually, plus 8% of any future income from xerographic copying as patent royalties. To promote the new technology, Haloid coined the word "xerography" from Greek roots as a more palatable alternative to the awkward term "electrophotography." In 1958, the company even changed its name to "Haloid-Xerox," highlighting its commitment to the technology. By 1961, it had dropped the "Haloid" name entirely, keeping only the invented word "Xerox" as its brand name. The official Chinese translation became "施乐" (Shīlè). Regardless of historical period, we will henceforth refer to the company as Xerox.

Carlson's invention had outlined the basic principles and workflow of xerography, but a technical blueprint alone doesn't sell—it had to be turned into a practical product. As discussed earlier, since acquiring Rectigraph, Xerox had been deeply involved in the automation of copying machines, with many of its patents from the 1930s onward focusing on paper feeding and cutting mechanisms. However, Carlson's technology used plain paper, unlike the specialized paper used in traditional photostats. How could Xerox build an entirely new copier around Carlson's technology? The company had to start from scratch.

In 1950, Xerox introduced the first copier using xerographic technology: the Model A. It was not user-friendly, and the market response was lukewarm. For nearly a decade afterward, Xerox poured extraordinary effort into developing a user-friendly copier. Carlson's foundational patent expired in 1957, but even by then, Xerox had yet to produce a viable product. Fortunately, the company had filed several key patents during that period, which helped delay competitors from catching up.

From 1947 to 1960, Xerox invested a staggering $75 million in developing a new copier—roughly double the company's cumulative operating profit during that time. The company borrowed heavily and encouraged employees to buy company stock. John Dessauer even mortgaged his own house to raise funds. When John Rutkus—later a key figure in the invention of the new copier—joined Xerox in 1955, his first task was to build himself a desk, because Xerox was so cash-strapped it couldn't even provide basic office furniture.[2] The 1950s were a painfully difficult time for

[2] David Owen. (1986). Copies in Seconds Nobody Thought are Needed Xerography until We Had it. The Atlantic February 1986 Issue. Retrieved March 31, 2024 from: https://www.theatlantic.com/magazine/archive/1986/02/copies-in-seconds/665716/ Also available at: https://cdn.theatlantic.com/media/archives/1986/02/257-2/132614921.pdf

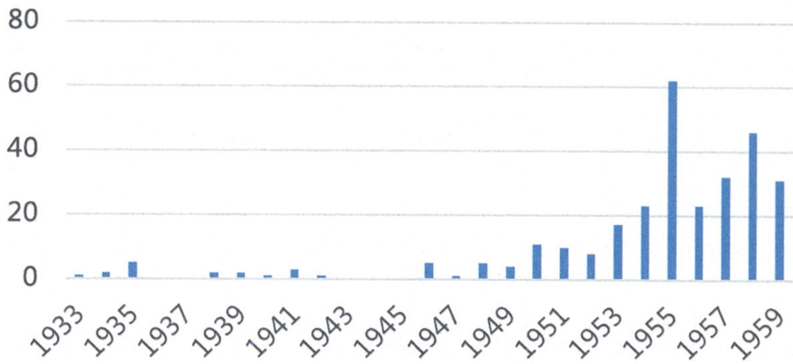

Fig. 43.4 Patent filing trends at Xerox during the 1930s, 1940s, and 1950s

Xerox, with an atmosphere of pessimism clouding the entire company. Reflecting on that era in an interview with *The New Yorker*, Dessauer recalled, "*Various members of our own group would come in and tell me that the damn thing would never work.*"[3]

As shown in Fig. 43.4, Xerox's early patent filings were relatively sparse—at most, only a few per year—hardly the output of a typical tech company. During World War II, although the company received a surge in orders and reaped handsome profits, its patent activity stagnated. However, after purchasing Carlson's patents, Xerox saw a counterintuitive spike in patent filings, even though it had no stable profit engine at the time. This trend reveals the company's bold, all-in gamble in the face of crisis.

After a long and arduous development process, Xerox finally created a large, bulky, and costly copier by the late 1950s. The model took 9 years to develop and consumed $12.5 million in R&D costs. With this machine, Xerox—dormant for a decade—found its historical turning point and soared.

This copier, named the Model 914, was officially released on September 16, 1959, and became one of the most successful products in Xerox's history. Its appearance is roughly depicted in Fig. 43.5 (taken from US Patent No. 2945434A). The model number "9" and "14" refer to the paper size it used: 9 inches by 14 inches (229 mm × 356 mm, slightly larger than A4 paper). Visually, the Model 914 resembles modern office copiers in size and form, measuring approximately 110 cm in length, width, and height. It consisted of 1260 components and weighed about 294 kilograms. As shown in Fig. 43.6, its operating principle involved scanning the

[3] Gross, D. Betting the Company: Joseph Wilson and the Xerox 914 from Forbes Greatest Business Stories of All Time. Retrieved March 31, 2024 from: https://www.stephenhicks.org/wp-content/uploads/2012/01/forbes-xerox.pdf

Fig. 43.5 Drawing from US2945434A

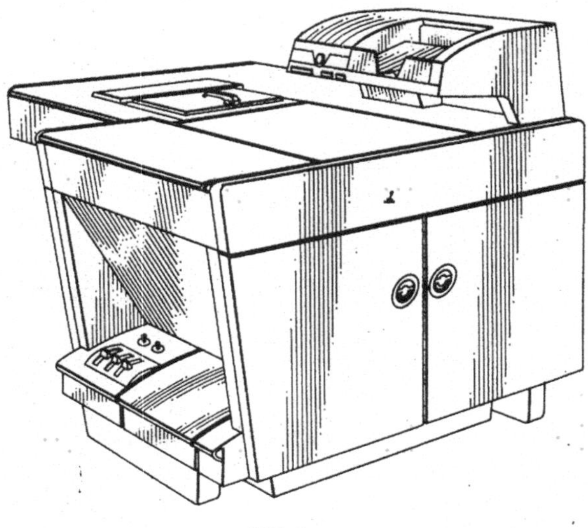

FIG. 1

document with light, then projecting the reflected image onto a photoconductive drum. Static electricity on the drum formed a pattern that mirrored the original image. Toner would then adhere to the charged areas on the drum. Finally, the drum passed over a sheet of blank paper, where a heated pressure roller fused the toner to the page, producing a finished copy.

The Model 914 could produce 7 copies per minute and up to 100,000 copies per month. At the time, traditional copiers cost around $300 and were notoriously inconvenient to use. Most users copied only 15 to 20 pages a day, with 90% of users copying fewer than 100 pages daily.[4] Compared to these machines, the Xerox 914 represented a jaw-dropping leap—a quantum advance into an entirely different dimension.

The Model 914's patent innovations focused mainly on **paper feeding mechanisms**. One of xerography's key breakthroughs was the use of **plain paper** rather than photographic paper. Plain paper is light and readily adheres to the statically charged drum, enabling toner transfer. But this created a problem: how to remove the paper from the drum afterward? Mechanical grippers might scratch the drum's surface, and vacuum systems risked sucking up toner along with the paper.

Xerox engineers explored many solutions, but **John Rutkus** had a breakthrough while inflating his child's bike tires. He placed a sheet of paper tightly against the roof of his car—mimicking how paper sticks to a charged drum—then used the air pump's nozzle to blow air at the sheet's edge. Just as he suspected, the paper dislodged easily once a bit of air entered the gap between the sheet and the roof.

[4] Chesbrough, H., & Rosenbloom, R. S. (2002). The role of the business model in capturing value from innovation: evidence from Xerox Corporation's technology spin-off companies. *Industrial and corporate change*, *11*(3), 529–555.

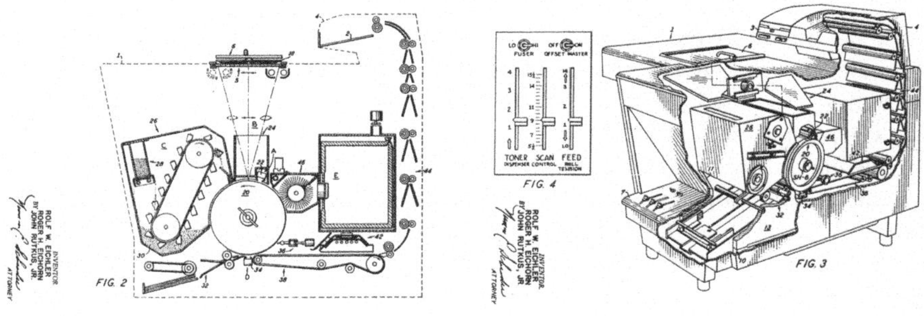

Fig. 43.6 Figure 2 and 3 from U.S patent US2945434A

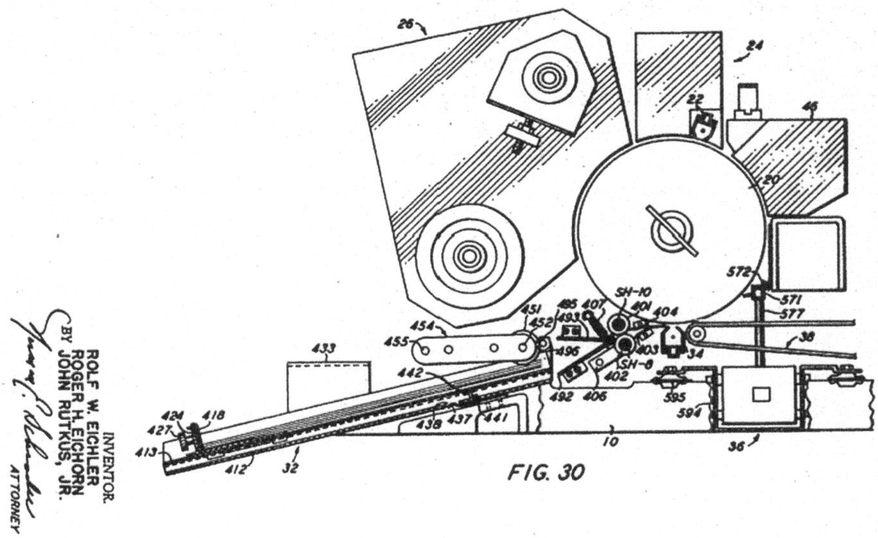

Fig. 43.7 Figure 30 from US2945434A, with component 36 as the air-jet paper pick-off mechanism

Xerox incorporated this solution into **US2945434A**. In Fig. 43.7, device number **36** at the bottom right is the **paper pick-off mechanism**, which uses a jet of air to separate the paper from the drum. Once the paper's edge is lifted, gravity causes it to fall onto a conveyor. Xerox also filed a separate patent for this innovation (**US824658A**).

Paper handling posed another novel challenge. An old American saying goes: *"You can't push a rope."* The implication is clear—ropes (and by extension, paper) must be pulled, not pushed. Xerox physicist **Robert Gundlach** recalled that his friend, during physics exams, would repeatedly chant: *"Force equals mass times acceleration. You can't push a rope. That's all there is to physics."* Inspired by this,

Gundlach realized that, like ropes, paper is too limp to be pushed from behind. It must be **pulled from the front**. The Model 914's **paper feeding system**, as described in **US2945434A**, used **separator rollers** and **feed rollers** with gentle friction to grasp and pull the leading edge of the paper—a principle still used in modern printers and copiers.

Despite its advantages, the 914 had serious drawbacks. By today's standards, it was **unsafe, unstable**, and prone to **overheating**. Xerox even included **a small fire extinguisher** with every 914 unit sold. In fact, outright **sales of the Model 914 were limited**—it was too expensive to manufacture, and mostly the **U.S. government** bought it (due to legal restrictions preventing the government from leasing such machines). The government's procurement price reached as high as **$29,500 per unit**.

Having completed development, Xerox now faced a daunting new challenge: **production**. The R&D had nearly bankrupted the company, and despite its long history in **photographic paper**, Xerox **had no facilities** capable of manufacturing the massive Xerox 914 copier.

Xerox approached IBM in hopes of a commercial partnership, aiming to leverage IBM's powerful, well-established corporate sales team to promote the Xerox 914 copier to large enterprise clients. IBM took the proposal seriously and commissioned the renowned consulting firm **Arthur D. Little and Co.** to conduct a thorough market analysis. After careful research, the consultants concluded that the Xerox 914 copier was already extremely expensive, and its consumables equally so. No customer, they predicted, would be interested. They estimated that **5000 Xerox 914 machines** would be enough to meet the **entire U.S. market demand**. This conclusion was reminiscent of IBM founder **Thomas J. Watson Sr.'s** famous 1943 remark: *"I think there is a world market for maybe five computers."* In its final report, Arthur D. Little summarized:

> *The Xerox 914 copier is impressive for specialized reproduction tasks, but it has no future in the office copying equipment market.*

Based on this conclusion, IBM declined Xerox's partnership offer. Kodak and General Electric also turned Xerox away for similar reasons. Thus, Xerox had to find its own way to market the new product.

On **September 26, 1959**, Xerox officially launched the **Xerox 914 copier**, introducing a new business model—**leasing**. By 1965, the monthly lease for a Xerox 914 was **$95**, which included **2000 prepaid copies**. The copier had a built-in **counter** to track total usage. If users exceeded the 2000-copy quota, they would pay **$0.04 per page** thereafter. Xerox also provided **maintenance and servicing**. This was a gamble—if customers didn't consistently make over 2000 copies a month, Xerox wouldn't turn a profit.

But once again, Xerox **bet right**: after launch, customers were averaging **2000 copies per day**, not per month—equivalent to roughly **$80 of daily usage**!

Xerox copiers not only worked fast, they delivered **excellent print quality**. One iconic TV commercial featured a little girl helping her father use the Xerox 914 to copy a document. Afterward, she handed him both the original and the copy, and the father, unable to tell them apart, asked which was which. The girl scratched her head and said, *"I forgot!"* The ad underscored two key selling points of the Xerox 914: **ease of use** (even a child could operate it) and **flawless print quality** (copies indistinguishable from originals).

At the beginning of this chapter, we saw a 1913 ad for the **Photostat** machine, where the mere claim of "no errors" was a selling point. Decades later, Xerox's ad promised **true indistinguishability**—a revolutionary leap. Today, those qualities are taken for granted and no longer appear in advertisements. Back then, however, Xerox's bold marketing even drew accusations of false advertising from competitors—allegations the Xerox 914's actual performance easily silenced.

With Carlson's foundational patents now expired, companies like **3M** and **Eastman Kodak** tried to break into the copier market using xerography. But their machines couldn't match the Xerox 914's ease of use and still required **specialized paper**.

Once the Xerox 914 hit the market, the company experienced **explosive growth**. In **1959**, Xerox posted revenues of **$33 million** and profits of just **$2 million**. By **1961**, revenue had **doubled** to **$66 million**, with profits at **$5.3 million**. In **1962**, profits **doubled again** to **$13.9 million**. In **1963**, sales shot up to **$176 million**, with profits reaching **$22.6 million—more than ten times** what they were 4 years earlier.[5]

By **1965**, Xerox's revenue reached **$392 million**, with over **60,000 units of the Xerox 914** leased or sold, accounting for **62% of the company's total revenue**.[6] In **1966**, revenue surpassed **$500 million**.

But revenue alone didn't tell the whole story. In terms of **net profit**, Xerox ranked **63rd in the United States** in 1966. Its **profit margin** was estimated to be in the **top 10 nationally**. Its **stock value soared**, increasing more than **60 fold between 1959 and early 1967**, minting a generation of millionaires. Even Xerox's top executives had been blindsided. When shares were trading for mere cents, some had cautiously advised friends and family not to buy in—a warning they would come to regret as the stock skyrocketed.[7]

During his early days working on xerography, Chester Carlson had hired an assistant named Kornei out of his own pocket. Unconvinced of the invention's future, Kornei resigned and relinquished all future rights. Years later, as Xerox stock soared, Carlson gifted him a small number of shares. By 1972, those shares were worth $1 million.

---

[5] Xerox Corporation. (n.d.). *The story of xerography* [PDF]. Retrieved March 31, 2024 from https://www.xerox.com/downloads/usa/en/s/Storyofxerography.pdf

[6] Gross, D. Betting the Company: Joseph Wilson and the Xerox 914 from Forbes Greatest Business Stories of All Time. Retrieved March 31, 2024 from: https://www.stephenhicks.org/wp-content/uploads/2012/01/forbes-xerox.pdf

[7] Brooks, J. (2014). *Business adventures: Twelve classic tales from the world of wall street*. Open Road Media.

In 1968, *Fortune* magazine named Carlson one of the wealthiest individuals in America, estimating his net worth at over $150 million. He passed away that same year and was inducted into the National Inventors Hall of Fame in 1981.

Yet Xerox's wild success drew regulatory scrutiny. In 1972, the Federal Trade Commission (FTC) launched an antitrust investigation, citing Xerox's 86% market share in office copiers. After 3 years of litigation, Xerox reached a settlement similar to IBM's earlier Consent Decree: it agreed to freely or cheaply license many of its hard-earned patents to competitors.

Xerox's rise is a textbook "American Dream" story and was famously featured in the book *Business Adventures: Twelve Classic Tales from the World of Wall Street* under the chapter titled "Xerox Xerox Xerox Xerox." In 1991, Warren Buffett gave this book to Bill Gates, who later wrote on his blog that Xerox's story was his favorite, and that the chapter's title was "*alone belongs in the Journalism Hall of Fame.*"[8]

Flush with cash, Xerox was far from complacent. It invested heavily in research and development—not only in its core business of copying and printing, but also in the promising field of computing. Yet to this day, Xerox remains a company focused primarily on printing, and continues to wrestle with the unintended consequences of having its brand name, "Xerox," become a generic term.

According to then-CEO **Joseph C. Wilson**, the name "Xerox" was designed to mimic the visual symmetry of **Kodak**—both names possess a pseudo-palindromic quality, appearing similar when read from either end. Initially, Xerox worried the brand name might lack impact, especially since it sounded like "zero," which some businesspeople might associate with bad luck. But these concerns quickly vanished as the brand proved to be too powerful. In English-speaking countries, "xerox" rapidly evolved into a verb, turning Xerox's trademark into a victim of its own success. This, perhaps, was the curse of having such a narrow product line.

In 1965, Xerox released the Xerox 2400 copier. Consistent with the company's naming convention, the model number reflected the machine's performance—it could produce 2400 copies per hour. The leasing model remained in place, and per-copy fees gradually dropped to around half a cent, yet profits continued to flow. The Xerox 2400 was even larger than the original 914 and was among the first to feature an automatic document feeder (ADF), paper cutter, hole punch, and collator/sorter—all advancing the automation of document reproduction.[9] Soon after, Xerox introduced the Xerox 3600, which, as the name implies, could produce 3600 copies per hour.

By the mid-1970s, Xerox launched the Xerox 9200 copier (patent number US4054380A, see Fig. 43.8). It boasted an astonishing speed of two pages per second, or 7200 copies per hour—double the output of the 3600.

[8] Richard Feloni. (2014). There's A Great Xerox Growth Story In Bill Gates' Favorite Business Book. Retrieved April 8, 2024 from https://www.businessinsider.com/xerox-growth-in-bill-gates-favorite-book-business-adventures-2014-7

[9] Xeroxnostalgia.com. (2015, February 26). *Xerox 2400*. Retrieved March 31, 2024 from https://xeroxnostalgia.com/2015/02/26/xerox-2400/

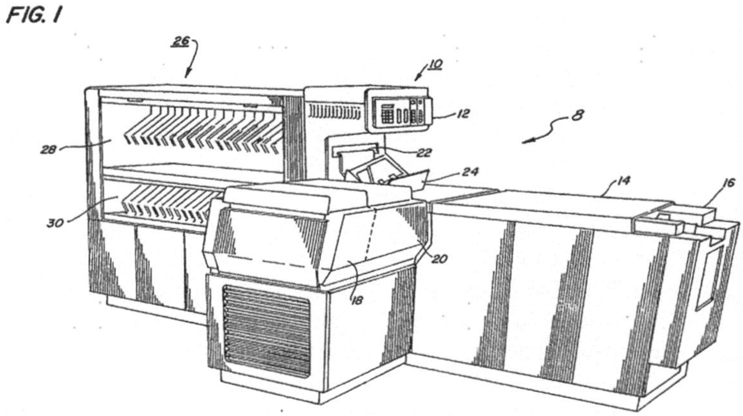

Fig. 43.8 Patent drawing from U.S patent US4054380A

A key innovation of the Xerox 9200 was its **auxiliary paper tray**.[10] Today, most household printers only have a single tray. When paper runs out, users typically pause printing, reload paper, and resume printing after checking for proper alignment to avoid jams. In practice, however, many users just slap a new stack of paper onto the tray without pausing—because the risk of jamming is low and, even if it happens, it's a minor inconvenience.

But in enterprise settings—where printers and copiers handle thousands of pages at a time—adding paper without pausing the machine drastically increases the risk of paper jams. And when a jam occurs, it can bring the entire workflow to a halt. To address this, the Xerox 9200 included an auxiliary paper tray. When the primary tray ran low, users could load fresh paper into the auxiliary tray and simply press a button to switch trays—no interruption required. This innovation enabled non-stop high-volume printing, solving a major pain point for business users.

In addition, the Xerox 9200 supported jam recovery, interrupt printing, and other advanced features. All these technologies were documented in patent US4054380A.

Henry Chesbrough, a scholar at UC Berkeley, identified Xerox's development strategy as focused on copying speed and continuous operation time. He argued that Xerox's leasing model resembled a razor-and-blade strategy: the machines were leased cheaply, while profits came from consumables. To drive up consumables sales, Xerox prioritized faster speeds and longer uptime. A close look at US4054380A supports Chesbrough's claim—features like the auxiliary paper tray, jam recovery, and job interruption functions all aimed to maximize uptime.

[10] Xeroxnostalgia.com. (2013, October 27). *Xerox 9200*. Retrieved March 31, 2024 from https://xeroxnostalgia.com/2013/10/27/xerox-9200/

Chesbrough made a thought-provoking point: *"A mediocre technology pursued within a great business model may be more valuable that a great technology exploited via a mediocre business model."*[11]

This model allowed Xerox to extract high service and consumables fees from enterprise clients. However, it also introduced **strategic risks**—Xerox neglected smaller businesses and the low-end copier market. As CEO Wilson once said: *"…our profits came from how many copies were made on those machines. If a copier was slow in generating copies, that was money plucked out of our pocket"*.[12]

Xerox's relentless innovation yielded a formidable patent portfolio, especially in feeding and binding technologies, creating robust **technological moats**. The company also had a **world-class enterprise sales team**. Its direct leasing model bypassed dealers, ensuring that sales reps knew both the products and their customers' needs intimately. Large corporate clients valued **service reliability** over price sensitivity. Xerox built a **24/7 service network** to support them—at a price, of course. These service contracts became a major revenue stream.

With its portfolio of patent technologies, strong client relationships, and robust after-sales service, Xerox had erected a formidable fortress around its core business—virtually impervious to attack.

As previously mentioned, when Xerox launched the **Xerox 914 copier**, it had sought partnerships with **IBM** and **Kodak**, but was rebuffed by both the Blue and Yellow Giants. However, after Xerox single-handedly created the commercial copier market, IBM and Kodak were eager to get in on the action, hoping to claim a share of this lucrative new industry. IBM possessed deep financial reserves and strong trust from enterprise clients. Its mainframe business—like Xerox's copier business—relied on a leasing model and had already built a comparable after-sales service network by the 1960s. Leveraging these strengths, IBM deployed over **80,000 units** of its **Copier II** model globally by **1977**, capturing **10% of the world market**—but this was still a far cry from challenging Xerox's **dominant 70%+ share**.[13]

In response to IBM's challenge, Xerox quickly retaliated with patent litigation. In **1970**, just as IBM was releasing its first-generation **Copier I**, Xerox sued IBM for patent infringement in a **New York District Court**, citing **22 different patents**.[14]

In fact, IBM had previously received a limited license from Xerox to use xerographic technology in certain "computer peripheral devices" and had received some trade secrets as technical support. However, Xerox had **not** granted IBM permission

---

[11] Chesbrough, H. (2010). Business model innovation: Opportunities and barriers. *Long Range Planning, 43*(2–3), 354–363. https://doi.org/10.1016/j.lrp.2009.07.010

[12] Chesbrough, H. (2010). Business model innovation: Opportunities and barriers. *Long Range Planning, 43*(2–3), 354–363. https://doi.org/10.1016/j.lrp.2009.07.010

[13] Smith, D. K., & Alexander, R. C. (1999). *Fumbling the future: How Xerox invented, then ignored, the first personal computer*. iUniverse. p182

[14] William Smith. (1970). I.B.M. Challenges Xerox in Copier Field. Retrieved April 8, 2024 from https://www.nytimes.com/1970/04/22/archives/ibm-challenges-xerox-in-copier-field-copymachine-maker-counters.html

to use those technologies in the **Copier I**. In retaliation, IBM counter-sued Xerox, accusing it of monopolistic practices.[15] When IBM launched the **Copier II**, Xerox once again filed a new lawsuit for patent infringement. The legal battle between the two giants dragged on for **8 years**. In **1978**, they reached a settlement: **IBM paid $25 million** to Xerox, and the companies engaged in cross-licensing agreements. But by then, the **golden era of large commercial copiers** was drawing to a close. In **1988**, IBM exited the copier business altogether, selling it to **Kodak**.

Kodak, for its part, tried a different approach—touting its renowned **"Kodak Quality"**. In the 1970s, it launched copiers that surpassed both Xerox and IBM in reproduction quality, winning a loyal customer base. But despite this, Kodak captured only a small slice of the market and never posed a serious threat to Xerox.

Xerox had successfully defended its core territory. But in the end, what truly shook Xerox's dominance in the copier industry was neither IBM nor Kodak, but the Japanese companies that began rising in the 1970s—especially those who adopted a "rural encircling the city" strategy by targeting the low-end market.

Among these Japanese challengers, Canon emerged as Xerox's most formidable rival. Canon's strength lay in its core technologies: microelectronics developed from its calculator business and optics and imaging technologies from its camera division. In 1968, Canon invented the New Process (NP), replacing dry toner with liquid toner. This innovation significantly reduced the size of key copier components and lowered production costs. Though somewhat inconvenient to use, the NP technology bypassed Xerox's patent barriers.[16] Further weakening Xerox's position, in 1975, as part of its settlement with the Federal Trade Commission (FTC), Xerox voluntarily licensed its patents to competitors, effectively dismantling its once-formidable patent wall.

It's worth noting that Xerox wasn't forced into this by the FTC—it had a choice. It could have restructured its business model or relinquished its patent rights. Ultimately, Xerox chose to retain its business model and control over its subsidiaries, sacrificing its patents in the process. C. Peter McColough, then CEO of Xerox, later remarked in a New York Times interview that patents had been vital when Xerox was "small and weak," but their importance diminished once the company became an industry giant. Ironically, Canon made the same mistake—it did not adequately protect its liquid toner technology and ended up licensing it to over 20 competitors. The first Japanese brand to truly break through in the low-end U.S. copier market wasn't Canon, but Ricoh.

In 1970, Canon introduced the NP-1100, a plain-paper copier based on its NP technology. It soon followed with the world's first liquid-dry hybrid copier, the NP-70. At the time, Xerox didn't even consider Canon a serious rival. But Canon had set its sights firmly on Xerox's turf. Under the leadership of Takeo Maeda, Canon formally designated the copier business as its core strategic focus.

---

[15] Xerox Corp. v. International Business MacH. Corp., 399 F. Supp. 451 (S.D.N.Y. 1975) Retrieved April 8, 2024 from https://law.justia.com/cases/federal/district-courts/FSupp/399/451/1489539/

[16] Gomes-Casseres, B., & McQuade, K. (1991). *Xerox and Fuji Xerox*. Harvard Business Review Press (China Case Studies).

By the late 1970s, Canon's compact copiers could print 8 to 10 pages per minute, and were priced between $700 and $1200. In contrast, Xerox's large commercial copiers were capable of 90 to 120 pages per minute and, despite their steep prices, still appealed to large corporate clients. Flagship Xerox models sold for $80,000 to $130,000 apiece. Canon's strategy was clear—it bypassed Xerox's entrenched enterprise clients and focused on small and medium-sized businesses.[17] Using ordinary office supply dealers to distribute its copiers, Canon effectively avoided direct confrontation.

SMEs that couldn't afford Xerox leases increasingly purchased compact copiers from Canon, Ricoh, and other Japanese firms—instead of making trips to Xerox's centralized copying centers.

As Canon tightened its grip, Fuji Xerox, Xerox's Japanese subsidiary, proposed in 1978 that the company should enter the low-end copier market to counter Canon's expansion in the U.S. However, Xerox headquarters rejected the idea.

Early copiers were notoriously fragile, but this wasn't a major issue for Xerox, which operated on a leasing model—it could always collect additional maintenance service fees from customers. Canon, targeting individual consumers, knew it couldn't count on users to keep paying for repairs. Therefore, Canon focused on developing "service-free" machines. To accomplish this, Canon bundled all critical components—the drum, charging unit, toner, and cleaning unit—into a replaceable cartridge. This meant that even users with no technical knowledge could easily replace the cartridge and handle basic maintenance themselves. This innovation was not only key to unlocking the personal copier market, but also paved the way for laser printers to dominate later on.

Canon executive **Hajime Mitarai** once said, "*It's a small thing … if you have to pick a single factor that contributed most to the success of the laser printer, it's this cartridge idea, which was our invention.*"[18]

Unlike Xerox, Canon had neither the means nor the desire to build a 24/7 service network. Instead, it focused on enhancing the machine's reliability. Each cartridge could produce roughly 2000 copies, and Canon designed its machines to be as simple as possible, enabling staff at ordinary office supply stores to carry out repairs.

In Fig. 43.9, you can see the black-line sections of patent diagram US4540268A, which depict Canon's cartridge. It's clear that Canon's innovation largely involved miniaturizing and consolidating traditional components into a removable housing. While the core combination—developer, cleaner, drum, and corona charger—didn't change much, this new configuration gave rise to a transformative "razor-and-blade" business model. Replaceable cartridges became Canon's main profit driver, while service costs plummeted. This new model ultimately shook Xerox's dominance.

Was a company as long-established and wealthy as Xerox—one with decades of mechanical expertise—really incapable of designing a smaller copier?

---

[17] Van der Heijden, K., Bradfield, R., Burt, G., Cairns, G., & Wright, G. (2009). *The sixth sense: Accelerating organizational learning with scenarios*. John Wiley & Sons. p34

[18] *Canon, lone wolf.* (1994, October 1). *Wired.* Retrieved March 31, 2024 from https://www.wired.com/1994/10/canon/

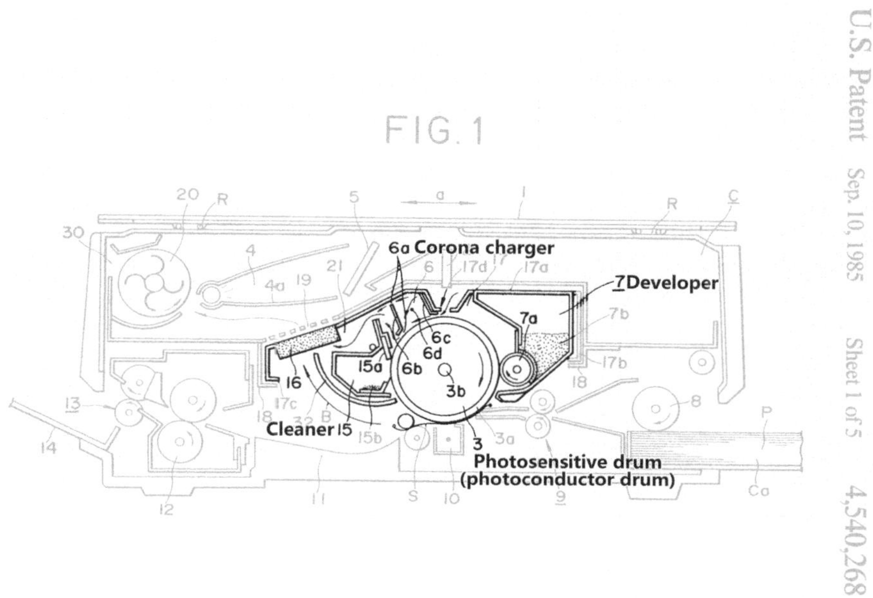

Fig. 43.9 Patent drawing of the Xerox 914 photocopier (top) and patent drawing of U.S. Patent No. 4,540,268 (Canon's removable ink cartridge, bottom)

By 1985, Canon had overtaken Xerox to become the world's largest copier manufacturer (though not yet the largest printer maker). Only after this did Xerox try to enter the mass market—but by then, the best window of opportunity had closed. Over the next decade, the rise of IBM-compatible PCs fueled a new era of economic growth in the U.S., and the personal printer market boomed in its wake. Xerox could only watch from the sidelines, overshadowed by competitors like HP, who dominated the field (a topic explored later).

Beyond its liquid toner innovations, Canon also explored other low-cost solutions, including the now-familiar inkjet printing.

Inkjet printing is fundamentally different from xerography—it does away with the drum entirely, allowing for inexpensive printers that cost little more than their ink cartridges. Like many great inventions, this concept originated in academia. IBM was the first to attempt commercializing it with the IBM 4640, but the print quality and market response were poor. It wasn't until the 1980s that Japanese companies successfully brought the technology to market.

Canon and HP independently developed thermal bubble inkjet technology during this period. Canon's version was a serendipitous discovery—a researcher noticed that heating a soldering iron near a syringe caused ink to spray out, inspiring the BubbleJet system. The principle was simple: a heater near the nozzle generated steam bubbles, which expanded and burst, forcing ink through the nozzle. Canon filed for a Japanese patent (JP61059911B) and then expanded to other jurisdictions like the UK, France, Germany, Australia, and Canada (e.g., GB2007162A).

HP, around the same time, discovered a similar approach—legend has it that one of its engineers got the idea while watching a drip coffee machine. This shared inspiration led both companies to develop inkjet printers.

Canon's BubbleJet technology allowed it to branch into color printing, a domain previously dominated by expensive, specialized equipment. Thanks to the low cost of the BubbleJet line, Canon cracked open the low-end market. Xerox, however, failed to compete. In 2001, Xerox officially withdrew from the consumer market, choosing to focus on high-end commercial printing instead.

Today, Xerox still maintains strong profits in the high-end segment—but its overall scale is a shadow of its former self.

Chapter 44
Xerox's Expansion at the Frontier: Tried, Struggled, and Failed

In *Fumbling the Future: How Xerox Invented, Then Ignored, The First Personal Computer*, bestselling American author Douglas K. Smith writes:

> *The scientists at Xerox PARC created more than a personal computer. They designed, built, and used a complete system of hardware nd software that fundamentally altered the nature of computing itself. Along the way, an impressive list of digital "firsts" came out of PARC. In addition to the Alto computer, PARC inventors made the first graphics-oriented monitor, the first hand held 'mouse' inputting device simple enough for a child, the first word processing program for non expert users, the first local area communications network, the first object-oriented programming language, and the first laser printer.*[1]

Charles Geschke, co-founder of Adobe and a former member of PARC, once made a sharply insightful comparison between PARC's work and IBM's research, noting a core distinction in their vision: the computers at PARC *were used "not to calculate but to communicate.*"[2]

The PARC mentioned here refers to the Palo Alto Research Center, established by Xerox in 1968 in Palo Alto, California. At the time, computers were far from essential tools for engineers, and even Xerox's executives were unsure about the future role of computing. But with deep pockets and bold ambition, Xerox managed to recruit Bob Taylor, who had served as director of the Information Processing Techniques Office at the Pentagon, to lead the new research center. Although not a computer scientist himself, Taylor had remarkable skill in managing researchers and directing projects. He was one of the key figures behind ARPANet, the precursor to the internet, earning him the moniker "father of the internet." He had a unique eye for talent in the computing field.

Taylor went on to recruit some of the brightest minds in computer science. Xerox poured significant resources into PARC and gave it considerable autonomy. But this very freedom kept PARC somewhat detached from Xerox's core operations. Under

---

[1] Smith, D. K., & Alexander, R. C. (1999). *Fumbling the future: How Xerox invented, then ignored, the first personal computer*. iUniverse. p14

[2] Pfiffner, P. (2002). *Inside the publishing revolution: The Adobe story*. Adobe Press. p18

Q. Yue, F. Yang, *Governing Innovation: A Global History of Patent-Driven Technological Progress*, https://doi.org/10.1007/978-981-95-5570-3_44

Taylor's leadership, PARC's experts fulfilled their mission with distinction. They designed a multifunctional personal computer known as the "Alto," producing over 40 working prototypes. To make it user-friendly for non-specialists, they invented the mouse, developed a graphical user interface, created WYSIWYG (what-you-see-is-what-you-get) word processing software, and pioneered Ethernet, a technology that allowed computers to communicate with each other.

By 1979, though PARC had yet to produce a commercial success, it had already earned a solid reputation in the tech community. During a business discussion with Xerox, Apple's Steve Jobs requested a visit to PARC, which Xerox granted. There, a researcher named Larry Tesler demonstrated the Alto computer for him. Tesler would later become Apple's Chief Scientist and is best known as the inventor of the now-familiar "cut-copy-paste" functionality.

Jobs was enthralled by the graphical user interface and intuitive design of the Alto. He asked Tesler, "Why isn't Xerox selling this? You'd blow everyone away!" A few months later, Tesler joined Apple, and the company began developing the Lisa computer. In 1983, the Lisa became the first Apple computer to feature a graphical user interface.

When Microsoft later launched its Windows operating system with a GUI, Jobs was furious and sued Microsoft, initiating a legal battle that lasted for years. In the final judgment, the court explicitly stated: "*Apple cannot get patent-like protection for the idea of a graphical user interface, or the idea of a desktop metaphor which* ***concededly came from Xerox***."[3]

To complement the GUI, researchers at PARC had introduced several innovations, including the use of the mouse and onscreen cursor.

The mouse was not originally a Xerox invention. In 1968, Dr. Douglas C. Engelbart of the Stanford Research Institute created the first mouse—a wooden shell housing two heavy steel rollers. Positioned perpendicular to each other, these rollers allowed the cursor to move horizontally and vertically on the screen. Engelbart showcased this invention at an academic conference, hoping to demonstrate that computers could interact with humans in multiple intuitive ways. Stanford Research Institute filed a patent for the invention (US3541541A), but it was never commercialized.

Several of the computer scientists Bob Taylor recruited to PARC had been Engelbart's students. They improved upon his rudimentary mouse design, secured patents, and developed an international patent portfolio.

Engelbart's original mouse was crude, consisting of two cylindrical rollers that restricted cursor movement to strictly horizontal and vertical directions. Xerox's US3835464A patent introduced significant improvements: replacing the two rollers with a trackball and four miniature rollers positioned around it, enabling movement in any direction. The later US3987685A patent built upon this by replacing the four rollers with a pair of perpendicular shafts, and adding a spring-loaded biasing

[3] Apple Computer, Inc. v. Microsoft Corp., 35 F.3d 1435 (9th Cir. 1994) Retrieved April 8, 2024 from https://law.justia.com/cases/federal/appellate-courts/F3/35/1435/605245/

element to keep the trackball in contact with the shafts. This design became the basis for most mechanical mice in the 1990s. Xerox's three-button mouse, developed in the 1970s, closely resembled today's mouse in both function and appearance. Figures 44.1 and 44.2 show Engelbart's wooden mouse and Xerox's early 1970s three-button model, respectively.

Mechanical mice from the 1990s to early 2000s generally followed the design of US3987685A (see left side of Fig. 44.3), where a biasing element presses a trackball against a pair of perpendicular shafts. Xerox sought international protection for this design, filing patents in the UK, Canada, Germany, France, the Netherlands, Sweden, Japan, and Switzerland. However, its applications in Germany and France were later withdrawn, and the Swiss patent (CH607053) was revoked in 1988 due to non-payment of fees. Whether this lapse was intentional or not remains unclear, but what is certain is that one of the world's most renowned mouse manufacturers—Logitech—is a Swiss company. The right side of Fig. 44.3 shows the internal structure of Microsoft Serial Mouse 2.0A mechanical mouse from the 1990s.

Xerox's mouse cost approximately $300 to produce. After witnessing it at PARC, Steve Jobs immediately contacted a supplier to design a simpler version. He laid out two demands: the design must be simple, with only one button; and the cost must be under $15.

In 1981, Xerox released the Star 8010 system, which included its three-button mouse—the first commercially available mouse on the market. Though Apple's Lisa, released in 1983, generated more attention, its $10,000 price tag limited its success. It was not until 1984, with the launch of the Macintosh at $2500, that the graphical user interface and mouse gained widespread adoption. In a deliberate move to force users to adopt the mouse, Jobs even removed the arrow keys from the Macintosh keyboard.

Thanks to Jobs's charisma, both the Lisa and Macintosh were hailed as revolutionary products—especially the Mac line, which lives on to this day. Meanwhile, Xerox's Star system has largely faded into obscurity.

Ethernet was another groundbreaking invention to emerge from PARC. In 1973, with a number of Alto computers already operational within PARC, engineer Robert

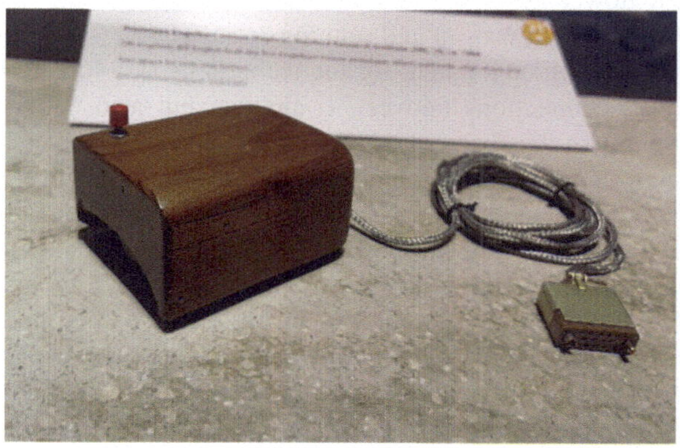

Fig. 44.1 Engelbart's wooden mouse. (Photograph by the author at the Computer History Museum)

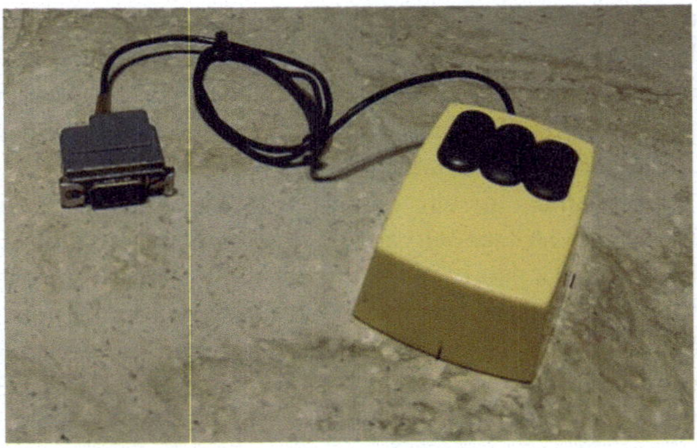

Fig. 44.2 Xerox's early 1970s three-button mouse. (Photograph by the author at the Computer History Museum)

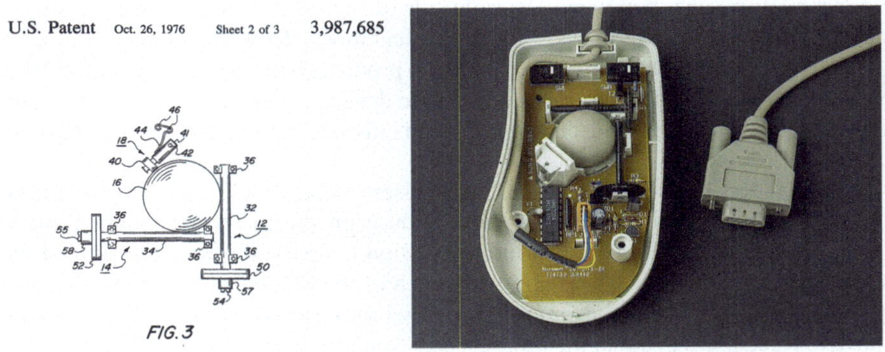

Fig. 44.3 Left—Xerox's US3987685A patent drawing; Right –Microsoft Serial Mouse 2.0A mechanical mouse (1990s) (https://commons.wikimedia.org/wiki/File:Microsoft_Serial_Mouse_2.0A-8524.jpg)

Metcalfe devised a standard for short-range computer networking. On May 22 of that year, he scribbled the phrase "Alto Ethernet" in his notebook along with a simple flowchart—thus, Ethernet was born. By November, the first Ethernet-based network system was up and running.

Xerox filed a patent for the technology (US4063220A). Ethernet not only managed connections between computers but also between computers and printers, hinting at the shape of the future in the PC era. Despite this, Xerox never commercialized Ethernet as a standalone product. Frustrated, Metcalfe briefly left Xerox but was persuaded to return—only to find that upper management still refused to greenlight commercialization of his Ethernet innovation. In 1979, having lost all faith in Xerox, Metcalfe paid $1000 out of his own pocket to secure a license for US4063220A and struck out on his own. With support from Intel and DEC, he founded 3Com to manufacture and sell Ethernet adapter cards.

Metcalfe proved prescient. Initially, he intended to emulate IBM and Xerox by selling directly to large enterprise clients in the UNIX workstation market. But then the IBM PC exploded onto the scene, giving rise to an entirely new PC market. Metcalfe pivoted sharply, abandoning the IBM/Xerox model in favor of selling Ethernet gear through distributors. In 2009, HP acquired 3Com for $2.7 billion in cash—by comparison, Xerox's total assets in 2017 amounted to just $15 billion.

The computer scientists at PARC fully understood the significance of printing technology to Xerox. In 1974, they completed Bravo, a word processing program designed for non-specialists. Thanks to the graphical user interface of the Alto, Bravo became the first software to deliver true "what you see is what you get" printing. One of Bravo's key developers, Charles Simonyi, not only coined the term WYSIWYG but would later gain fame as the father of Microsoft Word and Excel. In 1981, on Metcalfe's recommendation, Simonyi pitched himself to Bill Gates and left Xerox for Microsoft, becoming a core figure there. Long one of the highest-paid software developers in the industry, Simonyi ranked 161st on the 2020 Forbes list of America's richest people, with a net worth of $4.5 billion. Although he held many important patents at Microsoft, Simonyi had no patents to his name during his time at Xerox.[4]

Despite PARC's abundance of revolutionary ideas, the 200+ scientists in this team, largely isolated from Xerox's corporate core, consistently failed to get their innovations accepted by company leadership. Although Bob Taylor had a sharp eye for talent, he was four management layers removed from Xerox's top executives. The entire PARC operation was far removed from upper management, even farther from marketing departments, and utterly disconnected from end users. These computer geniuses had no direct understanding of what ordinary users truly wanted. Many commercialization proposals—such as for Metcalfe's Ethernet—were blocked long before they ever reached decision-makers.

To make matters worse, PARC's most fertile period coincided with a critical moment when Xerox was being challenged by IBM, Kodak, and Japanese firms in the copier market. Focused solely on its core business, Xerox's leadership had neither the capacity to pursue new markets nor the foresight to realize they held the keys to the next technological era.

It wasn't until 1981 that Xerox released the Xerox Star 8010 system, which incorporated many of Alto's innovations. Yet, as with the Alto, the Star floundered—its creators had little understanding of customer needs, and the machine simply couldn't compete. Priced at a staggering $16,595, it was far beyond the reach of ordinary users.[5] By contrast, Apple's 1977 Apple II sold for just $1298 for the 4K model and $2638 for the 48K version. The IBM PC, launched that same year and hailed as a revolutionary force, was priced under $3000.

With the advent of the IBM PC, the era of PC clones took off, eventually dethroning mainframes as the dominant tool for office work. Alongside this shift, small and

---

[4]Andrew Pollack. (1990). Most of Xerox's Suit Against Apple Barred.Retrieved April 8, 2024 from: https://www.nytimes.com/1990/03/24/business/most-of-xerox-s-suit-against-apple-barred.html

[5]*The Xerox Star 8010 "Dandelion"*. (n.d.). DigiBarn Computer Museum. Retrieved March 31, 2024 from https://www.digibarn.com/collections/systems/xerox-8010/index.html

mid-size printers—once Xerox's domain—became essential business devices. Xerox, once a market titan with the key to the PC age in its hands, failed to capitalize on personal computing and was overtaken in the printing and copying market by Canon and HP. In the new era, Xerox was toppled from its throne.

This story is one well worth contemplating.

The story of Xerox serves as a reminder that markets are like territory—if a company doesn't claim them, its competitors will. In the 1970s, both the low-end copier segment and the nascent personal computer industry were within easy reach for Xerox. Yet for various reasons, the company failed to act, squandering its clear advantage and allowing emerging rivals to surge ahead. By 2000, Xerox was reeling from massive financial losses. A scandal involving accounting fraud at its Mexico division sent its stock plummeting, and the company teetered on the brink of bankruptcy. It was only under a new CEO, who decisively abandoned the low- and mid-range markets, cut costs, laid off staff, and refocused on premium clients, that Xerox managed a turnaround. Today, it remains a multi-billion-dollar corporation—but given the immense possibilities it held in the 1970s, one cannot help but lament the lost potential.

In 1990, U.S. President George H. W. Bush visited Hewlett-Packard. In his remarks, he enthusiastically thanked HP engineers for "inventing the laser printer," praising them for their contribution to technological advancement. It was the sort of gaffe more commonly associated with his son, George W. Bush, yet it vividly illustrates HP's dominance and enduring impact in the laser printer market. In reality, the laser printer was not only invented by Xerox but also successfully commercialized by the company. It secured a strong market foothold—yet even that success faded into obscurity amid the crushing failures of a new era.

Today, most copiers, printers, and scanners are integrated into multifunction devices, and many users may not realize that copiers and computer printers hail from two distinct technological eras. A copier produces its output by scanning a physical original; as mentioned earlier, the principles of reprographic duplication were already established by the early twentieth century. The process involved photographing documents and required neither computers nor digital technology. In contrast, the ancestor of the computer printer was the typewriter, which actually predates the copier. But it wasn't until advancements in word processing, image rendering, and network communication protocols that printers could be tightly integrated with computers. And as noted, these were precisely the technological domains where PARC excelled. The rise of the printer market was inseparable from the spread of personal computers. By the late 1970s, Xerox had decades of experience in designing and manufacturing copiers, while PARC had developed a suite of computer technologies. If the company had simply combined these strengths, even if it failed to dominate the personal computer market, it could have secured leadership in printing.

Many of PARC's inventions were overlooked by Xerox. But as a company born from the copier business, Xerox was never going to miss the laser printer. It was the perfect fusion of PARC's innovations and Xerox's core expertise. A close examination of its patents reveals that this product could have served as Xerox's beachhead in the digital age.

Gary Starkweather, widely recognized as the inventor of the laser printer, joined Xerox in 1964. At first, his team worked on a project related to data transmission—specifically, linking two copiers so that one could scan and the other reproduce the image remotely. During this work, Starkweather discovered that laser scanning yielded far more precise images. This insight sparked a bold idea: What if the copier were taken out of the equation entirely? What if a computer could control a laser to produce a print directly—creating a printer, not a copier?

His supervisor wasn't impressed. At the time, Xerox had no intention of entering the computer business. Even its image transmission projects were geared toward enhancing copiers and fax machines. Starkweather's concept hinged on computers, which placed it far outside the company's strategic focus. Compounding this, laser hardware was still prohibitively expensive. Even if Starkweather's vision could be realized, the resulting product would be costly and unlikely to generate meaningful profit.

But Starkweather remained convinced that laser technology would eventually become affordable and that laser printing was the way of the future. He cordoned off a small area in his office with a curtain and continued his work in secret. When his boss found out, he repeatedly tried to persuade him to give it up, even threatening to fire him. But Starkweather stood firm. Around 1969, he succeeded in modifying a Xerox copier into a prototype laser printer—but at that time, there was still no computer system capable of interfacing with it.

In 1970, the Palo Alto Research Center (PARC) was established. Upon hearing the news, Gary Starkweather lobbied hard to transfer from Xerox headquarters to PARC—and was granted approval.[6] By 1971, with the support of PARC's computer scientists, Starkweather completed the design of an entire laser printing system.

It's worth clarifying that laser printing does not involve engraving letters onto paper with a laser. Instead, it uses a laser to scan a photoconductive drum, forming an electrostatic image on its surface. Toner is then attracted to this image and fused onto paper through high heat. This process is fundamentally similar to Chester Carlson's original xerographic technique. Achieving high-speed, high-precision laser scanning on the drum couldn't rely on simply moving the laser emitter rapidly. Starkweather's solution was a 24-faceted rotating mirror, about the size of a doughnut, mounted on a central spindle driven by a motor. As the spindle rotated, the mirror turned, reflecting the laser beam across the drum to perform the scan. Xerox filed a patent for this rotating polygon mirror (US3867571A; see Fig. 44.4). Half a century later, this polygonal mirror remains a core component in modern laser printers.

Yet this mirror was not the most important innovation in PARC's laser printing system. The true significance was embedded in the system's name: EARS. The

[6] Cade Metz. (2020). Gary Starkweather, Inventor of the Laser Printer, Dies at 81. Retrieved April 8, 2024 from https://www.nytimes.com/2020/01/15/technology/gary-starkweather-dead.html

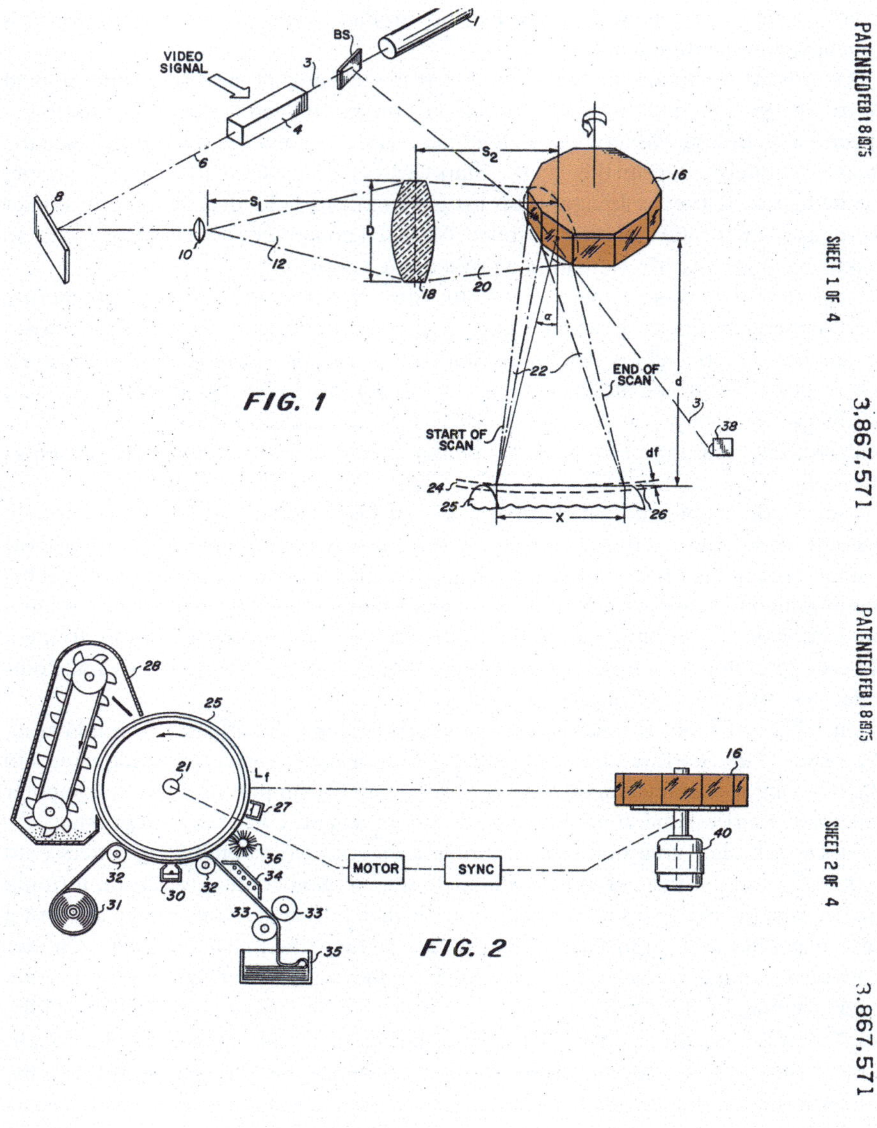

Fig. 44.4 Patent drawing from U.S Patent US3867571A. The white section represents the polygon mirror

acronym stood for Ethernet (E), Alto (A), Research Character Generator (R), and Scanned Laser Output Terminal (S). The "R" component, the Research Character Generator, or RCG, was responsible for converting digital information into signals readable by the printer. It was designed primarily by Butler Lampson, who, along with Charles Simonyi, developed the Bravo word processor mentioned earlier—enabling WYSIWYG document creation and printing. The "S," or Scanned Laser Output Terminal, was Starkweather's original name for the laser printer. Together, Ethernet linked the computer and printer, the Alto offered a graphical interface, Bravo allowed for intuitive document editing, and the laser system completed the printing process. In short, the EARS name represented a convergence of some of PARC's most groundbreaking technologies. The resulting laser printer became one of Xerox's blockbuster products in the late 1970s.

This single printer incorporated a host of epoch-defining technologies. The designer of its printing protocols would later leave Xerox to found Adobe. Yet what Xerox brought to market was merely a printing system—not the broader computing revolution it had in its hands.

Xerox productized the EARS system by integrating it with other in-house technologies, ultimately resulting in the imposing Xerox 9700 laser printer, launched in 1977. Staying true to Xerox's tradition of high performance, the 9700 boasted a resolution of 300 DPI, could print two pages per second, and up to 120 pages per minute. Its automation was impressive: it retained auxiliary paper trays, with the main tray holding 2500 sheets and a secondary tray handling 400 sheets. On the output end, it featured two stackers, each with a capacity of 1500 pages. It came with a CRT display and keyboard control console and could interface with IBM mainframes via parallel channel.

The Xerox 9700 sold for $350,000 and remained one of the fastest, highest-capacity printers on the market for years. Unfortunately, this success coincided with the rise of the personal computing era, led by IBM, Apple, and Microsoft. Offices were filling up with personal computers, and users disliked trekking to central print rooms to output a few pages. Bill Gates envisioned "a computer in every home," while HP and Canon imagined compact desktop printers in every office. For Xerox, the 9700 was one of its few profitable ventures related to computing.

In 1977, Xerox filed 365 patents—but only 13 included the keyword "comput*," three of which were design patents. Patents specifically tied to the 9700 were limited, including US3884408A, which covered the machine's stapling and paper handling mechanisms. Many of the transformative innovations from the EARS system—like the graphical interface or WYSIWYG software—weren't protected under these patents.

There's no denying that PARC researchers had a real need for laser printing, but they mainly saw it as an output device for Ethernet-connected workstations.[7] They didn't design with the broader needs of white-collar workers in mind. For instance, contemporary printers lacked features now considered standard, such as feedback mechanisms for users. In 1980, MIT's AI Lab purchased a Xerox 9700. Richard Stallman, who worked there at the time, was used to modifying printers to notify users about current print jobs or paper jams. However, he had no access to the 9700's driver source code and couldn't make these changes.

---

[7] Strassmann, P. A. (2008). *The Computers Nobody Wanted: My Years with Xerox*. Lulu.com.

Later, he learned that Carnegie Mellon University had the source code. He asked to see it, assuming—as was typical in the programmer community—that such requests were routine and usually granted. But CMU declined, citing confidentiality agreements. What now seems perfectly reasonable struck Stallman as outrageous: "I paid for this machine—why shouldn't I be able to modify a few lines of code?" Outraged, he launched the free software movement, started the GNU Project, and founded the Free Software Foundation.[8] Thus, the Xerox 9700 became a catalyst for one of computing's most influential ideological movements.

If 1977 marks the birth of the laser printer, the question then becomes: what were Xerox's competitors doing at the same time?

[8] Williams, Sam (2002). *Free as in Freedom: Richard Stallman's Crusade for Free Software*. O'Reilly Media. ISBN 0-596-00287-4. Chapter 1.

Chapter 45
Hewlett-Packard's Boundary Expansion: Instruments, Computers, and Printers

Hewlett-Packard Company, or HP, was founded in the 1930s in a garage by its two founders—a tale that has since become one of Silicon Valley's most iconic "garage startup" origin stories. From its inception, HP's primary business focus was the design, manufacture, and sale of various measurement and testing instruments. Like Harold Corporation, HP secured numerous military contracts during World War II, amassing substantial capital. By the 1960s, HP's product lines included medical instruments, gas chromatography analyzers, mass spectrometers, and laser measurement and ranging devices. Lasers were one of HP's areas of technical expertise, and it held far more laser-related patents than Xerox.

Following the invention of the transistor at Bell Labs, HP entered the emerging semiconductor industry. While applying transistor technology to improve its measurement equipment, HP also actively explored new possibilities enabled by semiconductors.

Figure 45.1 illustrates HP's U.S. patent filings involving transistors throughout the twentieth century. HP filed its first transistor-related patent in May 1958 (US3441833A), applying the technology to a voltage regulator. The following year, it filed two more patents for transistor voltage regulators (US3105188A and US3101442A), as well as a patent for using transistors in a sampling oscilloscope (US3011129A). This demonstrated HP's ability to ride the wave of technological advancement in its core domain. In the 1960s, HP filed a total of 71 transistor-related patents, making up 16.6% of its total 429 U.S. patents for that decade. Transistor amplifiers became a key focus of HP's patent activity. Pursuing technological breakthroughs at the scientific frontier—this was one of the reasons behind HP's success.

Looking at HP's patents from the 1960s, we can see the company's earliest forays into computer technology. In August 1968, HP filed US3530440A for a "typical basic computing system"—likely the company's first computer-related patent. That same year, it filed a patent for a "digital calculator system for computing square roots" (US3576983A), followed by three additional calculator-related patents in 1969 (US3641328A, US3668461A, and US3623156A).

Q. Yue, F. Yang, *Governing Innovation: A Global History of Patent-Driven Technological Progress*, https://doi.org/10.1007/978-981-95-5570-3_45

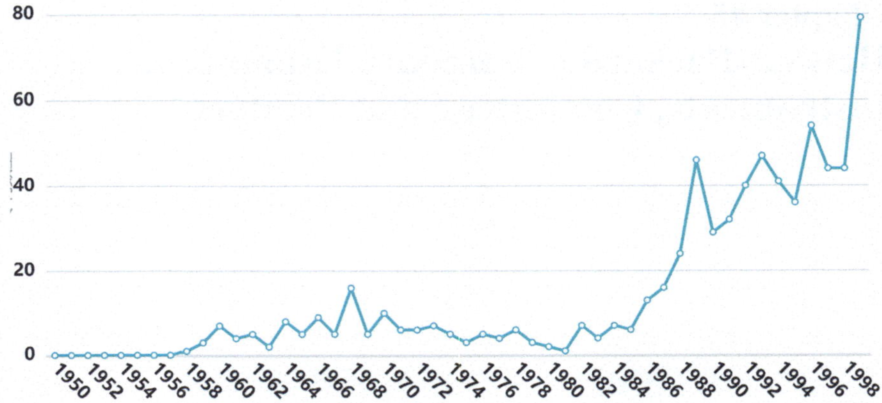

Fig. 45.1 Number of HP U.S. patent filings involving transistors during the twentieth century

On the product side, HP launched the HP 2116A minicomputer in 1966, with a price range of $25,000 to $60,000—marking its entry into the computer industry.

Why did HP choose to venture into computing? After the invention of the transistor, computing technology advanced at a breakneck pace. Many corporations, universities, and research institutions began purchasing or leasing IBM mainframes—HP included. Researchers often had to manually input data collected with HP instruments into IBM computers for analysis. In conversations with HP's sales reps, they began requesting smarter instruments—devices that could feed data directly into a computer, eliminating the need for manual data entry.[1]

HP's founders, Bill Hewlett and Dave Packard, took note of these customer needs and made the strategic decision to develop computing products. During this process, the design team proposed integrating computers with HP's other products. Packard recalled: "Their [the team's] vision was a complete system, with HP instruments connected to our printers and plotters, all controlled by HP computers." He believed that computers could serve as the "glue" binding HP's other offerings together and propelling the company forward.[2] Hewlett and Packard studied the products of DEC and Wang Laboratories, ultimately choosing to build the 2116A based on DEC's architecture.

The 2116A was an "instrumentation computer," designed primarily for engineers working with measurement devices—it was still very much aligned with HP's core business. Initially, HP expected universities and research institutions to be the primary customers for the 2116A. But to the company's surprise, the machine became a runaway success in the industrial sector, leaving Hewlett and Packard astonished. In hindsight, HP realized the 2116A had a hidden advantage: durability. Having

[1] Malone, M. S. (2007). *Bill & Dave: How Hewlett and Packard built the world's greatest company*. Penguin. p173

[2] Malone, M. S. (2007). *Bill & Dave: How Hewlett and Packard built the world's greatest company*. Penguin. p175

spent decades building instruments tough enough for harsh conditions—from oil fields and workshops to the open sea and even outer space—HP's engineers had unconsciously carried these rugged design standards over into the new product. Reflecting on this, Hewlett said, "(The 2116A) had an unexpected bonus we hadn't fully realized at first: our instruments had to exceed 'B-grade' shock and vibration standards… So naturally, our computers did too."

At the time, most other minicomputers were delicate and fragile. HP's machines, by contrast, could operate in higher-temperature, more rugged environments. The 2116A's first customer, the Woods Hole Oceanographic Institution, installed the computer on a research vessel where it performed reliably amid constant rocking and vibration—helping HP establish a strong reputation for reliability.

The robust quality that HP had developed in response to demanding use cases turned out to be a critical factor for success in new markets. Many companies aspire to expand their boundaries, but intention alone is not enough. Capital and technological expertise developed in core areas, experience across varied scenarios, and high-quality products and services—these are the true enablers of successful expansion.

In 1965, mathematician Malcolm McMillan, building on Jack Volder's CORDIC (COordinate Rotation DIgital Computer) algorithm, co-developed a scientific calculator. McMillan later approached HP to pitch this invention. HP's engineers admired the calculator's software design but deemed its hardware a mess. Coincidentally, an independent inventor named Tom Osborne had just developed a compact electronic calculator with a keyboard and display (see Fig. 45.2), which he brought to HP. Hewlett was so taken with Osborne's elegant design that he couldn't put it down. HP eventually combined McMillan's software with Osborne's hardware, launching the HP 9100A calculator in 1968.

The HP 9100A was considerably larger than the portable calculators Texas Instruments was releasing at the time, but the two were not meant to serve the same market. For years, HP insisted on calling the 9100A a "programmable calculator" rather than a computer. All of its patent filings used the term "programmable calculator." Yet decades later, some IT professionals and media outlets looking back would hail the 9100A as the first personal computer in history.[3] Bill Hewlett explained the naming decision as follows: "*If we had called it a computer, it would have been rejected by our customers' computer gurus because it didn't look like an IBM. We therefore decided to call it a calculator, and all such nonsense disappeared.*"[4]

At the time, IBM's mainframes sold for hundreds of thousands of dollars and were so large they could fill several office rooms. Despite being called "minicomputers," HP's own machines were anything but miniature—each

[3] Aaron, C. (2000, December). *Wired 8.12: Must read: The first PC. Wired.* Retrieved March 31, 2024 from https://web.archive.org/web/20151004045031/http://archive.wired.com/wired/archive/8.12/mustread.html?pg=11

[4] Hewlett-Packard. (1968). *History of the 9100A desktop calculator.* Hewlett-Packard. Retrieved March 31, 2024 from http://www.hp.com/hpinfo/abouthp/histnfacts/museum/personalsystems/0021/0021history.html

Fig. 45.2 Tom Osborne's electronic calculator design. (Source: Daderot. (2015, March 27). *Electronic calculator prototype, made by Thomas E. Osborne for Hewlett-Packard, 1964, view 2 – National Museum of American History* [Image]. Wikimedia Commons. https://commons.wikimedia.org/wiki/File:Electronic_calculator_prototype,_made_by_Thomas_E._Osborne_for_Hewlett-Packard,_1964,_view_2_-_National_Museum_of_American_History_-_DSC00028.jpg. This image is in the public domain (CC0 1.0 Universal))

about the size of a large wardrobe. So from the perspective of the era, it was difficult to associate the $5000 HP 9100A with the idea of a 'computer'.

The team behind the HP 9100A came from various departments. Dave Cochran, an engineer who had been designing voltmeters 1 day, found himself at a calculator development meeting the next, listening to discussions about algorithms. His first question to the presenter was, "What's an algorithm?" Even Dick Monnier, the project lead for the HP 9100A, had no prior experience in computing—his previous project was an oscilloscope.

Bill Hewlett had a specific vision for the HP 9100A: it needed to be compact enough to fit in an office environment, about the size of a typewriter. At one meeting, he laid out the requirement that it must sit neatly on a typewriter stand and be small enough to slide into the kneehole of his walnut office desk.

In 1967, the design team completed a prototype of the HP 9100A. While Hewlett happened to be out of the office, several members of the team barged into his workspace to test it. The prototype fit perfectly on the typewriter stand, but when they tried to slide it into the desk's kneehole, it got stuck. Thinking quickly, one engineer ran downstairs to fetch a saw, trimmed away an eighth of an inch from the desk's opening, and got it to fit. Legend has it that Hewlett soon noticed the alteration—but always pretended not to.[5]

[5] Malone, M. S. (2007). *Bill & Dave: How Hewlett and Packard built the world's greatest company*. Penguin. p188

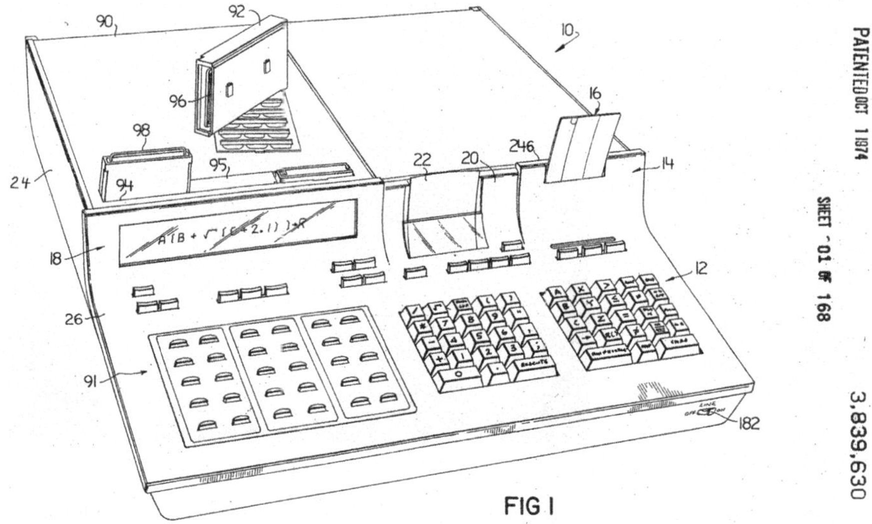

Fig. 45.3 Drawing from patent US3839630A

HP continued refining this "programmable calculator" and filed a patent at the end of 1971 for a "programmable calculator using algebraic language" (US3839630A). Over the next few years, HP filed more than 20 related patents. Notably, while Xerox was still indecisive about its computing strategy, HP's programmable calculators were already outfitted with basic printing capabilities. In Fig. 45.3 from the patent, component 20 is the print unit, while component 22 holds the paper. One of the divisional patents, US4152771A, specifically protected a "display mechanism that signals to the user when the printer is out of paper." Fig. 45.4 shows a real-world image of the HP 9100A calculator.

It's worth noting that this wasn't HP's first attempt at print-enabled devices. As mentioned earlier, HP's core business was measurement instruments, and "the palest ink is better than the best memory." No matter the type of instrument, users ultimately needed to record the results. Over time, HP engineers grew tired of jotting down data by hand and devised a printing solution for one of their signal generators. This resulted in the 1957 release of the Model 560A Digital Recorder.[6]

The term "digital" here did not refer to digital technology in the modern sense—it simply meant "numerical." According to HP's 1956 patent (US2864307A), which covered this "printer," the 560A Digital Recorder functioned like a printing press, using rotating drums with engraved digits to imprint numbers onto paper. From today's perspective, calling this machine a "printer" might be generous—it had a print speed of just five lines per second, each line containing 11 digits. Still, this was

[6] Fleming, L. (2002). Finding the organizational sources of technological breakthroughs: the story of Hewlett-Packard's thermal ink-jet. Industrial and Corporate Change, 11(5), 1059–1084. https://doi.org/10.1093/icc/11.5.1059

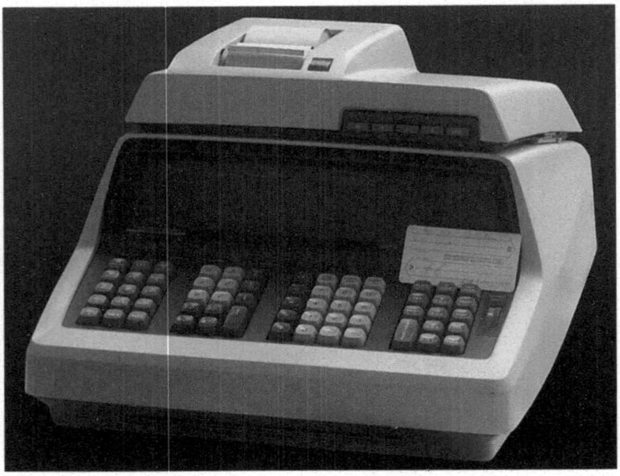

Fig. 45.4 HP 9100A Programmable Calculator. (Source: Rama & Musée Bolo. (2011, October 25). *HP 9100 A computer, on display at the Musée Bolo, EPFL, Lausanne* [Image]. Wikimedia Commons. https://commons.wikimedia.org/wiki/File:HP9100A-IMG_0393.jpg. This image is shared under the Creative Commons CC BY-SA 2.0 France license)

far faster than any human could manage and made it possible to record measurements that changed multiple times per second. The device was a hit upon release and laid an unintentional foundation for HP's later ventures into computer printing. After all, if measuring devices required printers to output results, wouldn't computing machines need the same?

Thus, equipping programmable calculators with printing capabilities was a logical evolution. In hindsight, US2864307A and the 560A Digital Recorder represent HP's first steps into the printer market (Fig. 45.5).

In its early years, HP focused its printing patents primarily on inkjet technologies. It wasn't until around 1988 that the company began filing laser printer-related patents, such as US4847641A.

In 1981, HP released its first laser printer, the HP 2680A, which printed 45 pages per minute and was priced at $108,500. Compared to Xerox's 1977 Xerox 9700 laser printer, the HP 2680A was more than 50% slower but significantly smaller and cheaper, making it one of the most cost-effective options on the market at the time. Universal print protocols didn't exist yet, and the HP 2680A was only compatible with HP's own HP 3000 minicomputer, which sold for around $120,000. However, the HP 3000 offered extensive software support for the printer, including a $7500 "Interactive Design System" and a $5000 "Interactive Formatting System." Additionally, HP developed one of the earliest WYSIWYG (What You See Is What You Get) presentation software programs, called *Bruno*. Today we would classify it as slide presentation software—Microsoft PowerPoint being the most recognizable. Bruno was later renamed *HP-Draw* and could be installed on HP 3000 systems.

HP's printer division had high hopes for the HP 2680A, expecting to sell 75 units in the first month. In reality, not a single unit was sold. Despite the

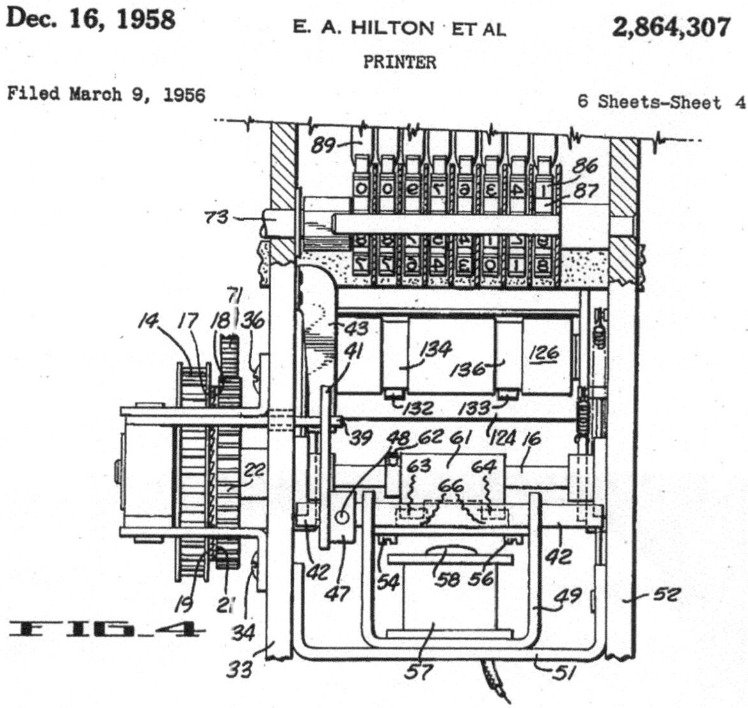

Fig. 45.5 Drawing from HP's 1956 "printer" patent (US2864307A)

rough start, HP kept the product line alive, and by 1987, the printer had sold a total of 1500 units.[7]

While the HP 2680A itself wasn't a commercial hit, it provided HP's printing team with invaluable early experience and foreshadowed the eventual convergence of computers and printers as a new growth frontier. In the 1980s, there were 2348 U.S. patent applications that included both the keywords "comput*" and "print*," with HP holding at least 30 of those—trailing only IBM and Pitney Bowes among U.S. companies (see Fig. 45.6). As personal computers and printers became new business domains for HP, they began to interact in a way that created synergies far exceeding their individual contributions.

In May 1984, HP released its first desktop laser printer, the **HP LaserJet** (see Fig. 45.7). Compact enough to sit on a desk and priced at $3500, it delivered impressive print quality and became a hit. The LaserJet series and HP's PCs helped fuel each other's sales success.

[7] *Title of the page*. (n.d.). *The Museum of HP Calculators*. Retrieved March 31, 2024 from http://www.hpmuseum.net/display_item.php?hw=442

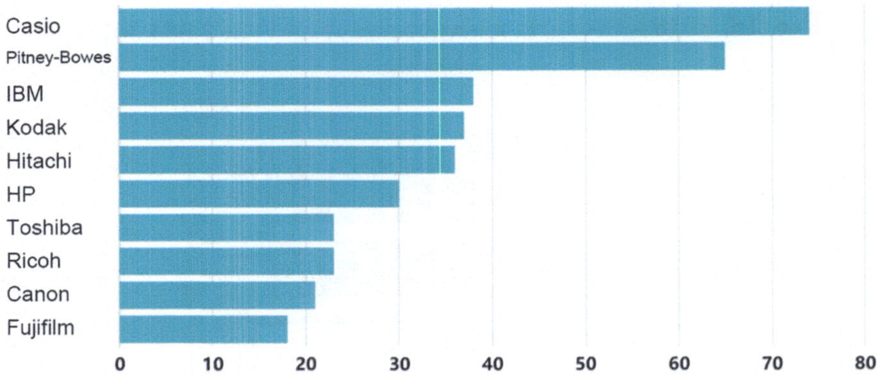

Fig. 45.6 Top 10 applicants in 1980s U.S. patents containing both "comput*" and "print*"

Fig. 45.7 1984 first-generation HP LaserJet 1. (Source: Atomic Taco. (2013, September 19). *HP LaserJet I* [Image]. Wikimedia Commons. https://commons.wikimedia.org/wiki/File:HP_LaserJet_I_%2812935740974%29.jpg. This image is shared under the Creative Commons CC BY-SA 2.0 license)

Jim Hall, then R&D manager in HP's printer division, recalled that in the early 1980s, HP's main priority was PCs, and the PC division held sway over resources and influence with distributors. After LaserJet development concluded, the printer team begged the PC sales unit to include the LaserJet in their presentations to dealers. The PC team reluctantly agreed but relegated the printer to the end of the agenda. By that point, many senior attendees had already left the room. But as soon as the LaserJet demo began, the remaining staff rushed out to bring their superiors back in. In the end, every distributor present opted to purchase both PCs and printers.[8] Later, some even committed to HP-150 and Vectra PC orders just to secure printer supply.[9] Within a year, HP captured 85% of the desktop laser printer market. The first-generation LaserJet sold 500,000 units, far surpassing Xerox's Xerox 9700 in influence.

During this era, HP's PC business moved from the periphery into its core business, while the printer division became a new frontier. This interplay between the core and frontier drove HP's phenomenal success in the 1980s and 1990s.

Interestingly, between the "laser printer year zero" of 1977 and the release of the first LaserJet, HP had not filed a single patent specifically on laser printers. HP had 12 U.S. patent applications related to "laser," covering topics like semiconductor lasers, electro-optic circuits, and photolithography methods—but none explicitly mentioned printing. Of its 35 "printing" related patents, 11 dealt with thermal printers for programmable calculators, and most of the rest focused on inkjet printing. At the time, HP's leadership had more confidence in inkjet printing as the company's pathway into the mid-to-high-end market. In terms of laser printing technology alone, HP's contributions were minimal compared to Xerox's foundational innovations. HP's laser printer engines were sourced from Japanese suppliers: the HP 2680A used a Canon engine, later replaced by Ricoh, and eventually switched back to Canon's **CX engine**—known for its replaceable cartridge design.

HP's real technological strength and patent activity in desktop printing lay in **inkjet printers**, where the company achieved significant consumer-market success. As noted in the "Rise and Fall" chapter, Antonio Pérez, later named one of the "Worst CEOs of 2011" for his time at Kodak, once led HP's inkjet division. During his 25-year tenure at HP, Pérez believed inkjet printing's future lay in the consumer market, not commercial use. Laser printer prices were falling fast, rendering inkjet non-competitive in the mid-to-high-end sector. Large corporate clients still trusted IBM and Xerox and wouldn't pay a premium just for the HP brand. But the booming consumer PC market created a new opportunity for affordable, low-end inkjet printers. Pérez urged HP to pivot toward simplicity and ease of use.[10]

In 1984, HP launched the **HP 2225** inkjet printer. Though it had many flaws—such as requiring special paper, having delicate and expensive printheads, and even including instructions on fixing the printhead with a paperclip—it was incredibly compact, even smaller than most of today's inkjets. The 2225 came in multiple variants offering different connectivity:

- HP 2225A: HP-IB interface and built-in power supply
- HP 2225B: HP-IL interface and battery-powered

[8] Hall, J. HP LaserJet–The Early History. Retrieved April 8, 2024 from https://www.hpmemoryproject.org/an/pdf/HP_LaserJet_early_history.pdf

[9] Malone, M. S. (2007). *Bill & Dave: How Hewlett and Packard built the world's greatest company*. Penguin.

[10] Lawrence Fisher. (1996). How Hewlett-Packard Runs Its Printer Division. Retrieved April 8, 2024 from https://www.strategy-business.com/article/12217

- HP 2225C: parallel port and built-in power supply
- HP 2225D: serial port connectivity

At a time when PC and commercial computing standards were still evolving, these options showcased the printer's versatility. Another standout feature was its quiet operation, which earned it a strong presence in libraries.

Inkjet printer sales began soaring in 1990. HP dominated the early '90s as the market leader, although it started to fall behind competitors by the mid-1990s. The inkjet printer followed a classic **razor-and-blade model**: the printer was inexpensive, while the ink cartridges were pricey. Cartridges would often clog if unused for a while, turning them into a key revenue stream for HP.

HP developed its own **Printer Control Language (PCL)** and, with backing from Microsoft, cemented a position at the heart of the personal computing revolution. During the fierce battle of the 1980s between IBM-compatible PCs and Apple's Macintosh, HP's integration of personal computers, printer protocols, WYSIWYG (What You See Is What You Get) software, and laser printers formed a powerful synergy—making HP a crucial pillar of the PC camp. It's important to remember that many of these technologies had their origins in the 1970s at Xerox.

HP released its **third-generation LaserJet** printer in 1990, featuring a new **Resolution Enhancement technology (REt)** for dramatically improved print quality. The accompanying **PCL 5** language offered advanced font-scaling capabilities. In 1991, the **LaserJet IIISi** debuted with built-in Ethernet support, becoming the first desktop laser printer to offer this feature. Ironically, the "E" in Xerox PARC's original **EARS** system had stood for Ethernet—yet it was HP that ultimately brought Ethernet-enabled laser printing to market success.

HP's LaserJet journey wasn't without competition. One of its earliest challengers was Apple's **LaserWriter**—a printer many may not even realize Apple once made. Released in 1985, the LaserWriter printed at 8 pages per minute and was priced at $6995. Like HP's LaserJet, it used the **Canon CX engine**. Though expensive, the LaserWriter supported **up to 16 connected Macs**, bringing the effective cost per machine down to about $450—cheaper than a LaserJet. Positioned between Xerox's expensive commercial printers and HP's budget-friendly models, the LaserWriter found initial success and even made Apple the world's largest printer company for a time during the 1980s.[11] But as HP's prices fell and the PC platform triumphed over the Mac, Apple's printer sales declined, and the company exited the market by the end of the twentieth century.

At the heart of the LaserWriter was **PostScript**, Adobe's foundational product. This technology was deeply linked to Xerox's legacy: Adobe's founders, **John Warnock** and **Charles Geschke**, had both worked at PARC and were instrumental in developing Xerox's **Interpress** printing protocol. After failing to convince Xerox to commercialize Interpress, the pair left to start Adobe.

Initially, Warnock and Geschke intended to build a hardware workstation for printing. But in 1983, **Steve Jobs** intervened. He added the LaserWriter to Apple's

---

[11] Pfiffner, P. (2002). *Inside the publishing revolution: The Adobe story*. Adobe Press. p50

development agenda and tapped the newly founded Adobe for software support. After seeing Adobe's demo, Jobs declared: *"I don't want your computer. I don't want your printer. I want your software."*

Jobs paid Adobe a **$15 million licensing fee** for PostScript and invested another **$25 million** to acquire **20% of the company**—shares that Apple would later sell for **$87 million** six years down the line. With Apple's support, Adobe pivoted from hardware to focus solely on software development.[12]

Though the **PC vs. Mac war** ended in a decisive victory for the PC side, Apple's alliance with PostScript carved out a lasting niche in the **design industry**. Even today, many still view the Mac as the platform of choice for creatives—an enduring legacy of Adobe's early support.

While PCs and Macs fought tooth and nail, **Xerox stood on the sidelines**, watching as its high-end commercial printer market was steadily eroded. It failed to enter the new consumer printing segment and had long since ceded the PC battlefield.

By **1992**, HP's printer sales had reached **$3 billion**, accounting for over **one-fifth of total revenue**.[13] In **1999**, HP spun off its legacy instrumentation business—its original cornerstone—into a separate entity, **Agilent**, which also included the medical devices division. In **2001**, HP reorganized into three business units: **Computer Systems**, **Imaging and Printing Systems**, and **IT Services**. That year, printing brought in **$19.447 billion** in net revenue, surpassing the **$17.771 billion** from computing and the **$7.59 billion** from services—making printing the **core pillar** of HP's business.[14] Michael S. Malone, a prominent American journalist, once called HP "*the world's greatest company*." in his best-selling history of the firm. He considered HP's invention of the **inkjet printer** as its **last truly great technological breakthrough of the twentieth century**.[15]

To protect its lucrative consumables business, HP began aggressively leveraging patents. In **1997**, HP sued third-party supplier **RoT (Repeat-O-Type Stencil Manufacturing Corporation)** for infringing multiple ink cartridge patents. From **2001** onward, HP began explicitly citing the disruptive impact of third-party suppliers in its annual reports.

This story offers a crucial lesson: in expanding technological frontiers, the **synergy between different domains** is essential. At HP, for example, personal computers and printers reinforced each other, supported by internal software development and later bolstered by Microsoft. Similarly, Apple built a virtuous cycle among personal computing, networking protocols, and Adobe's printing technologies. In hindsight, we realize what **Xerox truly missed out on**—the graphical user interface operated via a mouse, Ethernet, the Interpress printing protocol, and WYSIWYG

---

[12] Pfiffner, P. (2002). *Inside the publishing revolution: The Adobe story*. Adobe Press. p34

[13] Malone, M. S. (2007). *Bill & Dave: How Hewlett and Packard built the world's greatest company*. Penguin.

[14] Hewlett-Packard Company. (2001). *Annual report 2001* [PDF]. Retrieved March 31, 2024 from http://www.hp.com/hpinfo/investor/financials/annual/2001/text_only_10k.pdf

[15] Malone, M. S. (2007). *Bill & Dave: How Hewlett and Packard built the world's greatest company*. Penguin.

word processors. All of these were Xerox inventions. Had Xerox integrated these technologies and decisively entered the PC and computer printer markets, who would have emerged the victor of the 1980s?

The **Xerox 9700 laser printer** could have been a perfect transitional platform, combining Xerox's legacy in copying with new advances in laser imaging, networking, and software. But while Xerox dreamt of machines capable of printing hundreds of pages in minutes, HP had already sold over a million compact printers. Once an opportunity is missed, it rarely comes back.

Most people know about the **epic GUI battle** between Apple and Microsoft in the 1980s, which even caused a lasting rift between the two. Fewer realize that Xerox also joined the fray later, hoping to reclaim some of its legacy.

In **1989**, Xerox filed a lawsuit against Apple in the **U.S. District Court for the Northern District of California**.[16] Notably, Xerox did **not** sue for patent infringement. That's because back in the 1970s, when Xerox made many of its groundbreaking innovations, software patents were still in their infancy. Programmers and patent attorneys alike were just beginning to navigate this new legal terrain. Moreover, PARC, as a research-oriented institution, encouraged open publication rather than aggressive patenting.[17]

On **October 28, 1985**, Xerox filed several **design patents** for desktop computer icons—possibly the earliest icon-related patents in history. These included **USD295631S**, **USD295632S**, **USD295633S**, **USD296705S**, and **USD297243S**, and were among the first to use the word *icon* in the title. Xerox is particularly proud of the now-ubiquitous "hamburger icon"—three stacked horizontal lines—designed by **Norman Cox** in the 1970s. This minimal icon, representing a menu, now appears in virtually every mobile interface. Cox also created the dog-eared document icon.

Despite their pioneering nature, these design patents were largely ineffective—there's no record of lawsuits or invalidation proceedings involving them, and they had little industry impact. After Xerox, only a few Japanese firms pursued icon-related design patents. Apple didn't begin filing them until 1992, and Microsoft only joined in **1999**.

In truth, the concept of using icons in user interfaces wasn't entirely original to Xerox. Some researchers[18] attribute the idea to a **1971 academic paper**.[19] Regardless, Xerox had minimal patent presence in GUIs during the 1970s and only began filing software-related GUI patents in 1986—like **US4899136A**, which wasn't granted until 1990. Another was the 1988 application for "virtual and simulated objects in the user interface of a display processor." Clearly, Xerox had come to appreciate the

[16] Xerox Corp. v. Apple Computer, Civil Action No. C89-4482-CALENE, USDC NDCA (Dec. 14, 1989)

[17] Stephens, K. (1990). Xerox Finally Wakes Up, But Is It Too Late?. *Santa Clara Computer & High Tech. LJ*, *6*, 407.

[18] Keith Stephens, Xerox Finally Wakes Up, But Is It Too Late?, 6 Santa Clara High Tech. L.J. 407 (1990). Retrieved April 8, 2024 from http://digitalcommons.law.scu.edu/chtlj/vol6/iss2/9

[19] Huggins, W. H., Iconic Communications, IEEE Transactions on Education, Vol. E14, No. 4, pps. 158–63 (Nov. 1971)

GUI's commercial value, but by then it was **too late**—and Xerox was no longer in the personal computing game.

Apple, by contrast, filed its first GUI-related patent in **1985—US4786893A**, covering a method for generating RGB color signals from composite digital video. Even this patent had limited impact; Apple's real legal actions focused on **copyright infringement**.

Xerox's lawsuit against Apple included several claims. First, Xerox argued that Apple's registered software copyrights for the **Lisa** and **Macintosh** systems were **derivative works** of Xerox's **Smalltalk** and **Star** systems, and therefore invalid. However, based on a 1986 court precedent, such challenges had to go through administrative channels first—the Copyright Office—before being brought to court.

Second, Xerox claimed that Apple's series of copyright infringement lawsuits against Microsoft and HP constituted **unfair competition**. According to Xerox, these high-profile cases led the public and industry to mistakenly believe Apple was the originator of GUI technology. This misunderstanding caused tech companies to seek licenses from Apple, not Xerox, thereby diminishing Xerox's own licensing opportunities. Xerox contended that Apple had unjustly enriched itself through these licenses and that **Xerox deserved a share of that revenue**.

Apple's attorney famously mocked Xerox's legal claims with a biting courtroom quip: On the Hoover Dam, said the lawyer, there are two beavers. One asks the other, "Did you build that dam?" The other replies, "No, but it's based on an idea I had."

Beavers, known for damming rivers with sticks and mud, are nature's engineers—but the Hoover Dam is one of the modern world's greatest feats, listed by the BBC as one of the "Seven Engineering Wonders." Moviegoers might recall its majestic presence in *Transformers* (2007), where it served as the backdrop to the first major battle and the containment site for Megatron. Next to this colossal structure, even a ten-meter-tall Decepticon looks like a mere beaver.

To Apple's lawyer, the **Mac and Lisa computers** were Hoover Dam-level achievements. Xerox's work at PARC? Just a pile of sticks and mud. The subtext was clear: Xerox's influence was primitive, raw, and indirect at best.

Xerox, for its part, waited until much later to confront HP. By **1997**, HP dominated the printer market with a 51% share and $43 billion in revenue, while Xerox held just 2% and earned $18.2 billion. It wasn't until then that Xerox filed its **first patent infringement lawsuit against HP**, citing patents like **US74437122A** (filed 1981) and **US5030971A** (filed 1989).[20] The case, filed on **May 14, 1998**, in Rochester, New York, targeted multiple HP printers and cartridges. HP countersued on **May 27**, invoking a touchscreen interface patent; Xerox followed up with a resolution enhancement patent, and HP responded with a cartridge patent. In **1999**, HP

[20] Wired. (1998). Xerox Sues HP over Patents. Retrieved April 8, 2024 from https://www.wired.com/1998/05/xerox-sues-hp-over-patents/

Fig. 45.8 Handwriting input method designed in Palm's U.S. Patent US6493464B1

added an image-enhancement patent to the fray. The legal battle lasted less than two years, ending in a **settlement in March 2000**.[21]

Xerox also sued 3Com and Palm in 2001, invoking US5596656A.[22] As noted earlier, Palm's Graffiti handwriting input system had become a standout innovation in the pre-smartphone PDA era. Though Palm held its own patents (e.g., US6493464B1, see Fig. 45.8), Xerox argued Graffiti fell within the scope of their broader handwriting recognition patent (see Fig. 45.9).

These lawsuits reflected a shift in Xerox's identity: from an innovator protected by patents to a company relying on patents as its last lifeline. As we've seen, the lifeblood of high-tech firms is not just innovation but competitive products. Once a company ceases to offer compelling products, even the strongest patent portfolio becomes powerless—a stagnant reservoir with no fresh water inflow. When that happens, the company risks devolving into a non-practicing entity (NPE), extracting revenue through litigation instead of technology.

By 2001, Xerox was in deep trouble. After a series of lawsuits and product flops, it posted massive losses. In April 2002, the SEC accused Xerox of accounting fraud and levied substantial fines. Xerox admitted it had overstated revenue by $6.4 billion and pre-tax profit by $1.4 billion, causing a market uproar and pushing the

[21] Carolyn Koo. (2000). Xerox and Hewlett-Packard Settle Litigation. Retrieved April 8, 2024 from https://archive.nytimes.com/www.nytimes.com/library/tech/00/03/biztech/articles/21tsc-xerox.html

[22] Xerox Corp. v. 3Com Corp., 198 F. Supp. 2d 283 (W.D.N.Y. 2001), Retrieved April 8, 2024 from:https://law.justia.com/cases/federal/district-courts/FSupp2/198/283/2513355/

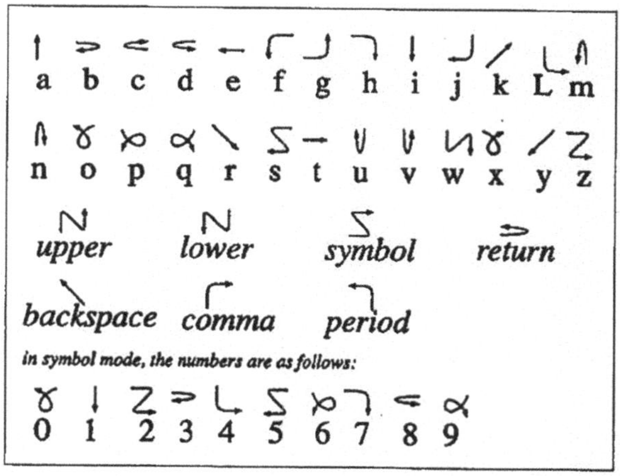

Fig. 45.9 Handwriting input method designed in Xerox's U.S. Patent US5596656A

company to the brink of bankruptcy. It survived only after a sweeping reorganization, shedding non-core assets and focusing on imaging systems.

But the saga didn't end there. In 2019, Xerox, now worth just $6.7 billion, attempted a hostile takeover of HP, whose market cap had ballooned to $31.3 billion. HP resisted with massive stock buybacks and a "poison pill" defense. The *Los Angeles Times* lamented:

> *For anyone who has grown up in the computer technology era, the recent histories of HP (that is, Hewlett Packard) and Xerox must elicit nothing but sadness…What's saddest of all is the idea that merging these two doddering old hulks may be the only way to save them.*[23]

The pandemic ultimately derailed the effort. In April 2020, Xerox cited COVID-19 as the reason for abandoning the acquisition.

[23] Michael Hiltzik. (2019). HP and Xerox: Can two dinosaurs carry each other into the high-tech future? Retrieved April 8, 2024 from https://www.latimes.com/business/story/2019-11-15/hp-and-xerox-dinosaurs

Chapter 46
Building a Core of Light and Shadow

Through the stories of Bosch, Xerox, and HP, we've aimed to clarify our framework of core zones, borderlands, transitional zones, and frontiers. These cases show how companies use their accumulated technological strength in the core zone to expand outward. In this chapter, we turn to the examples of Disney and Pixar to explore another kind of support that the core can provide to its boundaries—one that is not merely technological, nor solely based on brand equity or IP rights, but rooted in the soft power and core philosophy of a company's main business.

It may seem implausible that a company could pursue a century-spanning R&D strategy. Yet after reviewing hundreds of patents, we are convinced that both Disney and Pixar have long adhered to a consistent technical ethos. Eventually, the two companies merged and together became the most successful film studio in cinematic history.

We live in a three-dimensional world, where spatial distance is crucial to our daily perception. Yet the human eye has not evolved to measure depth with lasers—it merely captures light reflected off objects and sends two-dimensional images to the brain. From there, the brain must rely on imagination and inference, combining the slightly offset perspectives of the left and right eyes into a coherent three-dimensional experience. This ability, unique to predatory animals, is called stereoscopic vision.

So what distinguishes two-dimensional from three-dimensional imagery? The answer is simple: **depth**—the visual cue that reveals spatial relationships between near and far. In the fields of painting, photography, and cinematography, this quality is often referred to as "depth of field."

Achieving a sense of depth in 2D imagery is no easy task. For millennia, artists have experimented with techniques to re-create the 3D world on flat surfaces—paper, walls, canvas. It wasn't until the Renaissance that humanity developed a reliable solution: **linear perspective**, a method that brought a new level of realism to art. Early perspective treated foreground, figures, and backgrounds more or less equally, resulting in depictions that, while sophisticated, still lacked the subtlety of natural human perception.

Q. Yue, F. Yang, *Governing Innovation: A Global History of Patent-Driven Technological Progress*, https://doi.org/10.1007/978-981-95-5570-3_46

It was a genius of singular vision who captured this subtlety: Leonardo da Vinci. His masterpiece *Mona Lisa*, which gained worldwide fame after its sensational theft in 1911, employed a groundbreaking technique. Behind the subject lies a blurred landscape; the foreground is sharp, the background soft, creating a powerful illusion of depth.

To today's viewers, this effect is familiar—portrait mode on smartphones uses a wide aperture to blur the background and isolate the subject. In English, the term **bokeh**, borrowed from Japanese, describes this effect. In painting, the technique is known as **aerial perspective**, and it has become fundamental. Even in simple sketches or comic art, closer elements are drawn with bolder, darker lines, while distant ones are rendered with finer, lighter strokes to simulate depth.

Yet this chapter is not about da Vinci. It is about two modern geniuses who solved the problem of "depth" in animation. As creatures of a 3D world, humans are instinctively drawn to sweeping vistas and layered mountain ranges. These two artists succeeded in conveying the vastness of plains and the layering of peaks with uncanny realism on a flat screen—leaving an indelible mark on the history of visual art.

In 1914, American director and cartoonist **Earl Hurd** invented a revolutionary method for producing animation: using **transparent materials** instead of paper to layer characters and backgrounds. The process involved first drawing static backgrounds—like scenery, streets, or interiors—on one layer, then drawing the moving characters on clear sheets. These transparent layers were then placed over the backgrounds during photography. This meant artists didn't need to redraw the background for every frame; they could reuse the same background layers repeatedly.

This technique saved an enormous amount of labor and became the backbone of the animation industry throughout the twentieth century. Because the transparent material was typically made from **celluloid**, the technique came to be known as **cel animation**. Hurd patented his innovation in 1915 (US Patent No. **1143542A**; see Fig. 46.1), and his system even allowed animators to separate different parts of a character's body onto distinct layers, optimizing effort.

For those who grew up in the 1980s and 1990s, most of the cartoons they watched—especially Japanese animation—were made using Hurd's cel animation technique. Even in the early twenty-first century, traditionalists like Hayao Miyazaki continued to resist computer animation, clinging instead to hand-drawn techniques on celluloid.

Hurd's patent holds monumental importance in animation history—and it made him a wealthy man. Many people assume patents are only useful in industrial contexts, but they play a vital role in cultural industries as well. After inventing cel animation, Hurd partnered with his boss, John Bray, to form the Bray-Hurd Patent Trust. Until the patent expired in 1932, any American company wishing to produce animated films had to pay royalties to the trust.

John Bray was also a towering figure in animation history, credited as the first person to industrialize the animation production process. Beyond the cel animation patent, Bray held a suite of foundational patents covering animation techniques. Notoriously combative, he was litigious to the end. When he passed away in 1977,

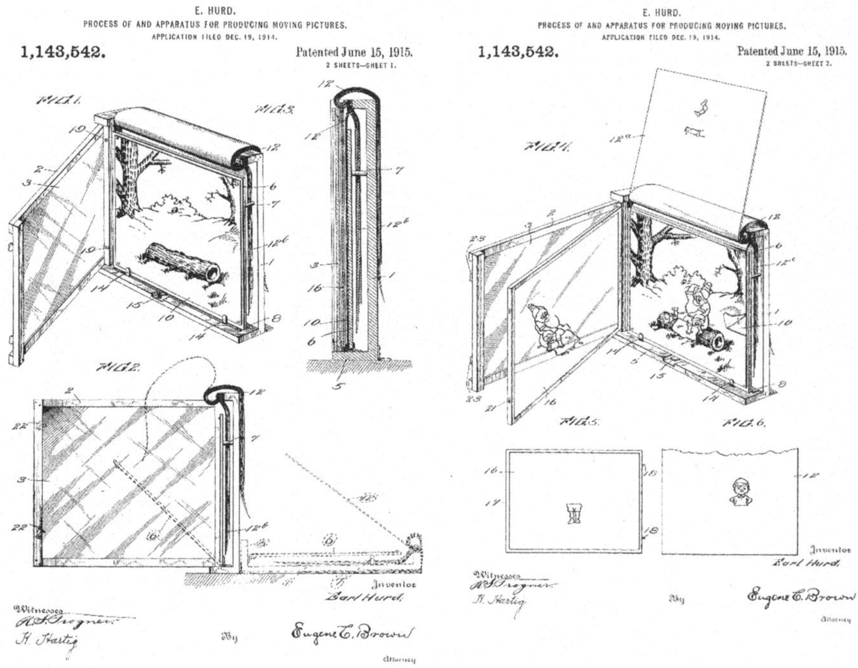

Fig. 46.1 Patent drawing of U.S. Patent US1143542A

some animators reportedly sighed with relief, remarking that they could finally make cartoons again.[1]

Earl Hurd, meanwhile, applied his cel animation patent to produce the pioneering "Bobby Bumps" short films—among the earliest works in animation history. In his later years, Hurd joined Disney and worked on such classics as *Snow White and the Seven Dwarfs*. That film, though grounded in Hurd's cel animation, achieved immortality by layering on Disney's own groundbreaking patent innovations.

Like other animators of his time, Walt Disney used cel animation and paid licensing fees to Bray and Hurd. But Disney, ever the perfectionist, found the flatness and lack of spatial depth in traditional two-dimensional animation unsatisfying.

Creating depth on a flat medium with linear and atmospheric perspective is not particularly difficult—if the image is static. But as soon as motion enters the picture, everything becomes exponentially harder. Even children understand the parallax effect: as one moves forward, nearby objects appear to move quickly in the opposite direction, while distant objects shift slowly, and the sun and moon seem to travel

[1] Sito, T. (2006). *Drawing the line: the untold story of the animation unions from Bosko to Bart Simpson*. University Press of Kentucky, p51

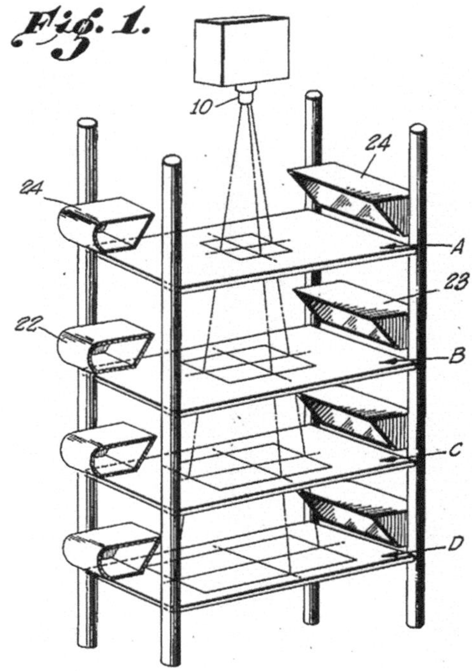

Fig. 46.2 Patent drawing of U.S. Patent No. US2281033A

alongside. Depicting this in dynamic animation, using traditional techniques, would require redrawing scenes from multiple angles—an insurmountable labor burden.

At Disney's insistence, engineer **William Garity** developed a massive, complex piece of camera equipment and patented it in 1937 (US Patent No. **2281033A**). This would become one of Disney's earliest and most important core patents: the **multiplane camera**. See Fig. 46.2 for the patent illustration.

From the outside, the machine looked like a table with four stacked platforms, each rigged with lights. At the top sat a downward-facing camera. Transparent cel sheets bearing artwork were mounted at different levels. According to Claim 1 of the patent:

> *In a method of producing an effect of depth in photographic images of pictorial representations, the steps of: placing pictorial representations in two parallel spaced planes along the optical axis of a camera lens, said pictorial representations being complementary, the representation closer to said camera lens being carried by a transparent member, illuminating the camera side of each of said representations, illuminating the rear representation with light of sufficiently higher intensity than the illumination of the front representation to compensate for absorption losses of light passing through the front representation and its carrying medium, and moving the camera lens and the rear representation along the camera lens axis and in the same direction, between periods of photography by said camera.*

Today, to zoom in or out on an image, we simply pinch a touchscreen or adjust parameters on software. But a century ago, for animators to scale an image, redrawing it was often the only option. The multiplane camera offered a physical solution based on the principle that **objects closer to the lens appear larger**. This technique remains in use today—**Peter Jackson's *Lord of the Rings*** used it to make Gandalf appear towering next to hobbits, simply by adjusting actor proximity to the camera.

However, changing the distance between a cel layer and the light source affects illumination. Disney engineers solved this by installing individual light fixtures on each platform—an innovation embedded in the multiplane camera patent.

To be fair, the patent doesn't feel so much ingenious as **brute-force engineering**—solving a simple problem with an enormous machine. Each shot captured with the multiplane camera cost tens of thousands of dollars and consumed massive amounts of electricity. A Disney executive once joked that running it nearly "blackened out the entire Burbank power grid."

Yet the machine delivered exactly what Walt Disney wanted.

Even with the multiplane camera, Disney's animators worked under harsh conditions. The early Disney studios were sweatshops in every sense, operating on rigid production lines. Just like today's animation pipelines, artists began with black-and-white line drawings, which were passed on to colorists. Every sheet came with time codes—from A to E—dictating urgency. "A" required completion in 10 minutes, "E" allowed up to 40. Artists needed hall passes to leave their desks—even to use the restroom. In the ink-and-paint division, these passes remained in use until the 1980s.

Without the multiplane camera, Disney's workforce would have had to produce hundreds or even thousands more background layers—impossible even for their large team. In short, **without this machine, the animated films we know today might have looked entirely different**.

The United States Patent and Trademark Office honored the multiplane camera as a "Celebrity Patent" on its official website, explaining that the device featured seven different layers and a movable camera filming vertically from above, which created an illusion of depth that had not previously been possible in animation. It noted that the first animated short to test this camera was *The Old Mill* from the *Silly Symphony* series, which won the Academy Award for Best Animated Short in 1937. After that success, Walt Disney applied the technique in *Snow White and the Seven Dwarfs* and in all subsequent productions. The office also pointed out that although depth-of-field effects in animation are now achieved through computers, the multiplane camera remains one of the most important innovations in animation history.[2]

As the USPTO noted, depth effects are now created with computers, and since the earliest days of 3D computer animation, the challenge of representing depth has been addressed inherently by the technology. Compared to da Vinci and Disney, the man who pioneered depth in computer graphics is far less known—but no less important. That man is **Ed Catmull**, the co-founder of Pixar and a recipient of the Turing Award. In the 1970s, when 3D computer animation was still nascent, Catmull

---

[2] U.S. Patent and Trademark Office. (n.d.). *Behind the curtain. Inventors Eye*. Retrieved March 31, 2024 from https://www.uspto.gov/learning-and-resources/newsletter/inventors-eye/behind-curtain

introduced the concept of the **z-buffer** (or "depth buffer"), a foundational technique for storing depth information.

Thus, da Vinci, Disney, and Catmull—spanning oil painting, animation, and computer graphics—represent a lineage of genius converging across centuries to solve the same aesthetic challenge: rendering *depth* on a two-dimensional surface.

From the earliest cave paintings in Lascaux, artists have instinctively used lines to depict figures and animals. More than ten thousand years later, Impressionist painters rejected line, opting instead for vibrant fields of color. As Édouard Manet put it: *"There are no lines in nature, only areas of color, one against another."*

Returning to *Mona Lisa*, the painting's celebrated "mysterious smile" owes much to da Vinci's mastery of **sfumato**, a technique he described as "*without lines or borders, in the manner of smoke or beyond the focus plane*."[3] He softened the contours of Mona Lisa's eyes, mouth, and background to create a subtle, lifelike presence. Some scholars believe da Vinci achieved this through painstaking layering—repeatedly applying and reapplying pigment with his fingers over years of work.

Walt Disney, too, had his own artistic aversion to line. In animation, he detested visible outlines. To mute their prominence, he required artists to match the ink lines to the color of the object they outlined.[4] Early Disney animations show this clearly: compare, for instance, 1950s *Looney Tunes* with 1961's *One Hundred and One Dalmatians*. Bugs Bunny is often outlined in hues that blend with the background, while the Dalmatians are rendered with stark black contours—sharp and graphic.

Though the new technique used in *One Hundred and One Dalmatians* lowered production costs and earned solid box office returns, Disney himself disliked the resulting look. Determined to minimize visible outlines, Disney filed a 1939 patent to address this artistic concern. The patent, **US2254462A**, granted in 1941, was titled "Method of producing depth and texture effects in animation."

In the background section, the inventors wrote:

> A small animal, such as a rabbit or chipmunk, is typically furry. But in animation, such animals are outlined with continuous dark lines, which contrast with the color of the figure itself. These unbroken lines divide elements like a mosaic or stained-glass panel. In contrast, our method renders fur with a soft, undefined edge, creating roundness, depth, and texture—more realistic, artistic, and visually pleasing.

The technical method was simple: add a transparent overlay atop the cel sheet bearing the character. This extra layer carried only the shadows and edge treatments. The result was a figure with softened contours and subtle shading. Claims 2, 3, and 6 of the patent specifically note that shadow elements "partially obscure the

[3] Art History Project. (n.d.). *Sfumato*. Retrieved March 31, 2024 from https://www.arthistoryproject.com/subjects/sfumato/

[4] Pallant, C. (2011). *Demystifying Disney: a history of Disney feature animation*. Bloomsbury Publishing USA. p70: *Anderson notes how Disney inherently hated lines, stressing how he pushed the Studio 'into cel-paint ink lines, where the ink line is the same colour as the area it is encompassing ...*

outline," blending the lines into surrounding colors—an effect straight out of Manet's color-block philosophy.

Disney applied this principle rigorously, especially in *Snow White and the Seven Dwarfs*, where shadow lines often melt seamlessly into the background. The patent reflects a deep artistic fixation: **a war on the hard line**.

Just as two eyes give humans depth perception, our **two ears** enable 3D hearing. Sound travels through air at 340 m/s, and the minute delay between its arrival at each ear lets our brains locate the source. British physicist **John Strutt** first described this phenomenon in 1896 as the **"binaural effect."** To us, sound, like vision, is experienced in three dimensions.

In 1931, British engineer **Alan Blumlein** noticed something odd while watching a film with his wife: all the sound seemed to come from the same spot on the movie screen, breaking the immersive illusion. Turning to his wife, he declared that he would find a way to fix this. On December 14 that year, Blumlein filed a patent in the UK titled *"Improvements in and relating to sound transmission, sound recording and sound reproduction systems"*, based on John Strutt's binaural theory. The patent was granted in 1933 under number **GB394325A**. Employed by the broadcasting company EMI, Blumlein would later design radar systems for the British military during World War II. Ironically, his invention of stereo sound was first deployed not in cinemas, but in long-range anti-aircraft radar systems. He died tragically in a plane crash during the war and never lived to see his invention adopted for civilian use.

Disney's foray into stereo sound came even earlier—before Blumlein's invention had entered public use. By the late 1930s, Mickey Mouse was losing popularity, and some within the Walt Disney Company even advocated abandoning the character. But Walt Disney refused to give up on his creation. To revive Mickey's stardom, he poured lavish resources into a short film titled *The Sorcerer's Apprentice*, spending $120,000—three times the cost of a typical Disney short. It was evident the film would never recoup its costs.

On the advice of renowned conductor **Leopold Stokowski**, Disney added six more animated segments to *The Sorcerer's Apprentice*, compiling them into a feature-length film called *Fantasia*, with Stokowski conducting the soundtrack. These additional segments were more abstract and less narrative-driven, relying on color and form to visualize the music. Their simpler production helped offset the cost of the ambitious central piece. The result was a film where music took center stage—*Fantasia* became a movie about sound.

But when the film was previewed, Disney was deeply dissatisfied with the theater's sound system. Blumlein's stereo patent had not yet been commercialized.

To address this, Disney engineers led by William Garity developed a **multi-channel audio system** called **"Fantasound"**. This system made *Fantasia* the world's first commercial **stereophonic** motion picture. The first claim of the Fantasound patent describes:

> A method of generating new audio effects in a theater in timed relation with the projection of a motion picture, comprising: placing loudspeakers along the walls, ceiling, and behind the screen of the theater to simulate an auditory environment surrounding the audience;

> projecting the film; simultaneously reproducing multiple program sound recordings via reproducing pickups, along with a control track via an intelligent control pickup; selectively directing audio oscillations from one or more of said program recordings to designated loudspeakers based on data in the control track; and dynamically adjusting the amplitude response of said loudspeakers via intelligent control.

But the system's complexity proved to be its greatest obstacle. *Fantasia* required each screening venue to install dozens of custom speakers and employ a dedicated operator. No studio was willing to undertake such an effort. Disney had to self-distribute the film using a traveling roadshow model.

By April 1941, after 11 roadshow stops, *Fantasia* grossed $1.3 million. However, each theater setup cost $85,000, and rental fees added more overhead. With World War II raging across Europe, *Fantasia* could not be released overseas—nearly half of its anticipated revenue vanished. Despite its artistic brilliance, the film was a commercial failure—at least initially.

Yet *Fantasia* was no ordinary cartoon. With little in the way of traditional plot, it was essentially a visual symphony, featuring top-tier musicians and orchestras. For music lovers, it offered more enduring appeal than most narrative films—and heavily relied on immersive sound. After stereo sound systems became commercially viable, *Fantasia* was re-released and remastered multiple times. By the 1990s, its cumulative box office exceeded $83 million.

Interestingly, Walt Disney's insistence on high-fidelity sound led the company to purchase advanced audio testing tools, giving a Palo Alto startup its first major contract. In 1938, Disney ordered eight **HP 200B** audio oscillators at $71.50 apiece—more than $500 total. This was the first significant sale for the company that would become **Hewlett-Packard (HP)**, incorporated the following year in July 1939. The story mirrors how Texas Instruments' portable radios helped catalyze the market for miniaturized electronics—underscoring how a single technology contract can spark a chain reaction across an entire industry. Scientific breakthroughs may begin in the lab, but it's often the entrepreneur or product specialist who delivers the final push into history.

It's poetic that the first company to pioneer synchronized sound and stereo in cinema wasn't a live-action studio—but **an animation house**. Though its characters were "flat" on screen, Disney insisted that they be three-dimensional in appearance, in environment—and in sound. This pursuit of **realism in the imaginary**, of immersive sensory experience in a fictional world, lies at the heart of Disney's early patent philosophy.

As early as 1928, during the production of *Steamboat Willie*—Disney's first sound cartoon—Walt Disney had already demonstrated a remarkably high standard for sound quality and synchronization. At the time, major studios like Warner Bros., Fox, and RCA had launched their own sound film systems: Warner's Vitaphone, Fox's Movietone, and RCA's Photophone. After evaluating these systems, Walt Disney found none of them satisfactory. In a letter to his brother Roy Disney, he criticized RCA's system, saying, "It's nothing more than an orchestra playing in the background, adding nothing but noise. The dialogue makes no sense and the synchronization is nonexistent." Back then, most filmmakers didn't even consider the issue of syncing sound to visuals.

Determined to achieve perfect audiovisual synchronization, Disney rolled up his sleeves. One of his earliest patents, **US1941341A**, describes a *"method and apparatus for synchronizing moving pictures with accompanying sound."* Its purpose was to ensure that animated characters' movements were precisely aligned with sound. Claim 1 of the patent reads:

> A method for preparing a motion picture synchronized with an accompanying musical score, comprising the steps of: producing a chart associating a predetermined musical score with a sequence of progressive action steps, preferably illustrated in pictorial form; assigning the designated images on said chart to successive frames of a film to be produced, indicating them thereon; assigning and indicating on the chart the musical notes to be synchronized with the designated frames; drawing images for each action step terminating with the frame associated with the assigned note; photographing said drawn images in the order specified on the chart and associated with said successive frames; separately recording the musical score; timing the musical recording according to the note frequencies assigned on the chart; and combining the sequence of photographed images with the recorded sound of the score.

Much like the multiplane camera, this invention gives off a "brute-force brilliance" vibe: it involved meticulous charting and timing of every visual and musical cue, including a mechanical system to signal instructions to orchestra players. But this fastidiousness became a hallmark of Disney's approach. From that point forward, American theatrical and even some television animation would aim for flawless synchronization between lip movements, gestures, and sound—a stark contrast to Japan's fast-and-frugal animation model, which emphasized "key frames." In the Japanese system, artists drew the key visuals, while assistants filled in the transitional frames, reducing costs.

Readers of this book are likely familiar with *Tom and Jerry*, produced by Metro-Goldwyn-Mayer. In the episode *The Cat Concerto*, Tom plays *Hungarian Rhapsody No. 2* with astonishing precision—his animated piano fingerwork matches the music note-for-note. Numerous fans have attempted to recreate this scene using real pianos, and many of their performances are available on platforms like Bilibili and YouTube.

So what's Disney's secret to enduring success? As these patents show, it lies in the relentless pursuit of subtle artistic effects often hidden in the details. The saying goes, "The devil is in the details," but Disney's pursuit goes far beyond mere meticulousness.

Looking across Disney's twentieth-century patents, we see that for a company known for fantasy and whimsy, its most important guiding principle wasn't dazzling sound and light—it was **realism**. Even though every child knows that animated films are not "real," Disney's mission was precisely to pursue **realism within illusion**. We call this philosophy *"truth within fantasy"*.

- Disney invented the power-hungry multiplane camera to recreate the **depth of field**—because the real world is three-dimensional.

- Patent **US2254462A** added a separate shadow layer to animation to blur and mask outlines—because the real world has **no hard lines**.
- Disney's method of marking film frames to synchronize action and music aimed for **perfect audio-visual timing**—because in real life, motion and sound happen **simultaneously**.
- The company re-engineered theaters and built new audio systems to achieve **true stereophonic sound**—because sound in the real world comes from **every direction**.

Most of Disney's early patents were guided by this core principle: *realism within fantasy*. Under this philosophy, Disney not only revolutionized animation but went on to create **Disneyland and Disney World**, becoming a giant in theme parks and hospitality. In the twenty-first century, Disney acquired **Pixar** and **Lucasfilm**, pushing the boundary of realism in fantasy worlds even further. And notably, for a media company, Disney is now one of the most focused players in robotics—particularly in making robots that **look, move, and feel like humans**.

Next, we'll explore how Disney has extended this guiding philosophy across its other business ventures, making magic not just through storytelling, but through technological mastery.

Chapter 47
Blurring the Lines Between Illusion and Reality

As previously mentioned, American companies like Harold, Hewlett-Packard, and Texas Instruments all profited directly from World War II—but Disney did not. With the outbreak of war, Disney lost access to the European market entirely, and its headquarters were requisitioned by the U.S. military to protect the nearby Lockheed factory. Walt Disney could only eke out modest revenue by producing training films for the armed forces. After the war, animation studios like MGM and Warner Bros. rose rapidly, with MGM's *Tom and Jerry* series alone earning seven Academy Awards for Best Animated Short Film, posing a formidable challenge to Disney. In response, Disney invested heavily in the production of *Cinderella*. Like earlier Disney films, *Cinderella* was exquisitely crafted—and extremely expensive to produce, costing more than twice what Warner Bros. spent on their animations.[1] The financial risk was immense, and the pressure weighed heavily on Walt Disney, both mentally and physically.

On his doctor's advice, Disney took up a hobby to relieve stress. He began building small handmade gadgets and soon completed a scale model of a steam locomotive. He became obsessed with trains, purchasing model trains and enlisting the help of company engineer Roger Broggie to construct a miniature railroad in his office. Eventually, Walt Disney created an entire model railway layout in his office—large enough to fill a two-car garage—which attracted many curious employees.[2]

Disney had long entertained the idea of creating a park. Many fans wrote in expressing a desire to visit Disney, but in reality, there was little of interest for visitors to see at the studio. Combining this feedback with his growing passion for model trains, Disney found inspiration for the company's future.

Construction of Disneyland began in July 1954, and the park officially opened in July 1955. It quickly became one of California's most iconic tourist attractions. In

[1] Gennawey, S. (2013). *The Disneyland Story: The Unofficial Guide to the Evolution of Walt Disney's Dream*. Unofficial Guides.

[2] Gennawey, S. (2013). *The Disneyland Story: The Unofficial Guide to the Evolution of Walt Disney's Dream*. Unofficial Guides.

Q. Yue, F. Yang, *Governing Innovation: A Global History of Patent-Driven Technological Progress*, https://doi.org/10.1007/978-981-95-5570-3_47

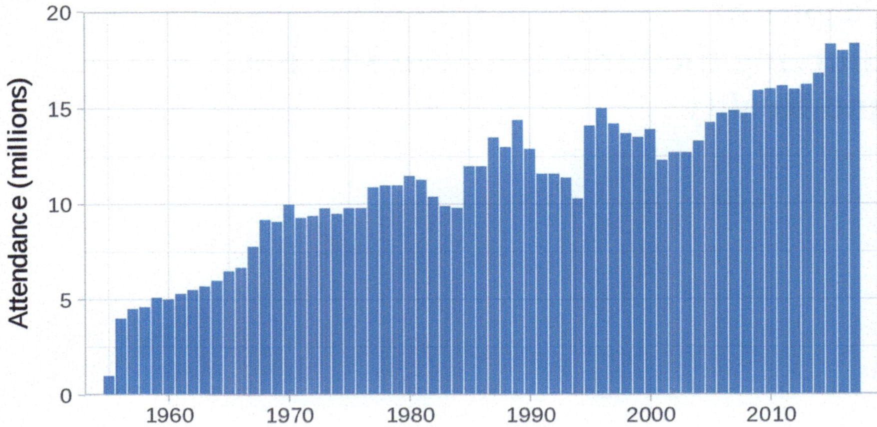

Fig. 47.1 Number of visitors to Disneyland, 1955–2017 (vertical axis in millions of visitors)

its first year, Disneyland generated $10 million in revenue—one-third of Disney's total income that year.[3] Over the following years, the park continued to grow steadily. According to *Time* magazine, in 1957, "the average visitor spent $2.72 on rides and admission, $2 on food, and 18 cents on souvenirs—including Disneyland flags, maps, and Donald Duck hats. Attendance in 1957 was up 11% from 1956, exceeding 4.5 million visitors, and Disneyland grossed over $11 million".[4] As shown in Fig. 47.1, Disneyland's annual visitor numbers continued to climb, providing a steady stream of cash flow even during the company's downturns. After Walt Disney's death in 1966, the park became the backbone of the company—Disneyland in California and Disney World in Orlando together contributed nearly three-quarters of the company's revenue.

From a patent perspective, Disney had been granted 21 patents before 1955 (see Table 47.1), all related to animated films.

From 1955 onward, Disney began filing patents for **physical amusement rides**, and over the next 40 years, these surpassed the number of patents related to animation and film—consistent with the critical role Disneyland played during the company's downturns. According to IPC classification, Disney obtained about **46 patents** for animation and film (see Table 47.2) and **62 patents** for physical amusement facilities (see Table 47.3). After the 1995 release of *Toy Story*, Disney filed more patents in areas such as computer animation, film effects, and streaming media—aligning with its expansion into new fields in the twenty-first century. Disney's success in these areas underscores the indispensability of patent-supported R&D.

(See Tables 47.2 and 47.3 for full patent breakdowns by IPC classification.)

---

[3] Gennawey, S. (2013). *The Disneyland Story: The Unofficial Guide to the Evolution of Walt Disney's Dream*. Unofficial Guides.

[4] Jennifer Latson. (2015). How Disney Proved That No Price Was Too High for Fun. Retrieved April 6, 2024 from https://time.com/3949988/disneyland-history/

Table 47.1 Disney Patents Granted Before 1955, All Related to Animation

Publication Number	Filing Date	IPC Main Classification	Title
US2298618A	1940-07-31	H04S3/00	*Sound reproducing system*
US2280988A	1939-11-14	C09D13/00	Pencil Composition
US1964909A	1933-07-14	G03B31/00	*Timer*
US2281033A	1939-05-08	G03B15/08	*Method of producing animated photoplays*
US2362980A	1941-09-02	B44F1/06	*Cartoon cell and method of making same*
US2449705A	1944-11-14	G03B21/54	Sound Picture Magazine Projector of the Continuously Moving Film Type
US2418943A	1944-12-08	G03B21/46	*Register pin mechanism*
US2312158A	1942-04-24	G03B15/08	*Apparatus for use in photographing miniatures*
US2449702A	1944-12-08	G03B15/08	*Animating device*
US2602668A	1946-05-03	G11B3/08	*Automatic phonograph resetting apparatus*
US2198006A	1938-11-16	G03B15/00	*Control device for animation*
US2247554A	1939-04-12	G11B5/02	*Method of recording and reproduction of sound and apparatus therefor*
US2313867A	1940-07-31	H04R3/12	*Sound recording system*
US2269813A	1940-07-31	G01J1/42	*Photometer system*
US1941341A	1931-04-02	G03B31/00	*Method and apparatus for synchronizing photoplays*
US2254462A	1939-11-21	G03B15/08	*Method of creating depth and texture effects in animated cartoons*
US2260092A	1939-12-26	G03B15/08	*Method of creating cartoon effects*
US2314629A	1939-12-18	G03B15/08	*Method of creating illusions*
US2201689A	1936-09-01	G03B15/08	*Art of animation*
US2246796A	1940-07-31	G11B7/28	*Apparatus adapted for scoring and rerecording sound*
US2348983A	1942-05-26	B44F1/10	*Method of simplifying animation*

Table 47.2 Patents Related to Animation and Film (45 items)

IPC Main Category	Number of Patents
G03—Photography; Cinematography; Optical Systems or Apparatus (analogous to photography); Electrography; Holography	13
G06—Computing; Calculating; Counting	9
H04—Electric Communication Technique	7
G02—Optics	6
G09—Education; Cryptography; Display; Advertising; Seals	3
G01—Measuring; Testing	2
G05—Controlling; Regulating	1
G10—Musical Instruments; Acoustics	1
G11—Information Storage	1
H01—Basic Electric Elements	1
C01—Inorganic Chemistry	1

Table 47.3 Patents Related to Amusement and Entertainment Equipment (62 items)

IPC Main Category	Number of Patents
A63—Sports; Games; Amusements	24
B61—Railways	9
A01—Agriculture; Forestry; Animal Husbandry; Hunting; Trapping; Fishing	4
F41—Weapons	4
B05—Spraying or Atomising in General; Applying Liquids or Other Fluid Materials to Surfaces, in General	3
A47—Furniture; Domestic Articles or Appliances; Coffee Mills; Spice Mills; Suction Cleaners in General	2
F42—Ammunition; Blasting	2
A42—Headwear	1
A44—Haberdashery; Jewelry	1
A62—Life-Saving; Fire-Fighting	1
B01—Physical or Chemical Processes or Apparatus in General	1
B24—Grinding; Polishing	1
B25—Hand Tools; Portable Power-Driven Tools; Handles for Hand Implements; Workshop Equipment; Manipulators	1
B60—Vehicles in General	1
B62—Land Vehicles for Travelling Otherwise than on Rails	1
B63—Ships or Other Waterborne Vessels; Related Equipment	1
B65—Conveying; Packing; Storing; Handling Thin or Filamentary Material	1
E04—Building	1
E03—Water Supply; Sewerage	1
F16—Engineering Elements and Units; General Measures for Producing and Maintaining Effective Functioning of Machines or Installations	1
F23—Combustion Apparatus; Combustion Processes	1

We believe Disneyland represents a highly successful case of boundary expansion. It fully leveraged Disney's intellectual property, developed a new business model, and achieved extraordinary commercial success. What's truly admirable is that Disney's R&D efforts consistently adhered to its core philosophy: creating realistic fantasy worlds.

Among Disney's many theme park innovations, the patent US5403238A, titled "Amusement Park Ride," stands out as particularly influential. The term "ride" here primarily refers to the track-guided vehicles and the surrounding themed scenery. According to the patent's description, the invention aims to give passengers a *"realistic moving ride vehicle experience that is actually not happening."* This ambition is perfectly aligned with the core concept we've previously discussed: the pursuit of realism within fantasy.

Claim 1 of the patent reads:

> *... a ride vehicle adapted to follow said path, said ride vehicle including a passenger supporting structure, a motion base supporting said passenger supporting structure that articulates said passenger supporting structure in a plurality of degrees of freedom relative to and independent of motion of said ride vehicle along the path, and a control system that controls said motion base to articulate said passenger supporting structure.*

Building on this, **Claim 9** describes three actuators that allow the ride vehicle to simulate multiple axes of movement—pitch, roll, and elevation—enabling the vehicle to respond dynamically to the surrounding scenery. For instance, when the vehicle passes a faux obstacle like a fallen tree (as in Fig. 47.2), it might first rise and then descend. Animal-shaped ride vehicles can even mimic behaviors like rearing up or bowing down (see Fig. 47.3).

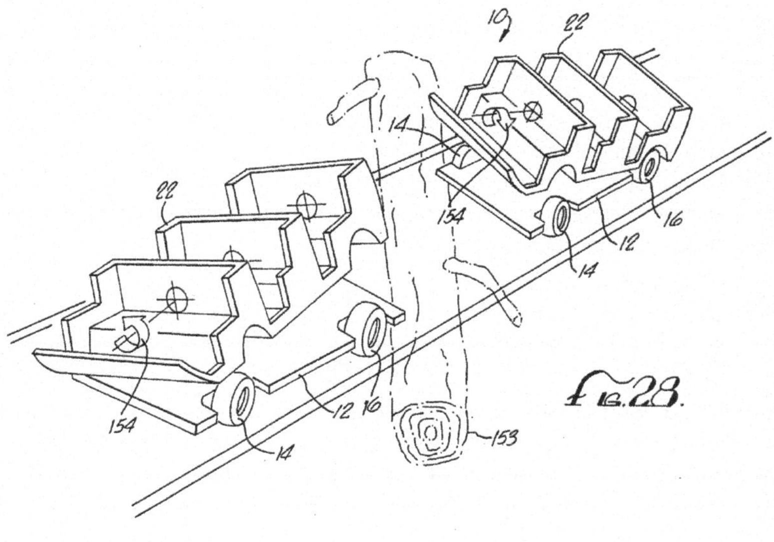

Fig. 47.2 Patent drawing 28 of U.S. Patent No. US5403238A, amusement train simulating traversal over an obstacle

Fig. 47.3 Patent drawing 2 of U.S. Patent No. US5403238A, amusement vehicle simulating a pitching motion

Though children riding these track-based vehicles at Disneyland likely realize they're just oversized toys—not real animals or locomotives—the point is not deception. It's the insistence on evoking **a convincing sense of realism**, even with clearly artificial props. While technological limitations at the time prevented perfectly lifelike motion, Disney focused on choreographing **key, expressive movements** to simulate the most delightful and dramatic motions—a philosophy entirely consistent with Disney's animation patents.

In 1994, Disney took the concept even further with **US5595121A**, a patent designed to add *"back-push"* and *tremor effects* for roller coaster passengers. According to the patent's description:

> The ride vehicle of the present invention is capable of delivering stimulation previously unobtainable in traditional coasters or powered rides, including reverse propulsion and highly precise control during reverse motion… Due to the high accelerative capacity, tighter horizontal curves and shorter hills can be used to achieve equivalent or enhanced tremors.

Disney's 1994 **US5551920A** patent (see Fig. 47.4) marks one of the company's earliest ventures into virtual reality. While today's VR efforts from tech giants and startups focus heavily on visual simulation, Disney's invention concentrated on **kinesthetic realism**—how the human body perceives motion and gravity. The patent observes:

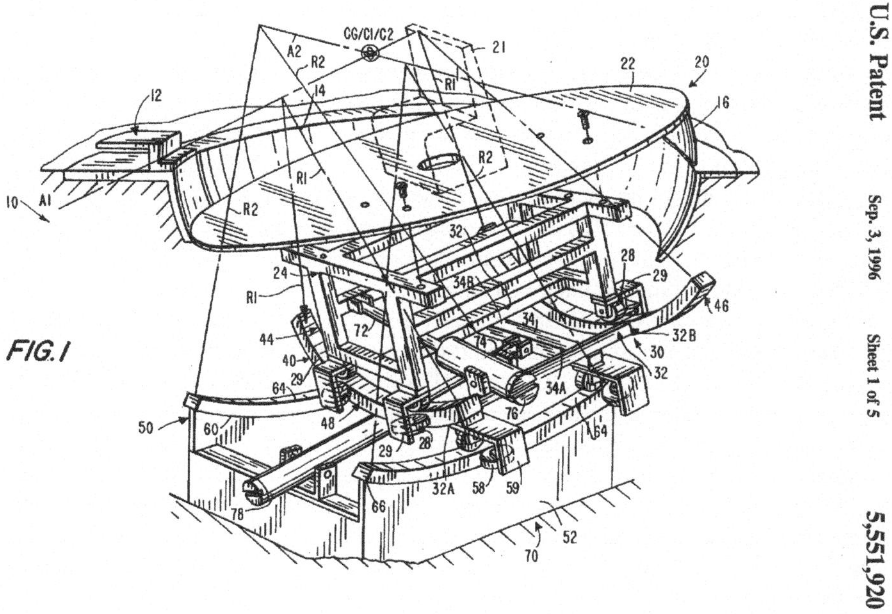

Fig. 47.4 Patent drawing of U.S. Patent No. US5551920A

> *...if rotational motion is imparted to the user, the center of rotation of the apparatus may not be coincident with the user's center of gravity. As a result, such an apparatus can cause nausea or other motion sickness for the user.*

Disney's goal was to align the rotational center with the user's center of gravity and, in attractions like the "magic carpet," ensure the ride's motion closely matches the user's inputs in a *realistic* way.

Across all these patents, the unifying design principle remains the same: to let users **experience the illusion of reality**, even when they know the vehicle or environment is not real.

Audio-Animatronics: Giving Life to Machines

As noted earlier, Disney may well be the most committed company in the world to making robots feel human. As early as the 1960s, Disney began developing what it called **Audio-Animatronics (AA)**—a groundbreaking technology for creating lifelike robotic figures.

In 1964, Disney was commissioned by companies like General Electric to create a robotic stage show for the New York World's Fair. The result was *The Carousel of Progress*, a mechanical performance portraying the evolution of the American family, which received widespread acclaim. After the fair, the attraction was relocated to Disneyland and renamed *The Carousel of Progress*. Today, updated AA technology remains central to major Disney attractions such as *Star Wars* and *Pirates of the Caribbean*.

Much like the company's fastidious approach to synchronizing voice and movement in animated characters, Disney demanded the same fidelity from its animatronic figures—mouth movements had to match speech exactly. Early figures, constrained by primitive technology, could only open and close their mouths vertically and could not blink, leading to the notorious **"uncanny valley"** effect. But Disney persisted. By the 1970s, it had designed robotic figures capable of realistically mimicking human mouth movements, culminating in **US4177589A**, a 1977 patent that protected this technical advance.

The specification of **US4177589A** clearly points out the limitations of traditional mechanical figures, whose mouths could only open and close vertically, unable to simulate rounded vowels like "O" and "U" or consonants like "F" and "V." This patent, however, offers a technological breakthrough: the animatronic figure's lips could now perform more nuanced movements. They could form circular shapes to pronounce vowels such as "O," stretch into a wide "A" sound smile, or extend the upper lip while retracting the lower one to form "F" consonants. Additionally, the eyes of the figures could blink intermittently. To mimic the look and feel of human skin, these figures were clad in a special flexible material—**polyvinyl chloride plastic (PVC)**.

The expressive motions of these Audio-Animatronics (AA) figures were powered by pneumatic or hydraulic actuators. As illustrated in Fig. 47.5 (bottom left), most of the cylinders are piston assemblies, while line 129 denotes the air supply tubes. The structure labeled 14 in the upper right enables the figure to stretch its mouth sideways, part 12 forms rounded vowel shapes, and part 11 controls blinking.

The entire mechanical structure is governed by an **audio encoding/decoding system** that responds precisely to specific phonemes such as "F," "A," "O," and "U." While the figures do not produce sound themselves, live voice acting or recorded audio is used during performances. As the audio plays, the system simultaneously feeds the phonetic signal into the decoder, which detects the relevant sounds and triggers corresponding facial movements.

For example, when the animatronic figure's mouth is open and the system detects an "O" or "U" sound, the circular-mouth mechanism (12 in the upper right) activates. Air flows through tube 70, pushing arm 43 forward, compressing a spring and causing the lips to form a round shape. When an "A" sound is detected, the smiling mechanism (structure 14) is triggered—air flows through tube 208, lifting rod 203 to pull back the T-shaped piece 205, stretching the mouth corners backward and upward.

The show *Carousel of Progress*, which features AA figures and originated at the 1964 World's Fair, remains a staple at Disneyland today. It exemplifies Disney's long-term dedication to developing human-like robots. In fact, among today's major tech companies, **Disney may be the only one with a direct commercial imperative to make humanoid robots as realistic as possible**—a trait well worth observing.

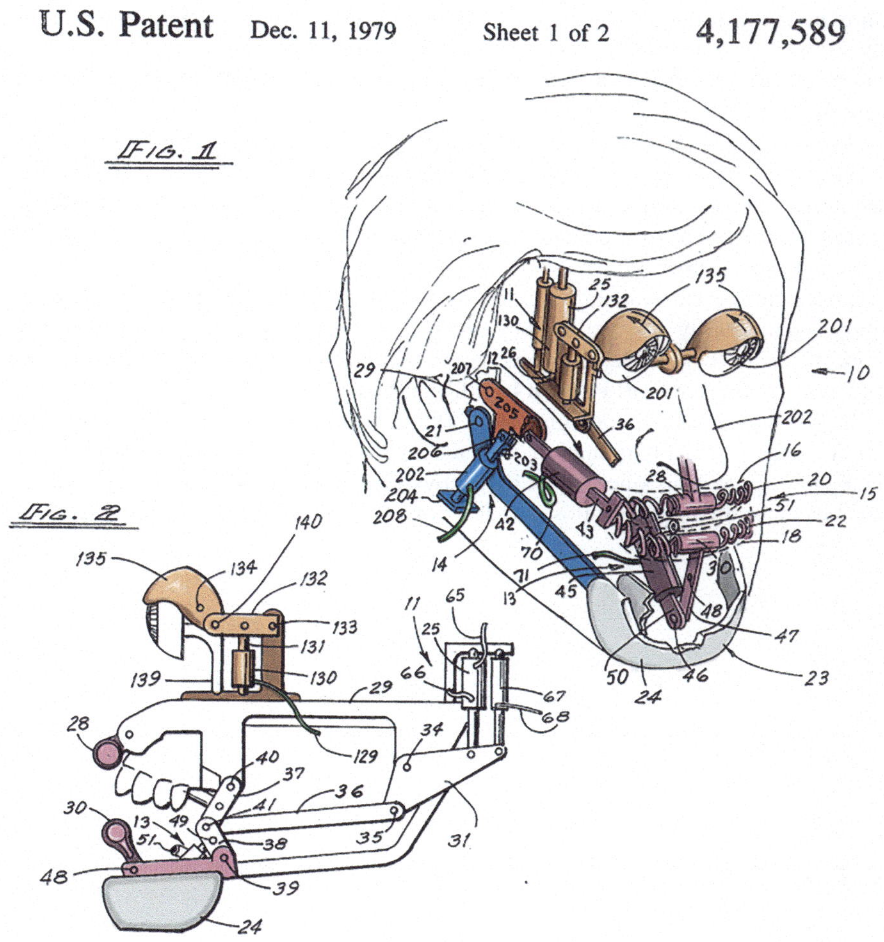

Fig. 47.5 Mechanisms from US4177589A

For instance, **US8052185B2** describes a robotic hand with "fingers that move and behave more like human hands." The patent notes that early robotic hands typically consisted of two-finger grippers or claw-like mechanisms, limited to grasping or pinching motions. But for Disney, utility or cost-efficiency isn't the priority—**verisimilitude is.** So they designed a five-fingered robotic hand, each finger with three joints, capable of accurately mimicking human gestures.

US9162720B2 presents a method for replicating human movement in robots. Based on a human's center of mass (COM) trajectory during motion, it uses inverse kinematics to calculate a robot's corresponding motion. **US9044859B2** focuses on gait generation and control for bipedal walking robots. **US7740953B2** covers an artificial skin system composed of elastic materials with varying hardness to simulate the texture of real skin. And **US8651916B2** tackles robotic eyes that emulate human eye behavior. It uses a small display to project eye movement sequences, which are then viewed through a convex lens—creating a lifelike eye with realistic gaze motion.

As previously noted, attention to detail and pursuit of perfection are not uniquely Disney traits. Many great companies, products, and patents show a similar dedication to user experience at a granular level. But what sets Disney apart is a **deep-seated commitment to "realism"**—crafting experiences with an internal logic that feels tangible, even within an overtly fantastical world.

This omnipresent realism helps users—whether they're children, teenagers, or adults—temporarily forget that they are in a fictional universe. When designers work to build believable illusions, users instinctively cooperate, **choosing to immerse themselves rather than break the illusion**.

Disney's principle of "pursuing truth within fantasy" has evolved from the two-dimensional world of animation into the tangible, three-dimensional reality of theme parks—and has since expanded to toys, games, and other derivative products. The legacy of this design philosophy is a powerful example worth emulating.

Chapter 48
Legacy and Innovation in the Digital Age

From the 1970s through the 1990s, digital technology surged forward, and computer animation rose from obscurity to prominence. During this period, Disney experienced both highs and lows, successes and setbacks in new ventures, yet its core philosophy and business foundations remained intact.

At the heart of this continuity was Roy Disney (1930–2009), nephew of Walt Disney. After Walt's death, Roy joined the company board as the Disney family's representative. Reverent toward his uncle's legacy, Roy was a staunch conservative who exerted tremendous influence over the board. He orchestrated the ousting of two CEOs and drove out longtime executive Jeffrey Katzenberg, an event that directly led to the founding of DreamWorks. Roy's leadership was most famously embodied in the two "Save Disney" campaigns of 1984 and 2003.

Disney's 1977 animated film *The Rescuers* was well-received, but it seemed to deplete the company's creative momentum. For the next 18 years, Disney failed to produce another first-class animated hit. In 1979, famed animator Don Bluth and 17 senior animators defected to start their own studio, which became Disney's chief rival throughout the 1980s and early 1990s. During this time, Disney's animation division floundered. As noted, the theme parks contributed nearly three-quarters of the company's revenue during this slump.

Between 1981 and 1982, Disney's films floundered at the box office, with only *The Fox and the Hound* performing decently. In 1983, mounting losses from its film division and the costs of launching the Disney Channel caused profits to plummet below $100 million. The stock price fell from $84 to $49.[1] Roy Disney resigned from the board amid internal discord, further destabilizing the company.

This turmoil caught the attention of financier Saul Steinberg, who saw Disney's depressed stock as a golden opportunity. Steinberg calculated that if Disney's assets—including film rights and 17,000 acres of undeveloped land near

[1] Taylor, J. (1987). Storming the Magic Kingdom: Wall Street, the raiders and the battle for Disney. *(No Title)*. p42–p52

Q. Yue, F. Yang, *Governing Innovation: A Global History of Patent-Driven Technological Progress*, https://doi.org/10.1007/978-981-95-5570-3_48

Disneyland—were liquidated, the company was worth at least $100 per share. He organized a consortium to buy up shares, intending to break up and sell the company piece by piece. He even found buyers for everything except the theme parks, which he wanted to keep.

Disney's finance team noticed unusual trading activity, with daily volumes jumping from 200,000 to 900,000 shares by mid-March 1984. Initially suspecting Rupert Murdoch, they were blindsided when Steinberg revealed himself at the end of March, filing a disclosure with the SEC. On April 11, his firm Reliance held 9.3% of Disney stock. By the end of the month, Steinberg filed an amended declaration expressing intent to acquire 25% for "investment purposes," though few believed his benign explanation.[2]

At this critical juncture, Roy Disney returned, rallied executives and investors, and launched a successful counteroffensive to buy back the stock. Among the key allies was the Bass family of Los Angeles, owners of the Lakers. After the successful "Save Disney" campaign, Sid Bass became the company's largest individual shareholder. Roy rejoined the board, promptly fired CEO Ron Miller, and brought in Michael Eisner from Paramount as CEO. Eisner, in turn, hired Jeffrey Katzenberg to chair Disney's film division.

Eisner and Katzenberg initially focused on live-action films. Katzenberg spearheaded the Touchstone Pictures label, releasing family-friendly comedies like *The Emperor's New Clothes* and *Good Morning, Vietnam.* Disney's film division rose from last place in 1984 to the top by 1987. However, in making room for live-action, the animation department was displaced, wandering through warehouses and garages until settling in Glendale. In 1985, the animated feature *The Black Cauldron* flopped at the box office, further shaking Eisner's confidence in feature animation. Disney then pivoted to establishing a television animation unit.

During this tumultuous period, Disney's core—its technology, its artistic philosophy—stood on the verge of collapse.

We know that George Lucas's 1977 production *Star Wars* ignited the success of special effects–driven science fiction films for the first time. The 1980s saw a surge of critically and commercially successful sci-fi blockbusters in the U.S., including *Superman* (1978), *Blade Runner* (1982), and *The Terminator* (1984). Lucas also established Industrial Light & Magic, a pioneering visual effects company that became a cornerstone of the American special effects industry. These films redefined "fantasy" by portraying imagined characters and environments with an unprecedented degree of realism. Beginning in 1982, Lucasfilm also began filing patents related to computer systems, riding the wave of the digital age. In contrast, Disney had little to show during this period in terms of movie effects or computer animation patents. Without the ability to present visually plausible fantasies, what could set Disney's live-action films apart from other special effects–driven productions?

---

[2] Taylor, J. (1987). Storming the Magic Kingdom: Wall Street, the raiders and the battle for Disney. *(No Title)*. p64

It was at this critical moment that Roy Disney stepped in, persuading Michael Eisner to allow him to personally oversee the animation division.

Under Roy Disney and Jeffrey Katzenberg's leadership, Disney animation began a dramatic resurgence starting in 1989, producing one masterpiece after another that achieved both box office success and critical acclaim. Because many of these films drew upon classic fairy tales or traditional stories, the media dubbed this period the "Disney Renaissance." This era brought forth timeless works such as *The Little Mermaid* (1990), based on the Danish fairy tale; *Beauty and the Beast* (1991); *Aladdin* (1992), adapted from *One Thousand and One Nights*; and *The Lion King* (1994), inspired by *Hamlet*.

During the Disney Renaissance, Katzenberg served as head of the entire Disney studio, while Roy Disney directly managed the animation division. Tensions between them ran high. Katzenberg relished the media spotlight and often received the credit for Disney's creative revival—something Eisner did little to prevent. Many considered Katzenberg the primary architect of the Disney Renaissance, which infuriated Roy Disney.[3]

During this time, Don Bluth, who had previously left Disney, remained a formidable competitor. Meanwhile, Hayao Miyazaki's Studio Ghibli was gaining global recognition. The healthy rivalry among these three animation powerhouses ushered in a new golden age for animated film. Miyazaki's *Lupin III* series drew inspiration from Disney's *The Great Mouse Detective*, and Glen Keane of Disney acknowledged Miyazaki's influence on his own work.

Katzenberg was ambitious and actively vied for the position of second-in-command at Disney—a role then held by COO Frank Wells. CEO Michael Eisner had reportedly promised Katzenberg the role if Wells ever left. In 1994, when Wells tragically died in a helicopter crash, Katzenberg moved quickly to claim the position. But Roy Disney vehemently opposed his promotion and threatened Eisner with an all-out battle if he proceeded. Eisner ultimately yielded to Roy Disney, prompting Katzenberg to leave in anger. Following a series of internal clashes, Roy Disney dismissed Katzenberg, who responded by suing Disney and ultimately won a $250 million settlement. Their rift was final. Katzenberg went on to co-found DreamWorks with Steven Spielberg, establishing Disney's fiercest rival for years to come.

Former CEO and board chairman Robert Iger was openly critical of Roy Disney in his memoirs. He wrote that Roy was an incorrigible drinker who frequently sent ill-advised and damaging emails after drinking, creating substantial headaches for Disney's executive leadership.

Nonetheless, Iger acknowledged Roy Disney's unwavering loyalty to his uncle's vision and legacy. He recalled an incident where Disney introduced a batch of multicolored Mickey Mouse plush toys. Roy Disney was incensed and wrote to Iger

[3] Drew Taylor. (2020). The Disney Renaissance Didn't Happen Because of Jeffrey Katzenberg; It Happened In Spite of Him. Retrieved April 8, 2024 from: https://collider.com/jeffrey-katzenberg-disney-renaissance-impact-influence-explained/

demanding their removal: "Black, white, red, and yellow—those are Mickey's colors. You can't change them!"

So who was truly the driving force behind the Disney Renaissance—Katzenberg or Roy Disney? As patent professionals, we're neither inclined nor equipped to dissect office politics or Hollywood family drama. But in our domain of expertise, we can point to one essential contributor: the breakthrough digital technology embodied by patent US5091849A.

A critical enabler of the Disney Renaissance was the CAPS (Computer Animation Production System), co-developed with Pixar. Pixar, originally the computer animation division of Lucasfilm, may have conceived the early version of CAPS under George Lucas's umbrella. Ed Catmull recounts in his memoir:

> At George's request, we developed a video editing system that allowed editors to use computers to edit film. But this move drew a lot of resistance. George envisioned a program that allowed for easier storage and organization of footage and significantly improved editing speed over traditional methods… The technical challenge was significant, but it paled in comparison to the entrenched resistance to change. The old method required slicing film into pieces and taping it back together—editors had mastered this technique. Switching to a new method could temporarily slow them down, so they were reluctant to adopt it. Their pushback severely hampered the rollout of the system, even though George was a strong proponent.

We fully understand the editors' reluctance. But we also clearly recognize that the wheels of time cannot be stopped. No matter how editors may feel, technological progress will march on. In just the brief period covered by this chapter, how many industries have been crushed under the wheels of new technology, never to rise again? Expert film cutters, ten-fingered typists, master photo colorists—one after another, entire professions have vanished from history. The only way not to become relics of the past is to actively embrace new technology.

In 1983, George Lucas divorced his wife and needed to pay a $50 million settlement. At the same time, Steve Jobs had been ousted from Apple by John Sculley, temporarily thwarting his ambitions to change the world. Jobs had ample cash on hand, and Lucas needed money urgently. Thus, Lucas spun off the animation division into a new company, Pixar, and sold it to Jobs. After Pixar became independent, the CAPS system continued development and was eventually refined and deployed at Disney, where it blossomed and bore fruit.[4]

Once independent, Pixar entered negotiations with Disney regarding the CAPS system. The two parties held discussions for 18 months. Beginning in 1987, they jointly developed the system. This was Pixar's first major business deal after Jobs acquired the company. When Disney's engineers took over, CAPS was still a half-finished product and had yet to be tested.[5] The collaboration saw Pixar focus on

---

[4] Mike Seymour. (2012). Alvy Ray Smith: RGBA, the birth of compositing & the founding of Pixar. Retrieved April 8, 2024 from https://www.fxguide.com/fxfeatured/alvy-ray-smith-rgba-the-birth-of-compositing-the-founding-of-pixar/

[5] Robertson, Barbara. (1994). "Disney Lets CAPS Out Of The Bag". *Computer Graphics World*. 17 (7): 58–64. ISSN 0271-4159.

graphics and the user interface, while Disney took responsibility for system architecture and filed US Patent 5091849A for their improvements.

According to the patent's description, earlier systems designed for TV animation existed, but they couldn't handle the massive data volumes required for 35 mm film resolution. The system in US5091849A includes two Global Area Networks (GANs): the first transmits control data to processing devices (workstations), while the second delivers digital image data to at least one processing device. Instead of relying on a centralized processor, the system distributed data loads across multiple processors, thus solving the problem of computational power.

CAPS didn't replace the fundamental process of traditional animation. Starting in 1989, all characters—including animals—were still drawn by hand with pencils, frame by frame. However, those drawings were then scanned into computers and processed using the CAPS system.[6]

Back in the 1930s, the multiplane camera could support a maximum of five layers, with frame dimensions limited to just a few meters in width and height. Operating the camera was complex and expensive. CAPS removed the size constraints entirely. By the mid-1990s, the system could handle 2K resolution images. Within the digital environment, each layer could be freely moved, zoomed, or resized—eliminating the need to manually reposition heavy camera rigs or set up individualized lighting for each plane. Animators could color images with ease, add shading or blending effects, and manipulate them at will.

Like the multiplane camera, CAPS was never intended to cut costs—it was meant to improve quality. According to Eisner, CAPS did not save Disney any significant money.[7]

Disney continually enhanced the system. *The Little Mermaid* included only three multiplane shots—nearly the maximum its budget and processing power could afford at the time. By the time of *The Lion King*, the pinnacle of 2D animation, Disney's artists could render hundreds of multiplane scenes using CAPS. Each scene could comprise 50 to over 100 layers. The vast, majestic landscapes of Africa—skies, plains, jungles, trees, and rocks—each occupied their own layer. As the camera panned, these layers moved in relative motion, at varying speeds and in logical alignment with real-world depth, creating rich, immersive environments with stunning detail.

For a long time, Disney kept the CAPS system a closely guarded secret. Iconic films like *Beauty and the Beast*, *The Lion King*, and *Aladdin* all used CAPS. Audiences sensed a difference: characters, foregrounds, backgrounds, and skies appeared with distinct layers of depth. Scene transitions were smooth, and effects like simulated focus used high resolution, while out-of-focus areas were softened. Viewers could feel Disney's 2D animation had reached new heights, even if they couldn't explain exactly how.

---

[6] Robertson, Barbara. (1994). "Disney Lets CAPS Out Of The Bag". *Computer Graphics World*. 17 (7): 58–64. ISSN 0271-4159.

[7] Pallant, C. (2011). *Demystifying Disney: a history of Disney feature animation*. Bloomsbury Publishing USA. p96

The Lion King was not only a story of royal vengeance on the savanna—it also showcased the CAPS system's remarkable evolution: animators commanding hundreds of layers to depict depth and perspective with unprecedented realism. Lacking flashy visual effects, the film's hand-drawn brilliance and technical finesse remain captivating even today.

In the computer science community, the acknowledged "father of computer graphics" is Ivan Edward Sutherland. In 1967, at the invitation of fellow pioneer David Evans, Sutherland left Harvard to join the University of Utah. Together, they transformed its computer science department into the cradle of modern computer graphics.

Among Sutherland and Evans's students were a remarkable number of future tech luminaries, including Adobe co-founder John Warnock, Silicon Graphics and Netscape founder Jim Clark, and Atari founder Nolan Bushnell. Ed Catmull, one of Pixar's founders, was also one of Sutherland's most gifted protégés.

In 1973, Sutherland recommended the young Catmull to visit Disney and showcase the computer animation technology being developed at the University of Utah. According to Catmull's memoir, the meeting turned out to be a disappointment. Disney showed no interest in Sutherland's outreach initiative, and the spirit of Walt Disney, who had once embraced new technology, had long since vanished. Although Catmull delivered his presentation with enthusiasm, he recalled that the audience remained expressionless and that, to them, computers and animation seemed to have no connection at all.

Catmull recalled that Disney's aversion to computer animation stemmed mainly from the disappointing results of their 1971 live-action/animation hybrid film *Bedknobs and Broomsticks*. Disney had hoped to use computers to generate scenes with millions of bubbles, but the technology was still too immature and failed to deliver. While Disney executives warmly invited Catmull to join their Imagineering division, they showed no real interest in computer animation. They reportedly said, "If computer animation can't even do bubbles properly, then its time hasn't come." Catmull, unwilling to give up his dream of computer animation, declined their offer.

In 1979, George Lucas established a computer division within Industrial Light & Magic and hired Catmull to lead it. Catmull eagerly accepted. As previously noted, this division later became Pixar, which was eventually acquired by Steve Jobs.

For the first 9 years after acquiring Pixar, Jobs focused most of his time and energy on NeXT, visiting Pixar only once a year. During this early period, Pixar's main business was hardware, with occasional video ads. Located in the San Francisco Bay Area—rather than the Hollywood film industry hub of Los Angeles—Pixar maintained more of a tech-company culture.

In traditional animation and film companies like Disney, production crew members are not necessarily employees of the company. Most film workers are freelancers assembled into temporary teams for a project, then disbanded once the film is complete. Not so at Pixar: all production staff were salaried employees under contract. After *Toy Story* was completed, all crew members still reported daily to the Pixar offices.

Pixar's distinctiveness lay in this approach: if there was no movie to make, Jobs would assign the team to R&D rather than waste their time on small-scale projects like short films or ads. This strategy enabled Pixar to accumulate a trove of important patents in the field of 3D animation.[8]

We believe that many of Pixar's patents in computer animation reflect a continuation of Disney's "realism within fantasy" philosophy. Several key examples illustrate this legacy.

At the end of 2000, Pixar filed a patent for an "inertial field generator," describing a method for "controllably coupling kinematic properties to dynamically simulated elements." According to the patent, this generator could influence dynamic elements such as hair, clothing, and steam.

This technology does not aim for blind realism, but rather for realistic effects within exaggerated or fantastical scenes. The patent explains that when a character falls from a height, animators often make the fall unnaturally fast to heighten drama. In reality, the character's T-shirt might billow upward and obscure the face. Similarly, a furry character swinging its arm rapidly might have hair whipping in all directions. However, this overly "real" effect might distract from the animation's emotional impact.

To solve this, Pixar's invention classified objects into "moving bodies" and "dynamic attachments." With the inertial field generator, dynamic attachments would still move in sync with the body's acceleration but be tempered by simulated counterforces—so even if the character moved at exaggerated speed, hair or clothing would react in a more realistic manner.

One year after the patent was filed, Pixar released *Monsters, Inc*. One of its main characters, Sulley, was a monster covered in blue and purple fur. Another key character was a young human girl named Boo. This marked the first time a 3D animated protagonist had over two million individually rendered hairs. While humans had previously played minor roles in Pixar films, Boo was a central figure in this one—a major milestone in computer animation history.

In 2005, Pixar filed a patent based on two 2004 provisional applications for "variable motion blur" technology, which was granted in 2008 (US7348985B2). Prior to this, motion blur could be applied to computer animation, but the blur level was uniform across all objects in a frame. Pixar aimed to assign different motion blur levels to different objects—or even to different parts of the same object—better matching how humans perceive motion. This effect is particularly evident in the *Cars* movie series.

In March 2010, Pixar filed provisional applications for two technologies: "redirecting dynamic hair attributes" and "separating shaft and body dynamics in hair." These were granted in 2015 as patents US8698810B2 and US9098944B2. Then in February 2012, they filed another patent on "artistic simulation of curly hair," which was granted in 2016 as US9449417B1.

[8] M. Holson, Laura (2003). Animated Film Is Latest Title to Run Aground At DreamWorks. Retrieved March 31, 2024 from The New York Times Website: https://www.nytimes.com/2003/07/21/business/animated-film-is-latest-title-to-run-aground-at-dreamworks.html

These patents focused on hair simulation—specifically, long curly hair. Pixar modeled a strand of curly hair like a beaded necklace: the central "shaft" acted like a chain, while the curls resembled attached springs or beads. When a character's head moved, both the shaft and curls moved together. To avoid a stiff, wire-like appearance, the shaft and curls were assigned different masses.

In June 2012, Pixar released *Brave*, featuring Princess Merida as the protagonist—famous for her wild, voluminous curly red hair. Though previous Pixar films had featured long-haired female characters, none had hair as bold and prominent as Merida's, which occupied nearly a quarter of her overall figure and interacted frequently with her body. To make Merida's hair movement appear as natural as possible—without looking overly fluffy—Pixar set the gravitational effect on her hair to approximate that of the moon's gravity, achieving the desired visual effect.

Jobs was very proud of Pixar's technology in computer animation: "*We have 10 years of proprietary software systems that you cannot buy anything close to in the marketplace. You have to build them yourself.*"[9] Starting in 1989, Pixar provided software support and technology licensing to film studios such as Disney and Sony, contributing to productions like *Jurassic Park*. This became Pixar's main source of revenue in its early years.

Pixar held a robust portfolio of foundational patents in computer animation, including US4897806A, which addressed stochastic sampling; US5175808A on non-affine image warping; and US5307452A concerning image creation, manipulation, and display. Many of these were essential, difficult-to-work-around patents in the field of computer graphics.

In the mid-1990s, Steve Jobs was informed by his CFO that key technologies used in Pixar's RenderMan system might be at risk of infringement—especially its motion blur technology. One notable example was US4897806A, filed in 1985, one of Pixar's earliest patents and the first to claim "motion blur" in computer animation. At that time, any company developing rendering technology—especially giants like Microsoft and Silicon Graphics—would inevitably use some of Pixar's foundational techniques.

Jobs was excited by the news—finally, an opportunity to sue his old nemesis, Microsoft. But Larry Levy advised against going to war. *Toy Story* was nearing its final production deadline, and Levy cautioned that a lawsuit could distract the engineers at a critical time. Moreover, Microsoft and Silicon Graphics operated in markets different from Pixar's and posed no direct threat to its business. The two agreed to resolve the matter via licensing, hoping to secure much-needed cash.

Jobs initially hoped for $50 million, but Levy cautioned that such a high demand would likely trigger a lengthy legal dispute. He suggested a more pragmatic approach, emphasizing that patent licensing was not intended as Pixar's business model but rather as a short-term financing tactic, something to be used only

---

[9] Alcacer, J., Collis, D. J., & Furey, M. (2010). The Walt Disney Company and Pixar Inc.: To Acquire or Not to Acquire?. *Harvard Business School Case*, 709–462.

occasionally in difficult times. According to him, licensing could provide temporary relief but could not deliver long-term success.

Jobs accepted Levy's counsel. After 3 months of negotiation, Pixar secured $6.5 million from Microsoft. Talks with Silicon Graphics took a full year and yielded a slightly higher sum. As a supplier, Silicon Graphics also provided Pixar with free graphics workstations.[10] Time would prove Pixar's strategy wise. Patents are critical strategic tools, but as the saying goes, "Weapons are ominous instruments, not to be used lightly." If a company over-relies on litigation or licensing, it risks diverting engineers from innovation and alienating its audience.

Just as Pixar was thriving, Disney faced another crisis.

After 2000, seeing the rise of computer animation, CEO Michael Eisner began to lose faith in traditional animation—much to Roy Disney's dismay. In an interview with the *Los Angeles Times*, Eisner compared 2D animation to black-and-white films, declaring it was nearing its end. He planned to shut down Disney's historic Florida studio—which had produced *Who Framed Roger Rabbit* and *Mulan*—and replace the studio's annual feature release schedule with more frequent but less ambitious productions. He even considered releasing some quality titles directly to DVD. In 2003, Eisner laid off a large number of traditional animators and shifted fully to computer animation (many of whom Pixar would later rehire after the acquisition). To Roy Disney, this was a betrayal of the company's legacy. It was also one of the reasons behind his second "Save Disney" campaign.[11]

Around 2003, Roy Disney's conflict with Eisner reached its peak. Roy called on the board to remove Eisner, who in turn cited a company bylaw stating board members must be under 72—effectively forcing Roy to resign from the board.

After his forced departure, Roy launched a media campaign titled "Save Disney," outlining seven accusations against Eisner and urging shareholders to vote against him. The fourth accusation especially revealed Roy's values:

> All our shareholders believe… the company has become soulless in its pursuit of profit, always chasing the 'quick buck' while neglecting long-term value. This has eroded public trust in Disney.

The sixth accusation was: "Failure to maintain good relationships with key partners—most notably, Pixar."

The campaign eventually led to Eisner's resignation. Roy returned to the board as a non-voting honorary member, and the diplomatic Robert Iger took over as Disney's CEO.

While Disney and Pixar had long enjoyed a good relationship, Jeffrey Katzenberg and Steve Jobs did not. In 1991, when Katzenberg was still head of Disney Studios, Jobs visited Disney to pitch NeXT's NeXTcube and NeXTstation workstations. Jobs passionately claimed his products could empower every American household

[10] Levy, L. (2017). *To Pixar and beyond: My unlikely journey with Steve Jobs to make entertainment history*. Simon and Schuster.

[11] Telotte, J. P. (2010). *The mouse machine: Disney and technology*. University of Illinois Press. p209

to make animations like Disney. The comment unexpectedly infuriated Katzenberg. He coldly retorted:

> I own animation. Nobody touches it. That's like some punk asking to date my daughter. I've got a shotgun, and if anyone touches her, I'll blow his balls off.

Though NeXT never landed a major deal with Disney, Pixar's experimental short films caught Katzenberg's eye. He commissioned Pixar to make a feature-length animated film about toys—what would become the revolutionary *Toy Story*.

In the early 1990s, Jobs was struggling. NeXT's sales were disappointing, while Microsoft was dominating. Customers defected en masse to Windows NT and Sun Microsystems. Pixar had burned through $60 million of Jobs' personal fortune and was still not profitable. *Toy Story* became Pixar's one and only lifeline.

Katzenberg was only willing to allocate a small share of box office revenue to Pixar, plus a fixed production fee. Jobs, however, demanded a higher percentage of the box office, as well as future profits from DVD and VHS sales, and software rights. Their negotiations erupted into a heated dispute. In the end, Pixar received just 10% to 15% of the box office revenue, while Disney retained the copyrights—including the right to produce sequels. Katzenberg told Jobs that Disney never paid more than $15 million to external production partners. Jobs reluctantly accepted these terms. In reality, Katzenberg had lied: *Beauty and the Beast* had also been outsourced, and Disney had paid a staggering $32 million for it. Once Jobs realized he had been duped, Katzenberg had already left Disney, and there was nothing he could do.

After *Toy Story*'s massive success, Jobs sought to renegotiate the deal and secure greater rights for Pixar, but Eisner refused to budge. By 2003, however, the tables had turned—Jobs had returned to prominence at Apple and no longer had to defer to Disney. Relations between the two companies were on the verge of collapse.

In 2005, Disney CEO Robert Iger visited Hong Kong Disneyland and noticed something unsettling: every character in the park's parade from the past decade had come from Pixar, not Disney. He presented this fact to the board and emphasized, "The future of Disney animation is the future of Disney itself." He proposed acquiring Pixar.

In the cases of Kodak, Sony, and Xerox, we often lament missed opportunities: if Kodak had embraced digital imaging sooner, if Sony had adapted to the mobile internet age, if Xerox had promptly commercialized PARC's innovations… But for these companies, history offered no second chances. Fortunately, while Disney did not immediately leap into computer animation, it was granted a rare opportunity to recover.

Iger first set out to repair Disney's relationship with Jobs. At the time, Apple had just launched the fifth-generation iPod. During a meeting, Iger agreed to allow ABC (American Broadcasting Company) shows to be distributed on the iPod. Eventually, Jobs agreed to sell Pixar to Disney in a stock-swap deal. In January 2006, Disney acquired Pixar for $7.4 billion, with each share of Pixar common stock exchanged for 2.3 shares of Disney stock. Jobs, who held a 49.65% stake in Pixar, became Disney's largest individual shareholder with a 7% stake—worth $3.9

billion—surpassing former CEO Michael Eisner (1.7%) and Roy Disney (1%). Jobs also joined Disney's board. Roy Disney passed away in 2009, and Jobs himself died not long afterward. Today, Disney's largest individual shareholder is not a member of the Disney family but Steve Jobs's widow. John Lasseter was appointed Chief Creative Officer of Disney, overseeing both Disney Studios and Pixar Animation. Ed Catmull remained president of Pixar and also assumed leadership of Disney Animation Studios.

With the integration of Pixar and Industrial Light & Magic (ILM), Disney launched a renewed campaign in the live-action film arena.

As mentioned earlier, Disney had made several attempts to break into live-action films—some of them successful—but true triumph only came in recent years. In 2008, Disney films accounted for 10.5% of the U.S. box office. By 2018, that share had risen to 26.3%, and by 2019 it hit 33%, with a record-breaking global box office revenue of $11.9 billion—$3.8 billion from the U.S. and $8.1 billion internationally—surpassing all competitors.[12] This was due in part to Disney's acquisitions of Marvel, Pixar, and Lucasfilm, but the company's formidable technological prowess also played a key role.

As shown in Fig. 48.1, according to a June 2021 analysis by patent intelligence firm GreyB,[13] Disney holds the most patents in the field of animation. The second and third largest categories are "Broadcasting and Streaming" and "User Information and Data Processing," respectively. These two areas largely reflect Disney's technical foundation in streaming. In 2017, Disney acquired a 75% stake in BAMTech, a

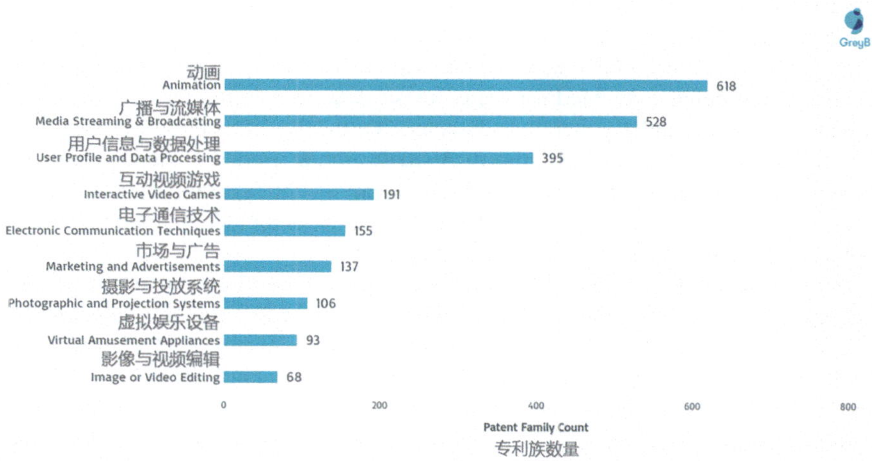

Fig. 48.1 Patent analysis of Disney's portfolio by GreyB, June 2021

[12] Nancy Tartalione. (2019). Disney Crossing Record $10 Billion at Worldwide Box Office; First Studio in History to The Mark. Retrieved from Deadline's Website: https://deadline.com/2019/12/disney-crosses-10-billion-worldwide-box-office-new-all-time-record-1202803824/

[13] Grey B. (2023). What's Inside Disney Patent Portfolio? Retrieved March 31, 2024 from GreyB Website: https://insights.greyb.com/disney-patent-portfolio/

streaming technology company, and renamed it Disney Streaming Services to support its ESPN+ platform. In November 2019, Disney launched its own streaming service, Disney+. On its first day, Disney+ attracted 10 million subscribers; it has since surpassed 100 million and become Netflix's fiercest competitor.

ILM's iMoCap system was first used in Disney's *Pirates of the Caribbean: Dead Man's Chest*. As the pioneering force in sci-fi visual effects, Industrial Light & Magic became part of the Disney family in 2012 when Disney acquired Lucasfilm.

In addition to incorporating the technological achievements of Industrial Light & Magic and Pixar, Disney established the **Disney Research Center** in 2008, opening a research lab in Zurich, Switzerland, to pursue cutting-edge research in **artificial intelligence**, **machine learning**, and **visual computing**. To date, the Zurich lab has contributed at least 254 patents, most falling within the **G06** and **H04** categories. These patents span 3D animation technologies, as well as innovations in **3D printing**, **drones**, and **robotics**.

Disney Research's most renowned achievement in recent years is the **Medusa Performance Capture System**, co-developed with Industrial Light & Magic—currently the most advanced facial performance capture system on the market.

Unlike traditional facial capture systems, which require actors to wear headgear with markers, cameras, and lights, the Medusa system eliminates the need for such markers. Instead, actors undergo a high-precision facial scan before filming to build a comprehensive expression library. During scanning, actors make a series of expressions as directed, and the Medusa system captures the corresponding facial geometry with astonishing detail—up to millions of data points on a single face, capturing even the smallest wrinkles and contours. The team then reviews the video to select the optimal moments for each expression. After live-action filming, these scans are used to enrich the characters' digital performances. ILM's **Muse system** precisely manipulates the Medusa scans' mesh data, translating them into lifelike expressions for computer-generated characters. Of the five films nominated for **Best Visual Effects** at the **2019 Academy Awards**, three used the Medusa system. Disney Research received a **2019 Sci-Tech Oscar** for the breakthrough. The most recent *Avengers* films also relied on Medusa, with characters like Hulk and Thanos showcasing expressive realism thanks to this technology.

Disney's recent handling of IP and franchise sequels has drawn mixed reactions. The new *Star Wars* films alienated many longtime fans, while the *Mandalorian* TV series won widespread acclaim and numerous awards. *Avengers: Endgame* shattered box office records in China, but *Shang-Chi and the Legend of the Ten Rings* faced heavy backlash even before release due to casting controversies. We don't aim to judge Disney's IP decisions. No matter how its stories evolve, **human beings will always dream**, and Disney continues striving to make those dreams real. As long as that vision is sustained through technological innovation, Disney has the potential to keep expanding its frontiers.

Leaving aside the brand and IP advantages supporting new ventures, **from a patent perspective**, Disney has undergone **three major expansions**:

1. **The creation of Disneyland**, backed by a strong portfolio of physical attraction patents. From the start, Disneyland transitioned from a peripheral project to a core revenue stream, becoming Disney's second-largest business. Walt Disney fulfilled his childhood dream of creating a fantastical playground for children and adults alike. Originally, Disney's customers were primarily children; Disneyland brought in couples and parents, broadening the audience.
2. **The acquisition of ABC and ESPN** in the 1990s marked Disney's move into television. Years of technical accumulation have since supported the launch of **Disney+** and **ESPN+** streaming services. As mentioned earlier, streaming has become one of Disney's most patent-heavy domains. With its vast reservoir of IP, Disney stands as Netflix's strongest competitor. While the COVID-19 pandemic hit Disney's theme park and film divisions hard, it provided a perfect opportunity for its streaming services to grow. The audience base has expanded beyond families to include sports fans and general TV viewers.
3. **In the twenty-first century**, through acquisitions of Pixar and Lucasfilm (and ILM), and the forward-looking efforts of Disney Research, the company has remained a global leader in **computer animation** and **visual effects**, winning over audiences with live-action, action-packed, and sci-fi films.

After these three expansions, **Disney's audience reach may be the widest among all the companies we've analyzed**. Sony and Apple products often have price thresholds; HP focuses on enterprise markets; Bosch mainly targets automakers and faces stiff competition from cheaper alternatives. In contrast, **Disney's customers are anyone who loves to dream**, regardless of age. As we've emphasized throughout this book, the core of Disney's technological philosophy is crafting **realistic fantasy worlds**, aligning perfectly with the aspirations of dreamers everywhere. Our repeated reminder—**"The people are most important; the nation comes second; the ruler is least"**—underscores that it is the **people's expectations**, or **consumer needs**, that should drive innovation and business expansion.

Part IV
Governance and Administration

Chapter 49
Introduction

The title of this book is *Governing Innovation.*

Innovation is the foundation of all patent work and the reason the patent system exists. Innovation is also intrinsic to human nature—from stone tools to iron, from caves to houses, from slash-and-burn farming to an understanding of the solar terms, innovation long predates the birth of patent law and has always been closely tied to human well-being. We believe that such a magnificent aspect of human nature cannot be arbitrarily restricted or forcibly managed; it can only be guided and shaped in accordance with its natural course. This process, we refer to as **governance**.

> *Nature, rather than religious theology, provided the metaphors for many key concepts in early Chinese philosophy... Water nourishes life, wells up from underground, flows naturally, levels itself when still, filters out impurities, clarifies itself, yields to external force but ultimately wears down even the hardest stone, turns to ice as hard as rock, or vanishes into steam—it is a model for philosophical reflections on the nature of the cosmos.*[1]

The still, level surface of water represents the impartiality of law. In fact, the Chinese character for "law" (法) is derived from this very image—something every Chinese law student learns in their first lecture.

Professor Wang Renbo of China University of Political Science and Law, in *Water: The Core Metaphor of Chinese Legal Thought,*[2] also argues that "early Chinese philosophers did not rely on concepts or logic to reflect on law, but instead drew inspiration from the observation, imagination, and contemplation of water. The imagery provided by water became a prototype for Chinese legal thought". In ancient times, mastering the governance of water was regarded as a supreme political achievement. Great governance was the ultimate pursuit. In modern times, the rule of law (fazhi) is the goal, and the vision is one of "governance in China." The

[1] Allan, S. (2002). *The way of water and the beginnings of Chinese philosophy* (水之道与德之端) (H. Zhang, Trans.). Shanghai People's Publishing House.

[2] 王人博. (2010). 水:中国法思想的本喻. 法学研究, 032(003), 186–194.

Q. Yue, F. Yang, *Governing Innovation: A Global History of Patent-Driven Technological Progress*, https://doi.org/10.1007/978-981-95-5570-3_49

character *zhi* (治, governance) has remained a guiding ideal across millennia—this is why we are so fond of this word.

The character *li* (理), on the other hand, originates from something even more intimately connected to Chinese culture: jade. According to *Shuowen Jiezi*[3] under the section for jade, "Li means to shape jade—to follow its natural grain and cleave it accordingly," that is, to carve the rough stone into a refined artifact. Some scholars argue that from this root meaning, *li* developed in two directions: from the grain of jade came concepts like "principle of things" and "cosmic order," which stand alongside *dao* (the way) and *yi* (righteousness); from the act of working jade came meanings like "to govern" or "to manage,"[4] which ties it back to *zhi* (治). In *Xunzi: Cultivating the Self*, we find: "A few governed is called order (li), too many governed is called chaos (hao)."

Zhi (治理, governance) is both action and aspiration. *Li* represents methodology—governing in accordance with the natural order, as well as with reason and emotion. For these reasons, we chose the term "governance" (治理) as both the title of this book and this final section. Though the patent system is an imported construct, it has taken root in our soil. It should now be reexamined, critiqued, and governed in accordance with ideas that suit our context, so it can better serve society. The first three parts of this book dealt with stories from abroad. But we do not advocate swallowing foreign ideas wholesale. Our goal is to understand history, seek truth, and identify models and patterns that are truly instructive.

The truth of history is often obscure, hidden in a thicket of records. To faithfully reconstruct these stories, we spent considerable effort gathering materials—seeking to verify what lies behind folklore, news articles, and academic papers. We are not historians of technology, nor management scholars, nor investigative journalists. We lacked access to primary sources. As historian Fu Sinian wrote in *On the Mission of the Institute of History and Philology*, "Progress lies in direct research of source material. When one only studies the interpretations or systems created by others without thoroughly engaging with the underlying facts, one regresses".[5] From the beginning of this project, our goal was to find first-hand materials to compare against common narratives in news, academia, and industry lore.

As patent professionals, our most familiar tool—**the patent database**—quickly became our most important archive of primary material. Patents clearly trace the development and dissemination of technologies. They can be cross-verified with product launches, company trajectories, and legal disputes. When combined with court statements and judicial summaries, they become a priceless trove of first-hand history.

In his foundational treatise *Shitong*, Tang-dynasty historian Liu Zhiji distinguished between "records written at the time" and "narratives written afterward." The former, he argued, were more valuable as historical sources. By that measure,

[3] Xu, S. (n.d.). *Shuowen jiezi* (《说文解字》).

[4] 李翔宇. (2008). 汉语"管理"词源考. 管理观察, (6), 109–111.

[5] 傅斯年. (1928). 歷史語言研究工作之旨趣. 中央研究院歷史語言研究所集刊, 3–10.

patents are undoubtedly "records written at the time" for the stories we explore. Liu also identified ten forms of anecdotal history—some honest, others embellished. Corporate accounts of R&D often fall into the latter category, glossing over inconvenient truths. Media reports and online commentary, too, frequently sacrifice accuracy for impact. Liu championed the method of "gathering diverse perspectives and sifting through them meticulously," aspiring to this ideal: "If future generations of historians continue their work, then even though people pass away and time moves on, their deeds will shine on like the stars in the sky." Today, few tech companies would appoint a Chief History Officer, but patent databases, combined with journalistic and scholarly sources, can help us preserve the true arc of technological history—before it vanishes into myth.

In the "Rise and Fall" section of this book, I used this method to challenge the notion that "first-rate companies sell standards, second-rate companies sell patents." I critiqued IBM's licensing-driven patent strategy. In the "Disruption and Revolution" section, I analyzed how Sony and Philips clung to outdated standards, neglected consumer needs, and were ultimately outpaced by more agile competitors. These companies were chosen deliberately. When I entered this profession 20 years ago, many colleagues were in awe of IBM's legendary $1 billion in licensing revenue, or envious of the royalty income earned by CD and DVD patent pools. Even I, as a newcomer, once admired these "patent overlords." But after working with them, I came to see things differently.

Sure, IBM's $1 billion licensing revenue is impressive. But while IBM clung to patents, Amazon, Google, and Microsoft dominated the cloud computing space—IBM's supposed core market. Even after acquiring Red Hat, IBM saw little real progress. A B2B company unable to outmaneuver its true competitors resorted to targeting consumer-oriented internet firms that posed no threat. This kind of "victory" is hollow—much like the Jin dynasty in the thirteenth century, which, unable to fend off the Mongols in its heartland, decided to attack the Southern Song in the hopes of "compensating northern losses with southern conquest." Even if such efforts yield tactical gains, they can never redeem a lost core.

In the "Disruption and Revolution" section, we advocated that enterprises should boldly break through their own boundaries, embrace self-revolution with the resolve of a warrior severing his own arm, and create new worlds. In the "Frontiers" section, we emphasized that companies should first build a stable core, then expand outward to establish new strongholds in emerging technological fields, advancing steadily and fearlessly. We believe this is the royal road to enterprise management.

The "royal road" we champion—or the path of governance—is not merely a path to profit. If an enterprise truly follows the royal road, serves the public, and builds century-spanning businesses through new technologies and industries, then wealth will naturally follow. As the saying goes, "If one rejoices with the world and worries with the world, yet does not rule it, such a person has never existed." As for how much money one can earn, or how soon one achieves financial freedom, our answer is simply: "Why must a ruler speak of profit? There is only benevolence and righteousness."

Our criticism of IBM, Sony, and Philips is not because their revenues declined or market value shrank, or that they failed to make money—but because they held the world's top-tier technologies and talent yet could not stand at the forefront of humanity's exploration of the stars and seas. From the standpoint of a long-time patent attorney serving enterprises, it pains me to see the fruits of countless R&D professionals, earned through years of dedication, wasted in obscurity and failing to benefit the public. A company's lifespan is short; a lawyer's or manager's career even shorter. For individuals or small firms, a few big successful cases may suffice to make a fortune and secure a comfortable life. But such petty maneuvers fall outside the scope of this book.

In September 2018, a U.S. company called "Shipping & Transit" filed for bankruptcy. This notorious patent troll had filed over 500 patent lawsuits during its existence.[6] In 2016 alone, it initiated 107 lawsuits—making it the most litigious company in that year.[7] At the time of its bankruptcy filing, "Shipping & Transit" held 34 U.S. patents and 29 international patents. According to procedural law, it had to declare the value of all remaining assets. Thus, we arrive at Fig. 49.1, where the company valued all 34 of its patents at a grand total of $2—$1 for its U.S. patents and $1 for its non-U.S. patents.

The company's former name, "ArrivalStar," hints at its operations in transportation and logistics—and its core activity was patent extortion within this industry. Around 2012, "Shipping & Transit" filed lawsuits in multiple U.S. states targeting local transportation agencies, including transit authorities, traffic departments, and port commissions. According to a 2012 report, the company typically demanded

Part 10: **Intangibles and intellectual property**

59. Does the debtor have any interests in intangibles or intellectual property?

☐ No. Go to Part 11.

■ Yes Fill in the information below.

	General description	Net book value of debtor's interest (Where available)	Valuation method used for current value	Current value of debtor's interest
60.	**Patents, copyrights, trademarks, and trade secrets**			
	United States Patents (see attached list)	$1.00	Litigation	$1.00
	Worldwide Patents (see attached list)	$1.00	Litigation	$1.00

Fig. 49.1 Excerpt from Shipping & Transit's bankruptcy filing on September 6, 2018, page 7. (The file is available at: https://www.eff.org/files/2018/10/30/shipping_transit_bankruptcy_petition.pdf)

[6] Daniel Nzer. (2018). Stupid Patent of the Month: How 34 Patents Worth $1 Led to Hundreds of Lawsuits. Retrieved April 8, 2024 from https://www.eff.org/deeplinks/2018/10/stupid-patent-month-how-34-patents-worth-1-led-hundreds-lawsuits

[7] Lex Machina. (2016).Lex Machina's Fourth Annual Patent Litigation Year in Review Report Shows 22 Percent Decline in Patent Filings in 2016. Retrieved April 8, 2024 from https://lex-machina.com/media/press/report-shows-22-percent-decline-in-patent-filings-in-2016/

settlements of $50,000 to $75,000.[8] A 2013 report from the American Public Transportation Association (APTA) revealed that at least 11 transportation agencies had settled with ArrivalStar to avoid costly litigation. For example, a transit division under King County, Seattle, lacking any patent litigation knowledge, paid $80,000 to settle a claim.[9]

In June 2013, APTA filed for invalidation of 13 ArrivalStar patents. The company immediately backed down, agreeing not to sue any APTA member. By 2018, it could no longer profit and filed for bankruptcy—shamelessly admitting that all its patents were worth just $2.

Using worthless patents valued at $2 to extort millions over more than a decade—if wealth is the only goal, what business compares to the dark art of patent trolling? But as a patent attorney, I believe whether one admits to holding $1 garbage patents, or uses such patents to extort money, both are stains on a professional career. We have spent years studying both technology and law, researching cases and drafting arguments—not to exploit legal loopholes or drain society of its wealth. Such behavior contributes nothing of value, wastes judicial and administrative resources, and is the very antithesis of our concept of governance.

In 1928, Alexander Fleming happened to notice a mold that inhibited bacterial growth, leading to the discovery of penicillin. He made this discovery public in a brief paper. Ten years later, Howard Florey's team at Oxford University was the first to grasp the significance of Fleming's discovery. After years of labor, they succeeded in purifying penicillin. Though Fleming, for technical and legal reasons, did not patent his discovery, Florey's method of purification was fully patentable. Team member Ernst Chain suggested they file a patent, but Florey, adhering to the ethical norms of British medical science at the time, declined.[10]

In August 1942, penicillin was successfully used to treat a patient, attracting media attention. Fleming gave interviews and quickly became a celebrity. Florey, however, refused all interviews and barred his team from engaging with the press, letting Fleming take all the credit. In 1945, Fleming, Florey, and Chain were jointly awarded the Nobel Prize, creating a memorable chapter in scientific history.

History is full of figures like Florey—people who shunned fame and fortune to unconditionally gift their discoveries to society. Their actions are worthy of admiration, but we do not ask readers to forgo patents and become saints like Florey. James Watt, whose inventions fueled the First Industrial Revolution over 200 years ago,

[8] Joe Mullin. (2012). A new target for tech patent trolls: cash-strapped American cities. Retrieved March 31, 2024 from ArsTechnica Website: https://arstechnica.com/tech-policy/2012/03/a-new-low-for-patent-trolls-targeting-cash-strapped-cities/

[9] Joe Mullin. (2012). A new target for tech patent trolls: cash-strapped American cities. Retrieved March 31, 2024 from ArsTechnica Website: https://arstechnica.com/tech-policy/2012/03/a-new-low-for-patent-trolls-targeting-cash-strapped-cities/

[10] BUD, ROBERT (1998). Penicillin and the new Elizabethans. The British Journal for the History of Science, 31(3), 305–333. doi:10.1017/s0007087498003318

rightly observed that invention stems from three motives: *the desire to doing good to society, the desire of fame and the hope of increasing his private fortune.*[11]

This is akin to the classical Chinese stories of Zigong ransoming a slave and Zilu accepting a cow: Zigong paid out of his own pocket but refused state compensation, prompting Confucius to criticize him for imposing moral burdens on others. Zilu saved a person's life and accepted a thank-you gift, earning Confucius' praise. The patent system, by institutional design, acknowledges human desires and puts them to use. We do not preach from moral high ground. The governance we advocate is not about everyone emulating Zigong—it's perfectly fine to be like Zilu: accept praise and rewards with a clear conscience.

Philo Taylor Farnsworth was a prodigious inventor hailed as a teenage genius. He began conducting independent research at the age of 14 and filed a patent for a "television system" at 21, which was granted in 1930 (U.S. Patent No. US1773980A). Later, the corporate giant RCA invested heavily in television R&D, only to find that it could not circumvent Farnsworth's patent. This led to a prolonged legal battle over television technology. Though the lawsuit lasted many years, Farnsworth ultimately prevailed and received a substantial licensing fee from RCA. However, the long legal struggle left him mentally and physically exhausted.[12]

After the litigation ended, Farnsworth was ready to unleash his full potential—but the timing could not have been worse. The United States entered World War II, and the government suspended all television-related research. When the war ended and Farnsworth resumed work in the field, he found himself unable to compete with the much larger RCA. He was eventually forced to sell his company. RCA, for its part, refused to acknowledge Farnsworth's contributions. In 1956, it released a documentary titled *The Story of Television*, which exclusively chronicled RCA's development of the medium—as if Farnsworth had never existed.

This deeply wounded Farnsworth, who grew to despise television in his later years. He once told his children, "There's nothing on it worthwhile, and we're not going to watch it in this household, and I don't want it in your intellectual diet."[13] Yet, in the year he passed away, Farnsworth finally saw something on TV that he believed was worth it. On July 21, 1969, Apollo 11 landed on the moon, and for the first time, humanity set foot on another celestial body. Watching this historic event

---

[11] Murfitt, S. (2017). *The English patent system and early railway technology 1800–1852* (Doctoral dissertation, University of York). "There seems to be only three motives that can excite a man to make improvements in the arts, the desire to doing good to society, the desire of fame and the hope of increasing his private fortune." James Watt, letter to Lord Loughborough, July 8, 1785, Boulton and Watt Papers, MS 3147/3/85. Birmingham City Reference Library.

[12] Erik Gregersen. (2024). Philo Farnsworth. Retrieved April 8, 2024 from Britannica Website: https://www.britannica.com/biography/Philo-Farnsworth

[13] Pennsylvania Historical and Museum Commission. (n.d.). *Philo T. Farnsworth historical marker*. Pennsylvania Historical Markers. Retrieved March 31, 2024, from https://share.phmc.pa.gov/markers/

unfold on the screen, Farnsworth turned to his wife and said, "This has made it all worthwhile."[14]

When I first read Farnsworth's story and came across that line, I couldn't help but feel a deep resonance. We often speak of humanity's journey to the stars and seas—but how many of us will ever actually go there? For ordinary people, the most we can hope for is to contribute in some small way to that grand endeavor. If, in our lifetime, we can do something—however minor—that supports China's spacecraft soaring through space, then indeed, "made it all worthwhile."

The path of governance we advocate is not one of patent extortion, nor the Machiavellian strategy of building patent pools to "tax" entire industries, nor the saintly act of selflessly abandoning all patent rights. Rather, it is the simple principle that allows us, at the end of our lives, to say, "This has made it all worthwhile."

Throughout this book, we've introduced numerous inventors and entrepreneurs, most notably from nineteenth- and twentieth-century America and Germany. Max Weber, in his seminal work *The Protestant Ethic and the Spirit of Capitalism*, offers deep insights into their values. A nineteenth-century American entrepreneur or inventor might work around the clock and accumulate vast wealth—yet live a life of abstinence, avoiding smoking, drinking, indulgence, and luxury. Religion taught them to labor diligently; wealth was seen as divine reward for righteous conduct. Laziness, greed, gluttony, and lust were original sins to be resisted. We Chinese also value diligence and thrift and are not opposed to borrowing from others' ideas—but the mindset and lifestyle of nineteenth-century Puritan America are too far removed from us. We must seek our own philosophy of governance and innovation from within Chinese tradition.

So, how should innovation be governed according to traditional Chinese thought? Though ancient scholars did not explicitly discuss innovation governance, they wrote extensively on state governance. Perhaps we can draw analogies from these teachings.

The Confucian classic *The Great Learning* offers this wisdom:

> The ancients who wished to illuminate illustrious virtue throughout the world would first govern their states; to govern their states, they would first regulate their families; to regulate their families, they would first cultivate themselves; to cultivate themselves, they would first rectify their hearts; to rectify their hearts, they would first be sincere in their intentions; to be sincere in their intentions, they would first extend their knowledge; the extension of knowledge lay in the investigation of things. Only when things are investigated is knowledge extended; only then are intentions sincere, hearts rectified, selves cultivated, families regulated, states governed, and the world pacified.

Later scholars summarized this as the "Eight Steps" (*Ba Mu*): investigation of things, extension of knowledge, sincerity of intention, rectification of the heart, self-cultivation, regulation of the family, governance of the state, and pacification of the world.

---

[14] Television Academy. (n.d.). *Hall of Fame: Philo Farnsworth*. Academy of Television Arts & Sciences. Retrieved March 31, 2024, from https://televisionacademy.com/

"Governing the state and pacifying the world" was the highest aspiration of the ancients. Among our peers, a few have reached such heights. Albert Einstein, a former Swiss Patent Office examiner, experienced a burst of creativity in his 4th year on the job, publishing four groundbreaking papers that "revolutionized humanity's understanding of the physical world"[15]—what became known as his "miracle year." Herman Hollerith, a former U.S. patent examiner, invented the tabulating machine in his spare time, laying the foundation for one of the most significant tech enterprises in history. Chester Carlson, once a patent engineer, toiled away after hours to invent the photocopier, benefiting billions and greatly boosting productivity. Then there was Ivar Giaever, who worked at the Norwegian Patent Office before winning the Nobel Prize 18 years later. These accomplishments surely merit the title "pacifying the world" and serve as inspiration for us all. While such feats are rare, to simply witness them and contribute in even the smallest way would make a life well lived.

If we cannot reach the heights of "governing the state and pacifying the world," we can at least start with the basics. The foundation of the "Eight Steps" is the investigation of things and extension of knowledge. By beginning with this, and progressing through sincerity, rectification, and cultivation, we may 1 day contribute meaningfully to society, to the governance of China, and to the peace of the world. What greater joy could there be?

To practice the Confucian principle of *gewu zhizhi*—"investigating things to extend knowledge"—Wang Yangming famously spent 7 days observing bamboo in his courtyard, attempting to uncover its underlying principles. In the end, he fell ill and gained no clear insight. Yet, this perseverance and commitment ultimately led to his enlightenment in Longchang, making him one of the greatest sages in Chinese history.

We modern people are far too busy to spend seven straight days contemplating a single object. But for patent professionals, patents are well within the domain of our expertise. If we reflect on them day and night, echoes will surely come. As long as we are diligent and proactive in our investigations, insights into innovation will inevitably arise.

As patent professionals, we work daily with a wealth of resources: patent databases, academic databases, and case law repositories. These are intellectual treasure troves—well-organized, uniformly formatted, and easily searchable—painstakingly built by predecessors and ready to sustain a lifetime of study. Those of us who specialize in prior art searches roam through these vast archives; those who write and prosecute patents contribute new bricks to the edifice.

These two tasks may seem unrelated, but they share a deeper commonality: human beings and tools evolved together, and our affinity for tools is embedded in our DNA. When we find the right tool, we instinctively want to use it—push it to its fullest potential. Faced with such a remarkable tool as the patent database, how

[15] 杨振宁. (2005). 爱因斯坦: 机遇与眼光——第 22 届国际科学史大会大会报告. 科学文化评论, (4), 10–16.

could we possibly walk away empty-handed? Entering a treasure trove daily, we should not leave without gems in hand.

The next step after *gewu zhizhi* is to rectify the mind and be sincere in one's intentions. We firmly believe that patent professionals should uphold sound values. It is a fact that people have desires; the pursuit of wealth and a better material life is not inherently wrong. But ideals and desires should not be confused. Proper use of patent law can help us uncover truths, validate genuine innovations, and contribute to humanity's collective knowledge. It should not be twisted into a tool for personal gain, textual manipulation, patent fabrication, frivolous litigation, and the squandering of public resources.

Enterprises should aim to serve millions through exceptional products. Patents are meant to protect such products—they are a means, not an end. The quantity of patents, their strategic layout, litigation outcomes, and licensing deals should all support the product and the market, safeguarding the enterprise's right to serve the public. When patents are reduced to gimmicks for capital markets or metrics for executive bragging rights, priorities are inverted—this is the wrong path and one that should be rejected.

Once we have investigated things, rectified our intentions, cultivated ourselves, and brought harmony to our homes, we may not govern a state—but we can govern innovation in our own way. The essence of "governance" (*zhili*) lies in following the flow of water and the grain of jade. To govern innovation, one must identify its laws, patterns, trends, or methodologies—and act accordingly. Follow the natural order, serve the people.

With this purpose in mind, we completed the first three parts of this book. Though we claim no grand discoveries, we have gleaned some insights from the history of patents. Next, we will distill the core ideas of this book into a simple model.

Chapter 50
Innovation and Patents: A Fundamental Connection

In our imagination, human knowledge and technology resemble a vast, Earth-centered space (see Fig. 50.1), which has been continually expanding since early humans first discovered fire and crafted stone tools. This expansion is driven by human innovation. From stone tools to bows and arrows, wheels, and pottery, every historically significant innovation has profoundly transformed human life, making it better—and it is through these transformations that humanity came to recognize the immense value of innovation.

Thus, within the boundaries of innovation, humanity established a man-made dividing line—intellectual property—bounded by time, to distinguish between new and old technologies. The operating rules of the patent system best reflect the fundamental principles of this intellectual property regime. Both old and new technologies nourish human society, but in general, new technologies offer greater satisfaction, make life more convenient, provide more employment opportunities, and create more wealth. In short, new technologies offer richer nourishment to human society. As a result, we grant inventors of new technologies exclusive rights to implement them for a limited period. This incentivizes further innovation and, through patent disclosure and licensing, promotes the wide dissemination and implementation of new technologies—ultimately enriching society as a whole.

Innovation in different fields is like a set of parallel universes enveloping the physical world. The outer boundaries of these parallel realms are defined by humanity's most recent innovations, and they continually expand, evolve, and shift under the propulsion of innovation. These realms are separated by a membrane referred to as the "field of the art," with wormholes connecting the various spaces. The individuals featured in this book are like innovators existing within distinct parallel universes—fierce competition exists within each field, but across fields, innovation tends to proceed peacefully, almost without rivalry. For example, IBM in the computing domain, Sony and Philips in the music playback space, and Bosch in mechanical manufacturing. Foundational scientific principles serve as wormholes that pierce through all these realms, providing the essential tensile strength that supports each. Yet the specific innovations within each realm are built upon different

Q. Yue, F. Yang, *Governing Innovation: A Global History of Patent-Driven Technological Progress*, https://doi.org/10.1007/978-981-95-5570-3_50

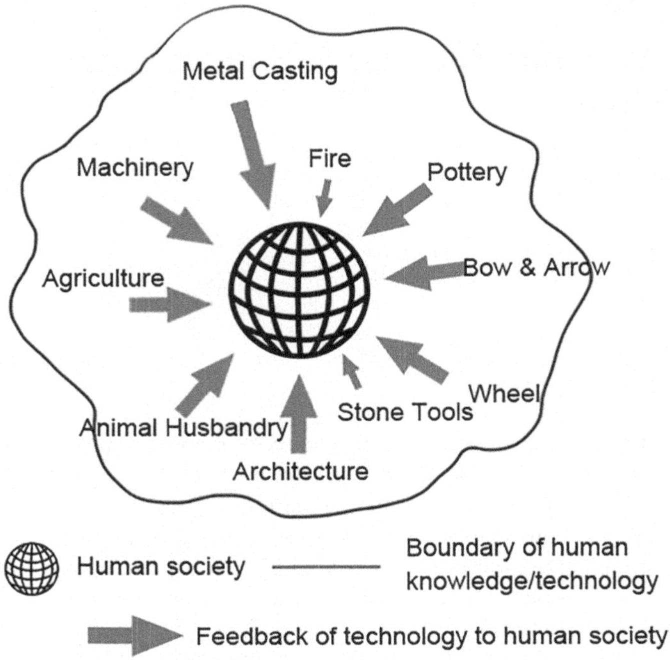

Fig. 50.1 A foundational model of how technology and innovation nourish human society

scientific foundations and operate independently, feeding back into and nourishing the physical world (see Fig. 50.2).

Within a single parallel realm, competition among innovators is fierce, and they use patents to mark out jagged, interlocking territories (see Fig. 50.3). These borders are surveyed and delineated by patent attorneys and judges under the rules of the patent system. The core of this surveying effort lies in determining whether an invention is novel—patent invalidation systems, for instance. Under the law, true innovation boundaries can be recognized and protected through injunctions, rulings, and decisions. However, how to govern innovation and encourage its implementation is a matter that lies beyond the scope of law.

The law can determine the boundaries of patents, but it cannot extend the frontiers of human knowledge. When patent boundaries encroach into old technology or stretch into areas beyond the current knowledge frontier, it benefits neither innovation nor society.

The domain of new technology is far from peaceful. Conflicts over patent boundaries are constant; battles rage on, draining societal resources. Are these battles worthwhile? Do they bring us to the point where we can say, "It was all worth it"? We believe that since the ultimate aim of technology is to benefit society, then the purpose of these battles should be to protect the right to implement innovation that enriches society. If the battle is to defend one's right to use innovation for public

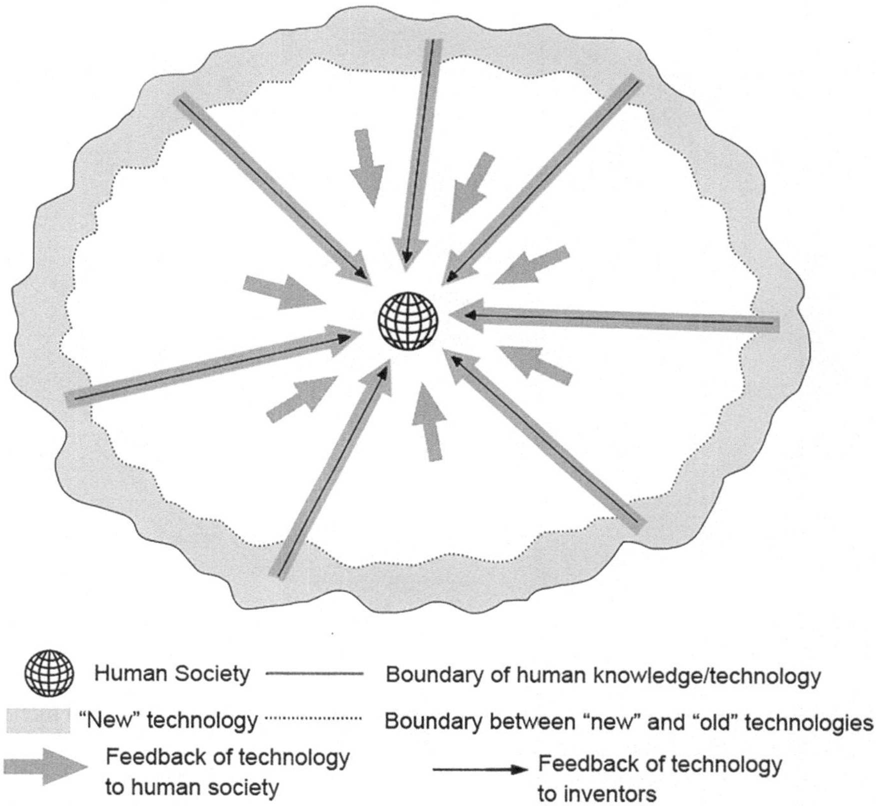

Fig. 50.2 The place of the modern intellectual property system within the foundational model

benefit, then it is undoubtedly a fight worth waging. But if the purpose is to deprive others of that right—or worse, to parasitize those who implement innovation for personal gain, thus obstructing the dissemination of technology—then whether that battle is worthwhile deserves deep reflection.

Of course, the domain of old technologies is not monolithic either. We divide it into *useful old technologies* and *obsolete dead technologies*. Dead technologies offer little to no value to society and are economically worth nearly nothing (see Fig. 50.4).

With the rapid advance of technology, the realm of dead technologies is also expanding. The faster innovation moves, the faster these obsolete zones grow. Take music storage as an example: from vinyl records to tapes, from reel-to-reel to cassette tapes, then to audio CDs, MDs, and finally internet-based music—all within less than a century. Technologies like Sony's MD (MiniDisc), once a culmination of its audio prowess, had a shorter lifespan than even reel tapes and quickly fell into disuse, becoming a dead technology. While vinyl retains artistic and cultural value,

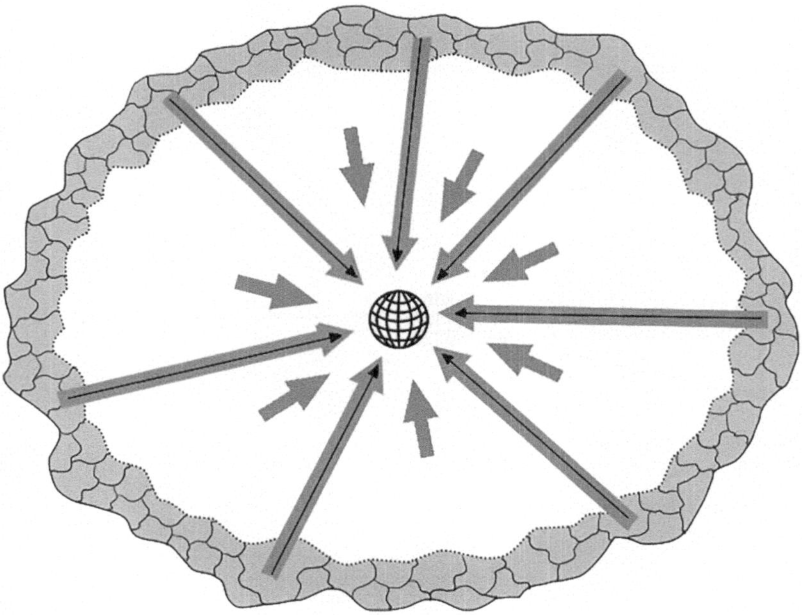

Fig. 50.3 Jagged patent borders shaped by fierce competition among innovators

reel-to-reel and cassette tapes have little remaining economic worth outside of niche applications—this is the “dead zone” in our model.

In the earlier chapters on rise and decline, we discussed the value of self-reinvention. During the era of cassette tapes and CDs, Sony and Philips stood as giants, transforming how humanity experienced music. But in the digital music era, Sony clung to its proprietary formats—MD, Memory Stick, ATRAC3—hesitating in the face of the internet music tidal wave. In the blink of an eye, MD was consumed by the dead zone, reduced to obsolete tech. These sleek discs, once the source of fond memories, lost all ability to serve society—this marked a symbolic moment in Sony’s decline (see Fig. 50.5).

We can think of *dead technologies* as deserts, *old technologies* as jungles, *new technologies* as fertile farmlands, and the *unknown* as uncultivated land. Deserts are not completely barren—there are oases and mineral treasures hidden within. Likewise, within the realm of discarded technologies, there are still valuable gems. Even in the age of digital imaging, some directors and enthusiasts remain devoted to the texture of film; and even today, some governments are still struggling to phase out 3.5-inch floppy disks. But for a forward-looking tech company, being half-buried in the sands is no way to thrive. An outstanding tech company must constantly race alongside the brightest and fastest innovators, advancing with them into unexplored territory to secure a strong strategic position.

As discussed in the chapters on disruption and expansion, companies must proactively shed the relics swallowed by the desert. When a company is firmly

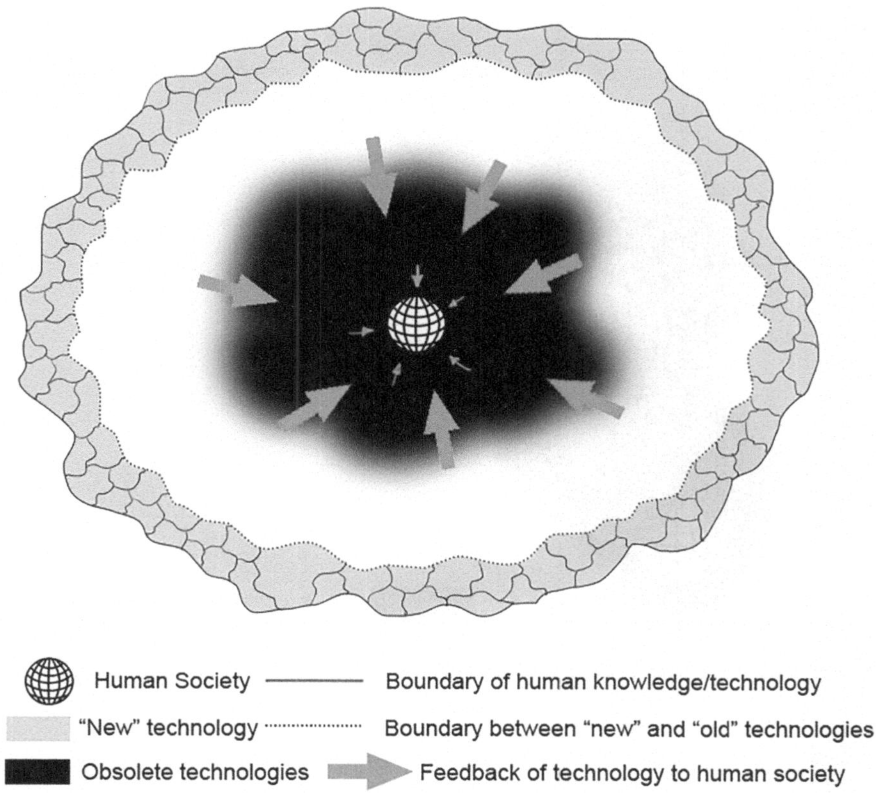

Fig. 50.4 Obsolete, low-value, or entirely useless dead technologies

established, it should look ahead to the vast uncultivated land stretching before it, stand atop rich and vital farmland, and rely on the thick jungle behind to resist the desert's encroachment. Always moving forward and embracing self-renewal—this is not only the path to survival but also the only true path to better serve humanity.

In the preface to *Records of the Grand Historian*, Sima Qian quotes Confucius: "Rather than carrying it in empty words, it is better to show it through vivid and concrete deeds." So now, let us share a few short stories to explore these humble insights with you, dear reader.

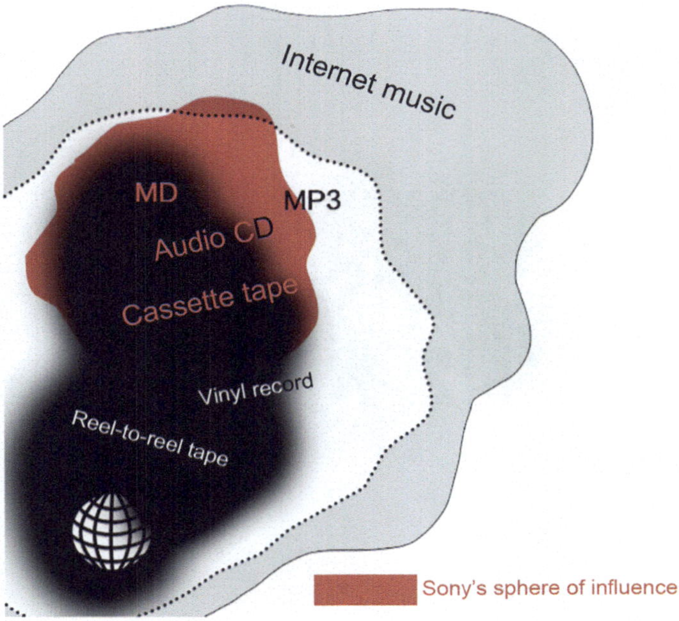

Fig. 50.5 Sony as a case study in the evolutionary model of dead, old, and new technologies

Chapter 51
Innovation Protection: Evolution of Rules

It is widely known that the patent system first originated in Venice, Italy, and was later introduced to England. In 1558, the legendary Queen Elizabeth I ascended the throne, ushering in the first golden era of modern British history. In 1585, war broke out between England and Spain, culminating in the defeat of the Spanish Armada at the Battle of Gravelines. From then on, England emerged as a dominant naval power.

As the saying goes, "Every cannon shot costs ten thousand pieces of gold." While the Anglo-Spanish War marked the rise of England, it also brought immense financial pressure upon the crown. Elizabeth I devised various methods to raise funds—she issued privateering licenses to pirates, essentially endorsing robbery on the high seas, and used monopolistic business rights, under the guise of "patents," as tools for royal enrichment. Over several decades, she created over 50 monopoly rights in traditional industries, selling most of them to her close courtiers. By 1601, with the exception of the grain trade, nearly all industries in England had been monopolized through these "patents." Ironically, true inventors during that period often found it difficult to obtain any patent protection at all.[1]

Much like Emperor Wu of Han's state monopoly on salt and iron, Elizabeth's patents brought in substantial revenue for both the crown and the rights holders, helping fund the war effort. However, they also caused soaring prices and public outrage.

In late 1601, the House of Commons convened a session to denounce Elizabeth's misuse of patents. Some members branded the patent holders as parasites feeding on the nation and demanded the complete abolition of the patent system. Yet many of these MPs were themselves beneficiaries of monopolies. They railed against monopolies in industries they didn't profit from, but were reluctant to give up their own.

[1] Mossoff, A. (2000). Rethinking the development of patents: an intellectual history, 1550–1800. *Hastings Lj*, *52*, 1255.

Q. Yue, F. Yang, *Governing Innovation: A Global History of Patent-Driven Technological Progress*, https://doi.org/10.1007/978-981-95-5570-3_51

One of the more famous MPs was the adventurer Sir Walter Raleigh, who held a monopoly on the tin-smelting industry. He argued that while other patents harmed the country, his was beneficial. "Before the granting of my Patent," he said, "the Poor Workmen never had above two shillings the week … But since my Patent … they have four shillings a week."[2] Raleigh did admit, however, that this increased wage didn't come from thin air—or his own pocket. Since he acquired the monopoly, the price of tin had risen from 17 to 50 shillings, and the profits went into the monopolist's coffers.

The debate ultimately went nowhere. At its climax, Raleigh declared, "if all others may be repealed, I will give my consent as freely to the cancelling of this (his patents)." With that, the chamber fell silent—no one dared speak again.[3]

As illustrated in Fig. 51.1, the benefits of existing technologies to society were already limited. When monopolists further siphoned off a share, ordinary people were left with even less. The lesson learned from the birth of the patent system was simple: existing technology should not be monopolized.

Though Parliament didn't explicitly demand a response, Queen Elizabeth quickly addressed the issue. A few days later, the aging monarch gave a speech in Parliament defending her use of patents as a necessity for the nation rather than a means of enrichment, and pledged to cease selling monopolies. This address would become known as the "Golden Speech," her final reflection on her reign. Two years later, she passed away, and James I ascended the throne.

In 1610, under parliamentary pressure, James I canceled all traditional industry monopolies, preserving only patents for genuine inventions. Between 1607 and 1617, the philosopher Francis Bacon served as Commissioner of Suits, overseeing the issuance of patents. During his tenure, 40 patents were granted for technological inventions.[4] But in 1622, James I backtracked and sold a soap monopoly to his favorites, causing the price of soap to skyrocket sixfold and reigniting parliamentary fury. In 1624, Parliament compelled him to sign the Statute of Monopolies, which declared that only inventors who introduced new industries through technological innovation could be granted patents, limited to a term of 14 years.

Despite the progress made, England at the time was far from an ideal society. Corruption was rampant, bureaucracy prevailed, and the aristocracy lived in idle luxury while laborers toiled for 14 hours a day just to survive. The legal system remained convoluted and burdensome. Although some distinction was made between old and new technologies, there was still no formal mechanism to declare patents invalid.

---

[2] Kealey, T. (2022). The Industrial Revolution as a collective action problem: The House of Commons games patents of monopoly, November 1601. *Economic Affairs*, *42*(3), 418–441.

[3] Kealey, T. (2022). The Industrial Revolution as a collective action problem: The House of Commons games patents of monopoly, November 1601. *Economic Affairs*, *42*(3), 418–441.

[4] Kealey, T. (2022). The Industrial Revolution as a collective action problem: The House of Commons games patents of monopoly, November 1601. *Economic Affairs*, *42*(3), 418–441.

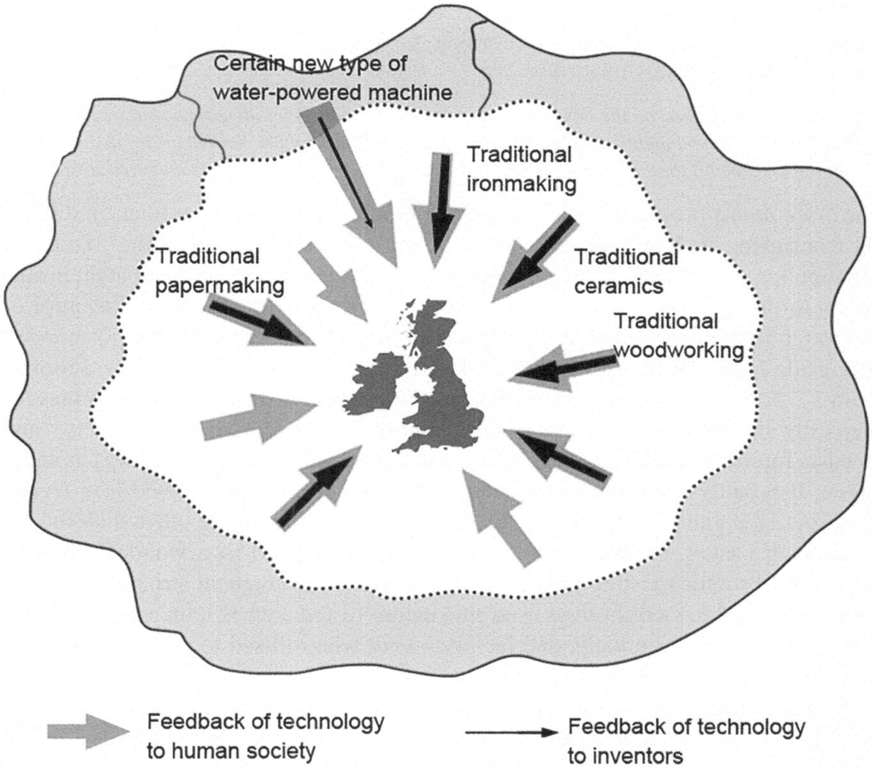

Fig. 51.1 Societal and proprietary return from technology during the First Industrial Revolution

Today, we understand patent rights as protections for intellectual achievements—rights that inventors deserve in exchange for their contributions to humanity.[5] But under the legal framework of Elizabethan England, patents were royal privileges, akin to grants of land, wealth, or honor. As such, only the monarch could revoke them.

If someone wanted to challenge a patent, they couldn't directly petition the courts. Instead, they had to appeal to the Attorney General, who would then initiate court proceedings on behalf of the Crown. There were four grounds for invalidation: lack of novelty, non-entitlement of the applicant, insufficient disclosure, and harm or inconvenience to the subjects of His Majesty. If any of these conditions were met, it was seen as deception against the Crown, and the royal privilege could be rescinded.

[5] Professor Zheng Chengsi, founder of intellectual property law studies in China, defined intellectual property as the exclusive legal rights over intellectual creations. Professor Liu Chuntian defined it as the collective term for the rights that creators of intellectual achievements and owners of commercial identifiers enjoy under the law.

This legal process was known as *Scire facias*, a Latin term meaning "to make known" or "to show cause."[6] According to *Yuan Zhao's Dictionary of Anglo-American Law*, it refers to:

> *A judicial writ issued on the basis of a legal document such as a judgment or letters patent, requiring the respondent to appear in court and explain why the holder of the document should not benefit from it, or why the document should not be annulled or declared invalid.*

The *Scire facias* procedure was so complex that it deterred most ordinary individuals from using it. It wasn't until more than a hundred years after the Statute of Monopolies was enacted that England witnessed its first major case of patent invalidation through *Scire facias*. The case involved Richard Arkwright, the inventor of the water frame, which helped inaugurate the modern factory production model. Alongside James Watt, Arkwright stood as a pioneer of the Industrial Revolution.

In 1775, Arkwright was granted a patent for "certain instruments or machines for preparing silk, cotton, flax, or wool for spinning." To protect his trade secrets—and to ease future enforcement—he drafted a patent specification that was notably vague. It broadly described components like a seed remover, toothed iron frame, yarn spools, spindles, a cloth spread with wool, flax, hemp, or other materials, a winnowing basket, cranks, cylinders, rollers, rollers fixed to a wooden frame, a cylindrical box for twisting the contents, associated mechanical structures, shafts, and flywheels.[7] Arkwright then used this patent to threaten Britain's earliest spinning mill owners. Eventually, one factory owner who refused to pay licensing fees initiated the arduous *Scire facias* process.

By the standards of the time, the Arkwright case was a landmark battle—comparable to today's patent wars between Apple and Samsung. Beginning in 1785, it went through three trials. Both parties hired top-tier legal counsel and expert witnesses. The plaintiffs challenged the patent on three grounds: they argued that Arkwright's description was too vague to enable replication of a working machine, and that the parts such as "the cylindrical box and corresponding mechanism, shaft, and flywheel" had no clear purpose—Arkwright's machines had never actually used these components. They also summoned numerous textile experts, including former employees of Arkwright, to testify that the machine lacked novelty.

Representing the Crown, the plaintiff's lawyer emphasized that the most important reason to revoke the patent was that "the grant caused harm and inconvenience to His Majesty's subjects," particularly by causing skilled workers to flee.

Arkwright responded with a powerful defense. He summoned high-profile expert witnesses including James Watt, his close friend Erasmus Darwin, and the widow

---

[6] Christopher Beauchamp. (2019). Repealing Patents (2019). 72 Vand. L. Rev. 647, Retrieved March 31, 2024, from: https://brooklynworks.brooklaw.edu/faculty/1052/

[7] Peter Wells. Tilaye Terrefe. (2020). A Brief History of the Evolution of the Patent of Invention in England. Canadian Intellectual Property Review Vol. 352,020. Retrieved March 31, 2024 from https://ipic.ca/_uploads/5f8da09982330.pdf

and son of James Hargreaves, inventor of the spinning jenny.[8] Watt and Darwin argued that the description was adequate.

After hearing all sides, the judge handed the specification to the jury and asked whether a mechanic could construct the machine based solely on Arkwright's description. The jury replied, "No." The judge then declared the patent invalid.

Watt's support for Arkwright was understandable: his own steam engine patent was also broadly written, lacking in technical details and vulnerable to attack. Those with a modest understanding of technological history will know that while Watt is commonly referred to as the inventor of the steam engine, this isn't entirely accurate. Thomas Savery and Thomas Newcomen had already developed practical steam engines before Watt (see Fig. 51.2), with Newcomen's engine achieving some adoption in mining.

Watt's major improvement was the introduction of a separate condenser, allowing steam to be drawn from cylinder B into the condenser for cooling. This maintained high temperatures in the cylinder and significantly increased efficiency. This innovation earned Watt his first patent in 1769. He followed this with a second patent in 1781 covering improvements in mechanical linkages involving gears and

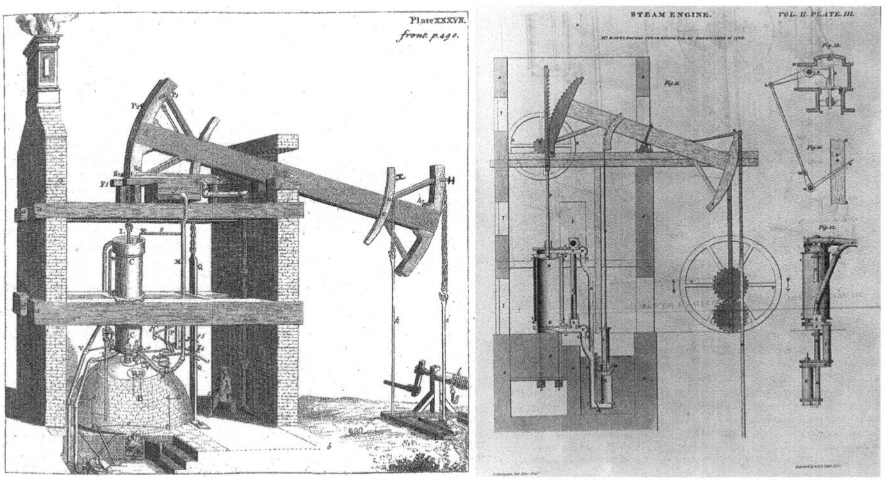

Fig. 51.2 Newcomen's steam engine (left) and Watt's steam engine (right). The two diagrams above are provided to help readers understand the principle, as Watt's steam engine patent did not preserve the original drawings. In the left figure, B is the boiler and C is the cylinder. Steam is injected from the boiler into the cylinder. The water pipe at Z allows cold water to be injected outside the cylinder, cooling it down and condensing the steam, thereby creating a vacuum and driving the piston to perform work

[8] Jeremy Norman. (2004). Invalidation of Arkwright's Patent Causes Explosive Growth of the Textile Industry in England. Retrieved from Jeremy Norman's HistoryofInformation.com: https://historyofinformation.com/detail.php?id=4679

rods, and a third in 1782 for the dual-action cylinder. These three core patents made the steam engine truly practical—and ensured that it was Watt's name, not Newcomen's, that entered the school textbooks.

Figure 51.2: Newcomen's steam engine (left) and Watt's steam engine (right). These are schematic representations, as Watt's patent included no illustrations. In the diagram, B is the boiler and C the cylinder. Steam is injected into the cylinder from the boiler. A pipe section labeled Z introduces cold water to the cylinder's exterior, condensing the steam and creating a vacuum that drives the piston.

Watt's 1769 patent (No. 913), still available in the European Patent Office database, is surprisingly concise—just three A4 pages in an 1855 printed edition (see Fig. 51.3). The technical portion spans only from line 5 on page 2 to line 10 on page 3. It outlines seven "principles," offering no details on assembly or operation, and includes no diagrams. Watt's decision to avoid technical depth was shaped by advice from his friend William Small, who urged him to file promptly, minimize detail, omit illustrations, and focus on principles. Small advised: "As for your principles, we think they should be described as broadly as possible… to protect you from infringement to the greatest extent allowed by the essence of your invention." Watt followed that advice.[9] Thus, the patent that changed human history wasn't titled

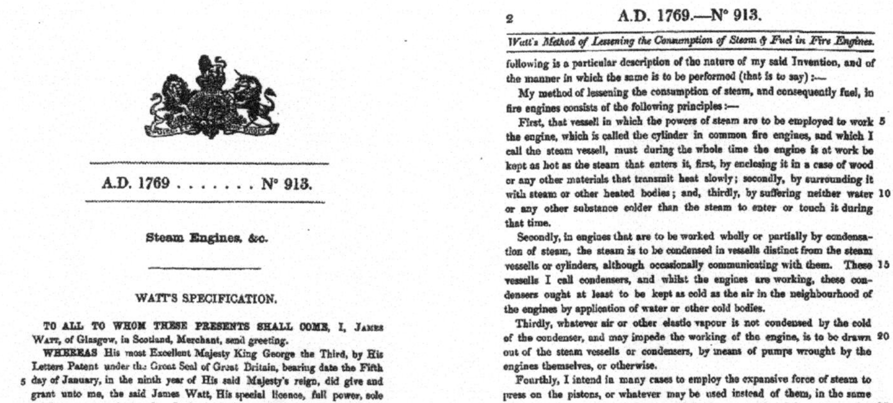

A.D. 1769 . . . . N° 913.

Steam Engines, &c.

WATT'S SPECIFICATION.

TO ALL TO WHOM THESE PRESENTS SHALL COME, I, JAMES WATT, of Glasgow, in Scotland, Merchant, send greeting.

WHEREAS His most Excellent Majesty King George the Third, by His Letters Patent under the Great Seal of Great Britain, bearing date the Fifth day of January, in the ninth year of His said Majesty's reign, did give and grant unto me, the said James Watt, His special licence, full power, sole priviledge and authority, that I, the said James Watt, my exors, admors, and assigns, should and lawfully might, during the term of years therein expressed, use, exercise, and vend, throughout that part of His Majesty's Kingdom of Great Britain called England, the Dominion of Wales, and Town of Berwick upon Tweed, and also in His Majesty's Colonies and Plantations abroad, my "NEW INVENTED METHOD OF LESSENING THE CONSUMPTION OF STEAM AND FUEL IN FIRE ENGINES;" in which said recited Letters Patent is contained a proviso obliging me, the said James Watt, by writing under my hand and seal, to cause a particular description of the nature of the said Invention to be inrolled in His Majesties High Court of Chancery within four calendar months after the date of the said recited Letters Patent, as in and by the said Letters Patent, and the Statute in that behalf made, relation being thereunto respectively had, may more at large appear.

NOW KNOW YE, that in compliance with the said provisoe, and in pursuance of the said Statute, I, the said James Watt, do hereby declare that the

2 A.D. 1769.—N° 913.

Watt's Method of Lessening the Consumption of Steam & Fuel in Fire Engines.

following is a particular description of the nature of my said Invention, and of the manner in which the same is to be performed (that is to say):—

My method of lessening the consumption of steam, and consequently fuel, in fire engines consists of the following principles:—

First, that vessell in which the powers of steam are to be employed to work the engine, which is called the cylinder in common fire engines, and which I call the steam vessell, must during the whole time the engine is at work be kept as hot as the steam that enters it, first, by enclosing it in a case of wood or any other materials that transmit heat slowly; secondly, by surrounding it with steam or other heated bodies; and, thirdly, by suffering neither water or any other substance colder than the steam to enter or touch it during that time.

Secondly, in engines that are to be worked wholly or partially by condensation of steam, the steam is to be condensed in vessells distinct from the steam vessells or cylinders, although occasionally communicating with them. These vessells I call condensers, and whilst the engines are working, these condensers ought at least to be kept as cold as the air in the neighbourhood of the engines by application of water or other cold bodies.

Thirdly, whatever air or other elastic vapour is not condensed by the cold of the condenser, and may impede the working of the engine, is to be drawn out of the steam vessells or condensers, by means of pumps wrought by the engines themselves, or otherwise.

Fourthly, I intend in many cases to employ the expansive force of steam to press on the pistons, or whatever may be used instead of them, in the same manner as the pressure of the atmosphere is now employed in common fire engines. In cases where cold water cannot be had in plenty, the engines may be wrought by this force of steam only, by discharging the steam into the open air after it has done its office.

Fifthly, where motions round an axis are required, I make the steam vessells in form of hollow rings or circular channels, with proper inletts and outletts for the steam, mounted on horizontal axles like the wheels of a water mill; within them are placed a number of valves that suffer any body to go round the channell in one direction only. In these steam vessells are placed weights, so fitted to them as intirely to fill up a part or portion of their channels, yet rendered capable of moving freely in them by the means herein-after mentioned or specified. When the steam is admitted in these engines between these weights and the valves, it acts equally on both, so as to raise the weight to one side of the wheel, and by the reaction on the valves successively to give a circular motion to the wheel, the valves opening in the direction in which the

Fig. 51.3 First two pages of Watt's Patent No. 913

[9] Murfitt, S. (2017). *The English patent system and early railway technology 1800–1852* (Doctoral dissertation, University of York).

"steam engine," but rather "a method for reducing the consumption of fuel in fire engines"—a method patent. Being a method, it faced looser requirements for illustration, enabling Watt to cover broad concepts with minimal specifics.

Watt, as Arkwright's ally and witness, was deeply dissatisfied with the case outcome. He wondered: if a patent had flaws, why wasn't it rejected at the outset? Why let inventors pay steep application fees, endure drawn-out lawsuits, and incur further legal expenses—only to have the patent revoked later?

In response, Watt penned *Thoughts upon Patent*, advocating for a reformed patent system with substantive examination. Yet Britain at the time offered no official response to his proposal.[10]

Watt's steam engine dramatically reduced fuel consumption and was widely adopted in mines throughout England. Engineers Edward Bull and Richard Trevithick also began manufacturing similar steam engines, which fell within the scope of Watt's patents. In 1792, Watt decided to file a lawsuit. His close friend Erasmus Darwin (grandfather of *On the Origin of Species* author Charles Darwin) advised him to let it go. On the eve of the industrial era, Erasmus Darwin had already pointed out that patent litigation was unreliable and served only to enrich lawyers. He left behind this memorable quote:

> *A lawsuit that lines a lawyer's pockets is like a snail inching its way up a pillar—climbing two or three inches only to slide back, until the whole pillar is covered in slime.*

At the time, patent law also suffered from severe uncertainty. In 1785, Watt's lawyer told him that although the Statute of Monopolies had been in place for over 150 years, there were no recorded case precedents—nothing to refer to for consistency.[11] Despite all this, Watt insisted on pressing forward with what would become a prolonged and costly legal battle. In the pivotal 1793 trial, Watt summoned eight expert witnesses. The defendants brought in seven—virtually every steam engine expert in Britain participated in this case. Among all the patents discussed in this book, Watt's invention arguably holds the highest historical stature—it led the entire industrial age. Even compared to the computer patents discussed in the "Boundary" chapter, this one outshines them. Novelty wasn't in question; the core dispute centered on whether Watt's disclosure was sufficiently enabling.

Although the judge leaned toward insufficiency, the jury ruled in Watt's favor. In the end, Watt and his business partner were awarded £30,000 in licensing fees. Records show that when Watt's key patent expired in 1800, after being in force for 31 years, it had brought him £200,000 in total revenue.[12]

While Watt won the case, Erasmus Darwin's warning still holds water over 200 years later. Moreover, Watt's patent war remains controversial. Edward Bull

---

[10] Christopher Beauchamp. (2019). Repealing Patents (2019). 72 Vand. L. Rev. 647, Retrieved March 31, 2024, from: https://brooklynworks.brooklaw.edu/faculty/1052/

[11] Osborne, R. (2013). Iron, Steam & Money: The Making of the Industrial Revolution. Random House.

[12] Brian Spear (2008). James Watt: The steam engine and the commercialization of patents., 30(1), 53–58. https://doi.org/10.1016/j.wpi.2007.05.009

and Richard Trevithick were also significant early steam engine inventors with unique contributions to high-pressure engines. Some scholars argue that Watt's victory delayed progress in steam engine development and thus slowed Britain's industrial revolution.[13]

From that point on, patent lawyers became the primary beneficiaries of patent wars. One might say that upholding due process and fighting for rights holders is the sacred duty of lawyers. But it is also essential to ask: Do these battles serve the interests of human society? Do they impede technological progress or obstruct the flow of benefits to humanity? In the legal community, someone must grapple with these questions.

As illustrated in Fig. 51.4, while patent-based technologies return value to society, some of that value goes to inventors—and some to patent professionals. But if patent practitioners take too much, that's not beneficial for society as a whole.

The invalidation of Arkwright's patent shocked industrial elites in Britain, including Watt, and its impact reached across the Atlantic. American scholar Christopher Beauchamp argues that the Arkwright case played a key role in shaping U.S. patent legislation.[14] At the time, one of America's founding fathers, Alexander Hamilton, was closely studying British technology and systems, hoping to import British machinery and attract British engineers to the U.S. Arkwright's technology was among his top priorities. The Arkwright case made clear that true innovation deserved patent protection, while fake innovation did not. As a result, the very first U.S. patent law included provisions for revoking patents.

In 1791—the year before Watt filed suit—a merchant named Benjamin Folger was granted a U.S. patent for a method of removing sediment from whale oil. Thomas Jefferson personally approved it in January 1792. A few months later, a merchant named Jonathan Jenkins filed suit in a New York district court seeking to invalidate the patent on the grounds of lack of novelty. According to Professor Beauchamp, this was the first genuine patent litigation in U.S. history, though the details are now lost. In the decades that followed, patent invalidation suits continued to occur.

In 1835, a judge named Joseph Hopkinson wrote in a court opinion:

> *A case is recorded of a patent for using the common stone coal in a common blacksmith's forge. The patentee went through the country exhibiting his parchment patent with the great seal of the department of state, and the signatures of the high officers of government*

---

[13] For example, Boldrin, M., Levine, D. (2013) 'What's Intellectual Property good for?'. Revue économique 64: 29–54, "Boulton and Watt's steam engine patent most likely delayed the industrial revolution by a couple of decades". There are also scholars who are against this view, for example Bottomley, Sean, 'Did James Watt's Patent(s) Really Delay the Industrial Revolution?', in Stephen H. Haber, and Naomi R. Lamoreaux (eds), *The Battle over Patents: History and Politics of Innovation* (New York, 2021; online edn, Oxford Academic, 21 Oct. 2021), https://doi.org/10.1093/oso/9780197576151.003.0004, accessed 16 Nov. 2023，以及Bottomley, Sean, 2014. "Patents and the first industrial revolution in the United States, France and Britain, 1700–1850," IAST Working Papers 14–14, Institute for Advanced Study in Toulouse (IAST).

[14] Christopher Beauchamp. (2019). Repealing Patents (2019). 72 Vand. L. Rev. 647, Retrieved March 31, 2024, from: https://brooklynworks.brooklaw.edu/faculty/1052/

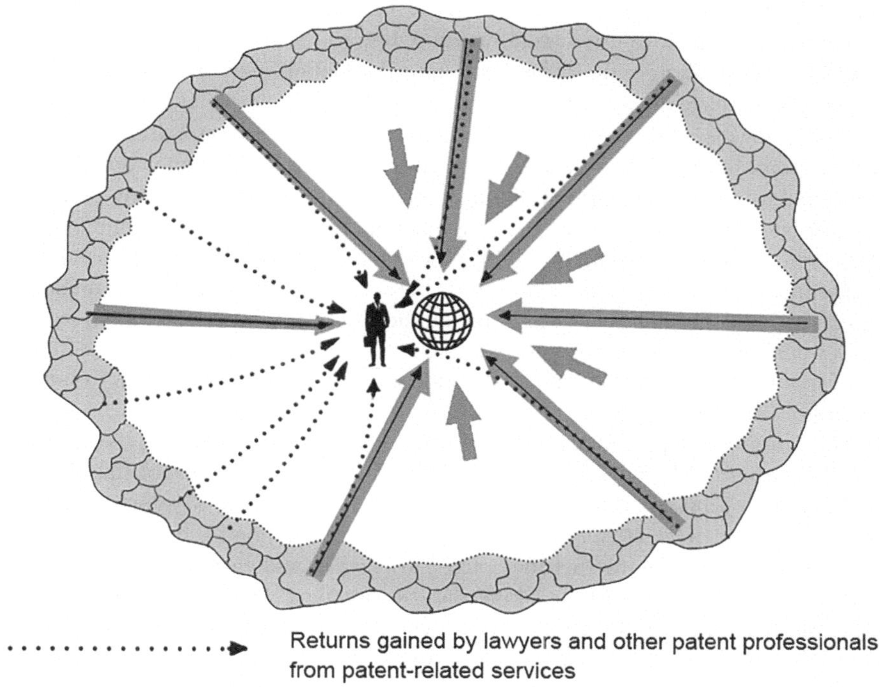

Fig. 51.4 Returns gained by lawyers and other patent professionals from patent-related services

> *appended to it. This would naturally alarm an ignorant smith, and as the patentee would sell him a right for two or three dollars, or for whatever he could get for it, a prudent man would prefer paying so small a sum, rather than go to law with an adversary apparently so well armed. To protect the public from such impositions, this tenth section was enacted, and gives the power to any person, interested or not in the discovery or the patent, to call upon the patentee for an examination of his right, and have it repealed.*[15]

This shows that some patent professionals were already manipulating the system—stretching patent claims into areas of prior art and extracting fees from old, low-value technologies. Fortunately, the system's designers responded in time.

These stories took place during the nascent stages of the modern patent system, before the great nineteenth-century debates over whether patents promote innovation. That is, in an age when Britain's feudal monarchy still held sway and America's legal system was still embryonic—before modern jurists devised intricate theories about patents—people had already recognized that patents should protect real innovation.

From that time on, more and more patent professionals have stood at the boundary between old and new technologies, seeking the truth and ensuring that patents

[15] Christopher Beauchamp. (2019). Repealing Patents (2019). 72 Vand. L. Rev. 647, Retrieved March 31, 2024, from: https://brooklynworks.brooklaw.edu/faculty/1052/

don't intrude on existing knowledge. This is the solemn duty of all patent professionals—and we believe it is a battle worth fighting.

Over 2800 years ago, the Phoenician king Pygmalion murdered his brother-in-law in order to seize the inheritance of his sister, Princess Dido. Fleeing with a small band of followers, Dido crossed the Mediterranean and landed on a stretch of North African coast in Libya. The locals were willing to trade with them and agreed to sell them land to settle. According to the agreement, Dido could choose any parcel of land, provided it did not exceed the area that could be covered by a single oxhide. Dido then ordered the hide to be cut into long, narrow strips, using them to encircle a large tract of land, including a hill beside the sea. The Libyans had no grounds for objection and could only watch as Dido and her followers established a city on the hill—founding what would become Carthage, the only North African power in history capable of rivaling European civilization, eventually contending with Rome for control of the Mediterranean.

This story first appeared in Greek records and spread widely during the colonial era. According to Zhang Qiaogui and Lin Jianyu (2021),[16] versions of the "oxhide land claim" legend exist in the United States, the Philippines, Myanmar, Taiwan, and Macau. The *Ming History* records: "At that time, the Franks, being aggressive, traded with Luzon. Seeing that the country was weak and easy to seize, they presented lavish gifts to the king, asking only for as much land as a cowhide could cover, promising to build merely a house. The king, unaware of the deceit, agreed. They then cut the hide into narrow strips, joined them together to form a perimeter stretching several thousand feet, and enclosed Luzon land as agreed. The king was alarmed but had already given his word and had no choice but to comply." The *Gazetteer of Taiwan Prefecture* includes this account: "In the first year of Tianqi, the Chinese pirate Yan Siqi recruited Japanese troops to camp at Tainan. After some time, a Dutch ship was blown ashore by a storm. Admiring the land, they sought to settle there. The aborigines refused, but the Dutch tricked them, saying, 'We ask only for the amount of land an oxhide can cover, and we'll pay generously.' Permission was granted. The Dutch cut the hide into threads and laid it around dozens of zhang; thus, they built Fort Zeelandia and settled in Taiwan."

These stories have been passed down for millennia, traveling from West to East, a testament to humanity's admiration for cleverness and to the grievances of those who were tricked. Using rules to one's advantage, parsing language to maximize benefit—is that not the very work of lawyers and patent professionals? Patent practitioners demarcate the boundaries of innovation; is this not akin to staking claims with an oxhide? Who wouldn't want to be like Princess Dido, using a small invention to carve out vast territory and build a legacy that lasts through the ages?

In patent history, the most renowned "land-claimer" was Samuel Morse, inventor of Morse Code. Originally a painter with no technical background, Morse became interested in electromagnetism after returning from a European art tour in 1832. On the ship home, he met scientist Charles Thomas Jackson, who conducted

[16]张桥贵, 林建宇. (2021). 传教士"牛皮圈地"传说流变与地域生成.《新华文摘》2021年03期

entertaining electromagnetic demonstrations and introduced Morse to the latest developments in the field. Inspired, Morse abandoned painting and enthusiastically dove into electromagnetic research. In October 1837, he filed a patent application for his telegraph invention, and in June 1840, received U.S. Patent No. 1647. The patent described a practical telegraph system, with Claim 3 protecting the now-famous Morse Code. Morse subsequently withdrew his applications twice to revise and broaden his claims. His final Claim 8 was extraordinarily expansive, covering electromagnetic circuits and mechanical structures used for transmitting information via signal and sound.

As mentioned, Morse was primarily a painter and had little commercial ambition. In 1845, he sold his patent to a group of partners and planned to retire. Among the new patent holders was Amos Kendall, a well-connected figure in American political and business circles, former U.S. Postmaster General, close friend of President Andrew Jackson, and an experienced lawyer. Kendall assured Morse that he would be free from legal or business troubles and founded the Magnetic Telegraph Company to manage the patent. As a pioneering telecommunications enterprise, it was similar to today's tech giants—engaged both in building infrastructure and licensing patents.[17] Among the company's partners was a businessman named O'Reilly, who purchased a license but expanded operations into states beyond the licensed territory.

In 1847, Kendall sued O'Reilly for breach of contract and sought an injunction. Although the court ultimately denied the injunction, O'Reilly was infuriated. Emboldened by victory, he aggressively sought more clients, deepening the rift with the Magnetic Telegraph Company. Aware of the patent risks, he also began acquiring other telegraph patents and refining his own technology.

In 1848, Kendall filed a new patent infringement suit in Kentucky circuit court—thus beginning the patent war between the two men, and drawing Morse into the fray. Like many accused of infringement today, O'Reilly argued both non-infringement and invalidity. For the latter, he brought in none other than Charles Jackson—the same scientist who had demonstrated electromagnetism to Morse aboard the ship—claiming that Morse had stolen the idea.[18]

What saddened Morse most was the betrayal of his longtime colleague, the renowned scientist Joseph Henry, who testified for O'Reilly. Henry was one of the

---

[17] Mossoff, A. (2014). O'Reilly v. Morse. Available at SSRN 2448363.

[18] Charles Jackson's motivations were unusual. A polymath renowned as a physician, chemist, and geographer, he was oddly obsessed with claiming credit for major discoveries. He contested Christian Friedrich Schönbein over the invention of guncotton, William Beaumont over digestive research, and even his own student William Morton over the anesthetic use of ether. Morton publicly demonstrated ether anesthesia at age 27 and filed a patent, becoming one of the pioneers of modern anesthesiology alongside Horace Wells. Jackson insisted the invention stemmed from his teachings, leading to bitter conflict. Wells claimed to have discovered inhalation anesthesia earlier, but a failed public demonstration discredited him. Courts ultimately ruled that ether anesthesia was a discovery, not an invention, and thus unpatentable. Morton's patent on ether mixtures was ignored, and many doctors bypassed it. Wells, disgraced, became addicted to chloroform and died by suicide in jail after attacking a prostitute. Jackson, meanwhile, died insane in an asylum.

most respected experts in electromagnetism at the time—the unit of inductance, the henry (H), is named after him. As previously mentioned, Morse had limited understanding of electromagnetic theory and had indeed received considerable help from Henry during his invention process. But when Morse's associates omitted Henry's name from a paper on the history of telegraphy, Henry felt slighted and ultimately sided against him. The lawsuit deepened their rift, and Morse later publicly denied Henry's contributions.

In September 1848, the court ruled in Morse's favor. Enraged, O'Reilly refused to accept defeat and even called on Congress to impeach the judge.

Yet Morse's patent litigation was far from over. The scope of disputes continued to expand. Not long after, inventor Alexander Bain filed a patent application for his own telegraph system. Leonard Gale, one of Morse's collaborators and a co-owner of his patent, had by then become a patent examiner. Citing Morse's patent, Gale rejected Bain's application. Bain sued, initiating an interference proceeding that once again dragged Morse into court. In 1849, the judge ruled that each invention had its own merits, and both patents remained valid.

But the storm was far from over. O'Reilly seized the opportunity to argue that his own technology originated from Bain, not Morse, and demanded a retrial—pulling Morse back into yet another infringement case. Meanwhile, another co-owner of the Morse patent, Francis O. J. Smith, also sued O'Reilly, and the case ballooned into a sprawling legal battle crossing state lines and counties. Smith's case alone generated 400 to 500 pages of testimony, while court records for another co-owner's lawsuit exceeded 1000 pages. During this protracted litigation, Smith and Morse fell out, with Smith ultimately siding with O'Reilly and testifying that Morse had plagiarized Jackson's idea.

In 1853, the Supreme Court recognized Morse as the true inventor of telegraphy. Though victorious, Morse emerged from the ordeal isolated and exhausted. Kendall had failed in his promise to shield Morse from legal and commercial troubles and could only offer this consolation: "*The troubles you encounter are but the tax a man has to pay for wealth and fame*."

In its final decision, the court upheld most of Morse's claims but declared Claim 8 invalid—making this one of the earliest landmark cases concerning "abstract ideas." The claim in question read:

> *I do not propose to limit myself to the specific machinery or parts of machinery described in the foregoing specification and claims; the essence of my invention being the use of the motive power of the electric or galvanic current, which I call electromagnetism, however developed, for making or printing intelligible characters, signs or letters at any distances, being a new application of that power, of which I claim to be the first inventor or discoverer.*

To claim the ability to transmit readable characters over a distance using electricity—what an audacious claim! Some scholars have suggested that Morse's Claim 8 was so broadly worded it could have covered later inventions such as the telephone, radio, fax machine, television, or even the internet.[19] The court wrote: "*(Morse)*

[19] Dratler Jr., J. (2005). Alice in Wonderland Meets the US Patent System. *Akron L. Rev.*, *38*, 299.

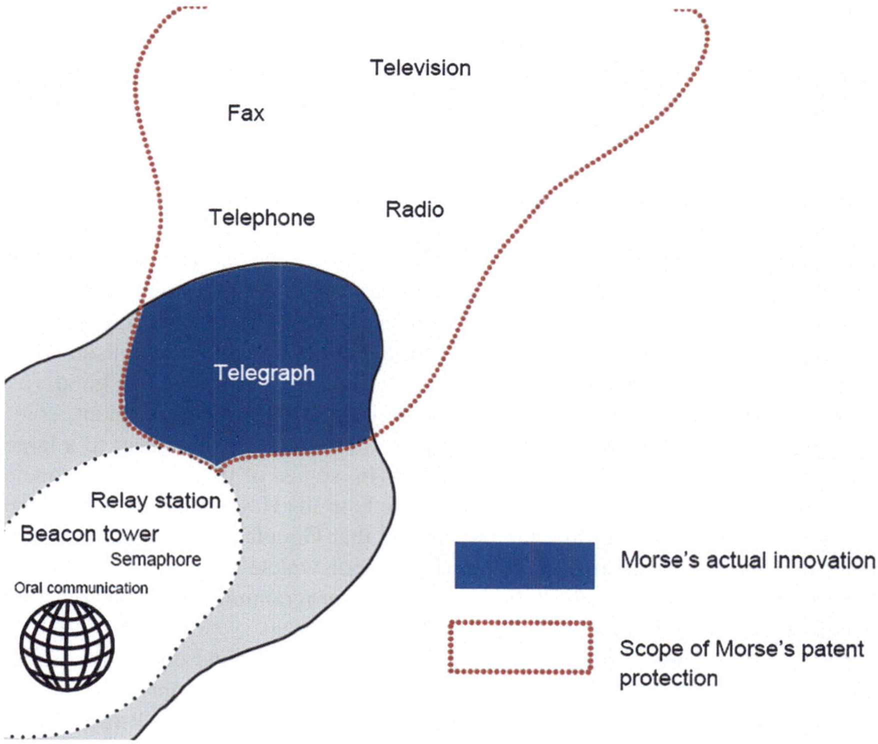

Fig. 51.5 A model of the actual boundary of innovation in telegraph-related technologies and the encroachment of Morse's patent protection into future technological domains

shuts the door against inventions of other persons… For he says he does not confine his claim to the machinery or parts of machinery, which he specifies; but claims for himself a monopoly in its use, however developed, for the purpose of printing at a distance… The court is of the opinion that the claim is too broad, and not warranted by law."[20]

There is no doubt that Morse's patent was a remarkable innovation, but it also resembled the oxhide enclosure tactic—staking claim not only to his own invention but also to future technologies in which he played no part. As illustrated in Fig. 51.5, he sought to monopolize beyond his contribution. If invalidation based on lack of novelty indicates that a patent has encroached on prior art, then the invalidation of Morse's Claim 8 shows that it encroached on the territory of future technology. Future technologies cannot transcend time and space to nourish human society in advance. Prematurely drawing boundaries merely closes the door to the inventions of others.

[20] O'Reilly v. Morse, 56 U.S. 62 (1853).

Whether it's an intrusion upon prior art or an encroachment upon the future, the core issue remains the same: the boundary defined by the patent fails to align with the boundary of actual human innovation. In our view, only when a patent's boundary matches the frontier of real innovation is it a proper patent—one whose boundary is worth defending. Otherwise, the rights holder inevitably takes something from either existing or future technologies that is not rightfully theirs.

It is the duty of patent professionals to seek truth from prior art and patent specifications, correcting the scope of patents so that they accurately reflect the true boundaries of human innovation. We believe this pursuit of truth is worthwhile. But we must remember: the ultimate goal of truth-seeking is to ensure that technology, through its implementation, can nourish human society. If we lose sight of this, treating the battle itself as the end goal—if the attention and resources consumed by the fight far exceed what's justified—then we risk doing more harm than good.

In January 2010, Oracle Corporation, a long-established software giant, completed its acquisition of Sun Microsystems, Inc., thereby gaining access to a large portfolio of Java software patents held by Sun. In August of the same year, Oracle filed a patent infringement lawsuit against Google in the U.S. District Court for the Northern District of California. Oracle alleged that Google's use of the Java language in its Android operating system infringed upon Oracle's software patents, and that Google had also copied 37 Java application programming interfaces (APIs) owned by Oracle, thus constituting copyright infringement.

Representing Oracle in this protracted legal battle was none other than David Boies, one of the top trial lawyers in the United States. Boies had previously represented George W. Bush in the landmark Supreme Court decision in Bush v. Gore during the 2000 presidential election, earning him the No. 2 spot on Time magazine's Person of the Year list—second only to Bush himself. In 2001, he represented the U.S. government in its antitrust case against Microsoft, putting immense pressure on Bill Gates and nearly resulting in the company's breakup. Boies, a man of extraordinary talents, had suffered from dyslexia as a child but possessed a photographic memory. Once he overcame his reading difficulties, he could recite lengthy passages with perfect accuracy, including citation and page numbers, which often left his courtroom opponents unnerved.

In the Oracle v. Google case, Boies and his legal team uncovered critical evidence from Google's internal emails. One of the most dramatic pieces was the now-infamous "Lindholm email."

This email, dated August 6, 2010, was written by Google software engineer Tim Lindholm to Android team lead Andy Rubin. In it, Lindholm mentioned that Google's founders, Larry Page and Sergey Brin, had tasked him with finding alternatives to Java for use in the Android and Chrome platforms. He wrote, "We've been through the alternatives… and they all suck. We conclude that we need to negotiate a license for Java under the terms we need." This email was sent less than a week before Oracle filed its lawsuit against Google, and it clearly reflected Google's internal response to Oracle's earlier communications regarding potential infringement.

Together with a 2005 email from Andy Rubin, which stated, "If Sun doesn't want to work with us, we have two options: 1) Abandon our current work and switch to MSFT CLR VM and C#; 2) Keep using Java, defend our decision, and face the consequences," the Lindholm email became a key piece of evidence suggesting willful infringement. Combined, the emails showed that Google was fully aware of Sun's patents and the potential risks of infringement, yet proceeded with its use of the Java-related technology.

To ensure the Lindholm email would be admissible, Boies's team took painstaking measures. Though the email was an internal company communication, it contained a line at the end stating "Highly Confidential – Attorney's Eyes Only." Normally, Google's legal team could have argued that this constituted privileged communication between client and counsel, and therefore was not subject to discovery or admissible as evidence.

So how did this email end up in Boies's hands? According to Florian Mueller, blogger at the prominent FossPatents site, the version obtained by Boies's team may not have been the final version, but rather one of nine drafts automatically saved on Google's servers.[21] The final version of the email bore the "Attorney's Eyes Only" designation and possibly other critical language that would have triggered Google's keyword filters and excluded it from discovery. However, the drafts likely did not include that designation and slipped through during document production.

Even so, if Google had raised an objection during the final trial, citing attorney-client privilege due to the designation on the final version, the judge might have excluded the email from evidence.

To avoid this possibility, Boies's team orchestrated a strategic ambush at a pre-trial hearing originally convened to discuss a motion filed by Google. Prior to the hearing, presiding Judge William Alsup indicated that he wanted to discuss issues related to willful infringement. Boies's team brought the Lindholm email to court, concealed within a stack of documents related to the motion. At the hearing, an Oracle attorney remarked that some of the materials were "sensitive" and requested not to address them in open court. Mistaking this for evasiveness, Judge Alsup insisted on seeing the documents. Oracle's lawyers seized the opportunity and handed over the Lindholm email.

After reading it, the judge recognized its importance and immediately questioned Google's counsel about the "alternatives" referenced by Page and Brin. Caught off guard and unaware of the email's privileged designation, Google's lawyers made a critical error. Instead of objecting, they attempted to explain the contents away: "… Oracle's representatives came and told us this lawsuit would be costly and that they were demanding billions. At that point, our CEO [referring to Larry Page] asked, what other alternatives do we have to avoid all this?" In effect, Google's lawyers tried to reinterpret "alternatives to Java" as "alternatives to litigation." Given the context of the email, this explanation appeared far-fetched and unconvincing.

---

[21] Florian Mueller. (2011). Google's Five Failed Attempts to Give Confidential Status to 'Damning' Email in Oracle Case. Retrieved March 31, 2024 from FossPatents: http://www.fosspatents.com/2011/11/googles-five-failed-attempts-to-give.html

After the hearing, Google's lawyers realized they had made a grave mistake: their courtroom response effectively confirmed that the "Lindholm email" was an internal company communication rather than a privileged exchange between attorney and client. This also gave the judge the impression that Google's lawyers were evasive and disingenuous.

A week later, Google filed a motion to exclude the Lindholm email from evidence, claiming it was privileged attorney-client communication and should not have been disclosed. They argued that Oracle's attorneys should have notified them in advance before presenting such documents. However, the seasoned David Boies had foreseen this. He never proactively introduced the email—instead, he only "reluctantly" presented it when the judge explicitly requested it. Before handing it over, Boies even made a point of noting that the email contained "sensitive content."

Although the Lindholm email was marked "Attorney's Eyes Only," its content clearly indicated it was a routine internal exchange between colleagues, not attorney-client communication. On that basis, the judge denied Google's motion, explicitly ruling that merely labeling a document as privileged or copying an attorney on it did not transform it into protected correspondence. Over time, both parties filed multiple motions concerning the email. Google threw substantial legal resources at the issue, attempting nine times to exclude the document from evidence—each effort failed.

In the 2012 trial, then-Google CEO Larry Page was compelled to take the stand and face David Boies's questioning about the Lindholm email. According to the Los Angeles Times,[22] Page—usually seen in casual attire—donned a rare suit and tie for the occasion. Wired magazine[23] reported that Page was slow in responding, evasive, and vague in his testimony. He avoided eye contact with Boies, often staring at the ceiling. The judge repeatedly had to prompt Page to answer with "yes," "no," or "not sure." Most of Page's responses were "not sure" or "don't recall." When asked to explain what Java was, Page stammered: "I think Java is a complicated thing, with many, many parts." Asked whether Android was a "crucial asset" to Google, the CEO shockingly replied, "I'm not sure."

When questioned about internal Google emails, Page behaved like an amnesiac, evading nearly every question Boies posed. Regarding all the emails shown in court, Page—once hailed as a Stanford prodigy—claimed not to remember any of them. He even downplayed his familiarity with Tim Lindholm, despite Lindholm's prominence as a star engineer who had been at Google since 2005. When Boies mentioned Lindholm by name, Page feigned ignorance and asked, "Which Tim?" CNBC mockingly called Page a "blank page"—a pun on his surname.

---

[22] Jessica Guynn. (2012). Google CEO Larry Page evasive in Oracle patent suit testimony. Retrieved March 31, 2024 from LA Times Website: https://www.latimes.com/business/la-xpm-2012-apr-19-la-fi-oracle-page-20120419-story.html

[23] Galeb Garling. (2012). Google Boss Takes Stand in Oracle Trial Dodges Android Questions. Retrieved March 31, 2024 from Wired Website: https://www.wired.com/2012/04/google-boss-oracle-trial/

This episode was dramatic and showcased the strategic prowess of America's top litigators, rivaling any legal drama. After enduring a devastating blow from Boies's textbook execution, did Google end up paying a record-breaking fine while Oracle cashed in handsomely?

In truth, Oracle's aggressive legal strategy achieved little. Of the seven Java patents it initially asserted, five were invalidated by the Patent Trial and Appeal Board (PTAB), and the remaining two (U.S. Patent Nos. US6061520A and USRE38104E1) went to trial. But the court found that Google had not infringed on any of the asserted claims—meaning no infringement, let alone willful infringement, had occurred. Despite the dramatic deployment of the Lindholm email, Boies's brilliant legal maneuvering ultimately made no difference in the outcome. The whole effort consumed massive legal fees and countless hours from Google's executives and engineering teams—for naught.

In one of the author's own U.S. patent cases, the opposing party also poured considerable effort into proving willful infringement—through interrogatories during discovery, depositions of third-party witnesses, and courtroom cross-examinations. Much valuable time was wasted quibbling over marketing material instead of debating patent validity, leaving the author deeply disappointed by what was supposed to be a showcase of elite legal argumentation.

As early as 2003, Professor Mark Lemley dubbed this phenomenon the "Willfulness Game" and called for an end to this excessive focus on proving willful infringement in patent litigation. Because of the high stakes associated with willfulness findings, many patent lawyers advise clients not to read competitors' patents during product development.[24] The logic is simple: if you keep your eyes shut and say nothing, you can avoid being accused of willful infringement. If a lawsuit does arise, just focus on proving non-infringement or invalidity and save on legal fees. Professor Robin Feldman also noted in a study that "many patent attorneys actively advise their clients not to review issued patents… which may trigger damages calculations and increase exposure to enhanced damages for willful infringement."[25]

While the author wholeheartedly supports punishing truly willful infringers, the way the U.S. system combines such penalties with discovery rules—and the ways lawyers exploit this—has completely deviated from the patent system's original purpose of "disclosure in exchange for protection." Encouraging companies to turn a blind eye and forgo using valuable patent databases is a perverse outcome of a system meant to promote knowledge sharing.

The vibrancy of U.S. patent litigation and patent monetization is deeply tied to its legal framework. In the days of Morse, O'Reilly, Edison, and Tesla, inventors frequently faced the rigors of interference proceedings. Under the early American "first-to-invent" system, inventors in interference cases had to present evidence proving they were the true original inventors of the patented technology. Losing

---

[24] Lemley, M. A, & Tangri, R. K. (2003). Ending Patent Law's Willfullness Game. UC Berkeley: Berkeley Law. Retrieved March 31, 2024 from https://escholarship.org/uc/item/6hx8d8tt

[25] Robin Feldman and Tom Ewing, The Giants Among Us, 2012 Stan. Tech. L. Rev. 1 (2012). Retrieved April 1, 2024 from: https://repository.uchastings.edu/faculty_scholarship/158

such a proceeding meant not only forfeiting the patent but also the honor of being recognized as the genuine inventor. To support their claims, inventors had to produce lab notebooks, experimental records, and testimonies from witnesses—all reviewed and rehearsed by attorneys. Because of this, U.S. patent litigation has been notoriously expensive since the nineteenth century. After being formally acknowledged as the inventor of the telegraph, Morse lamented the cost of litigation, remarking, "The law is too expensive," and that "being bled at will by lawyers is a bad thing; the best course is to accept it with grace."

Today, interference proceedings are a thing of the past in the U.S., but lab notebooks, experimental records, and testimony from R&D staff remain essential in patent litigation. Plaintiffs must not only prove the defendant's willfulness but also support their case with "secondary considerations" to uphold the patent's inventiveness. Plaintiffs often demand extensive discovery from defendants—emails, meeting notes, market research reports, and in the case of software patents, massive amounts of source code—driving litigation costs ever higher.

If Chinese patent monetization professionals wish to wield the same clout as their American counterparts and instill fear in industry circles, the current system is clearly inadequate. Compared to the U.S., China's patent invalidation and litigation procedures are far more cost-effective, with no extensive discovery process or heavy focus on proving willfulness before the patent's validity is established.

We believe the value generated by technology is ultimately finite: the more that goes to lawyers, the less is returned to society or to the inventors themselves. As discussed earlier, Lee De Forest and Edwin Armstrong wasted immense time and energy in patent litigation, only to end their lives in frustration. Nikola Tesla—often portrayed as Edison's archrival—also spent years embroiled in legal battles with Marconi over radio patents, dying penniless in the end.

A good patent system must protect legitimate rights and distinguish genuine innovation from false claims, but it should never become an end in itself. Litigation should serve to safeguard the fruits of innovation, to protect inventors' rights to serve the public, and to maximize societal returns—not merely to enrich the litigants. This is our vision for the patent system.

In our model, we liken a patent's scope to land. But land does not yield crops on its own. No matter how fertile it is, without diligent cultivation and weed control, it cannot nourish human society. We believe the life of a patent lies in its implementation.

In 1900, a reporter named Remsen Crawford somehow learned that seven of Edison's key patents were about to expire and believed he had stumbled onto a major scoop. On the night the patents expired, he rushed to Edison's office for an interview. Though it was late, Edison was still working in his lab. The security guard turned Crawford away, but after persistent pleading, agreed to relay a question: What were Edison's thoughts on his patents expiring? Edison responded via the guard: "Tell that fellow—tell him I said this: the expiration of those patents doesn't matter. Tell him Mr. Edison says he never had a monopoly on his inventions and doesn't expect to in this world. Patent expiration has no effect on an inventor's wealth."

Edison had a top-tier legal team and aggressively enforced his patents, famously driving the entire film industry westward, giving rise to Hollywood. Yet at heart, Edison was an industrialist. Licensing revenue was not the cornerstone of his empire—at least not in his own view.

Thirty years later, Crawford interviewed Edison again and asked why he wasn't the richest man in the world. By then in his 80s and battle-worn from decades of patent disputes, Edison had come to see the system clearly. He replied, "I've been granted 1,180 patents in my life. The money I spent on patents—including R&D and litigation—exceeded the royalties I received. I made my money as a manufacturer, by developing and selling products, not as an inventor."[26]

Edison's patent tactics have been both praised and criticized throughout history. But he was undeniably a doer, someone who brought great technologies to life and helped usher humanity into the electrical age. It's said that in his final moments, while slipping into unconsciousness, Edison's last words were, "It's very beautiful over there." Whether it was a deathbed vision or a glimpse of heaven, we cannot say—but it shows that this once-mighty figure found peace in the end.

Yet not all inventors of great technologies are so lucky to end their lives without regret.

On December 17, 1903, the Wright brothers successfully tested their powered manned aircraft, marking a giant leap for humankind into the age of aviation. By their time, many engineers and adventurers had already made countless attempts to achieve human flight, but none had succeeded—until the Wright brothers made a breakthrough based on experimentation, observation, and calculation. This invention would later receive what any inventor might dream of: near-perfect patent protection.

According to their patent specification, the technical problem the Wright brothers solved was overcoming the "disturbing forces" encountered by an aircraft in flight, as "these forces tend to cause the flying machine to deviate from the position it must maintain in order to produce the desired effect." Their test aircraft was a rudimentary glider: the wings were made of canvas stretched over a wooden frame. The pilot could pull on cables and pulleys to warp the wing edges upward or downward, thereby altering lift on each side to maintain balance. At the rear of the glider was a vertical rudder, referred to in the patent as a "vertical rudder," which could turn left or right and served as another key balancing mechanism. When the pilot warped the wings, the rudder would also pivot, maintaining stability and enabling directional control.

Of course, aircraft of that time had no cockpit. The Wrights had to lie prone on the aircraft, with their hands gripping a control lever to steer the horizontal elevator and ensure the plane faced into the wind (see Fig. 51.6). At the same time, they operated a T-shaped foot lever to warp the wings. Initially, they only considered wing warping. Although the aircraft had a vertical rudder, it was fixed and could not

[26] Jeff Nilsson. (2012). Edison and The Pirates: The Inventor's Solution to Copyright Theft. The Saturday Evening Post. January 28, 2012. Retrieved at: https://www.saturdayeveningpost.com/2012/01/pirates-patents-progress/

Fig. 51.6 The Wright brothers' early glider, with Orville Wright lying prone on the aircraft

move. After some unsuccessful trials, Orville suggested making the rudder moveable as well. Since their hands were already occupied with the forward controls and using feet was awkward, they redesigned the T-lever into a frame that rested beneath the pilot's body. This allowed the pilot to shift it using waist and leg strength. This frame was connected via cable and pulley to both the wing-warping mechanism and the vertical rudder. In the Wrights' view, using a single action and apparatus to control both the wing warp and vertical rudder made logical sense and didn't impose additional burden on the pilot. Ironically, this very feature would later become the only potential vulnerability in their patent.

As seen in Fig. 51.7, component number 18 is the central structure for controlling wing warping and rudder movement.

By shifting structure 18 with their waist and legs, the pilot pulled cables (15) to warp the wing edges upward or downward, while also turning the vertical rudder left or right. This technology—known as "wing warping"—left all contemporary aviation pioneers in the dust and significantly outpaced the existing progress in aerodynamics. In a 1906 letter to a friend, Wilbur Wright confidently wrote:

> *We are convinced that no one will be able to develop a practical flyer within five years. This opinion is based upon cold calculation. It takes into consideration practical and scientific difficulties whose existence is unknown to all but ourselves. Even you, Mr. Chanute, have little idea how difficult the flying problem really is. When we see men laboring year after year on points we overcame in a few weeks, without ever getting far enough along to meet the worse points beyond, we know that their rivalry & competition are not to be feared for many years.*[27]

In early 1903, the Wright brothers filed a patent application for their "wing warping" technology based on their glider tests. At the time, engines were already widely used in cars and motorcycles, so applying them to aircraft was a logical step. The Wrights hoped to include powered flight within their patent claims. However, as

[27] Giere, R. N. (Ed.). (1992). *Cognitive models of science* (Vol. 15). U of Minnesota Press.

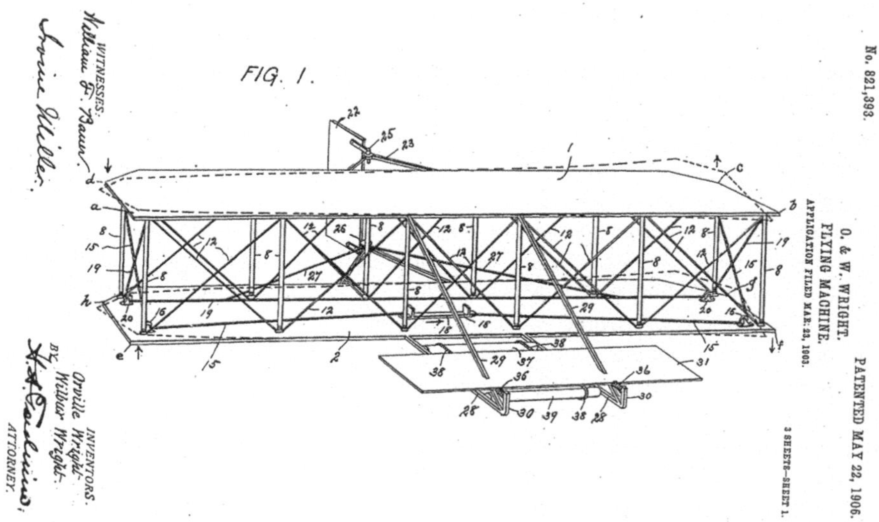

Fig. 51.7 Patent drawing of U.S. Patent No. 821393A

they had limited expertise in engines and had not yet conducted powered flight trials, they couldn't adequately address feasibility concerns raised by patent examiners. Their initial application was rejected.

After their historic flight in late 1903, they revisited the application. This time, they hired prominent patent attorney Harry Toulmin, at considerable expense. By then, American patent litigation had matured over decades, and patent lawyers had become skilled in legal precision. Toulmin ambitiously redrafted their patent claims. In claim 1, he deliberately omitted any mention of the pulley and cable mechanisms that warped the canvas wings. Instead, he focused solely on the vertical movement of the "lateral portions" of the wing. Claim 1 reads:

> *In a flying-machine, a normally flat aeroplane having lateral marginal portions capable of movement to different positions above or below the normal plane of the body of the aeroplane, such movement being about an axis transverse to the line of flight, whereby said lateral marginal portions may be moved to different angles relatively to the normal plane of the body of the aeroplane, so as to present to the atmosphere different angles of incidence, and means for so moving said lateral marginal portions, substantially as described.*

It's worth noting that drafting claims only became a legal requirement in U.S. patent law in 1870. Before that, many applicants voluntarily included claims, but they weren't yet a statutory necessity for defining patent scope. It was in this pre-claims context that Morse's famously expansive patent emerged. Before the courts fully adopted claims-based jurisprudence, judges would determine a patent's core inventive principle from the description. Technologies applying the same principle could be deemed infringing under the doctrine of equivalents.[28] But after 1870, as patent practice grew more sophisticated, courts began to distinguish between "pioneering" inventions and incremental ones. For truly "pioneering" technologies, courts might still invoke the doctrine of equivalents based on underlying principles—not merely the claim language—to determine the patent's protective scope.

In the 1898 case *Westinghouse v. Boyden Power Brake,*[29] the court formally defined a "pioneer patent" as one that protects "an invention that introduces a novel function or an entirely new device, or is so novel and important that it marks a distinct step in the progress of the art." The three quintessential examples cited were Howe's sewing machine patent, Morse's telegraph patent, and Bell's telephone patent. If an invention was deemed a "pioneer patent," it could receive broader protection than ordinary patents. Specifically, it could apply the doctrine of equivalents "uninhibitedly" to encompass all implementations operating on the same principle. This doctrine of pioneer patents lasted for over 80 years in American patent law until it was abolished in the 1980s by the Federal Circuit in the case *Texas Instruments v. United States International Trade Commission.*

However, such "pioneer patent" status was not determined by the Patent Office, but rather recognized by the courts during litigation—similar in some ways to how courts in China can recognize a trademark as "well-known" under judicial review. After the Patent Office granted their patent, the Wright brothers' patent did not receive "pioneer" status, but the brothers firmly believed that their patent was the cornerstone of aviation technology and never doubted its pioneering nature. At that point, they hadn't made a penny from their invention, yet they still had their lawyer, Harry Toulmin, file patent applications in the UK, France, Germany, Belgium, Russia, Italy, and even Australia.

The historic 1903 flight didn't cause much of a stir in the industry. Although the Wright brothers lacked experience in patent enforcement, they were highly aware of intellectual property and had executed rigorous secrecy. As mentioned, they believed their technology was years ahead of its time and their patent held immense value. Before their patent was published, they participated in some public competitions and demonstrations, earning modest recognition—but they kept the secret of wing-warping well hidden.

In 1906—the same year Wilbur famously declared that no one could develop a practical flying machine within 5 years—a young engineer named Glenn Curtiss reached out to the Wright brothers. Originally a bicycle specialist with his own workshop, Curtiss later built his own engines and converted bicycles into racing motorcycles. In 1903, he rode one of his motorcycles at 103 km/h for a mile, setting a world record that stood for decades. But by then, land vehicles could no longer satisfy his obsession with speed—he turned his eyes to the skies. After learning of

---

[28] Winans v. Denmead, 56 U.S. (15 How.) 330, 342 (1853); Evans v. Eaton, 20 U.S. (7 Wheat.) 356, 361 (1822) ("[I]f the two machines be substantially the same, and operate in the same manner, to produce the same result, though they may differ in form, proportions, and utility, they are the same in principle….").

[29] Westinghouse v. Boyden Power Brake Co., 170 U.S. 537, 561-62 (1898)

the Wright brothers' aircraft performance, Curtiss volunteered to design and install an engine for them.

Though not ready to collaborate with "outsiders," the Wrights agreed to meet Curtiss and invited him to their workshop. They got along well and admired each other's work. Whether the Wrights disclosed key technical information during this meeting—and whether Curtiss learned the secret of wing-warping—remains a contentious and unresolved mystery in the legal battles that followed.

In 1907, legendary inventor Alexander Graham Bell met Curtiss and invited him to join his *Aerial Experiment Association* to design aircraft engines. Bell, already in his 60s and a celebrated figure, remained passionately curious, with a particular interest in aviation at the time. That same year, Bell and Curtiss designed an experimental aircraft called the **White Wing**, which featured small triangular ailerons mounted at the edges of its dual canvas wings. These ailerons performed a function similar to the Wright brothers' wing-warping system.[30] The aileron design was credited to Bell himself. As noted earlier, Bell had extensive experience with patent litigation and played a pivotal role in establishing AT&T's dominance in the tech industry. By this time, the Wright brothers' patent had been published, and Bell had studied it closely. He believed that ailerons were sufficient to avoid infringing on their patent.

The *White Wing* eventually crashed due to pilot error in handling the ailerons.[31] In July 1908, Bell and Curtiss's team unveiled a new aircraft that successfully flew 1.6 km, winning the *Scientific American* Trophy and a $25,000 prize. Bell personally named it the **June Bug**. In April 1909, Bell, Curtiss, and others filed a patent application for their new aircraft (U.S. Patent No. US1011106A), focusing on the aileron design (see Fig. 51.8). Claim 1 reads:

> *In a flying machine, the combination of a supporting surface having a positive angle of incidence, a pair of lateral balancing rudders, one on each side of the medial fore and aft line of the structure and each of said rudders normally having a zero angle of incidence and connections between said rudders.*

In the specification, Bell and co. noted: "It has been suggested that a machine of this character might be made with flexible supporting surfaces, having lateral edges warpable or bendable so that the opposite ends of the supporting surfaces on either side of the center line of the machine present different angles of incidence." These flexible, warpable wings clearly referred to the Wright brothers' design. In contrast, Bell and Curtiss's aircraft had "rigid, non-bendable supporting surfaces." When the *June Bug* first attempted flight, wind passed through the canvas wings, causing insufficient lift and three failed takeoffs. Curtiss applied a sailor's technique by

---

[30] Aviastar.org. AEA Aerodrome No.2 White Wing. Retrieved March 31, 2024 from Aviastar.org: http://www.aviastar.org/air/usa/aea_whitewing.php

[31] Goldstone, L. (2015). Birdmen: the Wright Brothers, Glenn Curtiss, and the battle to control the skies. Ballantine Books.

Fig. 51.8 The White Wing aircraft

coating the canvas with a mixture of gasoline and paraffin, making it air-impermeable and turning the surface rigid.[32]

The specification further emphasized: "It is essential that the supporting surfaces be substantially rigid, and incapable of being distorted or bent to restore lateral balance when it is disturbed." Instead, at the edges of these rigid wings, there were "structures distinct from the supporting surface, in combination therewith, for raising the lower side of the machine and lowering the upper side thereof when its lateral balance is disturbed." In other words, while the Wright brothers used flexible wings and warping to maintain balance and control, Bell and Curtiss used a pair of ailerons—distinct from the wing surface—for the same purpose. From this perspective, their design more closely resembled modern aircraft, and the ailerons' distinction from the wing surface was a clear attempt to avoid infringing the Wright brothers' patent (Figs. 51.9 and 51.10).

The 1908 *Scientific American* Trophy required aircraft to take off under their own power. At that time, the Wrights' plane was not capable of this and did not participate. When they heard that Curtiss had won, they were outraged and immediately wrote him a letter: "Claim 14 of our patent 821,393 clearly covers the combination we hear you are using… We believe that it would be very difficult to build a successful machine without employing some features covered by our patent." That claim reads:

> *A flying-machine comprising superposed connected aeroplanes, means for moving the opposite lateral portions of said aeroplanes to different angles to the normal planes thereof,*

[32] Wright Brothers Aeroplane Company. (1999). The June Bug. Retrieved March 31, 2024, from Wright Brothers Aeroplane Company Website: https://www.wright-brothers.org/History_Wing/Wright_Story/Showing_the_World/June_Bug/June_Bug.htm

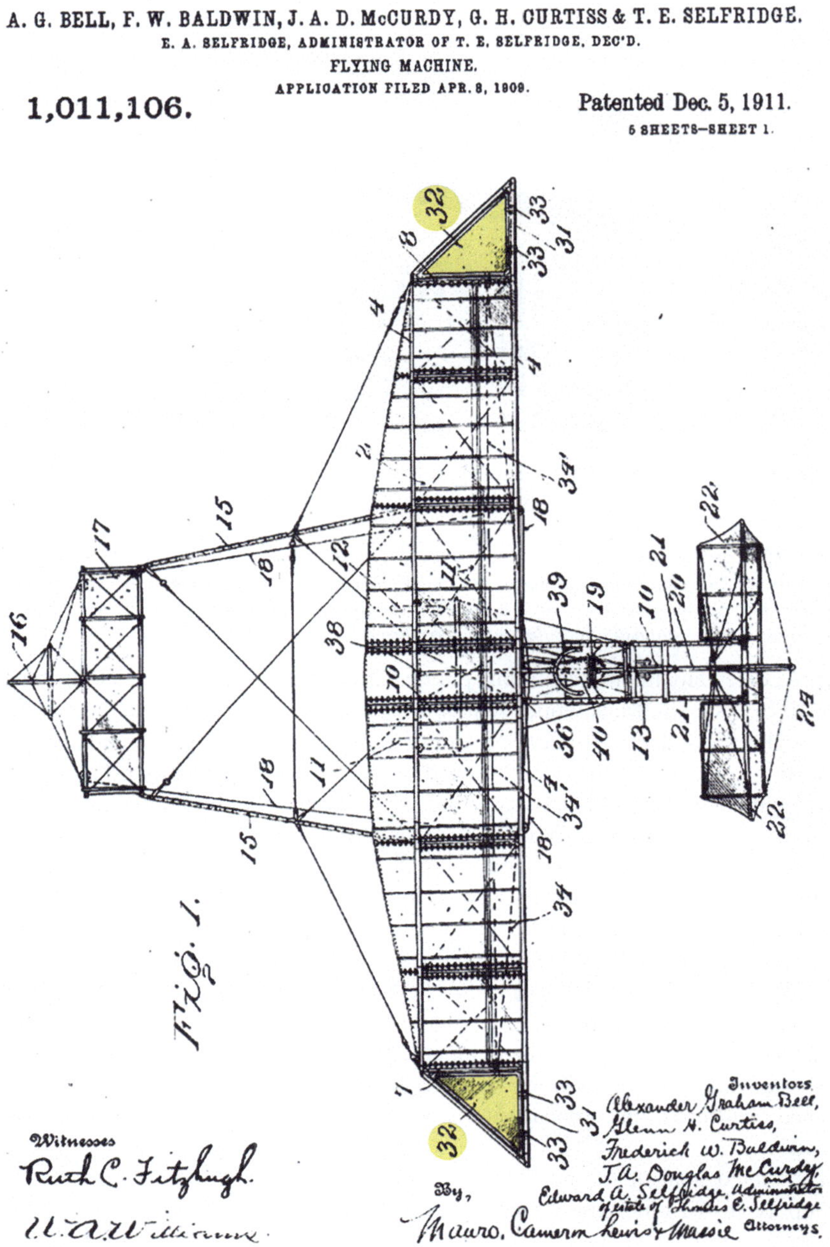

Fig. 51.9 Patent drawing from US1011106A. Component 32 is labeled a "lateral balancing rudder," later commonly known as an aileron

Fig. 51.10 The June Bug in flight. The triangular ailerons are visibly deflected at angles different from the main wing

> *a vertical rudder, means for moving said vertical rudder toward that side of the machine presenting the smaller angle of incidence and the least resistance to the atmosphere, and a horizontal rudder provided with means for presenting its upper or under surface to the resistance of the atmosphere, substantially as described.*

Curtiss denied any intent to continue public demonstrations and deflected the patent issue to Bell's team. Meanwhile, Wilbur Wright traveled to France for the brothers' first public flight demonstration, creating a sensation across Europe and receiving rockstar treatment. Until then, although their patent had been published and earned some reputation within the industry, their achievements hadn't been widely recognized due to strict secrecy. But this demonstration threw open the gates to public admiration: Just like the wings and tail feathers of birds, the warping or deflection of aircraft wing edges and vertical rudders were now clearly seen as the keys to conquering the skies—and the Wright brothers had already turned the key. Amid the cheers and accolades, the patent conflict was, for the moment, set aside.

In November 1909, the Wright brothers reorganized their company with the backing of a group of Wall Street investors—including the legendary financier J.P. Morgan himself, not just his company. With Morgan's support, the new company was valued at $1 million. The Wright brothers received $100,000 in cash, retained one-third of the company's shares, and were entitled to 10% of all licensing revenue.

The stated goal of the newly formed company was nothing short of absolute monopoly in the emerging American aviation market. A company spokesperson openly declared: "The objective of the Wright brothers and the board of directors is to protect and defend the Wright brothers' patents and oppose all new entrants...

One of the main purposes of forming a company with such a strong board is to assist the Wright brothers in all respects in defending their patent rights."[33]

While such a goal might sound somewhat unusual, the Wright brothers had at least found reliable partners who, for the most part, proved to be staunch supporters. Earlier that same year, Glenn Curtiss had also formed a new company—but his choice of partner would prove to be a disastrous one. His partner, Augustus Herring, was himself a former aircraft designer who had taken part in early flight experiments. In March 1909, he and Curtiss founded the Herring-Curtiss Company. To persuade Curtiss to partner with him, Herring falsely claimed he held patents capable of countering the Wright brothers' intellectual property. When Curtiss asked to see them, Herring claimed the patents hadn't yet been granted and had to remain confidential. In a stunning lapse of judgment, Curtiss believed him, paid the majority of the startup capital, and allowed Herring to hold a majority share in the new company. Meanwhile, Bell's Aerial Experiment Association formally disbanded in the same month the Herring-Curtiss Company was formed, having fulfilled its historical mission.

In 1909, the Wright brothers officially sued the Herring-Curtiss Company, seeking a preliminary injunction to prevent Curtiss from manufacturing aircraft. In addition to Claim 14, they asserted Claims 3, 4, and 15 of their patent. The relevant claims read as follows:

> ***Claim 3****: In a flying-machine, a normally flat aeroplane having lateral marginal portions capable of movement to different positions above or below the normal plane of the body of the aeroplane, such movement being about an axis ransverse to the line of flight, whereby said lateral marginal portions may be moved to different angles relatively to the normal plane of the body of the aeroplane, and also to different angles relatively to each other, so as to present to the atmosphere different angles of incidence, and means for simultaneously imparting such movement to said lateral marginal portions, substantially as described.*
>
> ***Claim 4****: 4. In a flying-machine, the combination, with parallel superposed aeroplanes, each having lateral marginal portions capable of movement to different positions above or below the normal plane of the body of the aeroplane, such movement being about an axis transverse to the line of flight, whereby said lateral marginal portions may be moved to different angles relatively to the normal plane of the body of the aeroplane, and to different angles relatively to each other, so as to present to the atmosphere different angles of incidence, of uprights connecting said aeroplanes at their edges, the uprights connecting the lateral portions of the aeroplanes being connected with said aeroplanes by flexible joints, and means for simultaneously imparting such movement to said lateral marginal portions, the standards maintaining a fixed distance between the parts which they connect, whereby the lateral portions on the same side of the machine are moved to the same angle, substantially as described.*
>
> ***Claim 15****: A flying-machine comprising superposed connected aeroplanes, means for moving the opposite lateral portions of said aeroplanes to different angles to the normal planes thereof, a vertical rudder, means for moving said vertical rudder toward that side of the machine presenting the smaller angle of incidence and the least resistance to the atmosphere, and a horizontal rudder provided with means for presenting its upper or under surface to the resistance of the atmosphere, said vertical rudder being located at the rear of*

[33] Goldstone, L. (2015). Birdmen: the Wright Brothers, Glenn Curtiss, and the battle to control the skies. Ballantine Books.

the machine and said horizontal rudder at the front of the machine, substantially as described.

That same year, Curtiss developed a new model called the **Golden Flyer** (or **Golden Bug**). This aircraft featured a new design in which the ailerons were placed between the upper and lower wings. This innovation was recorded in Curtiss's own patent application, U.S. Patent No. US1204380A (see Fig. 51.11 in the original document). From 1909 to 1913, the Wright brothers were consumed by litigation, shuttling between the U.S. and Europe, and did not file a single new patent application during that period.

Curtiss, meanwhile, was bursting with creativity. Despite the pressure of lawsuits and the meddling of an unreliable business partner, he managed to file a total of ten patent applications between 1909 and 1913 (see Table 51.1). Beyond his innovations in aileron design, Curtiss also succeeded in building the first American seaplane.

During the same period, aircraft design in Europe was flourishing with diverse innovation, far outpacing the United States, where progress was stifled under the shadow of the Wright brothers' patent litigation. In February 1907, French inventor Louis Blériot filed a patent (FR374494A) for a "universal joint controller," which used a joystick mounted on a gimbal to pull cables and control the ailerons (see Fig. 51.12). Moving the stick left caused the plane to turn left, and moving it right caused it to turn right. Aviation history would later confirm that this control method was the most effective for airplanes. Blériot first implemented this control system on the Blériot VIII, which made its maiden flight in 1908. The aircraft also featured

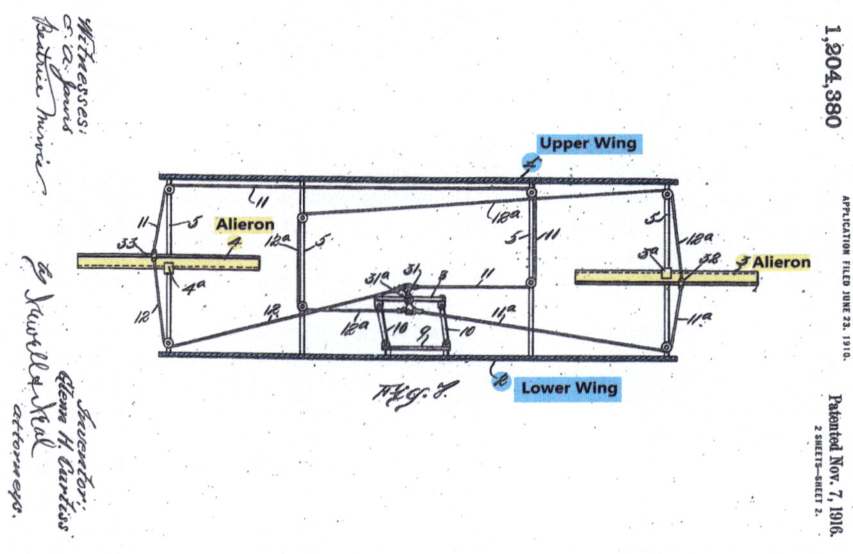

Fig. 51.11 Patent drawing of U.S. Patent No. 1,204,380A

Table 51.1 Patents filed by Curtiss, 1909–1913

Patent No.	Title	Filing Date
US1011106A	Flying-Machine	1909-04-08
US1204380A	Flying-Machine	1910-06-23
US1420609A	Hydroaeroplane	1911-08-22
US1027242A	Means for Launching Flying-Machines	1911-09-01
US1085575A	Controlling mechanism for flying-machines and the like.	1912-09-06
US1104036A	Flying-Machine	1912-09-14
US1108490A	Flying-Machine	1913-01-28
US1142754A	Flying-Boat	1913-06-04
US1223318A	Boat-Hull	1913-06-04
US1170965A	Hydroaeroplane	1913-09-16

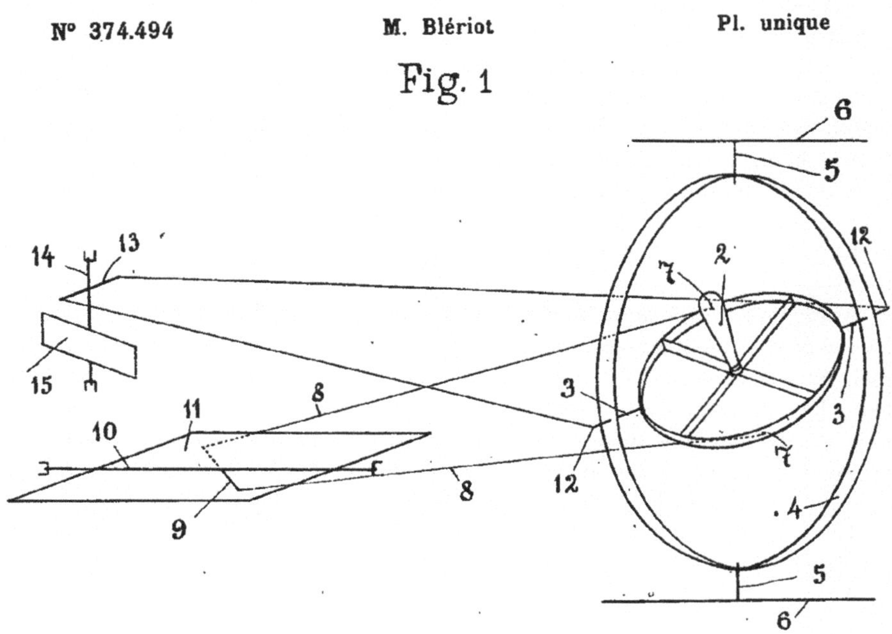

Fig. 51.12 Patent drawing of French patent FR374494A

foot pedals to control the rudder, forming the basic joystick-and-pedal configuration that became the standard in modern aviation.

The Wright brothers did not forget their European competitors. As previously mentioned, they had filed for patents in major European countries. In 1909, British media magnate Alfred Harmsworth offered a £1000 prize to the first pilot to fly across the English Channel and invited Wilbur Wright to participate. Concerned about the unpredictable weather over the Channel, Wilbur declined. Ultimately, Blériot completed the crossing and claimed the prize. His aircraft, a monoplane with flexible wings capable of warping but no separate ailerons, arguably fell within the scope of the Wright brothers' patent. Wilbur promptly sued him.

To pursue this litigation, Wilbur traveled to Europe multiple times, launching lawsuits in France, Germany, and the United Kingdom against various parties, with each stay lasting several months. However, the Wrights' legal efforts in Europe met limited success. Unlike the U.S., European legal systems did not recognize "pioneer patents." In 1912, a German court invalidated the Wrights' patent on the grounds that a friend of the Wrights had publicly discussed the wing-warping technique prior to their patent application. The German Supreme Court rejected their appeal and further restricted the patent's protection to the specific structure described in the specification.[34] By this time, most European aircraft were already using ailerons, effectively rendering the Wright patent irrelevant in Germany.

In the U.S., the Wright brothers also sued numerous pilots. At the time, civil aviation had not yet emerged as an industry, and airshows were one of the few ways to make money in aviation. Pilots became media stars, winning large prizes in flying competitions and capturing public attention.

When French aviator Louis Paulhan arrived in the U.S. for a competition, he was met at the dock by the Wright brothers' lawyers and U.S. marshals, who served him with a subpoena. Paulhan, who didn't speak English, was bewildered and thought he was being arrested. The next day, he received a court order requiring him to explain why he should be allowed to fly in the U.S. Before he could respond, an injunction followed, demanding a $25,000 bond or else he would be barred from performing. Enraged, Paulhan bought a ticket and sailed back to France the very next day.

At the time, the aviation industry had yet to fully emerge, and "pilot" was not a formal profession. Like the Wrights and Curtiss, many aircraft designers were also their own test pilots—both engineers and athletes. From a sporting perspective, using legal injunctions to eliminate rivals from competition was seen as unsportsmanlike. The lawsuits against Paulhan and Blériot severely damaged the Wright brothers' reputation in Europe. Major European aviation clubs declared they would not send pilots to the U.S. until the Paulhan case was resolved.

The U.S. had far fewer pilots than Europe at the time. Wilbur once proudly proclaimed, "Our goal is to cultivate enough aviation talent to end our reliance on

[34] Goldstone, L. (2015). Birdmen: the Wright Brothers, Glenn Curtiss, and the battle to control the skies. Ballantine Books.

foreign flyers, in case they decline our invitations." But under the Wright Company's terms, event organizers had to pay $1000 per day for a Wright-supplied pilot—of which only $50 would go to the pilot. Any prize money won by the pilot also had to be turned over to the Wright Company. By contrast, Curtiss allowed pilots to keep at least 50% of their winnings.[35] The Wrights also made an exclusive licensing deal with the Aero Club of America, which mandated that only sponsors holding a Wright patent license could receive official certification. Certified exhibitions were then required to pay 10% of gross revenue to the Wrights—a figure that exceeded the profit margins of many airshows and further stifled the growth of aviation in the United States.

At a large-scale airshow, American pilot John B. Moisant defeated British aviator Claude Grahame-White to win the grand prize—one of the few victories Americans had achieved in such events at the time. The American audience was moved to tears, and Wilbur Wright, who was present, cheered loudly with excitement. As the patent holder of the aircraft technology, the Wright Company earned $20,000 from ticket revenue sharing for the event, and its flying team received $15,000 in prize money. The company's board even awarded the Wright brothers a $10,000 raise. But the brothers were far from satisfied.

As previously mentioned, the Wrights had always demanded a cut of gross revenue. When they received the payout, they believed the amount was too low and suspected the organizers had skimmed off part of the gross revenue, shorting them by $15,000. They intended to sue. This time, even the company's senior partners couldn't stay silent. One early investor criticized the Wright brothers, saying, "While the company tries to promote the science and art of aviation, the Wright brothers care only for commerce and show little interest in real science," calling their behavior disgraceful.

Under pressure from the board, the Wrights dropped their case against the organizers—but not against the defeated British pilot Grahame-White. Just before boarding his ship home, Grahame-White was served with court papers. The Wrights demanded $50,000 in damages for patent infringement. They won the lawsuit and were awarded $1700.

But aviators were often nomadic, constantly on the move. Given the limited transportation and communication infrastructure of early twentieth-century America, it was nearly impossible to track down pilots like they had with Louis Paulhan at the dock—especially since many were foreigners. Harry Toulmin, the Wrights' patent attorney, advised them to stop suing individual pilots and instead target aircraft manufacturers and sponsors of airshows. Legally, this was a sound strategy. But from a business standpoint, it proved to be a disastrous misstep.

Aircraft manufacturers were at least direct competitors—suing them could be justified. But airshow sponsors were often major corporations, investors, or banks just beginning to show interest in the nascent aviation industry. Suing them

[35] Goldstone, L. (2015). Birdmen: the Wright Brothers, Glenn Curtiss, and the battle to control the skies. Ballantine Books.

effectively alienated potential allies and dampened their enthusiasm for aviation. The consequences of this approach were beyond what Toulmin, as a patent lawyer, could have foreseen.

Wealthy industrialist Harold McCormick organized an airshow in Chicago that drew an unprecedented crowd of 350,000. Orville Wright initially agreed to send the Wright Company's pilots. But upon learning that McCormick was allowing unlicensed pilots to compete, Orville changed his mind and ordered Toulmin to sue the organizing company. They later discovered that McCormick hadn't intended to profit from the event at all—the exhibition had cost \$195,000 to host and earned only \$142,000 in revenue. The organizing company, as the defendant, had no money left to pay the Wrights. In theory, the Wrights could have gone after McCormick personally, but their schedule was already overwhelmed with litigation, and they ultimately dropped the matter.

From America to Europe, the Wright brothers came under intense public criticism and engaged in heated debates with detractors. In the newly launched *Aircraft* magazine, Louis Paulhan wrote a piece condemning the Wrights, asserting that his aircraft design did not infringe on their patents. Toulmin responded with a rebuttal titled *Attacks on Wright Brothers Wholly Unjustified*, defending the Wrights not on technical grounds, but by emphasizing the hardships they had endured in building their business and their right to reap substantial rewards.

One lawyer, who supported Paulhan, published a counterargument. A passionate aviation enthusiast himself, this lawyer had designed and flown his own aircraft before a crash left him paralyzed from the waist down. In his article, he angrily decried the Wright patent as "owned by a greedy and wealthy company that could threaten the vitality of aviation, strangle national progress, and exclude the fruits of foreign advancement." He posed a scathing question: "Is the profit from aviation not enough to go around? Is it only enough for the Wright brothers?"

A naval officer at the time had flown both Wright and Curtiss airplanes. His assessment was:

> The Wright airplane has a hybrid control system… very unnatural. You have to move the control lever laterally while also shifting it forward and back to adjust altitude, and then operate a small controller on the stick… The Curtiss aircraft has better controls in some respects… You can tilt left or right to control the ailerons. The control wheel is mounted on a crossbar… to operate the elevators. Apart from managing power and throttle, you don't need to use your feet. The Wright aircraft was mounted on a glider, and required a falling weight to assist in takeoff. The Curtiss machine, on the other hand, had sufficient power to take off directly from the beach, ready for a powered ascent.

All signs pointed to the fact that the litigation-obsessed Wright brothers had fallen behind technologically. First, they were waging legal battles simultaneously in Europe and America against multiple opponents, with their schedules dominated by evidence gathering, cross-examinations, and court appearances. Second, in order to ensure their patent would be recognized as a "pioneer patent," the Wrights insisted that their original embodiment was unique and irreplaceable. If they had adopted ailerons, it would have undermined their own claims. But decades of aviation experience would prove that while the Wrights had indeed discovered the key to the skies, ailerons—given the material limitations of the era—were the superior solution.

While Europeans actively explored various aileron designs, in the U.S., only the stubborn Curtiss continued designing aircraft under immense legal pressure. The Wright brothers, in contrast, confined themselves within the rigid doctrine of flexible, warping wings.

Moreover, the Wright brothers showed little interest in the popular stunt flying shows of the time and refused to design aircraft specifically for aerial maneuvers. Orville once remarked, "I'm more interested in what ordinary people can do than in what daredevils or fools can do with a plane." That perspective wasn't necessarily wrong. But in reality, not only did the public love airshows—one of the few profitable avenues for the fledgling aviation industry—but also, extreme turns, dives, and climbs were essential techniques for early aerial combat.

Wilbur Wright, as the face of the company, threw himself into the legal battles with almost manic zeal. He scrutinized witness testimony personally, looking for inconsistencies. After a Curtiss witness testified, Wilbur would study the transcripts late into the night to identify flaws. He grew increasingly dissatisfied with Toulmin, their attorney. On one occasion, a Curtiss witness made a critical mistake during testimony that Toulmin missed entirely—but Wilbur caught it that night and immediately advised Toulmin to follow up. The next day, the witness admitted the error, contradicting other Curtiss testimony. Wilbur was ecstatic, even boasting about it in a letter to his brother.

According to one Wright Company manager:

> In the last two years of Wilbur Wright's life, he was wholly consumed by the patent war with Curtiss... Wilbur Wright died of typhoid fever after exhausting himself in the litigation.

He added,

> In the early days of the company... Orville Wright did all the actual work, because Wilbur had to devote himself entirely to the lawsuit.

In May 1912, Wilbur Wright suddenly contracted typhoid. Within days, he was critically ill and wrote what would be the last letter of his life—to a lawyer at the firm of Fish, Richardson & Neave. Even in his final days, patents dominated his thoughts. He complained that Toulmin had repeatedly agreed to procedural delays with Curtiss's lawyers, and lamented, "Now there are countless competitors entering the field, building their first truly airworthy aircraft. These planes cost half as much as ours." In his final moments, witnessing the blossoming and prosperity of the industry he had helped pioneer, Wilbur felt no joy—only bitterness and grievance.[36] His brother Orville would inherit that anger and continue the fight.

In 1913, the district court finally ruled in favor of the Wright brothers, upholding the validity of their patent and finding Curtiss guilty of infringement. More significantly, the Wrights' patent (US821393A) was officially recognized as a "pioneer

---

[36] Goldstone, L. (2015). Birdmen: the Wright Brothers, Glenn Curtiss, and the battle to control the skies. Ballantine Books.

patent." In his ruling, Judge John R. Hazel reviewed all known prior art before the Wrights' filing, praised their revolutionary contributions to aviation, called them "pioneers in the field of aeronautics," and declared that their disputed claims should be "interpreted broadly."

Under this doctrine, Curtiss's non-infringement defense fell apart. Curtiss argued that the vertical rudder on his aircraft was only for steering, not for "restoring balance." Many Curtiss pilots gave similar testimony. But one witness admitted using the rudder during an abnormal tilt and that the aircraft had regained balance as a result. Based on that testimony, the judge concluded that while Curtiss's aileron design differed from the Wrights' and "does not operate in every instance or under all conditions in the same way as the Wrights' principle," it could nonetheless "restore equilibrium in substantially the same manner." According to the prevailing case law at the time, such functional equivalence to a pioneer invention constituted infringement.

Judge Hazel was known to favor patent holders. As noted earlier, the infamous patent troll Sheldon had won his first lawsuit under Hazel's ruling, enabling him to launch a patent licensing empire that squeezed the entire automotive industry. At this very moment, Henry Ford was publicly railing against Hazel's decisions.

Curtiss appealed the ruling, but to no avail. In 1914, the Second Circuit Court of Appeals upheld Judge Hazel's decision, especially affirming the recognition of the "pioneer patent," stating:

> The patentee may fairly be regarded as the pioneer in actual flight of a heavier-than-air flying machine, and his claims are entitled to a broad construction.

The Wright Company's board was jubilant. They hoped to either drive all competitors out of the market or acquire them as subsidiaries. Orville Wright, however, was less aggressive. He proposed charging a 20% royalty on the sale price of every airplane manufactured in the U.S. At the time, a plane typically sold for around $5000. In 1914, over 1000 planes were produced in the U.S., meaning the Wright Company could easily pocket over a million dollars—a far more profitable venture than designing and building aircraft. Moreover, armed with the power of a "pioneer patent," the Wright Company had the right to claim royalties on virtually every aircraft built in the U.S. since 1906. Orville declared magnanimously that he would not pursue unintentional infringers—but Curtiss would not be spared. Every penny Curtiss had earned, Orville intended to reclaim.

By this time, the technological gap between the Wright Company and Curtiss had widened significantly. Not long after the 1914 appellate ruling, Alfred Harmsworth once again offered a £10,000 prize for the first pilot to successfully cross the Atlantic. Orville dismissed the challenge outright, deeming it impossible. But Curtiss, while loudly declaring he would appeal the decision, also announced his intention to build a plane capable of the transatlantic journey. On May 15, 1919, U.S. Navy Lieutenant Commander Albert C. Read piloted a Curtiss-built NC-4 seaplane from New York to Lisbon. After nearly 44 hours, he completed the first transatlantic flight in human history.

Following the 1914 verdict, Curtiss met Henry Ford. At the time, Ford had already defeated America's first patent troll, George Selden, and was hailed as a national hero. His assembly-line innovation was in full bloom, and he was at the height of both wealth and fame. Ford also harbored a deep disdain for patents. He offered Curtiss advice and introduced him to his own lawyer, Benton Crisp.

Crisp, a seasoned legal mind, believed there was little hope in appealing to the Supreme Court. The Court rarely heard patent cases, and the justices tended to focus on legal rather than factual issues—so a reversal was unlikely. Instead, Crisp recommended Curtiss attempt a design workaround. In the 1913 decision, Curtiss had argued that the three components of his aircraft—two ailerons and a rudder—did not operate in unison. The judge had rejected this claim, but subtly implied that if the components did not function together, the design might not infringe. Thus, Crisp advised Curtiss to allow pilots to operate the left and right ailerons independently, rather than simultaneously.[37] While the odds were still slim, as long as the Wright Company hadn't yet sued over the new model, Curtiss could keep building planes. It was a brazen strategy: Orville Wright, already exhausted by litigation, would have to file a whole new lawsuit just to get the redesigned aircraft barred.

Meanwhile, the previous lawsuit between the Wright and Curtiss companies was not yet fully resolved—while infringement had been established, the damages still had to go through a lengthy legal process. Orville lacked the resources and stamina to pursue two simultaneous lawsuits. Faced with a choice between profitability and monopolizing the aircraft market, Orville chose the latter. He wanted to prevent Curtiss from building even a single plane. So in 1914, the Wright Company sued Curtiss again.

That same year, the Sarajevo incident triggered the outbreak of World War I. The aviators who once pioneered the skies together now joined opposing armies, turning from comrades on the frontier of human exploration into bitter wartime enemies. At that moment, the Wright Company couldn't produce a single aircraft fit for combat, while Curtiss unveiled the Model J biplane in 1914—a design that quickly entered military service and became one of just 23 planes available to the U.S. Army Air Service at the time.

In 1915, Orville could no longer bear to continue. He sold his shares and patents to a Wall Street investment consortium, receiving nearly $1.5 million in cash. Though he retained the title of Chief Aeronautical Engineer, in practice, he had exited the company. In 1917, as the United States formally entered World War I, the government formed a committee—led by Curtiss's attorney, Benton Crisp—to ensure that American companies could continue aircraft production. The committee drafted a cross-licensing agreement that allowed all aircraft manufacturers to build planes under lower royalty fees, effectively ending all patent litigation in the U.S. aviation industry.

[37] Goldstone, L. (2015). Birdmen: the Wright Brothers, Glenn Curtiss, and the battle to control the skies. Ballantine Books.

Even so, the American aircraft industry had already fallen far behind Europe. Not a single American-built plane was capable of fulfilling military duties. In a bitter irony, the Wright brothers' company underwent several mergers and eventually combined with Curtiss's firm to form the Curtiss-Wright Corporation.

The Wright brothers' patent war severely hindered the implementation of their core patent, slowed the progress of the U.S. aviation industry, and greatly delayed the technological benefits aviation could have brought to society. Wilbur Wright had nearly sacrificed his life to litigation. Though Orville Wright lived until 1948, after Wilbur's death he made almost no further contributions to aviation. These two brilliant men birthed an entire industry from fragile gliders in just a few short years, only to strangle it with their own hands for the sake of personal gain—until their legacies, like shooting stars, faded from the sky.

We believe such a patent war was not worth it. The Wright brothers opened up a fertile new land but failed to cultivate it seriously. Instead, they patrolled its borders with excessive vigilance, frequently lashing out against others. Toward those who truly worked the land, they made exorbitant demands, seemingly wishing to turn the entire field into a no-man's-land. Naturally, such land could not yield proper returns to society, nor could it nurture a truly great enterprise. By the time of his death, Wilbur had all but become a professional patent litigator. But what were his legal victories compared to the towering legacy of founding an entire industry and benefiting billions of lives?

Chapter 52
Intellectual Property Rules: Tao of Application

Reflecting on the stories in the previous chapter, "Rise and Fall," we find that Herman Hollerith was a contemporary of the Wright brothers, and his groundbreaking invention was no less revolutionary. A true patent expert, Hollerith quietly refined his machine after obtaining his first patent. Aside from a brief episode of litigation after his initial patents expired, he exited the scene early, never becoming entangled in long patent wars. Ultimately, it was through the hands of the Watsons—father and son—that the great computer industry was born, propelling the technological leap of the late twentieth century and beyond.

IBM is great because it opened up new frontiers for humanity time and again. At the height of its power, it never used patent lawsuits to suppress competitors. Although its epoch-making System/360 left behind many patents, they now lie dormant in dusty archives. These patents left no tales for patent agents to brag about, no legendary courtroom battles for lawyers or judges. And yet, their contribution to humankind is immeasurable—far surpassing the handful of patents once boasted by Prodigy.

The many inventions of Sony and Philips have enriched countless lives, reshaped generations of lifestyles, and offered endless inspiration to artists. Today, when people speak of Sony and Philips' patents, they often think of the patent pools spanning from CDs to MP3s, or Philips' recent aggressive enforcement strategies. It's precisely for this reason that we've dusted off the forgotten early patents for tape recorders and the Walkman—those patents that left no litigation records, quietly nourishing human society in anonymity. These are the true foundation stones of Sony and Philips' legendary status.

In recent years, the cultivation of so-called "high-value patents"—a term promoted by the Chinese government and well known among Chinese patent professionals—has become a key topic in the patent world. But what defines a high-value patent? Many metrics have been proposed: courtroom victories, licensing revenue, stock market valuation, citation count, and so on. In our view, these quantifiable indicators are not the most crucial criteria. A truly valuable patent is one whose

Q. Yue, F. Yang, *Governing Innovation: A Global History of Patent-Driven Technological Progress*, https://doi.org/10.1007/978-981-95-5570-3_52

claims are precisely aligned with the scope of its innovation—no more, no less. And, as discussed earlier, its value must ultimately be reflected in implementation.

For example, in a past evaluation of China's Patent Awards, I encountered a patent from the State Grid Corporation—CN103730247A, titled "A Fully Shielded High-Voltage Isolated Voltage Transformer." It was used to calibrate high-voltage measurements and won an award at the 23rd China Patent Awards in 2022. It had no litigation history and only 10 citations. Yet according to the award documentation, this technology had already completed commissioning tests for all ultra-high-voltage substations in China and was used for field calibration in at least 11 overseas projects. These included Brazil's Belo Monte UHVDC transmission project and Pakistan's ±660 kV Matiari–Lahore transmission line.

Another example is CN102837105B, a gold-medal winner at the 22nd China Patent Awards, from China Railway Shanhaiguan Bridge Group, titled "A Welding Method for Q345qDNH Weathering Steel Used in Bridges." This little-known patent has been applied to the world's tallest bridge—the Beipanjiang Bridge—and the longest sea-crossing bridge—the Hong Kong–Zhuhai–Macau Bridge. It is one of the foundational patents behind China's architectural miracles of the twenty-first century.

These technologies rarely make headlines, have fought no battles in court, and attracted little attention in capital markets. Yet they've empowered the State Grid to scale mountains and cross deserts, bringing light, modern living, and life-changing opportunities to millions across Asia, Africa, and Latin America. They've enabled China Railway's engineers to carve paths through mountains and build bridges across waters, completing one awe-inspiring project after another. We wish we could unearth more stories behind such patents, but truth be told, there is little drama behind them—only the quiet toil of countless engineers. These patents may well be forgotten in time, but they are the ones that truly deserve to be called "high-value."

Though most of the stories told in this book are of the past, as we reach its conclusion, I hope readers will join me in looking ahead. We believe in the wisdom of humanity and in the power of a mature and healthy patent system—one that distinguishes genuine innovation from the false, enables the implementation of great inventions, and helps enterprises that stay on the righteous path to thrive for generations.

Some of the early issues with the patent system mentioned in this book—such as Watt's concern about whether patent specifications were sufficiently disclosed—are no longer a major concern for most patent holders today. According to the Chofn Research Institute's *2021 Statistical Analysis of Invalidation Decisions for Chinese Patents,*[1] of the 341 invention patents that were fully invalidated by the China National Intellectual Property Administration (CNIPA) in 2021, only a few were

[1] 超凡研究院. (2022). 2021年中国专利无效决定统计分析. Retrieved March 31, 2024 from Chofn Website: https://www.chofn.com/academy/620222c8c6659f6331f7a674/%E3%80%8A2021%E5%B9%B4%E4%B8%AD%E5%9B%BD%E4%B8%93%E5%88%A9%E6%97%A0%E6%95%88%E5%86%B3%E5%AE%9A%E7%BB%9F%E8%AE%A1%E5%88%86%E6%9E%90%E3%80%8B

due to insufficient disclosure. A total of 303 were invalidated for lack of novelty and inventiveness. Not long ago, the WeChat account "Enterprise Patent Watch" analyzed 52,000 patent invalidation decisions issued by CNIPA between 1987 and 2021, and found that only 65 patents were invalidated—wholly or in part—due to insufficient disclosure, accounting for just 0.1% of all decisions.[2]

In an interview cited earlier in this book, Edison remarked that judges' lack of technical understanding was the reason patent litigation was costly and inefficient. He once proposed: "A separate and special court. Take the whole business out of the regular judicial system.... I would appoint, to this special court for trying patent cases, judges from the faculties of colleges of technology, men who know something about science."[3] Today, that vision has been realized: intellectual property courts composed of judges with technical backgrounds now exist in many countries.

This book has consistently criticized the practice of patent monetization for the sole purpose of collecting licensing fees. In the "Rise and Fall" section, we posed this question: If patent monetization were the right path, then why, in the more than 200 years since the Industrial Revolution, has there never been a century-old enterprise built primarily on patent monetization? In reviewing patent history, we have found no such companies—but we have found numerous law firms that have thrived on patent services for over a century. The renowned law firm Fish & Richardson represented Alexander Graham Bell in his telephone patent disputes, Edison in the earliest lawsuits of the electrical age, and the Wright brothers in the first aviation patent wars. As we began writing this book, Fish & Richardson was representing iRobot in the first service robot patent case at the ITC. Bell and Edison's corporate legacies—AT&T and General Electric—are no longer what they once were; the Wright brothers' aviation company is now all but forgotten. But Fish & Richardson remains at the top of U.S. patent litigation firms.

In fact, China's current patent system does not favor the rise of NPEs or widespread patent monetization. China's invalidation procedure is relatively efficient, and legal fees are generally lower than in Western countries. This is a significant advantage of China's patent regime. Furthermore, for Chinese companies going global—whether targeting Western markets or emerging economies—we believe patent monetization should not be a primary goal. As long as companies remain customer-focused and grounded, when "Made in China" becomes unbeatable, China's patents will naturally become equally powerful. As the old saying goes, "To rejoice with the world and grieve with the world—who then would not be king?"

As this book goes to press, the long-anticipated Unified Patent Court (UPC) has officially launched in Europe, drawing widespread industry attention. Historically, Europeans have taken pride in their patent system, particularly in its resistance to patent trolls. One reason: enforcing patents across Europe has traditionally required

[2] 吴征高低. (2023). 深度 | 中国专利的十字路口. "企业专利观察"公众号. Retrieved March 31, 2024 from: https://mp.weixin.qq.com/s/eRjNkxc9_ENeJo-ZByfV_A

[3] Jeff Nilsson. (2012). Edison and The Pirates: The Inventor's Solution to Copyright Theft. *The Saturday Evening Post*. January 28, 2012. Retrieved March 30, 2024 from The Saturday Evening Post Website: https://www.saturdayeveningpost.com/2012/01/pirates-patents-progress/

litigation in multiple jurisdictions. Now, with a unified patent regime, some NPEs are eager to seize the opportunity, believing that the golden age of patent monetization in Europe is at hand. What impact the UPC will have on patent monetization in the coming years remains uncertain. But in the long run, any system that favors NPEs over real product developers is bound to be corrected—otherwise, it becomes a shackle on innovation. This is true for both the U.S. and Europe.

For all these reasons, we are optimistic about the future. Today's patent databases are larger and more accessible than ever before. Tools such as ChatGPT have demonstrated powerful reading, analysis, and summarization capabilities. Though not yet widely used in patent search, their rise is inevitable. The patent systems and tools of the future will only become more effective at distinguishing real innovation from the fake. Once that line is clear, patent litigation will no longer be dominated by trolls and frivolous lawsuits, and innovation will flourish.

From monarchic Britain, to fledgling America, to the era of corporate giants—and into our present day—the patent system has continued to evolve. Despite occasional setbacks, its trajectory has generally spiraled upward. Over more than two centuries, public attitudes toward monopoly and capital have shifted; standards for infringement and invalidity have varied; some eras favored patent holders, while others did not. What the next 10 or 20 years hold, we cannot say. But as long as the patent system exists, its core purpose—to encourage innovation—will not change. That core mission necessitates distinguishing true innovation from false and supporting the practical implementation of real breakthroughs, to benefit all of humanity.

Today, legislative trends around the world point toward faster, more efficient, and more economical invalidation procedures. That is the flow of the river; that is the grain of the jade.

Inventors expand the boundaries of human knowledge with sparks of genius. Patent engineers and agents define those boundaries. Patent attorneys and judges defend them in the name of the law. When rivals trespass, the system stands to protect the inventor's rights. We believe that only when the boundaries drawn by a patent align with the actual innovation can that patent be called genuine. That boundary is scientific, legal, and worth defending. Without real innovation behind it, the patent is unscientific, illegitimate, and undeserving of protection.

Balancing the rights of inventors with the public good, aligning patent boundaries with the true scope of innovation, and transforming invention into real-world value—this, in our view, is the governance of innovation.